Root Genomics and Soil Interactions

Root Genomics and Soil Interactions

Edited by
MARTIN CRESPI

WILEY-BLACKWELL
A John Wiley & Sons, Inc., Publication

Wiley-Blackwell is an imprint of John Wiley & Sons, formed by the merger of Wiley's global Scientific, Technical and Medical business with Blackwell Publishing.

Editorial offices: 2121 State Avenue, Ames, Iowa 50014-8300, USA
The Atrium, Southern Gate, Chichester, West Sussex, PO19 8SQ, UK
9600 Garsington Road, Oxford, OX4 2DQ, UK

For details of our global editorial offices, for customer services and for information about how to apply for permission to reuse the copyright material in this book please see our website at www.wiley.com/wiley-blackwell.

Library of Congress Cataloging-in-Publication Data

Root genomics and soil interactions / editor, Martin Crespi.
p. cm.
Includes bibliographical references and index.
ISBN 978-0-470-96043-1 (hardback : alk. paper)
1. Roots (Botany)–Physiology. 2. Roots (Botany)–Development. 3. Plant genomes. 4. Genomics.
5. Plant-soil relationships. I. Crespi, Martin.

QK644.R6523 2012
575.5′4–dc23 2012021109

A catalogue record for this book is available from the British Library.

Wiley also publishes its books in a variety of electronic formats. Some content that appears in print may not be available in electronic books.

Cover design by Modern Alchemy LLC

Set in 10.5/12 pt Times by Aptara® Inc., New Delhi, India
Printed and bound in Malaysia by Vivar Printing Sdn Bhd

1 2013

Contents

Color plate located between pages 144 and 145

Contributors

Pierre Abad INRA-CNRS-Université de Nice Sophia-Antipolis
Interactions Biotiques et Santé Végétale
Sophia Antipolis, France

Nicole Alloisio Centre National de la Recherche Scientifique
Ecologie Microbienne
Université de Lyon
Cedex Villeurbanne, France

F. Ariel Institut des Sciences du Végétal
Centre National de la Recherche Scientifique
Gif sur Yvette cedex, France

Leah Band Centre for Plant Integrative Biology
University of Nottingham
United Kingdom

F. Bardou Institut des Sciences du Végétal
Centre National de la Recherche Scientifique
Gif sur Yvette cedex, France

Thomas Baum Department of Plant Pathology and Microbiology
Iowa State University
Ames, Iowa, USA

Tom Beeckman Department of Plant Systems Biology and
Department of Plant Biotechnology and Genetics
Ghent University
Ghent, Belgium

Malcolm Bennett Centre for Plant Integrative Biology
University of Nottingham
United Kingdom

Didier Bogusz Equipe Rhizogenèse
Institut de Recherche pour le Développement
Montpellier, France

Thais L.G. Carvalho Laboratório de Biologia Molecular de Plantas
Instituto de Bioquímica Médica
Universidade Federal do Rio de Janeiro
Rio de Janeiro, Brazil

Siva K. Chamarthi International Crops Research Institute for the Semi-Arid Tropics
Hyderabad, India

Hyung-Taeg Cho Department of Biological Sciences and Genomics and Breeding Institute
Seoul National University
Seoul, Korea

Martin Crespi Institut des Sciences du Végétal
Centre National de la Recherche Scientifique
Gif sur Yvette cedex, France

Isabelle Damiani INRA-CNRS-Université de Nice Sophia-Antipolis
Interactions Biotiques et Santé Végétale
Sophia Antipolis, France

Pierre-Marc Delaux Laboratoire de Recherche en Sciences Végétales
Université de Toulouse
Castanet-Tolosan, France

José R. Dinneny Carnegie Institution for Science
Department of Plant Biology
Stanford, California, USA
and
Temasek Lifesciences Laboratory
Department of Biological Sciences
National University of Singapore
Singapore

Bruno Favery INRA-CNRS-Université de Nice Sophia-Antipolis
Interactions Biotiques et Santé Végétale
Sophia Antipolis, France

Paulo C.G. Ferreira Laboratório de Biologia Molecular de Plantas
Instituto de Bioquímica Médica
Universidade Federal do Rio de Janeiro
Rio de Janeiro, Brazil

Damien Formey Laboratoire de Recherche en Sciences Végétales
Université de Toulouse
Castanet-Tolosan, France

Florian Frugier Institut des Sciences du Végétal (ISV), CNRS
Gif-sur-Yvette cedex, France

Pooran M. Gaur International Crops Research Institute for the Semi-Arid Tropics
Hyderabad, India

Adriana S. Hemerly Laboratório de Biologia Molecular de Plantas
Instituto de Bioquímica Médica
Universidade Federal do Rio de Janeiro
Rio de Janeiro, Brazil

Tarek Hewezi Department of Plant Pathology and Microbiology
Iowa State University
Ames, Iowa, USA

Valérie Hocher Equipe Rhizogenèse
Institut de Recherche pour le Développement
Montpellier, France

Cyril Jourda Laboratoire de Recherche en Sciences Végétales
Université de Toulouse
Castanet-Tolosan, France

L. Krishnamurthy International Crops Research Institute for the Semi-Arid Tropics
Hyderabad, India

Laurent Laplaze Equipe Rhizogenèse
Institut de Recherche pour le Développement
Montpellier, France
and
Laboratoire Commun de Microbiologie
Centre de Recherche de Bel Air
Dakar, Sénégal

Philippe Laporte Laboratoire des Interactions Plantes Micro-organismes (LIPM) UMR CNRS-INRA
Castanet-Tolosan Cedex, France

Julien Lavenus Centre for Plant Integrative Biology
University of Nottingham
United Kingdom
and
Institut de Recherche pour le Développement
UMR DIADE
Montpellier, France

Mikael Lucas Centre for Plant Integrative Biology
University of Nottingham
United Kingdom
and
Institut de Recherche pour le Développement
UMR DIADE
Montpellier, France

Alistair Middleton Centre for Plant Integrative Biology
University of Nottingham
United Kingdom

and
Center for Biological Systems Analysis
Albert-Ludwigs-Universität
Freiburg, Germany

A.B. Moreno Institut des Sciences du Végétal
Centre National de la Recherche Scientifique
Gif sur Yvette cedex, France

Henry T. Nguyen Division of Plant Sciences
University of Missouri
Columbia, Missouri, USA
and
The National Center for Soybean Biotechnology

Andreas Niebel Laboratoire des Interactions Plantes Micro-organismes
UMR CNRS-INRA
Castanet-Tolosan Cedex, France

Philippe Normand Centre National de la Recherche Scientifique
Ecologie Microbienne
Université de Lyon
Cedex Villeurbanne, France

Boris Parizot Department of Plant Systems Biology and
Department of Plant Biotechnology and Genetics
Ghent University
Ghent, Belgium

Michaël Quentin INRA-CNRS-Université de Nice Sophia-Antipolis
Interactions Biotiques et Santé Végétale
Sophia Antipolis, France

Reyazul Rouf Mir International Crops Research Institute for the Semi-Arid Tropics
Hyderabad, India

Christophe Roux Laboratoire de Recherche en Sciences Végétales
Université de Toulouse
Castanet-Tolosan, France

Ive De Smet Division of Plant and Crop Sciences
School of Biosciences
University of Nottingham
Loughborough, United Kingdom

Mahendar Thudi International Crops Research Institute for the Semi-Arid Tropics
Hyderabad, India

Rajeev K. Varshney International Crops Research Institute for the Semi-Arid Tropics
Hyderabad, India
and

CGIAR-Generation Challenge Programme
DF, Mexico
and
School of Plant Biology
Faculty of Natural and Agricultural Sciences
The University of Western Australia
Crawley, WA, Australia

Laura M. Vaughn
Division of Plant Sciences
University of Missouri
Columbia, Missouri, USA
and
The National Center for Soybean Biotechnology
and
Department of Natural Resources and Environmental Sciences
University of Illinois

Elisabeth L. Williams
Division of Plant and Crop Sciences
School of Biosciences
University of Nottingham
Loughborough, United Kingdom

Michael Wilson
Centre for Plant Integrative Biology
University of Nottingham
United Kingdom

Preface

Roots and Their Soil Interactions: What We Can Learn from Genomics

Developmental plasticity allows higher organisms to adapt to their environment. In contrast to animals, plants exhibit a remarkable flexibility in their architecture and growth pattern in response to external conditions, due to the continuously active shoot and root meristems and their capability to generate new organs after embryogenesis. External cues influence plant growth by modulating hormone levels and signaling. The root architecture of the plant constitutes an important model to study how developmental plasticity is translated into growth responses under different soil conditions and plays an important role in water and nutrient acquisition. Indeed, primary root development and the formation of *de novo* meristems to generate lateral roots are conditioned by the soil environment. Lateral root growth and development is the main determinant of the shape of the root system, a trait controlled by internal cues and external factors. In addition to Arabidopsis, there are other relevant models where genomic information is becoming available, notably cereals and legumes. Both plants are able to develop symbiotic interactions with soil organisms, namely, mycorrhizal fungi and, for legumes, soil rhizobia. These interactions lead to further adaptation of root growth, the so-called mycorrhizal roots, and even to the formation of new organs, distinct from lateral roots, the nitrogen-fixing root nodules.

The diversity of root responses to biotic and abiotic stresses as well as symbiotic interaction can be analyzed at a genome-wide scale using transcriptomic and proteomic approaches. The advent of genomic technologies will open new perspectives for the analysis of how roots adapt to the soil environment. This work, mainly done in model systems such as Arabidopsis, uncover diverse regulatory genes (e.g., environmental sensors, protein kinases, transcription factors, and more recently, small regulatory RNAs) that participate in genetic programs, regulating root growth and architecture. Integration of these data with genomic approaches on different genetic backgrounds has already revealed, and will continue to reveal, critical regulatory networks and molecular hubs, whose orthologs could then be analyzed in crop plants to establish the generality of these mechanisms and impact agricultural practices.

This book contains 13 chapters from recognized experts in the field, which provide a comprehensive and integrated view of how root genomics can open new perspectives for root physiology and agriculture. The first six chapters deal with various novel areas where genomics, in combination with modeling, physiology, in-depth analysis of the transcriptome, and epigenetics, have revealed several regulatory networks controlling diverse aspects of root growth and development. Then, the remaining chapters describe genomic approaches being applied for the analysis of root responses to the soil environment, such as abiotic stresses, symbiotic interactions, or pathogenic nematode infections. The final chapter focuses on translational genomics and how genomics can guide crop improvement. I hope that this book will serve many, from plant researchers to plant and crop physiologists, breeders, graduate students, and their professors who want to have an overview of the highlights in root genomics and how this information could be screened and integrated without having considerable expertise in bio-statistics. While reading this book, the reader will realize how fascinating the actual global view of the genome is and how many complex mechanisms remain to be discovered to understand root growth and development. There are exciting agricultural challenges, such as the modulation of root architecture or drought adaptation, which may derive from the application of this new fundamental understanding of life principles to the control of major root traits.

Martin Crespi

Root Genomics and Soil Interactions

1 Genomics of Root Development

Boris Parizot and Tom Beeckman

Introduction

Roots: Rising from the Underground

Because of the different roles the root system plays in overall plant growth, root architecture is a fundamental aspect of plant growth and development. The root system especially acquires water and nutrients from the soil, anchors the plant in the substrate, synthesizes hormones and metabolites, interacts with symbiotic microorganisms, and insures storage functions. In light of these characteristics, more and more breeders turn their attention to this underground organ in order to increase yield. This requires a better understanding of the relation of this part of the plant with the environment and of its highly adaptive behavior (Lynch 2007; Gewin 2010; Den Herder et al. 2010).

Within the angiosperms, major differences in root architecture between dicotyledonous and monocotyledonous plants exist. Dicots develop a tap root system composed of a main primary root, already formed during embryogenesis, which grows vertically into the soil and gives rise to the emergence of numerous lateral roots extending the surface area. Monocots have a fibrous root system in which the embryonic primary root is only important for the early development of the plant (Feix et al. 2002) and in which an extensive postembryonic shoot-born root system is formed later on. Very little is known about the genetic and molecular mechanisms involved in the development and architecture of the root system in major crop species, generally monocotyledonous plants. Lack of insight is certainly a consequence of the difficulty to access and observe this organ in its natural habitat, namely the soil. Moreover, and probably because of this hidden character, the root has been neglected for a long time in crop improvement and in agricultural approaches aiming at increasing shoot biomass. Nevertheless, while most of the work has been donc on *Arabidopsis thaliana*, the awareness of the importance of the root system in modulating plant growth, together with progress in sequencing and new molecular techniques, has caused renewed interest in understanding molecular mechanisms in crop species (Hochholdinger and Zimmermann 2009; Coudert et al. 2010).

In the scope of root development and its interaction with the soil, in this chapter, we propose to focus on the mechanisms involved in root branching, which is a major determinant of root system architecture. The plasticity of the root system represents indeed an important potential for plants, being sessile organisms, to adapt to the heterogeneity of their environment. The soil is a complex mixture of solid, gaseous, and liquid phases, wherein nutrients are unequally distributed. Plants have therefore developed a highly sophisticated regulatory system to control their root architecture, in

Root Genomics and Soil Interactions, First Edition. Edited by Martin Crespi.

response to environmental cues, by modulating intrinsic pathways to optimize their root distribution in the soil and consequently guarantee an optimal uptake of nutrients necessary for growth and development (reviewed in Croft et al. 2012).

Primary Root Structure and Development: Lessons from the Arabidopsis Model

Branching of roots occurs through the development of new meristems inside the primary parent root. We therefore first discuss briefly the structure and development of the primary root in *Arabidopsis*, the model species in which major insights were obtained, thanks to its simple root architecture (Dolan et al. 1993; Malamy and Ryan 2001; Scheres et al. 2002; Casimiro et al. 2003; Casson and Lindsey 2003; Ueda et al. 2005; Iyer-Pascuzzi et al. 2009; Peret et al. 2009).

The root can be divided in three main zones. The most distal, at the tip of the root, is the meristematic zone, where the so-called initial cells give rise to the tissues constituting the root. The initial cells are kept in an undifferentiated state by the neighboring quiescent center (Van den Berg et al. 1997), a mitotically less active region, composed of few central cells in *Arabidopsis*. Higher up, in the elongation zone, cells progressively stop dividing and start to expand longitudinally. Finally, cells differentiate and acquire their final cell fate in the maturation zone (Truernit et al. 2006), which can be recognized by the appearance of the anatomical structures of the vascular tissues.

Distinct cell types are then composing the mature root (Figure 1.1). The outer layers, endodermis, cortex, and epidermis are organized in concentric layers and present a radial organization toward the longitudinal axis of the primary root (Dolan et al. 1993). The epidermis, which is the outermost layer of the root, is in direct contact with the soil and is often designated as rhizodermis. It is composed of two populations of cells: one producing root hairs and the other nonhair cells (Schneider et al. 1997). The root hairs are responsible for the major part of the nutrient uptake from the soil (Muller and Schmidt 2004) and also play other important roles such as the initial contact with certain symbiotic partners (Gilroy and Jones 2000; Perrine-Walker et al. 2011). Cortex and endodermis constitute the ground tissue and are derived from one single initial cell in *Arabidopsis* (Dolan et al. 1993; Scheres et al. 1994). The stele is situated internal to these layers and comprises the vascular cylinder, consisting of two bilateral poles of xylem alternating with two bilateral poles of phloem separated by procambium cells (Dolan et al. 1993). The stele also contains a heterogeneous layer, the pericycle, interfacing the vascular cylinder and the outer layers, playing a predominant role in root architecture and root branching (Parizot et al. 2008).

Root Branching

In dicotyledonous plants, such as *Arabidopsis*, elaboration of the root system occurs postembryonically by the formation of numerous secondary roots from the primary root that was formed during embryogenesis. These new roots are comparable to the primary root in structure and will be able to reiterate the branching process by in turn initiating tertiary roots. Roots of second, third, and higher order are defined as lateral roots. The plant can also produce adventitious roots, which initiate mostly at the base of the hypocotyl. Different markers related with cell identity show a similar pattern in the primary and lateral roots (Malamy and Benfey 1997b; Laplaze et al. 2005), indicating the possibility of a common developmental pathway. This hypothesis is supported by a high number of mutants affected in genes involved in root patterning, such as *SHORTROOT*, *SCARECROW*, and *LONESOME HIGHWAY*, showing similar defects in the primary and lateral roots (Helariutta

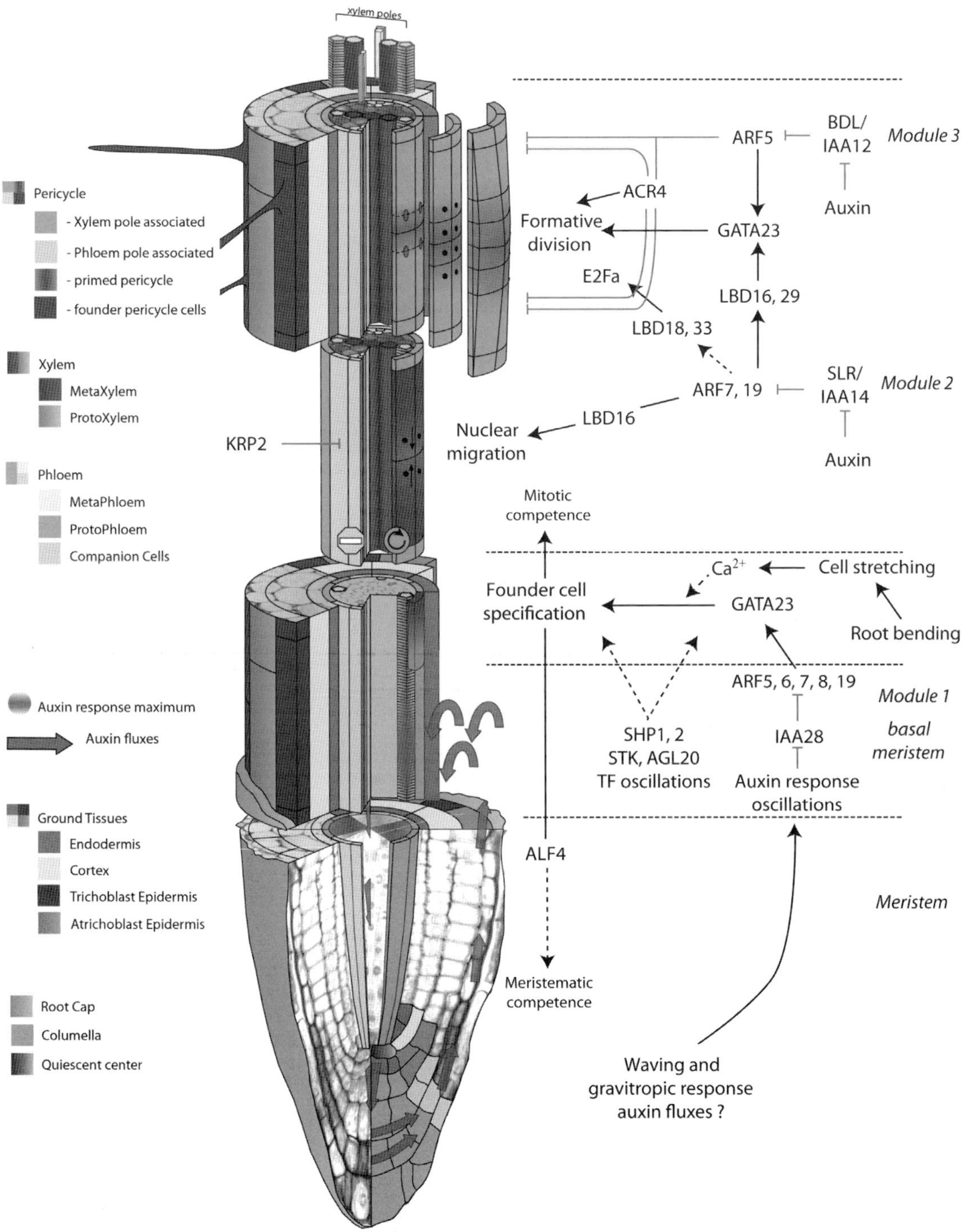

Figure 1.1 Structure of the primary root and different steps of lateral root initiation. See the text for detailed description. (For a color version of this figure, see the color plate section.)

et al. 2000; Wysocka-Diller et al. 2000; Parizot et al. 2008; Lucas et al. 2011). However, some differences can be observed in the behavior of the primary and the lateral roots toward external cues such as gravity and substrate nutrient concentrations (Zhang and Forde 1998; Mullen and Hangarter 2003; Bai and Wolverton 2011). A mutation in the gene *MONOPTEROS* impairs the apical–basal pattern formation of the embryo and leads to plants lacking a primary root, but that are still able to generate adventitious roots (Berleth and Jurgens 1993; Przemeck et al. 1996), indicating that early pathway(s) required for the embryonic formation of a root meristem are not required postembryonically. Also, a mutation in the gene *WOODEN LEG* has a major effect on the primary root development, with the suppression of the phloem elements and a drastic reduction in lateral root initiation (LRI), but does not affect the formation and branching of adventitious roots (Kuroha et al. 2006).

The monocots, such as maize, form different types of roots: primary, seminal, and adventitious roots, which can all form lateral roots. These root types also present similarities in their structures. However, mutants missing only a subset of these root types have been isolated, indicating that at least a part of the genetic program necessary for their formation is root-type specific (Woll et al. 2005; Hochholdinger and Tuberosa 2009).

Lateral Root Initiation

In *Arabidopsis* and most other dicotyledonous plants, lateral roots are formed from a restricted number of pericycle cells located in front of the xylem poles (Figure 1.1). The pericycle is a heterogeneous tissue composed of quiescent cells adjacent to the phloem poles and cells competent for LRI in front of the xylem poles (Beeckman et al. 2001; Parizot et al. 2008). Therefore, this layer presents a radial bilateral symmetry along the primary root, which reflects the diarch symmetry of the more internal vascular bundle as compared to the surrounding concentric radial layers of the outer tissues. The subpopulation of pericycle cells adjacent to the xylem poles can be considered as an extended meristem, as they conserve the ability to divide after leaving the root apical meristem (in contrast to the cells in front of the phloem poles), and give rise to the formation of a new organ (Beeckman et al. 2001; Casimiro et al. 2003). Although up to three adjacent pericycle cell files associated with each xylem pole are dividing during lateral root formation, cell lineage experiments have shown that only the central cell file will contribute significantly to the formation of the lateral root primordium (Kurup et al. 2005).

The first pericycle cell divisions that will give rise to a lateral root (i.e., formative divisions) can only be detected several millimeters above the primary root meristem, whereas in the lower part of a region named developmental window (Dubrovsky et al. 2006), it has been demonstrated that a subset of pericycle cells is already specified for LRI in a zone situated immediately above the primary root apical meristem, the basal meristem (De Smet et al. 2007; De Rybel et al. 2010b). The phytohormone auxin is most likely the signal triggering this priming, as auxin response recorded using the auxin response marker DR5 shows pulsations in the protoxylem cells of the basal meristem with a periodicity that can be correlated with the initiation of new lateral roots (Ulmasov et al. 1997; De Smet et al. 2007; De Rybel et al. 2010b; Moreno-Risueno et al. 2010). Up to now, different hypotheses have been proposed to explain the origin of these oscillating auxin response maxima in the protoxylem cells, and no consensus has been reached yet. Also the mechanism by which this auxin signal in the protoxylem cells is translated into the specification of founder cell identity in the neighboring pericycle cells is still unknown. Nevertheless, this intrinsic mechanism can be overruled, as the application of auxin on mature parts of the root above the basal meristem is still able to trigger LRI (Himanen et al. 2002), further reflecting the high plasticity of the root system.

The first morphological event preceding the division of two adjacent pericycle founder cells is the simultaneous migration of their nuclei to their common cell wall (De Smet et al. 2007). This migration is followed by an asymmetric anticlinal division of the pericycle cells, resulting in the formation of a core of small daughter cells flanked by larger cells (Dubrovsky et al. 2000). Successive anticlinal and periclinal divisions give rise to a lateral root primordium. Further divisions and elongation of the primordium cells result in the formation of a fully autonomous root, with a meristem similar to that of the primary root (Malamy and Benfey 1997b; Dubrovsky et al. 2001). Although the place of LRI differs between plant species, early patterning of the primordium is quite conserved (Casero et al. 1995; Malamy and Benfey 1997bb). The frequency of LRI in the *Arabidopsis* primary root can fluctuate in response to tropic and/or mechanical stimuli (De Smet et al. 2007; Ditengou et al. 2008; Laskowski et al. 2008; Lucas et al. 2008a). For example, a gravitropic stimulus applied to seedlings induces a lateral root at the place where the root bends to recover its normal growth angle (Lucas et al. 2008a).

Genomics of LRI

Most of the work on root development focused on the analysis of single mutants and allowed the discovery of many processes involved in the patterning of the different cell types within the primary root and in LRI. These studies show that root growth and development are complex processes with intricate pathways dealing with hormone biosynthesis, transport and signaling, tissue differentiation and dedifferentiation, nutrient sensing, cell divisions, and others (Iyer-Pascuzzi and Benfey 2009; Orman et al. 2011).

LRI has been increasingly studied over the last decade in the light of transcriptomics and proteomics. Material extraction for these analyses evolved from simple global root harvesting to more elaborated sampling techniques allowing a specific access to the tissues involved, such as laser capture microscopy (LCM; Woll et al. 2005) or cell sorting (De Smet et al. 2008). Also, the possibility to synchronize LRI circumvented the problems due to the discreteness of this event in plants grown in natural conditions (Himanen et al. 2002; Himanen et al. 2004). Different large-scale transcriptome and proteome studies have therefore been realized in different species, mainly *Arabidopsis* and maize, yielding information on various aspects of this de novo organogenesis: auxin response, asymmetric cell division, and pericycle tissue involvement. While these studies focused initially on the onset of lateral root development, a new era initiates now with the study of the formation and the patterning of the primordium after LRI and the emergence of the primordia from the parent root. Moreover, many other experiments are dealing with mechanisms related to LRI, such as meristem function, pericycle identity, and hormone treatment, and bring useful novel information, shedding light on this process. A list of omics experiments, directly or indirectly related to LRI is displayed in Table 1.1. A challenging task for the community will be to handle this wealth of data and search for appropriate system biology strategies to better understand the LRI process at the molecular level. To address this, a common effort of the biologists and the bioinformaticians is needed to design better experiments, rationalize and interpret the data, and make it accessible and understandable for the community. The most often characterized process in relation to LRI is the response to the hormone auxin.

Table 1.1 Omics Experiments dealing directly or indirectly with lateral root initiation. Publication year and reference, species, technique and experimental design, platform, tissues and preparation, pathways, treatments, and the biological process questioned by the experiment.

Year	Reference	Species	Omics technique	Experimental design	Platform / Technique
2003	Birnbaum et al.	Ath	Ts	Transcriptome atlas using FACS and GFP-reporter lines	Affymetrix ATH1
	Imin et al.	Mtr	Pr	Protein expression at the globular stages of somatic embryogenesis	Two-dimensional electrophoresis + MALDI-TOF-MS and MS/MS
	Menges et al.	Ath	Ts	Gene expression profiles during normal culture growth, synchronous cell cycle re-entry and synchronous cell cycle progression	Affymetrix ATH1
2004	Brinker et al.	Pta	Ts	Gene expression during adventitious root development by treating hypocotyls with auxin	Custom Pinus taeda microarray (2,178 cDNAs)
	Himanem et al.	Ath	Ts	Gene expression during lateral root inducible system (LRIS)	Spotted cDNA microarray (4k, Incyte Microarray Systems, Fremont, CA)
	Hochholdinger et al.	Zma	Pr	Comparison of protein accumulation between lrt1 mutant and wild type	Two-dimensional separation of proteins and MALDI-TOF mass spectrometry
	Nemhauser et al.	Ath	Ts	Gene expression upon treatment with Auxin or Brassinolide	Affymetrix ATH1
2005	Casson et al.	Ath	Ts	Gene expression in the apical and basal domains of the embryo at globular and heart stage	Affymetrix ATH1
	Nawy et al.	Ath	Ts	Tissue layer specific gene expression using FACS and GFP-reporter lines of the quiescent center and columella root cap	Affymetrix ATH1
	Okushima et al.	Ath	Ts	Comparison of gene expression upon auxin treatment between *arf7*, *arf19*, and *ar7 arf19* mutants and wild type	Affymetrix ATH1
	Vanneste et al.	Ath	Ts	Comparison of gene expression upon lateral root inducible system (LRIS) between *slr* mutant and wild type	Affymetrix ATH1
	Woll et al.	Zma	Ts	Comparison of gene expression between *rum1* mutant and wild type pericycle cells	Spotted cDNA microarray (12k , Generation II, Version A Iowa State MicroArray Facility)
2006	Che et al.	Ath	Ts	Gene expression during shoot, root and callus development in Arabidopsis tissue culture	Affymetrix ATH1
	Druka et al.	Hvu	Ts	Transcriptome atlas	Affymetrix Barley1 GeneChip
	Jiang et al.	Zma	Ts	Gene expression in the root cap, compared to the primary meristem and the quiescent center	Affymetrix Heterologous rice GeneChips
	Laskowski et al.	Ath	Ts	Gene expression during auxin treatment of seedlings roots	Custom Arabidopsis microarrays
	Lee et al.	Ath	Ts	Tissue layer specific gene expression using FACS and GFP-reporter lines of the phloem, xylem, and cortex	Affymetrix ATH1
	Levesque et al.	Ath	Ts	Comparison of tissue layer specific gene expression in *shr* mutant in comparison with wild type	Affymetrix ATH1
	Liu et al.	Zma	Pr	Comparison of protein accumulation between the mutant *rum1* and wild type	Two-dimensional electrophoresis + nanoHPLC-ESI-MS/MS massspectrometry
	Mouchel et al.	Ath	Ts	Comparison of gene expression in a natural allele variant of BRX gene with wild type, upon treatment with Auxin or Brassinolide	Affymetrix ATH1
2007	Brady et al.	Ath	Ts	Transcriptome atlas using FACS and GFP-reporter lines	Affymetrix ATH1
	Dembinsky et al.	Zma	TsPr r	Comparison of gene expression and protein accumulation between the stele, the pericycle and the ground tissues	Custom 12K microarray (GenII vA: GPL 4876) and 2D electrophoresis + nanoHPLC-ESI-MS/MS
	van Noorden et al.	Mtr	Pr	Comparison of protein accumulation between nodule initiation and auxin treatment in Medicago truncatula	Two-dimensional DIGE electrophoresis + MALDI-TOF/TOF MS

Table 1.1 *(Continued)*

Tissues	Tissue preparation					Pathways		Treatment								Biological process									
	Organ isolation	Micro dissection	LCM micro dissection	Cell Sorting	Other	Mutants	Biotic, abiotic factors	Auxin	NPA	Cytokinin	Brassinolide	Gibberellin	Stress	Biotic	Other	Lateral Root Priming	Lateral Root Initiation	Lateral Root Patterning	Root organogenesis	Embryo & Primary root	Pericycle	Primary root meristem	Callus & Regeneration	Tissue identity	Cell Cycle
Primary root sections and root cell layers																				D		D		D	
Embryonic cultures																			i	D					
Cell suspension															i		i								D
Hypocotyl segments								i									i	i	D						
Root section								i	i								D								
Primary roots						*lrt1*											D					i			
Entire seedlings								D			D						i								
Embryo apical and basal domains																			i	D		i			
Quiescent center and columella root cap																						i		D	
Entire seedlings						*arf7, arf19*		D									i								
Root segments						*slr*		i	i								D								
Pericycle cells						*rum1*											D				D				
Root segments								i		i							i	i				i	D		
Primary, seminal and crown roots																				i				D	
Primary meristem, quiescent center, root cap																						D		D	
Whole roots								D									D								
Phloem, xylem, and cortex																	i							D	
shr-2, pSHR::SHR:GR, *shr* root tips, sorted pWER::SHR:GFP						*shr*															D				
Primary roots						*rum1*											D								
Roots						BRX variant		D			D						i								
Primary root sections and root cell layers																	i	i		D	D	D		D	
Stele, Pericycle and ground tissues																	i				D			D	
Root segments						*sunn*	Nodulation	D						D			i	i	i						

Dicotyledonous species	
Ath	Arabidopsis thaliana
Mtr	Medicago truncatula
Gma	Glycine max
Pta	Pinus taeda
Ptr	Populus trichocarpa

Monocotyledonous species	
Zma	Zea maize
Osa	Oriza sativa
Hvu	Hordeum vulgare

Omics type	
Ts	Transcriptome
Tl	Translatome
Pr	Proteome
Mb	Metabolome

Information	
D	Direct
i	Indirect

(Continued)

Table 1.1 *(Continued)*

Year	Reference	Species	Omics technique	Experimental design	Platform / Technique
2008	Benedito et al.	Mtr	Ts	Transcriptome atlas	Affymetrix Medicago GeneChip
	De Smet et al.	Ath	Ts	Pericycle layer specific gene expression upon lateral root inducible system (LRIS)	Affymetrix ATH1
	Dineny et al.	Ath	Ts	Transcriptome atlas using FACS and GFP-reporter lines upon salt stress	Affymetrix ATH1
	Holmes et al.	Mtr	Ts	Comparison of gene expression between meristematic and non-meristematic root tissues	Affymetrix Medicago GeneChip
	Quesada et al.,	Ptr	Ts	Comparative analysis of the transcriptomes of Populus trichocarpa and Arabidopsis	42346 K 60-mer custom made microarray
2009	Gruber et al.	Mtr	Ts	Gene expression during salt-stress in root apexes	M. truncatula 16K+ microarrays (Mt16KOLI1)
	Jiao et al.	Osa	Ts	Transcriptome atlas using laser microdissection	70-mer custom made microarray (Qiagen/Operon)
	Mustroph et al.	Ath	Tl	Translatome atlas upon salt stress	Affymetrix ATH1
	Saleem et al.	Zma	Pr	Comparison of protein accumulation in immature embryo, cortical parenchyma and stele tissues between the mutant *rum1* and wild type	Two-dimensional electrophoresis + nanoHPLC-ESI-MS/MS massspectrometry
	Sena et al.	Ath	Ts	Gene expression during recovery of cell fates after cutting the root tip, and during the regeneration process	Affymetrix ATH-121501 full genome Arabidopsis microarray
2010	Libault et al	Gma	Ts	Transcriptome atlas using RNA-Seq	Illumina Solexa deep sequencing
	Liu et al.	Zma	Pr	Comparison of the pericycle layer specific protein accumulation between the mutant *rum1* and the wild type	Two-dimensional electrophoresis + nanoHPLC-ESI-MS/MS massspectrometry
	Matsuda et al.	Ath	Mb	Transcriptome atlas using metabolome data	LC-ESI-MS
	Moreno-Risueno et al.	Ath	Ts	Gene expression oscillations in the root tip	Affymetrix ATH1
	Muthreich et al.	Zma	Pr	Comparison of protein accumulation between the *rtcs* mutant and wild type immature embryos	Two-dimensional electrophoresis + nanoHPLC-ESI-MS/MS massspectrometry
	Saleem et al.	Zma	Pr	Protein accumulation in the cortical parenchyma and the stele of primary roots	Two-dimensional electrophoresis + nanoHPLC-ESI-MS/MS massspectrometry
	Severin et al.	Gma	Ts	Transcriptome atlas using RNA-Seq	Illumina deep sequencing
	Sugimoto et al.	Ath	Ts	Gene expression during in vitro callus induction	Operon Arabidopsis Genome Oligo Set Version 1.0 (Operon, Alameda, CA)
	Wang et al.	Osa	Ts	Transcriptome atlas on two different varieties	Affymetrix Rice GeneChip Genome Array
2011	Coudert et al.	Osa	Ts	Comparison of gene expression between *crl1* mutant and wild type stem bases	Affymetrix Rice GeneChip Genome Array
	Mathesius et al.	Gma	Pr	Comparison of protein accumulation between root apexes and root segments	Two-dimensional electrophoresis + MALDI-TOF/TOF MS
	Sekhon, R. S. et al.	Zma	Ts	Transcriptome atlas	Custom made microarray
	Takehisa et al.	Osa	Ts	Transcriptome atlas using laser microdissection	Agilent 4x44K microarray RAP-DB (G2519F#15241)
Unpublished	De Rybel et al.	Ath	Ts	Gene expression during lateral root initiation synchronization using a new chemical molecule, naxillin	Affymetrix ATH1
	Duclercq et al.	Ath	Ts	Auxin Cytokinin crosstalk and lateral root initiation	Affymetrix Tilling Microarray
	Middleton et al.	Ath	Ts	Gene expression during lateral root formation and emergence after synchronous induction of lateral root initiation	Affymetrix ATH1

Table 1.1 *(Continued)*

Tissues	Tissue preparation: Organ isolation	Micro dissection	LCM micro dissection	Cell Sorting	Other	Pathways: Mutants	Biotic, abiotic factors	Treatment: Auxin	NPA	Cytokinin	Brassinolide	Gibberellin	Stress	Biotic	Other	Biological process: Lateral Root Priming	Lateral Root Initiation	Lateral Root Patterning	Root organogenesis	Embryo & Primary root	Pericycle	Primary root meristem	Callus & Regeneration	Tissue identity	Cell Cycle
Roots, nodules and shoot tissues							Nodulation							D			i	i	i					D	
Xylem pole pericycle cells								i	i								D				D			D	i
Primary root sections and root cell layers							Salt stress						D							i		i		D	
Root tips and non meristematic root segments																		i		D		D			
Whole roots and shoot organs																						D		D	
Roots apexes							Salt stress															D			
Roots and shoot tissues																	i			D		D		D	
Roots							Hypoxy																		
Embryos, cortical parenchyma and stele						*rum1*											i			D				D	
Root tips								i		i								i	i	D			D		
Root tip, root , root hair, nodule, shoot tissues							Nodulation							D					i	D		D		D	
Pericycle cells						*rum1*											D				D				
Root, seed, shoot organs																						D		D	
Root segments corresponding to the oscillation zone																D									
Embryos						*rtcs*													D	D					
Cortical parenchyma and stele																	i							D	
Root, nodule, seed and shoot tissues							Nodulation							D					i	D		D		D	
Root, cotyledon, and petal explants								i		i							i	i	i			i	D	D	
Roots, callus, seed and shoot tissues								D		D		D		i						D			D	D	
Stem bases						*crl1*													D						
Root apexes and root segments																				i		D			
Root, seedling and shoot tissues																						D		D	
Sections, epi-exo, cortex and endo-stele tissues																	i	i				D		D	
Root segments								D							D	i	D								
Xylem pole pericycle cells								D		D							D				D				
Synchronized primordia																		D	D	i		i		i	

Dicotyledonous species	
Ath	Arabidopsis thaliana
Mtr	Medicago truncatula
Gma	Glycine max
Pta	Pinus taeda
Ptr	Populus trichocarpa

Monocotyledonous species	
Zma	Zea maize
Osa	Oriza sativa
Hvu	Hordeum vulgare

Omics type	
Ts	Transcriptome
Tl	Translatome
Pr	Proteome
Mb	Metabolome

Information	
D	Direct
i	Indirect

lrt1, lateral root 1; *slr*, solitary root 1; *arf*, auxin response factor; *rum1*, rootless with undetectable meristems 1; *shr*, short root; *wer*, werewolf; *brx*, brevis radix; *sunn*, supernumerary nodules; *rtcs*, rootless concerning crown and seminal roots; *crl1*, crown rootless 1.

IAA Proteins

Although little is known on the origin of the signals triggering the auxin maximum, which induces the priming of the pericycle cells or the migration of the nuclei leading to LRI, much more is known about the downstream auxin transduction pathways (Fukaki et al. 2007).

One of the first components of this pathway is *IAA14/SOLITARYROOT1* (*SLR1*), a member of the *Aux/IAA* gene family (Fukaki et al. 2002). Aux/IAA proteins are short-living nuclear proteins, most of which are induced early by auxin and act as active repressors of gene transcription. Accumulation of auxin causes the degradation of Aux/IAA proteins. Aux/IAA proteins are present all over the plant kingdom (Table 1.2). They are already described in monocots such as maize and rice (Jain et al. 2006; Wang et al. 2010b) and in dicots such as poplar (Kalluri et al. 2007), but are not found in animals (Riechmann et al. 2000). In the presence of auxin, Aux/IAA proteins bind to the F-box proteins TIR1, AFB1, 2, and 3 (Dharmasiri et al. 2005; Kepinski and Leyser 2005) and become targeted to the ubiquitin-dependent proteasome-degrading pathway (Zenser et al. 2001; Gray et al. 2003). Regulation of plant auxin sensitivity can be modulated by the control of TIR1 expression levels as it is the case during phosphate nutrient deprivation. Phosphate deprivation increases the expression of TIR1 and consequently causes Aux/IAA auxin response repressors to be degraded and LRI to be induced in *Arabidopsis* seedlings (Perez-Torres et al. 2008). Also, nitric oxide was recently shown to enhance TIR1–Aux/IAA interaction, which can explain how nitric oxide depletion blocks Aux/IAA protein degradation (Terrile et al. 2012).

A single point mutation in the conserved amino acid core sequence "GWPPV" in domain II of Aux/IAA proteins was shown to stabilize them (Ramos et al. 2001), leading to gain-of-function mutants. In the case of *IAA14/SLR1*, the resulting gain-of-function mutant *slr-1* fails to initiate formative divisions in the pericycle founder cells and consequently does not develop any lateral roots (Fukaki et al. 2002). Other gain-of-function mutants of some *Aux/IAA* gene family members, *IAA1/AXR5*, *IAA3/SHY2*, *IAA18/CRANE*, *IAA19/MSG2*, and *IAA28*, also show strong lateral root development phenotypes (Tian and Reed 1999; Rogg et al. 2001; Tatematsu et al. 2004; Yang et al. 2004; Uehara et al. 2008), indicating their involvement in this process. Nevertheless, none of these mutants totally block LRI as in the case of *slr-1*.

In addition, the loss-of-function *iaa14-1* mutant does not show any obvious phenotype (Okushima et al. 2005), as it is also the case for other loss-of-function *iaa* mutants (Rouse et al. 1998; Fukaki et al. 2002; Uehara et al. 2008), indicating a possible redundancy in the function of *Aux/IAA* genes. Recently, it was demonstrated that an OsIAA11 gain-of-function mutation caused the inhibition of lateral root development in rice (Zhu et al. 2011). Interestingly, on the basis of protein sequence, OsIAA11 is one of the closest homologs of IAA14 in *Arabidopsis* (Jain et al. 2006). Nevertheless, the mutation in OsIAA11 is semidominant for lateral root development, and the mutant phenotype differs from that of *Arabidopsis iaa14-1*, indicating that the auxin regulation pathways mediated by these two genes might be different (Zhu et al. 2011).

To unravel the pathways leading to the formative divisions downstream of SOLITARY ROOT (SLR), a comparative analysis was realized between the *slr-1* mutant and the wild type using the lateral root-inducible system (LRIS; Vanneste et al. 2005). It was observed that the mutation affected a number of cell cycle regulatory genes. The authors overexpressed the cell cycle regulator CYCD3;1 (which promotes G1 to S phase transition) in the *slr-1* background to rescue its rootless phenotype. Though inducing a few rounds of anticlinal divisions, this strategy failed in the formation of lateral root primordia, indicating that cell cycle activation in the pericycle cells of the mutant *slr-1* is not sufficient to get formative divisions and proper LRI (Vanneste et al. 2005).

Table 1.2 Plaza 2.5 output of LRI-related homologous genes in plant kingdom. The number of genes for each species and in each family corresponds to the result of a bioinformatics algorithm (http://bioinformatics.psb.ugent.be/plaza/). These results can be slightly different from the curated number of genes presented in the literature.

	Auxin Canonical Pathway					Cytokinin	GRAS Transcription Factors					MADS-Box Transcription Factors			
Green Plants Species	AUX	PINs	IAAs	ARFs	LOBs	IPT	GRAS	SHR	SCR	ALF4	ACR4	MADS	SHP1	SHP2	STK
Plaza Family Type	*HOM*	*HOM*	*HOM*	*HOM*	*HOM*	*HOM*	*HOM*	*ORTHO*	*ORTHO*	*HOM*	*ORTHO*	*HOM*	*ORTHO*	*ORTHO*	*ORTHO*
Plaza Family ID	*823*	*344*	*131*	*105*	*98*	*540*	*42*	*2280*	*3502*	*4876*	*7291*	*39*	*14121*	*13579*	10887
Land Plants															
Arabidopsis lyrata	6	9	33	23	39	7	41	1	1	1	1	64	1	1	1
Arabidopsis thaliana	4	9	28	25	37	8	37	1	1	1	1	60	1	1	1
Brachypodium distachyon	3	13	26	26	20	9	49	1	1	2	1	52			1
Carica papaya	4	9	21	21	31	6	46	1	1	1	1	93	1		1
Fragaria vesca	4	12	19	20	30	6	56	1	1	3	1	59	1		1
Glycine max	16	24	60	60	70	14	112	4	2	1	4	127	2		2
Lotus japonicus	5	9	17	40	38	6	52	1	1	2	1	53	1		1
Malus domestica	16	22	45	50	52	10	119	5	2	1	1	125	2	3	
Manihot esculenta	7	12	36	32	49	8	77	2	2	1	1	82	1		1
Medicago truncatula	3	9	20	40	28	18	66	2	1	5	1	62			1
Oryza sativa ssp. indica	5	20	32	33	36	10	63	1	2	1	1	69			1
Oryza sativa ssp. Japonica	5	15	31	31	30	9	58	1	2	1	1	55		1	1
Physcomitrella patens	7	6	2	20	25	3	38	2	3	2	1	17		6	
Populus trichocarpa	8	18	32	43	43	8	94	1	2	1	2	92			1
Ricinus communis	4	9	18	19	29	5	49	1	1	2	1	45			1
Selaginella moellendorffii	17	7	4	14	12	48	2	2	17			13			
Sorghum bicolor	5	13	25	27	27	8	78	2	1	1	1	55			1
Theobroma cacao	4	9	21	27	29	30	44	1	1	1	1	58	1		1
Vitis vinifera	4	13	23	30	37	6	42		1	1	1	51	1		1
Zea mays	5	16	39	43	35	10	88	3	1	1	1	81		1	1
Chlorophyta															
Ostreococcus tauri		1				1	1					1			
Ostreococcus lucimarinus		1				1	1					1			
Chlamydomonas reinhardtii		1					1					1			
Micromonas sp. RCC299		1		1		1	1					1			
Volvox carteri		1		1			1					1			

HOM, homologous families; ORTHO, orthologous families.

Auxin Response Factor Transcription Factors

Auxin response factors (ARFs) are transcription factors that act as repressors, except for a few of them (ARF4, 5, 6, 7, and 19) (Tiwari et al. 2003; Tiwari et al. 2004). They dimerize with Aux/IAA proteins (Guilfoyle and Hagen 2007), are present in most plant kingdom species (Table 1.2), are described in both monocots and dicots, such as rice and poplar (Wang et al. 2007; Kalluri et al. 2007), and are plant specific (Riechmann et al. 2000). The transcription factors NPH4/ARF7 and ARF19 have been shown to play a crucial role in LRI. They directly interact with IAA14/SLR, which causes their sequestration and prevents their positive transcriptional action (Fukaki et al. 2005). The degradation of IAA14/SLR leads to the release of these ARFs, which in turn activate the expression of their target genes. Consequently, the double mutant *arf7 arf19* presents a drastic LRI phenotype very similar to *slr-1* (Fukaki et al. 2002; Wilmoth et al. 2005). As this mutant is nevertheless still able to initiate some discrete roots, other ARFs may also be involved in this process (Fukaki et al. 2007).

A transcriptome experiment was performed to dissect further the SLR/ARF7/ARF19 pathway by comparing gene regulation upon auxin treatment in the mutants *arf7* and *arf19*, and in the double mutant *arf7 arf19* (Okushima et al. 2005). Several genes, which are not auxin-inducible anymore in the *arf7 arf19* double mutant compared to the wild type, such as lateral organ boundaries domain (LBD) proteins, are also impaired in the *slr-1* mutant background (Okushima et al. 2005; Vanneste et al. 2005). Among other results discussed in the next chapters, this observation indicates that even though the transcriptome experiment on the *arf7 arf19* mutant was performed with whole seedlings, it brings important information about specific processes such as LRI and formation. The functional analysis of genes concomitantly showing up in these transcriptome experiments greatly helped in understanding how a stabilized version of IAA14 blocks LRI and how ARF7 and ARF19 activate it (Okushima et al. 2007). Another auxin module, composed of MONOPTEROS/ARF5 acting downstream of BODENLOS (BDL)/IAA12, was recently reported to be activated during LRI. Overexpression of MONOPTEROS is able to rescue the phenotype of the *slr-1* mutant (De Smet et al. 2010), indicating that the auxin control on LRI is at least bimodal. It is most likely that other modules are involved in the control of root development as other ARFs have been shown to be implicated in adventitious and lateral roots (Tian et al. 2004; Mallory et al. 2005; Wang et al. 2005; Gutierrez et al. 2009).

Downstream of ARF Factors

LBD genes form a large family of plant-specific transcription factors that can be themselves the targets of ARF transcription factors. *LBD* genes are present all over the plant kingdom (Table 1.2) and are already described in dicots and monocots, such as *Arabidopsis*, maize, and rice (Yang et al. 2006; Taramino et al. 2007), but again do not exist in animals (Shuai et al. 2002). LBD16/ASYMMETRIC LEAVES2-LIKE18 (LBD16/ASL18) and LBD29/ASL16 appear to act immediately downstream of ARF7 and ARF19. They mediate auxin-regulated lateral root formation and their overexpression can rescue the *arf7 arf19* double mutant phenotype (Okushima et al. 2007; Lee et al. 2009). LBD16 was recently shown to regulate the polar migration of the nucleus and the asymmetric division of the lateral root founder cells (Goh et al. 2012). CROWN ROOTLESS1/ADVENTITIOUS ROOTLESS1 (CRL1/ARL1), which is among the closest rice homologs of AtLBD16 and AtLBD29, presents a conserved role in root development and regulates the formation of lateral and adventitious roots (Inukai et al. 2005; Liu et al. 2005). Similarly, ROOTLESS CONCERNING CROWN AND

SEMINAL ROOTS (RTCS) and RTCS-LIKE, which are also LBD proteins, are involved in the initiation and the formation of seminal and shoot-born roots in maize (Taramino et al. 2007).

Cell Cycle

LRI takes place above the differentiation zone, in a region of the root where cells hardly divide. The pericycle in front of the xylem poles makes an exception to this rule and divides asymmetrically to form a new lateral root. It was proposed, on the basis of cell cycle marker characterization and cell length measurements, that this peculiar behavior of the pericycle is insured by a particular mitotic status (Beeckman et al. 2001). Expression of the cyclin CYCA2;1 indicates the progression of the xylem pole pericycle cells toward the G2 phase of the cell cycle (Beeckman et al. 2001). The cyclin CYCB1;1 is expressed during the G2-M transition of the cell cycle, and because its transcription precedes the completion of cell division, it was selected and is now commonly used as a marker for LRI (Himanen et al. 2002; Himanen et al. 2004; Vanneste et al. 2005). Opposite to enhanced cell cycle progression at the xylem poles, expression of the cell cycle inhibitor KRP2 in the phloem pole pericycle cells indicates a blockage of these cells toward cell division (Himanen et al. 2002).

Several evidences indicate a role of the hormone auxin in the regulation of the cell cycle. Treatment of seedlings with *N*-1-naphthylphthalamic acid (NPA), which blocks the transport of auxin, triggers a general arrest of the cell cycle in the G1 phase and expression of KRP2 overall the pericycle (Himanen et al. 2002). Also, in the *slr-1* mutant, cell division in the pericycle is blocked, indicating a link between SLR/IAA14-dependent auxin signaling and cell cycle activation in the pericycle prior to LRI (Fukaki et al. 2002; Vanneste et al. 2005).

Nevertheless, the particular cell cycle fate of the pericycle does not seem to be controlled only by auxin. The nuclear protein ALF4 is necessary to maintain the mitotic competence of the xylem pole pericycle. A mutation in this gene blocks LRI, as observed with the mitotic marker CYCB1;1 : GUS, and causes on the other hand an increase in expression of the earlier cell cycle marker CDKB;1 : GUS. ALF4 is however not induced by auxin and might therefore control pericycle meristematic identity independently of the canonical hormone pathways (DiDonato et al. 2004).

Asymmetric and Formative Divisions

Cell cycle activation is naturally occurring in the xylem pole pericycle cells and leads to several rounds of asymmetric cell divisions, characteristic of the generation of new cell fates or organs (Scheres and Benfey 1999; De Smet and Beeckman 2011). In many plant species, LRI starts with an asymmetric and formative division of the founder cells (Casero et al. 1993; Casero et al. 1995; Malamy and Benfey 1997a). This notion is fundamental: the pericycle of the tomato (*Solanum lycopersicum*) mutant *diageotropica* (*dgt*) undergoes several rounds of divisions, which are not formative and fails to develop proper lateral roots (Ivanchenko et al. 2006). Similarly in *Arabidopsis*, bypassing the division blockage in the *slr-1* mutant by forcing cell cycle transition by means of CYCD3;1 overexpression gives a comparable phenotype, with stretches of pericycle cells undergoing proliferative divisions, but failing to form primordia (Vanneste et al. 2005).

A transcriptome experiment was performed especially to track regulation of gene expression occurring during the first asymmetric cell divisions of LRI. A transcript profiling of the xylem pole pericycle cells undergoing synchronous asymmetric divisions was performed making use of the LRIS in combination with cell sorting of the marker line J0121 (De Smet et al. 2008). Combining the list of genes significantly regulated during the treatment with previously published transcriptome experiments, the authors selected genes that are not involved in general cell cycle regulations, exception made for the G2-M phase transition (Menges et al. 2003), but which were already shown to be involved in LRI and to be dependent on the SLR/IAA14 auxin response pathway (Vanneste et al. 2005). This experiment allowed the discovery of the receptor kinase ARABIDOPSIS CRINKLY 4 (ACR4), which is expressed in the small daughter cells during the first asymmetric divisions in the xylem pole pericycle. This receptor was demonstrated to promote formative cell divisions in the pericycle and, once organogenesis has been started, to constrain their occurrence in the core cells of the future primordium by preventing divisions of the neighboring pericycle cells (De Smet et al. 2008). A comparable double role has been described in leaf development for the maize homolog CRINKLY4/CR4 (Becraft et al. 2001).

In *Arabidopsis*, asymmetric cell division is preceded by an auxin accumulation in the founder cells (Benkova et al. 2003; Dubrovsky et al. 2008; De Rybel et al. 2010b), which triggers the characteristic migration of two adjacent cell nuclei toward their common cell wall (De Smet et al. 2007; De Rybel et al. 2010b). The correct migration of the nucleus is dependent on signaling cascades involving IAA28, SLR/IAA14, ARF7, ARF19, and LBD16, highlighting the importance of auxin signaling in this process (De Rybel et al. 2010b; Goh et al. 2012). These observations are related to the observations made in the *gnom* mutant. The protein GNOM is a regulator of the intracellular vesicle trafficking involved in the recycling of the PIN1 auxin transporter to the membrane (Geldner et al. 2003). *gnom* mutants fail to express the marker ACR4 and to initiate proper formative divisions (De Smet et al. 2008). Another observation emphasizing the role of auxin is the possibility to rescue the phenotype of the *slr-1* plants overexpressing CYCD3;1 by auxin treatment, allowing to recover formative divisions and LRI (De Smet et al. 2010). Also recently, the transcription factor E2Fa was shown to be an essential component linking auxin pathways and asymmetric cell division triggering LRI. E2Fa expression is indeed regulated by the LBD18 / LBD33 dimer that is under direct control of the auxin-signaling pathway (Berckmans et al. 2011).

Digging into LRI, the Priming of the Pericycle

Although relatively much is discovered concerning the molecular pathways connecting the auxin signals driving the formative cell divisions of the pericycle and further lateral root development, very little is known on the origin of this signal and on the very early steps preceding LRI. Different hypotheses have been proposed to explain the origin of the recurrent auxin response oscillations observed in the basal meristem. One explanation would be that the auxin fluxes responsible for the gravitropic response in the root tip are driving LRI. Gravitropism and LRI are indeed coregulated in *Arabidopsis* (Lucas et al. 2008a). During gravitropic response, shoot-derived auxin is redistributed basipetally, relative to gravity, from the columella root cap cells to the epidermis cells, through the lateral root cap. Higher up in the region of the basal meristem, auxin is redirected to the stele by PIN1 and PIN2 transporters (Blilou et al. 2005; Leyser 2006; Ditengou et al. 2008). This feedback loop could explain the auxin response oscillations (Lucas et al. 2008b). A transcriptomic approach was recently performed to test the relationship between the auxin response maxima in the basal meristem and the control of root branching and bending (Moreno-Risueno et al. 2010). The authors tracked gene expression fluctuations in correlation to the auxin response oscillations and showed that these last ones were not sufficient to trigger the periodic specification of pericycle founder cells. This specification would rather be the consequence of the oscillating expression of transcription factors such as SHATTERPROOF1 and 2 (SHP1 and SHP2), SEEDSTICK (STK), and AGAMOUS-LIKE20 (AGL20) (Moreno-Risueno et al. 2010).

Another transcriptomic approach shed some light on the mechanisms involved in pericycle founder cells specification. The meta-analysis of different datasets related to LRI allowed the selection of genes that are differentially expressed between xylem and phloem pole pericycle cells, not expressed in other radial tissue layers, responsive to auxin, and that are predicted to have a role in asymmetric division but not in general cell-cycle phase transitions (De Rybel et al. 2010b; Parizot et al. 2010). Among the candidates, the authors characterized GATA23, a member of the GATA-type family of transcription factors that are known to have several regulatory roles in cell fate specification (Reyes et al. 2004). GATA23 is positively regulated by auxin treatment. It is expressed in a region above the basal meristem in correlation with the oscillating auxin response maxima, probably through an IAA28-controlled auxin response pathway, and induces founder cell identity (De Rybel et al. 2010b).

Apart from auxin, other hormones have been described to play an important role in LRI. LRI is inhibited in mutants with an increased level of ethylene (Nodzon et al. 2004; Ivanchenko et al. 2008; Prasad et al. 2010) and by abscisic acid treatment (De Smet et al. 2006). Cytokinins are known to inhibit LRI by blocking cell cycle progression in the founder cells and interfering with the auxin maxima (Li et al. 2006; Laplaze et al. 2007). Also, an increasing number of studies focus on the interplay of cytokinins with auxin pathways (Dello Ioio et al. 2008; Su et al. 2011; Zheng et al. 2011). Future experiments dedicated to this cross talk in relation to LRI should provide a lot of information on the synergies and antagonisms in these pathways (Duclercq et al. 2010). Furthermore, as it is the case for auxin, cytokinins are involved in multiple symbiotic processes (Hirsch et al. 1997; Barker and Tagu 2000; Lohar et al. 2004; Frugier et al. 2008). It has been shown that during symbiotic interactions, the bacteria can hijack the plant lateral root development program to form a nodule (Mathesius et al. 2000). The discovery of important genes commonly regulated in the different symbioses (Gherbi et al. 2008; Markmann et al. 2008), combined with histological similarities during development, such as divisions in the xylem pole pericycle (Péret et al. 2007), point to common regulatory mechanisms between lateral root and symbiotic nodule development pathways. Consequently, it may be valuable to compare these developmental processes at the transcriptomic or proteomic level, using published or upcoming datasets (Van Noorden et al. 2007; Benedito et al. 2008; Libault et al. 2010).

Primordium Patterning, Emergence, and Activation

A succession of stereotypic divisions accompanies the passage from pericycle founder cells to the emergence of a new lateral root through different well-characterized developmental stages (Malamy and Benfey 1997b). The first anticlinal and asymmetric divisions of the pericycle produce a single-layered primordium and are followed by periclinal divisions forming an inner and an outer layer. Further anticlinal and periclinal divisions yield a dome-shaped primordium that eventually emerges from the parent root and becomes autonomous. At this point, divisions of the different tissues formed in the primordium stop, and divisions of the initials cells that become responsible for root growth and elongation start (Malamy and Benfey 1997b).

In addition to its role in triggering LRI, auxin also plays an important role in the patterning and the emergence of the primordium. From the first stage on, an auxin gradient is formed in the central cells and is maintained in the tip of the primordium during its development (Benkova et al. 2003). Two-step models have been proposed for lateral root formation, and particularly for the establishment and the maintenance of this auxin gradient (Celenza et al. 1995; Laskowski et al. 1995; Sussex et al. 1995; Bhalerao et al. 2002). The AUX1 auxin influx carrier facilitates auxin uptake in the central core cells of the primordium as early as the first stage (Marchant et al. 2002). Later, when it reaches a size of about three to five cell layers, the primordium is able to synthesize its own auxin (Ljung et al. 2005). A combination of PIN polar auxin transport proteins helps in maintaining the gradient with a maximum at the tip. Mutations or mis-expression of these PIN proteins lead to abnormally shaped and/or nonemerged primordia (Benkova et al. 2003). It was recently shown that the nitrate transporter NRT1.1 not only transports nitrate but also facilitates auxin uptake in primordia and lateral roots, stimulating lateral root development and growth (Krouk et al. 2010). However, this NRT1.1-dependent auxin transport is inhibited by nitrate itself, defining another connection between nutrient and hormone signaling during organ development (Perez-Torres et al. 2008; Krouk et al. 2010). Downstream of auxin transport, the IAA14/SLR auxin response module was shown to be also involved during lateral root formation. Tissue-specific expression of a dominant negative form of the IAA14/SLR transcription factor in developing primordia leads to abnormal and/or arrested lateral root development (Fukaki et al. 2005).

Cytokinin hormones also play a crucial role in patterning the new lateral root primordium. Tissue-specific expression of a cytokinin biosynthesis gene, *IPT* (*ISOPENTENYL TRANSFERASE*) in the xylem pole pericycle, blocks lateral root patterning, especially by interfering with the correct expression of PIN genes and misleading a correct auxin gradient (Laplaze et al. 2007). Attempts to bypass this blockage by adding auxin to plants expressing IPT in the xylem pole pericycle, and therefore in the founder cells, caused cell proliferation but failed to induce the formation of patterned primordia. Cytokinin treatments of the root also perturb lateral root patterning and provoke abnormal and ectopic divisions, which give rise to a flattened primordium. However, the main effect of cytokinins appears to be prior to the formation of the primordium itself, as the expression of IPT in the developing primordium does not impair its development as much as in the case of earlier expression in the founder cells (Laplaze et al. 2007). Expression patterns of members of the IPT gene family are consistent with these observations; IPT3, which is expressed in the pericycle, presents higher expression in the phloem pole compared to the xylem pole pericycle, and IPT5 is expressed in young lateral root primordia (Miyawaki et al. 2004).

Next to its role in promoting formative cell divisions during early lateral root formation, ACR4 also plays a role in lateral root patterning. *acr4* mutants present altered expression of the boundary marker LBD5 and an abnormal auxin gradient as seen with the DR5 marker (De Smet et al. 2008). ACR4 encodes a membrane-associated receptor kinase, which indicates the possibility of novel signaling molecules, such as peptides, regulating LRI and primordium patterning. This vision is supported by the cell division inhibitory potential of ACR4 on cells where it is not expressed, indicating probable regulation by a noncell-autonomous signal (De Smet et al. 2008). Similar to observations in *acr4* mutants, mutation in the AP2 (APETALA2)/EREBP (ETHYLENE RESPONSIVE ELEMENT BINDING PROTEIN) transcription factor PUCHI leads to abnormal divisions in the lateral root primordium and ectopic divisions in the neighboring cells, giving rise to a flattened primordium (Hirota et al. 2007). Other genes belonging to *CUP-SHAPED COTYLEDON* and *LOB* (*LATERAL ORGAN BOUNDARIES*) gene families exhibit an expression pattern delineating the newborn organs. These families have been extensively described in the shoot of the plant and, among other roles, define the boundaries of emerging organs. Analogous expression patterns in the root suggest a similar role of these genes in restricting divisions in the cells neighboring the developing organ (Shuai et al. 2002; Vroemen et al. 2003; Laufs et al. 2004; Laplaze et al. 2005).

On its way to the outer world, the primordium still needs to overcome a last step, which consists in passing through the overlaying tissues, that is, endodermis, cortex, and epidermis. Although these tissues are composed of one layer of cells in *Arabidopsis* (Dolan et al. 1993), the cortex can be composed of up to 15 layers in other plant species such as maize (Hochholdinger et al. 2009). Recently, several cell wall remodeling (CWR) enzymes have been discovered in *Arabidopsis*, among which a pectate lyase (PLA2), a polygalacturonase (ADPG2), a pectin methyl-esterase (PME1), an expansin (EXP17), a xyloglucan:xyloglucosyl transferase (XTR6); and a glycosyl hydrolase (GLH17) are expressed in the overlaying tissues and most probably facilitate lateral root emergence by causing cell separation (Henrissat 1991; Neuteboom et al. 1999; Cosgrove 2000; Marin-Rodriguez et al. 2002; Vissenberg et al. 2005; Laskowski et al. 2006; González-Carranza et al. 2007; Swarup et al. 2008). Auxin also plays a role in this process and was recently reported to induce *CWR* genes and therefore facilitating cell wall separation (Swarup et al. 2008). SHY2/IAA3 was shown to mediate auxin response in the endodermis while the SLR/IAA14 / ARF7-ARF19 module induces expression of *LAX3* in the cortex and the epidermis (Tian and Reed 1999; Swarup et al. 2008). The expression of the LAX3 auxin influx carrier protein consequently increases the auxin concentration in the cortex, and later in the epidermis, in turn inducing CWR genes and facilitating primordium outgrowth.

The activation of the meristem hallmarks the transition between the primordium and a lateral root growing autonomously. It occurs during or soon after the protrusion through the epidermal layer. The importance of auxin homeostasis was again shown through the characterization of mutants. A mutation in the gene *ALF3* (*ABERRANT LATERAL ROOT FORMATION 3*) arrests lateral root growth soon after emergence but can be rescued by exogenous auxin application (Celenza et al. 1995). Mutation in the gene *MDR1* (*MULTIDRUG RESISTANCE 1*) impairs acropetal transport of auxin and causes reduced elongation and/or arrest of the lateral roots (Wu et al. 2007). Taken together, these observations indicate a step at which the lateral root becomes more, but not fully, independent from the primary root for the auxin supply necessary for its growth and elongation.

Rise of New Technologies to Understand Lateral Root Development

Inducing LRI

The problem of the discreteness of the LRI events along the main root could firstly be circumvented by the use of a combination of treatments: the plants are first grown on a medium containing NPA, which inhibits auxin transport, and later transferred to a medium containing 1-naphthaleneacetic acid, which synchronously and homogeneously induces LRI in the entire pericycle of *Arabidopsis* seedlings (Himanen et al. 2002). The LRIS was used in different transcriptomics experiments (Himanen et al. 2004; Vanneste et al. 2005; De Smet et al. 2007). However, a limit of this system is the use of auxin treatments, which trigger vast and pleiotropic changes in gene expression. This limitation could be elegantly bypassed by applying the same treatment to the mutant *slr-1* that does not form lateral roots under these conditions, in parallel to the wild-type plant, thus allowing the distinction of specific LRI genes from general auxin response pathways (Vanneste et al. 2005). In addition, chemical genomics, which makes use of small chemical compounds that can potentially interfere with metabolic pathways in planta allowed the discovery of new molecules capable of inducing LRI with a narrower effect on global gene expression (De Rybel et al. 2010a). Applied to the LRIS, such molecules should greatly help in deciphering the core pathways necessary and sufficient for LRI.

Mechanical external factors such as gravistimulation or bending of the root also induce LRI, on the external side of the bend (Ditengou et al. 2008; Lucas et al. 2008a; Richter et al. 2009).

Upon gravistimulation, the differential redistribution of auxin fluxes in the root tip creates an auxin maximum in the lower side of the root tip, inhibiting cell elongation and triggering a bend bringing the root tip back in the direction of the gravity vector (Ottenschlager et al. 2003; Swarup et al. 2005). Meanwhile, PIN relocation at the convex side results in an auxin maximum in the protoxylem cells, which can be the cause of LRI (Ditengou et al. 2008). During mechanical bending, it is hypothesized that the stretching of the cells results in cytoplasmic increases of Ca^{2+}, which in turns induces LRI (Richter et al. 2009). The possibility to synchronously induce LRI on a large population of plants by reorientation of the growing plates (i.e., gravistimulation) opens new perspectives for developmental studies. Among other possibilities, it will allow the production of transcriptome datasets following lateral root formation, from the initiation and through its development, without interfering massively with auxin pathways (Middleton et al. 2011).

Spatiotemporal Maps of Cell Types and Developmental Zones

A spatiotemporal gene expression map of individual root cell types and developmental zones in *Arabidopsis* was constructed by combining the use of tissue-specific marker lines with cell sorting (Birnbaum et al. 2005; Nawy et al. 2005; Lee et al. 2006; Brady et al. 2007). For the first time, these datasets made it possible to get precise in silico root expression profiles for a high number of genes. Furthermore, different studies can be envisaged based on tissue specificity, on radial and longitudinal expression patterns, and allow the construction of networks based on a combination of these observations (Brady et al. 2011). Cell sorting is however not easily transposable to all species, particularly due to the lack of tissue-specific marker lines but also due to the difficulties to prepare protoplasts of plants with a high number of cell layers and secondary cell wall thickening. This problem can be circumvented by using micro-dissection or LCM (Nelson et al. 2006; Ithal et al. 2007; Nelson et al. 2008). Different studies took profit of such approaches, with various degrees of resolution. In rice, the transcriptome of different cell types was analyzed throughout the plant (Jiao et al. 2009) and was recently ameliorated using better defined datasets specific for the root (Takehisa et al. 2012). The possibility to compare these transcriptome data between two different species provides the possibility, with an amelioration of the resolution for the root tissues, to isolate common regulatory genes responsible for changes in root architecture (Wang et al. 2010a).

Other studies comprising the root have been realized with a lower degree of resolution. These studies however contain information that can be valuable to increase our knowledge on primary and lateral root development. Transcriptome atlases have been realized in other monocotyledonous plants such as barley and maize (Druka et al. 2006; Sekhon et al. 2011) and also in dicotyledonous plants such as soybean and medicago (Benedito et al. 2008; Libault et al. 2010; Severin et al. 2010), which presents the advantage to contain different stages of nodulation, a process that can be compared to LRI and development (Mathesius et al. 2000). In addition, recent studies have taken advantage of this strategy and implemented the measurement of other regulatory mechanisms of gene expression, reflected in translatome, proteome, and metabolome analyses (Mustroph et al. 2009; Petersson et al. 2009; Matsuda et al. 2010).

ComparativOmics, the Future

While the importance of roots for nutrient acquisition and growth in crops is becoming increasingly clear, the molecular processes of LRI and development are very poorly understood in these species and remain mostly restricted to the model plant *Arabidopsis* (Den Herder et al. 2010; Gewin 2010). LRI is the major mechanism shared by angiosperms to control their root architecture. Consequently, it is very likely that it evolved from endogenous pathways present in a common ancestor and remained conserved throughout evolution. The conservation of gene families for which a role in LRI is known, among other anatomical and physiological observations, supports this theory (Table 1.2; Grunewald et al. 2007; Lau et al. 2008; Movahedi et al. 2011; Van Bel et al. 2012). Nevertheless, it is unclear whether orthology retained functionality through evolution. Genome duplication events and adaptation to environmental conditions may actually have led to gene diversity that highly complicates the task of looking for functional homologs only based on phylogenetic approaches (Van de Peer et al. 2009). In consequence, it is very complicated to define relevant orthologous genes from one species to another.

A lot of datasets have been generated during the last decade, which are directly or indirectly related to LRI (Table 1.2). In *Arabidopsis*, an effort to rationalize datasets together, in a compendium made accessible for the scientific community, allowed the discovery of new key regulators (De Rybel et al. 2010b; Parizot et al. 2010) and the generation of multiple networks (Lee et al. 2010; Brady et al. 2011). The possibility to compare the involvement of genes in different processes and in different species may allow to draw a link between orthologs and highlight the importance to realize "omics" studies in different species (Galbraith and Edwards 2010). A comparison of the transcriptional atlases in *Arabidopsis* and poplar already indicated the possibility to study organ-specific orthologs (Quesada et al. 2008). In this perspective, the rise of new sequencing technologies will be of great help in transferring root genomics knowledge to crop species (Varshney et al. 2009).

Acknowledgments

This work was supported by a grant from the Interuniversity Attraction Poles Programme (IAP VII/29), initiated by the Belgium Science Policy Office. Boris Parizot is supported by the Research Foundation Flanders (FWO, grant 3G002911). We thank Marlies Demeulenaere for comments that greatly improved the manuscript.

References

Bai, H. and Wolverton, C. (2011) Gravitropism in lateral roots of Arabidopsis pgm-1 mutants is indistinguishable from that of wild-type. *Plant Signaling and Behavior*, 6, 1423–1424.

Barker, S.J. and Tagu, D. (2000) The roles of auxins and cytokinins in mycorrhizal symbioses. *Journal of Plant Growth Regulation*, 19, 144–154.

Becraft, P.W. et al. (2001) The maize CRINKLY4 receptor kinase controls a cell-autonomous differentiation response. *Plant Physiology*, 127, 486–496.

Beeckman, T. et al. (2001) The peri-cell-cycle in Arabidopsis. *Journal of Experimental Botany*, 52, 403–411.

Benedito, V.A. et al. (2008) A gene expression atlas of the model legume *Medicago truncatula*. *Plant Journal*, 55, 504–513.

Benkova, E. et al. (2003) Local, efflux-dependent auxin gradients as a common module for plant organ formation. *Cell*, 115, 591–602.

Berckmans, B. et al. (2011) Auxin-dependent cell cycle reactivation through transcriptional regulation of Arabidopsis E2Fa by lateral organ boundary proteins. *Plant Cell*, 23, 3671–3683.

Berleth, T. and Jurgens, G. (1993) The role of the monopteros gene in organising the basal body region of the Arabidopsis embryo. *Development*, 118, 575–587.

Bhalerao, R.P. et al. (2002) Shoot-derived auxin is essential for early lateral root emergence in Arabidopsis seedlings. *Plant Journal*, 29, 325–332.

Birnbaum, K. et al. (2005) Cell type-specific expression profiling in plants via cell sorting of protoplasts from fluorescent reporter lines. *Nature Methods*, 2, 615.

Birnbaum, K. et al. (2003) A gene expression map of the Arabidopsis root. *Science*, 302, 1956–1960.

Blilou, I. et al. (2005) The PIN auxin efflux facilitator network controls growth and patterning in Arabidopsis roots. *Nature*, 433, 39–44.

Brady, S.M. et al. (2007) A high-resolution root spatiotemporal map reveals dominant expression patterns. *Science*, 318, 801–806.

Brady, S.M. et al. (2011) A stele-enriched gene regulatory network in the Arabidopsis root. *Molecular Systems Biology*, 7, 459.

Brinker, M. et al. (2004) Microarray analyses of gene expression during adventitious root development in *Pinus contorta*. *Plant Physiology*, 135, 1526–1539.

Casero, P.J. et al. (1995) Lateral root initiation by asymmetrical transverse divisions of pericycle cells in 4 plant species: *Raphanus sativus*, *Helianthus annuus*, *Zea mays*, and *Daucus carota*. *Protoplasma*, 188, 49–58.

Casero, P.J. et al. (1993) Lateral root initiation by asymmetrical transverse divisions of pericycle cells in adventitious roots of *Allium cepa*. *Protoplasma*, 176, 138–144.

Casimiro, I. et al. (2003) Dissecting Arabidopsis lateral root development. *Trends in Plant Science*, 8, 165–171.

Casson, S. et al. (2005) Laser capture microdissection for the analysis of gene expression during embryogenesis of Arabidopsis. *Plant J*, 42, 111–123.

Casson, S.A. and Lindsey, K. (2003) Genes and signalling in root development. *New Phytologist*, 158, 11–38.

Celenza, J.J.L. et al. (1995) A pathway for lateral root formation in *Arabidopsis thaliana*. *Genes and Development*, 9, 2131–2142.

Che, P. et al. (2006) Gene expression programs during shoot, root, and callus development in Arabidopsis tissue culture. *Plant Physiology*, 141, 620–637.

Cosgrove, D.J. (2000) Loosening of plant cell walls by expansins. *Nature*, 407, 321–326.

Coudert, Y. et al. (2011) Transcript profiling of crown rootless1 mutant stem base reveals new elements associated with crown root development in rice. *BMC Genomics*, 12, 387.

Coudert, Y. et al. (2010) Genetic control of root development in rice, the model cereal. *Trends in Plant Science*, 15, 219–226.

Croft, S. et al. (2012) Optimal root proliferation strategies: the roles of nutrient heterogeneity, competition and mycorrhizal networks. *Plant and Soil*, 351, 191–206.

De Rybel, B. et al. (2010a) Impaired IBA transport in the root tip is associated with enhanced lateral initiation. International Conference on Plant Growth Substances, Tarragone, Spain.

De Rybel, B. et al. (2010b). A novel aux/IAA28 signaling cascade activates GATA23-dependent specification of lateral root founder cell identity. *Current Biology*, 20, 1697–1706.

De Smet, I. and Beeckman, T. (2011) Asymmetric cell division in land plants and algae: the driving force for differentiation. *Nature Reviews Molecular Cell Biology*, 12, 177–188.

De Smet, I. et al. (2010) Bimodular auxin response controls organogenesis in Arabidopsis. *Proceedings of the National Academy of Sciences of the United States of America*, 107, 2705–2710.

De Smet, I. et al. (2007) Auxin-dependent regulation of lateral root positioning in the basal meristem of Arabidopsis. *Development*, 134, 681–690.

De Smet, I. et al. (2008) Receptor-like kinase ACR4 restricts formative cell divisions in the Arabidopsis root. *Science*, 322, 594–597.

De Smet, I. et al. (2006) A novel role for abscisic acid emerges from underground. *Trends in Plant Science*, 11, 434–439.

Dello Ioio, R. et al. (2008) A genetic framework for the control of cell division and differentiation in the root meristem. *Science*, 322, 1380–1384.

Dembinsky, D. et al. (2007) Transcriptomic and proteomic analyses of pericycle cells of the maize primary root. *Plant Physiology*, 145, 575–588.

Den Herder, G. et al. (2010) The roots of a new green revolution. *Trends in Plant Science*, 15, 600–607.

Dharmasiri, N. et al. (2005) The F-box protein TIR1 is an auxin receptor. *Nature*, 435, 441–445.

DiDonato, R.J. et al. (2004) Arabidopsis ALF4 encodes a nuclear-localized protein required for lateral root formation. *Plant Journal*, 37, 340–353.

Dinneny, J.R. et al. (2008) Cell identity mediates the response of Arabidopsis roots to abiotic stress. *Science*, 320, 942–945.

Ditengou, F.A. et al. (2008) Mechanical induction of lateral root initiation in *Arabidopsis thaliana*. *Proceedings of the National Academy of Sciences of the United States of America*, 105, 18818–18823.

Dolan, L. et al. (1993) Cellular organisation of the *Arabidopsis thaliana* root. *Development*, 119, 71–84.

Druka, A. et al. (2006) An atlas of gene expression from seed to seed through barley development. *Functional and Integrative Genomics*, 6, 202–211.

Dubrovsky, J.G. et al. (2000) Pericycle cell proliferation and lateral root initiation in Arabidopsis. *Plant Physiology*, 124, 1648–1657.

Dubrovsky, J.G. et al. (2006) Lateral root initiation in Arabidopsis: developmental window, spatial patterning, density and predictability. *Annals of Botany*, 97, 903–915.
Dubrovsky, J.G. et al. (2001) Early primordium morphogenesis during lateral root initiation in *Arabidopsis thaliana*. *Planta*, 214, 30–36.
Dubrovsky, J.G. et al. (2008) Auxin acts as a local morphogenetic trigger to specify lateral root founder cells. *Proceedings of the National Academy of Sciences of the United States of America*, 105, 8790–8794.
Duclercq, J. et al. (2010) Identification of the molecular components of the auxin–cytokinin interaction during lateral root organogenesis. International Conference on Plant Growth Substances, Tarragona, Spain.
Feix, G. et al. (2002) Maize root system and genetic analysis of its formation. In: *Plant Roots: The Hidden Half*, third edition. Marcel Dekker, New York, 239–248.
Frugier, F. et al. (2008) Cytokinin: secret agent of symbiosis. *Trends in Plant Science*, 13, 115–120.
Fukaki, H. et al. (2005) Tissue-specific expression of stabilized SOLITARY-ROOT/IAA14 alters lateral root development in Arabidopsis. *Plant Journal*, 44, 382–395.
Fukaki, H. et al. (2007) Auxin-mediated lateral root formation in higher plants. *International Review of Cytology*, 256, 111–137.
Fukaki, H. et al. (2002) Lateral root formation is blocked by a gain-of-function mutation in the *SOLITARY-ROOT/IAA14* gene of Arabidopsis. *Plant Journal*, 29, 153–168.
Galbraith, D.W. and Edwards, J. (2010) Applications of microarrays for crop improvement: here, there, and everywhere. *Bioscience*, 60, 337–348.
Geldner, N. et al. (2003) The Arabidopsis *GNOM ARF-GEF* mediates endosomal recycling, auxin transport, and auxin-dependent plant growth. *Cell*, 112, 219–230.
Gewin, V. (2010) Food: an underground revolution. *Nature*, 466, 552–553.
Gherbi, H. et al. (2008) SymRK defines a common genetic basis for plant root endosymbioses with arbuscular mycorrhiza fungi, rhizobia, and Frankia bacteria. *Proceedings of the National Academy of Sciences of the United States of America*, 105, 4928–4932.
Gilroy, S. and Jones, D.L. (2000) Through form to function: root hair development and nutrient uptake. *Trends Plant Sci*, 5, 56–60.
Goh, T. et al. (2012) The establishment of asymmetry in Arabidopsis lateral root founder cells is regulated by LBD16/ASL18 and related LBD/ASL proteins. *Development*, 139, 883–893.
González-Carranza, Z.H. et al. (2007) Expression of polygalacturonases and evidence to support their role during cell separation processes in *Arabidopsis thaliana*. *Journal of Experimental Botany*, 58, 3719–3730.
Gray, W.M. et al. (2003) Arabidopsis SGT1b is required for SCF(TIR1)-mediated auxin response. *Plant Cell*, 15, 1310–1319.
Gruber, V. et al. (2009) Identification of transcription factors involved in root apex responses to salt stress in *Medicago truncatula*. *Molecular Genetics and Genomics*, 281, 55–66.
Grunewald, W. et al. (2007) Developmental biology of roots: one common pathway for all angiosperms? *International Journal of Plant Developmental Biology*, 1, 212–225.
Guilfoyle, T.J. and Hagen, G. (2007) Auxin response factors. *Current Opinion in Plant Biology*, 10, 453–460.
Gutierrez, L. et al. (2009) Phenotypic plasticity of adventitious rooting in Arabidopsis is controlled by complex regulation of AUXIN RESPONSE FACTOR transcripts and microRNA abundance. *Plant Cell*, 21, 3119–3132.
Helariutta, Y. et al. (2000) The *SHORT-ROOT* gene controls radial patterning of the Arabidopsis root through radial signaling. *Cell*, 101, 555–567.
Henrissat, B. (1991) A classification of glycosyl hydrolases based on amino-acid-sequence similarities. *Biochemical Journal*, 280, 309–316.
Himanen, K. et al. (2002) Auxin-mediated cell cycle activation during early lateral root initiation. *Plant Cell*, 14, 2339–2351.
Himanen, K. et al. (2004) Transcript profiling of early lateral root initiation. *Proceedings of the National Academy of Sciences of the United States of America*, 101, 5146–5151.
Hirota, A. et al. (2007) The auxin-regulated *AP2/EREBP* gene *PUCHI* is required for morphogenesis in the early lateral root primordium of Arabidopsis. *Plant Cell*, 19, 2156–2168.
Hirsch, A.M. et al. (1997) The role of phytohormones in plant–microbe symbioses. *Plant and Soil*, 194, 171–184.
Hochholdinger, F. et al. (2009) The maize root system: morphology, anatomy, and genetics. In: *Handbook of Maize: Its Biology*. Springer, New York.
Hochholdinger, F. et al. (2004) Lateral roots affect the proteome of the primary root of maize (*Zea mays* L.). *Plant Molecular Biology*, 56, 397–412.
Hochholdinger, F. and Tuberosa, R. (2009) Genetic and genomic dissection of maize root development and architecture. *Current Opinion in Plant Biology*, 12, 172–177.
Hochholdinger, F. and Zimmermann, R. (2009) Molecular and genetic dissection of cereal root development. *Annual Plant Reviews*, 37, 175–191.
Holmes, P. et al. (2008) Transcriptional profiling of *Medicago truncatula* meristematic root cells. *BMC Plant Biology*, 8, 21.

Imin, N. et al. (2004) Proteome reference maps of *Medicago truncatula* embryogenic cell cultures generated from single protoplasts. *Proteomics*, 4, 1883–1896.

Inukai, Y. et al. (2005) *Crown rootless1*, which is essential for crown root formation in rice, is a target of an *AUXIN RESPONSE FACTOR* in auxin signaling. *Plant Cell*, 17, 1387–1396.

Ithal, N. et al. (2007) Developmental transcript profiling of cyst nematode feeding cells in soybean roots. *Molecular Plant–Microbe Interactions*, 20, 510–525.

Ivanchenko, M.G. et al. (2006) Mutations in the *Diageotropica* (*Dgt*) gene uncouple patterned cell division during lateral root initiation from proliferative cell division in the pericycle. *Plant Journal*, 46, 436–447.

Ivanchenko, M.G. et al. (2008) Ethylene–auxin interactions regulate lateral root initiation and emergence in *Arabidopsis thaliana*. *Plant Journal*, 55, 335–347.

Iyer-Pascuzzi, A. et al. (2009) Functional genomics of root growth and development in Arabidopsis. *Current Opinions in Plant Biology*, 12, 165–171.

Iyer-Pascuzzi, A.S. and Benfey, P.N. (2009) Transcriptional networks in root cell fate specification. *Biochimica et Biophysica Acta*, 1789, 315–325.

Jain, M. et al. (2006) Structure and expression analysis of early auxin-responsive *Aux/IAA* gene family in rice (*Oryza sativa*). *Functional and Integrative Genomics*, 6, 47–59.

Jiang, K. et al. (2006) Transcription profile analyses identify genes and pathways central to root cap functions in maize. *Plant Molecular Biology*, 60, 343–363.

Jiao, Y. et al. (2009) A transcriptome atlas of rice cell types uncovers cellular, functional and developmental hierarchies. *Nature Genetics*, 41, 258–263.

Kalluri, U.C. et al. (2007) Genome-wide analysis of *Aux/IAA* and *ARF* gene families in *Populus trichocarpa*. *BMC Plant Biology*, 7, 59.

Kepinski, S. and Leyser, O. (2005) The Arabidopsis F-box protein TIR1 is an auxin receptor. *Nature*, 435, 446–451.

Krouk, G. et al. (2010) Nitrate-regulated auxin transport by NRT1.1 defines a mechanism for nutrient sensing in plants. *Developmental Cell*, 18, 927–937.

Kuroha, T. et al. (2006) Cytokinin receptors are required for normal development of auxin-transporting vascular tissues in the hypocotyl but not in adventitious roots. *Plant Cell Physiology*, 47, 234–243.

Kurup, S. et al. (2005) Marking cell lineages in living tissues. *Plant Journal*, 42, 444–453.

Laplaze, L. et al. (2007) Cytokinins act directly on lateral root founder cells to inhibit root initiation. *Plant Cell*, 19, 3889–3900.

Laplaze, L. et al. (2005) GAL4-GFP enhancer trap lines for genetic manipulation of lateral root development in *Arabidopsis thaliana*. *Journal of Experimental Botany*, 56, 2433–2442.

Laskowski, M. et al. (2006) Expression profiling of auxin-treated Arabidopsis roots: toward a molecular analysis of lateral root emergence. *Plant Cell Physiology*, 47, 788–792.

Laskowski, M. et al. (2008) Root system architecture from coupling cell shape to auxin transport. *PLoS Biology*, 6, e307.

Laskowski, M.J. et al. (1995) Formation of lateral root meristems is a two-stage process. *Development*, 121, 3303–3310.

Lau, S. et al. (2008) The evolving complexity of the auxin pathway. *Plant Cell*, 20, 1738–1746.

Laufs, P. et al. (2004) MicroRNA regulation of the *CUC* genes is required for boundary size control in Arabidopsis meristems. *Development*, 131, 4311–4322.

Lee, H.W. et al. (2009) LBD18/ASL20 regulates lateral root formation in combination with LBD16/ASL18 downstream of ARF7 and ARF19 in Arabidopsis. *Plant Physiology*, 151, 1377–1389.

Lee, I. et al. (2010) Rational association of genes with traits using a genome-scale gene network for *Arabidopsis thaliana*. *Nature Biotechnology*, 28, 149–156.

Lee, J.Y. et al. (2006) Transcriptional and posttranscriptional regulation of transcription factor expression in Arabidopsis roots. *Proceedings of the National Academy of Sciences of the United States of America*, 103, 6055–6060.

Levesque, M.P. et al. (2006) Whole-genome analysis of the SHORT-ROOT developmental pathway in Arabidopsis. *PLoS Biology*, 4, e143.

Leyser, O. (2006) Dynamic integration of auxin transport and signalling. *Current Biology*, 16, R424–R433.

Li, X. et al. (2006) Cytokinin-mediated cell cycling arrest of pericycle founder cells in lateral root initiation of Arabidopsis. *Plant Cell Physiology*, 47, 1112–1123.

Libault, M. et al. (2010) An integrated transcriptome atlas of the crop model *Glycine max*, and its use in comparative analyses in plants. *Plant Journal*, 63, 86–99.

Liu, H. et al. (2005) ARL1, a LOB-domain protein required for adventitious root formation in rice. *Plant Journal*, 43, 47–56.

Liu, Y. et al. (2006) Comparative proteome analyses of maize (*Zea mays* L.) primary roots prior to lateral root initiation reveal differential protein expression in the lateral root initiation mutant rum1. *Proteomics*, 6, 4300–4308.

Liu, Y. et al. (2010) Regulation of the pericycle proteome in maize (*Zea mays* L.) primary roots by RUM1 which is required for lateral root initiation. *European Journal of Cell Biology*, 89, 236–241.

Ljung, K. et al. (2005) Sites and regulation of auxin biosynthesis in Arabidopsis roots. *Plant Cell*, 17, 1090–1104.

Lohar, D.P. et al. (2004) Cytokinins play opposite roles in lateral root formation, and nematode and rhizobial symbioses. *Plant Journal*, 38, 203–214.

Lucas, M. et al. (2008a). Auxin fluxes in the root apex co-regulate gravitropism and lateral root initiation. *Journal of Experimental Botany*, 59, 55–66.

Lucas, M. et al. (2008b). An auxin transport-based model of root branching in *Arabidopsis thaliana*. *PLoS One*, 3, e3673.

Lucas, M. et al. (2011) Short-root regulates primary, lateral, and adventitious root development in Arabidopsis. *Plant Physiology*, 155, 384–398.

Lynch, J.P. (2007) Roots of the second green revolution. *Australian Journal of Botany*, 55, 493–512.

Malamy, J.E. and Benfey, P.N. (1997a). Down and out in Arabidopsis: the formation of lateral roots. *Trends in Plant Science*, 2, 390–396.

Malamy, J.E. and Benfey, P.N. (1997b). Organization and cell differentiation in lateral roots of *Arabidopsis thaliana*. *Development*, 124, 33–44.

Malamy, J.E. and Ryan, K.S. (2001) Environmental regulation of lateral root initiation in Arabidopsis. *Plant Physiology*, 127, 899–909.

Mallory, A.C. et al. (2005) MicroRNA-directed regulation of Arabidopsis AUXIN RESPONSE FACTOR17 is essential for proper development and modulates expression of early auxin response genes. *Plant Cell*, 17, 1360–1375.

Marchant, A. et al. (2002) AUX1 promotes lateral root formation by facilitating indole-3-acetic acid distribution between sink and source tissues in the Arabidopsis seedling. *Plant Cell*, 14, 589–597.

Marin-Rodriguez, M.C. et al. (2002) Pectate lyases, cell wall degradation and fruit softening. *Journal of Experimental Botany*, 53, 2115–2119.

Markmann, K. et al. (2008) Functional adaptation of a plant receptor kinase paved the way for the evolution of intracellular root symbioses with bacteria. *PLoS Biology*, 6, e68.

Mathesius, U. et al. (2011) Comparative proteomic profiles of the soybean (*Glycine max*) root apex and differentiated root zone. *Proteomics*, 11, 1707–1719.

Mathesius, U. et al. (2000) Rhizobia can induce nodules in white clover by "hijacking" mature cortical cells activated during lateral root development. *Molecular Plant–Microbe Interaction*, 13, 170–182.

Matsuda, F. et al. (2010) AtMetExpress development: a phytochemical atlas of Arabidopsis development. *Plant Physiology*, 152, 566–578.

Menges, M. et al. (2003) Genome-wide gene expression in an Arabidopsis cell suspension. *Plant Molecular Biology*, 53, 423–442.

Middleton, A. et al. (2011) Regulatory networks in root development: an emerging story. Plant Growth Biology and Modeling Workshop, Elche, Spain.

Miyawaki, K. et al. (2004) Expression of cytokinin biosynthetic isopentenyltransferase genes in Arabidopsis: tissue specificity and regulation by auxin, cytokinin, and nitrate. *Plant Journal*, 37, 128–138.

Moreno-Risueno, M.A. et al. (2010) Oscillating gene expression determines competence for periodic Arabidopsis root branching. *Science*, 329, 1306–1311.

Mouchel, C.F. et al. (2006) BRX mediates feedback between brassinosteroid levels and auxin signalling in root growth. *Nature*, 443, 458–461.

Movahedi, S. et al. (2011) Comparative network analysis reveals that tissue specificity and gene function are important factors influencing the mode of expression evolution in Arabidopsis and rice. *Plant Physiology*, 156, 1316–1330.

Mullen, J.L. and Hangarter, R.P. (2003) Genetic analysis of the gravitropic set-point angle in lateral roots of Arabidopsis. *Advances in Space Research*, 31, 2229–2236.

Muller, M. and Schmidt, W. (2004) Environmentally induced plasticity of root hair development in Arabidopsis. *Plant Physiology*, 134, 409–419.

Mustroph, A. et al. (2009) Profiling translatomes of discrete cell populations resolves altered cellular priorities during hypoxia in Arabidopsis. *Proceedings of the National Academy of Sciences of the United States of America*, 106, 18843–18848.

Muthreich, N. et al. (2010) Regulation of the maize (*Zea mays* L.) embryo proteome by RTCS which controls seminal root initiation. *European Journal of Cell Biology*, 89, 242–249.

Nawy, T. et al. (2005) Transcriptional profile of the Arabidopsis root quiescent center. *Plant Cell*, 17, 1908–1925.

Nelson, T. et al. (2008) Plant cell types: reporting and sampling with new technologies. *Current Opinions in Plant Biology*, 11, 567–573.

Nelson, T. et al. (2006) Laser microdissection of plant tissue: what you see is what you get. *Annual Review of Plant Biology*, 57, 181–201.

Nemhauser, J.L. et al. (2004) Interdependency of brassinosteroid and auxin signaling in Arabidopsis. *PLoS Biology*, 2, E258.

Neuteboom, L.W. et al. (1999) A novel subtilisin-like protease gene from *Arabidopsis thaliana* is expressed at sites of lateral root emergence. *DNA Research*, 6, 13–19.

Nodzon, L.A. et al. (2004) The ubiquitin ligase XBAT32 regulates lateral root development in Arabidopsis. *Plant Journal*, 40, 996–1006.

Okushima, Y. et al. (2007) ARF7 and ARF19 regulate lateral root formation via direct activation of *LBD/ASL* genes in Arabidopsis. *Plant Cell*, 19, 118–130.

Okushima, Y. et al. (2005) Functional genomic analysis of the *AUXIN RESPONSE FACTOR* gene family members in *Arabidopsis thaliana*: unique and overlapping functions of ARF7 and ARF19. *Plant Cell*, 17, 444–463.

Orman, B. et al. (2011) EST-based approach for dissecting root architecture in barley using mutant traits of other species. In: A. Costa de Oliveira and R.K. Varshney (eds) *Root Genomics*. Springer, Berlin Heidelberg, Germany.

Ottenschlager, I. et al. (2003) Gravity-regulated differential auxin transport from columella to lateral root cap cells. *Proceedings of the National Academy of Sciences of the United States of America*, 100, 2987–2991.

Parizot, B. et al. (2010) VisuaLRTC: a new view on lateral root initiation by combining specific transcriptome data sets. *Plant Physiology*, 153, 34–40.

Parizot, B. et al. (2008) Diarch symmetry of the vascular bundle in Arabidopsis root encompasses the pericycle and is reflected in distich lateral root initiation. *Plant Physiology*, 146, 140–148.

Peret, B. et al. (2009) Lateral root emergence: a difficult birth. *Journal of Experimental Botany*, 60, 3637–3643.

Péret, B. et al. (2007) Auxin influx activity is associated with Frankia infection during actinorhizal nodule formation in *Casuarina glauca*. *Plant Physiology*, 144, 1852–1862.

Perez-Torres, C.A. et al. (2008) Phosphate availability alters lateral root development in Arabidopsis by modulating auxin sensitivity via a mechanism involving the TIR1 auxin receptor. *Plant Cell*, 20, 3258–3272.

Perrine-Walker, F. et al. (2011) Symbiotic signaling in actinorhizal symbioses. *Current Protein and Peptide Science*, 12, 156–164.

Petersson, S.V. et al. (2009) An auxin gradient and maximum in the Arabidopsis root apex shown by high-resolution cell-specific analysis of IAA distribution and synthesis. *Plant Cell*, 21, 1659–1668.

Prasad, M.E. et al. (2010) Arabidopsis RING E3 ligase XBAT32 regulates lateral root production through its role in ethylene biosynthesis. *Plant Physiology*, 153, 1587–1596.

Przemeck, G.K.H. et al. (1996) Studies on the role of the Arabidopsis gene *MONOPTEROS* in vascular development and plant cell axialization. *Planta*, 200, 229–237.

Quesada, T. et al. (2008) Comparative analysis of the transcriptomes of *Populus trichocarpa* and *Arabidopsis thaliana* suggests extensive evolution of gene expression regulation in angiosperms. *New Phytologist*, 180, 408–420.

Ramos, J.A. et al. (2001) Rapid degradation of auxin/indoleacetic acid proteins requires conserved amino acids of domain II and is proteasome dependent. *Plant Cell*, 13, 2349–2360.

Reyes, J.C. et al. (2004) The GATA family of transcription factors in Arabidopsis and rice. *Plant Physiology*, 134, 1718–1732.

Richter, G.L. et al. (2009) Mechanical stimuli modulate lateral root organogenesis. *Plant Physiology*, 151, 1855–1866.

Riechmann, J.L. et al. (2000) Arabidopsis transcription factors: genome-wide comparative analysis among eukaryotes. *Science*, 290, 2105–2110.

Rogg, L.E. et al. (2001) A gain-of-function mutation in *IAA28* suppresses lateral root development. *Plant Cell*, 13, 465–480.

Rouse, D. et al. (1998) Changes in auxin response from mutations in an *AUX/IAA* gene. *Science*, 279, 1371–1373.

Saleem, M. et al. (2009) Tissue specific control of the maize (*Zea mays* L.) embryo, cortical parenchyma, and stele proteomes by *RUM1* which regulates seminal and lateral root initiation. *Journal of Proteome Research*, 8, 2285–2297.

Saleem, M. et al. (2010) Specification of cortical parenchyma and stele of maize primary roots by asymmetric levels of auxin, cytokinin, and cytokinin-regulated proteins. *Plant Physiology*, 152, 4–18.

Scheres, B. et al. (2002) *Root development. In:* C.R. Somerville and E.M. Meyerowitz (eds) *The Arabidopsis Book*. American Society of Plant Biologists, Rockville, MD.

Scheres, B. and Benfey, P.N. (1999) Asymmetric cell division in plants. *Annual Review of Plant Physiology and Plant Molecular Biology*, 50, 505–537.

Scheres, B., et al. (1994) Embryonic origin of the Arabidopsis primary root and root meristem initials. *Development*, 120, 2475–2487.

Schneider, K. et al. (1997) Structural and genetic analysis of epidermal cell differentiation in Arabidopsis primary roots. *Development*, 124, 1789–1798.

Sekhon, R.S. et al. (2011) Genome-wide atlas of transcription during maize development. *Plant Journal*, 66, 553–563.

Sena, G. et al. (2009) Organ regeneration does not require a functional stem cell niche in plants. *Nature*, 457, 1150–1153.

Severin, A.J. et al. (2010) RNA-Seq atlas of *Glycine max*: a guide to the soybean transcriptome. *BMC Plant Biology*, 10, 160.

Shuai, B. et al. (2002) The lateral organ boundaries gene defines a novel, plant-specific gene family. *Plant Physiology*, 129, 747–761.

Su, Y.H. et al. (2011) Auxin–cytokinin interaction regulates meristem development. *Molecular Plant*, 4, 616–625.

Sugimoto, K. et al. (2010) Arabidopsis regeneration from multiple tissues occurs via a root development pathway. *Developmental Cell*, 18, 463–471.

Sussex, I.M. et al. (1995) Cellular and molecular events in a newly organizing lateral root meristem. *Philosophical Transactions of the Royal Society B: Biological Science*, 350, 39–43.

Swarup, K. et al. (2008) The auxin influx carrier LAX3 promotes lateral root emergence. *Nature Cell Biology*, 10, 946–954.

Swarup, R. et al. (2005) Root gravitropism requires lateral root cap and epidermal cells for transport and response to a mobile auxin signal. *Nature Cell Biology*, 7, 1057–1065.

Takehisa, H. et al. (2012) Genome-wide transcriptome dissection of the rice root system: implications for developmental and physiological functions. *Plant Journal*, 69, 126–140.

Taramino, G. et al. (2007) The maize (*Zea mays* L.) *RTCS* gene encodes a LOB domain protein that is a key regulator of embryonic seminal and post-embryonic shoot-borne root initiation. *Plant Journal*, 50, 649–659.

Tatematsu, K. et al. (2004) *MASSUGU2* encodes Aux/IAA19, an auxin-regulated protein that functions together with the transcriptional activator NPH4/ARF7 to regulate differential growth responses of hypocotyl and formation of lateral roots in *Arabidopsis thaliana*. *Plant Cell*, 16, 379–393.

Terrile, M.C. et al. (2012) Nitric oxide influences auxin signaling through *S*-nitrosylation of the arabidopsis transport inhibitor response1 auxin receptor. *The Plant Journal*, 70, 492–500.

Tian, C.E. et al. (2004) Disruption and overexpression of auxin response factor 8 gene of Arabidopsis affect hypocotyl elongation and root growth habit, indicating its possible involvement in auxin homeostasis in light condition. *Plant Journal*, 40, 333–343.

Tian, Q. and Reed, J.W. (1999) Control of auxin-regulated root development by the *Arabidopsis thaliana SHY2/IAA3* gene. *Development*, 126, 711–721.

Tiwari, S.B. et al. (2003) The roles of auxin response factor domains in auxin-responsive transcription. *Plant Cell*, 15, 533–543.

Tiwari, S.B. et al. (2004) Aux/IAA proteins contain a potent transcriptional repression domain. *Plant Cell*, 16, 533–543.

Truernit, E. et al. (2006) A map of *KNAT* gene expression in the Arabidopsis root. *Plant Molecular Biology*, 60, 1–20.

Ueda, M. et al. (2005) Stepwise understanding of root development. *Current Opinions in Plant Biology*, 8, 71–76.

Uehara, T. et al. (2008) Domain II mutations in *CRANE/IAA18* suppress lateral root formation and affect shoot development in *Arabidopsis thaliana*. *Plant Cell Physiology*, 49, 1025–1038.

Ulmasov, T. et al. (1997) Aux/IAA proteins repress expression of reporter genes containing natural and highly active synthetic auxin response elements. *Plant Cell*, 9, 1963–1971.

Van Bel, M. et al. (2012) Dissecting plant genomes with the PLAZA comparative genomics platform. *Plant Physiol*, 158, 590–600.

Van de Peer, Y. et al. (2009) The evolutionary significance of ancient genome duplications. *Nature Reviews Genetics*, 10, 725–732.

Van den Berg, C. et al. (1997) Short-range control of cell differentiation in the Arabidopsis root meristem. *Nature*, 390, 287–289.

Van Noorden, G.E. et al. (2007) Overlap of proteome changes in *Medicago truncatula* in response to auxin and *Sinorhizobium meliloti*. *Plant Physiology*, 144, 1115–1131.

Vanneste, S. et al. (2005) Cell cycle progression in the pericycle is not sufficient for *SOLITARY ROOT/IAA14*-mediated lateral root initiation in *Arabidopsis thaliana*. *Plant Cell*, 17, 3035–3050.

Varshney, R.K. et al. (2009) Next-generation sequencing technologies and their implications for crop genetics and breeding. *Trends in Biotechnology*, 27, 522–530.

Vissenberg, K. et al. (2005) XTH acts at the microfibril–matrix interface during cell elongation. *Journal of Experimental Botany*, 56, 673–683.

Vroemen, C.W. et al. (2003) The *CUP-SHAPED COTYLEDON3* gene is required for boundary and shoot meristem formation in Arabidopsis. *Plant Cell*, 15, 1563–1577.

Wang, D. et al. (2007) Genome-wide analysis of the auxin response factors (*ARF*) gene family in rice (*Oryza sativa*). *Gene*, 394, 13–24.

Wang, J.W. et al. (2005) Control of root cap formation by MicroRNA-targeted auxin response factors in Arabidopsis. *Plant Cell*, 17, 2204–2216.

Wang, L. et al. (2010a). A dynamic gene expression atlas covering the entire life cycle of rice. *Plant Journal*, 61, 752–766.

Wang, Y. et al. (2010b). Genome-wide analysis of primary auxin-responsive *Aux/IAA* gene family in maize (*Zea mays* L.). *Molecular Biology Reports*, 37, 3991–4001.

Wilmoth, J.C. et al. (2005) *NPH4/ARF7* and *ARF19* promote leaf expansion and auxin-induced lateral root formation. *Plant Journal*, 43, 118–130.

Woll, K. et al. (2005) Isolation, characterization, and pericycle-specific transcriptome analyses of the novel maize lateral and seminal root initiation mutant rum1. *Plant Physiology*, 139, 1255–1267.

Wu, G. et al. (2007) Mutations in Arabidopsis multidrug resistance-like ABC transporters separate the roles of acropetal and basipetal auxin transport in lateral root development. *Plant Cell*, 19, 1826–1837.

Wysocka-Diller, J.W. et al. (2000) Molecular analysis of *SCARECROW* function reveals a radial patterning mechanism common to root and shoot. *Development*, 127, 595–603.

Yang, X. et al. (2004) The IAA1 protein is encoded by *AXR5* and is a substrate of *SCF*(*TIR1*). *Plant Journal*, 40, 772–782.

Yang, Y. et al. (2006) Comparison and evolution analysis of two rice subspecies *LATERAL ORGAN BOUNDARIES* domain gene family and their evolutionary characterization from Arabidopsis. *Molecular Phylogenetics Evolution*, 39, 248–262.

Zenser, N. et al. (2001) Auxin modulates the degradation rate of Aux/IAA proteins. *Proceedings of the National Academy of Sciences of the United States of America*, 98, 11795–11800.

Zhang, H. and Forde, B.G. (1998) An Arabidopsis MADS box gene that controls nutrient-induced changes in root architecture. *Science*, 279, 407–409.

Zheng, X. et al. (2011) AUXIN UP-REGULATED F-BOX PROTEIN1 regulates the cross talk between auxin transport and cytokinin signaling during plant root growth. *Plant Physiology*, 156, 1878–1893.

Zhu, Z.X. et al. (2011) A gain-of-function mutation in *OsIAA11* affects lateral root development in rice. *Molecular Plant*, 5,154–161.

2 The Complex Eukaryotic Transcriptome: Nonprotein-Coding RNAs and Root Development

F. Ariel, A.B. Moreno, F. Bardou, and Martin Crespi

Genomic Approaches Reveal Novel Aspects of the Eukaryotic Transcriptome

Although the term "transcriptome" usually refers to the whole set of messenger RNA (mRNA) molecules in the cell, it is worth noting that mRNA only represents a fraction of all the transcripts that are actually transcribed. In fact, many transcripts are not translated into proteins, the so-called nonprotein-coding RNAs (npcRNAs), but still play a role in relevant cellular functions and complexes (e.g., ribonucleoproteins [RNP]). NpcRNAs are a class of RNAs with very poor protein-coding potential, but their function is associated to the RNA molecule itself. They form a heterogeneous group of RNAs that could be divided into three different classes according to their length and function. Generally, npcRNAs can range from 18 to 25 nucleotides for the families of microRNAs and small interfering RNAs (siRNAs) linked to posttranscriptional and transcriptional gene silencing, 20–300 nucleotides for small RNAs commonly found as transcriptional and translational regulators (e.g., small nuclear RNAs), or up to and beyond 10,000 nucleotides for medium and large npcRNAs mainly linked to the chromatin environment and epigenetic mechanisms. It is now known that there exist transcripts spanning chromosomal regions larger than 2 Mb (Carninci 2010). Indeed, transcripts originated and terminated in chromosomal regions that are as long as several megabases have been identified by RACE (Djebali et al. 2008). In the last years, bioinformatics and experimental strategies have revealed a remarkable number of novel npcRNA candidates in various model organisms from yeast or plants to *Homo sapiens* (Mattick and Makunin 2006; Yasuda and Hayashizaki 2008; Mercer et al. 2009). NpcRNAs are generally encoded in intergenic regions, and there exist a high number of antisense RNA transcripts, pseudogenes, and truncated transcripts in eukaryotes.

The transcriptome complexity involving npcRNAs has been hypothesized to exert a regulatory role that is required for the development and function of higher organisms (Carninci 2010). Indeed, certain npcRNAs have been implicated in different regulatory mechanisms in plant development (Brown et al. 2008; Voinnet 2009), in biotic interactions, and the response to environmental abiotic stress (Ben Amor et al. 2009; Jay et al. 2010; Sunkar 2010). Generally, npcRNAs are produced by RNA polymerase II and are capped and polyadenylated. Recent exploration of the poly(A)−/+ transcriptomes of HeLa and H9 cells revealed that while the majority of protein-coding transcripts are poly(A)+, a significant portion of them are either poly(A)− or bimorphic, being found in both the poly(A)+ and poly(A)− populations. Only stable excised introns were identified as a class of poly(A)− long npcRNAs (Yang et al. 2011). In plants, certain npcRNAs are produced

Root Genomics and Soil Interactions, First Edition. Edited by Martin Crespi.

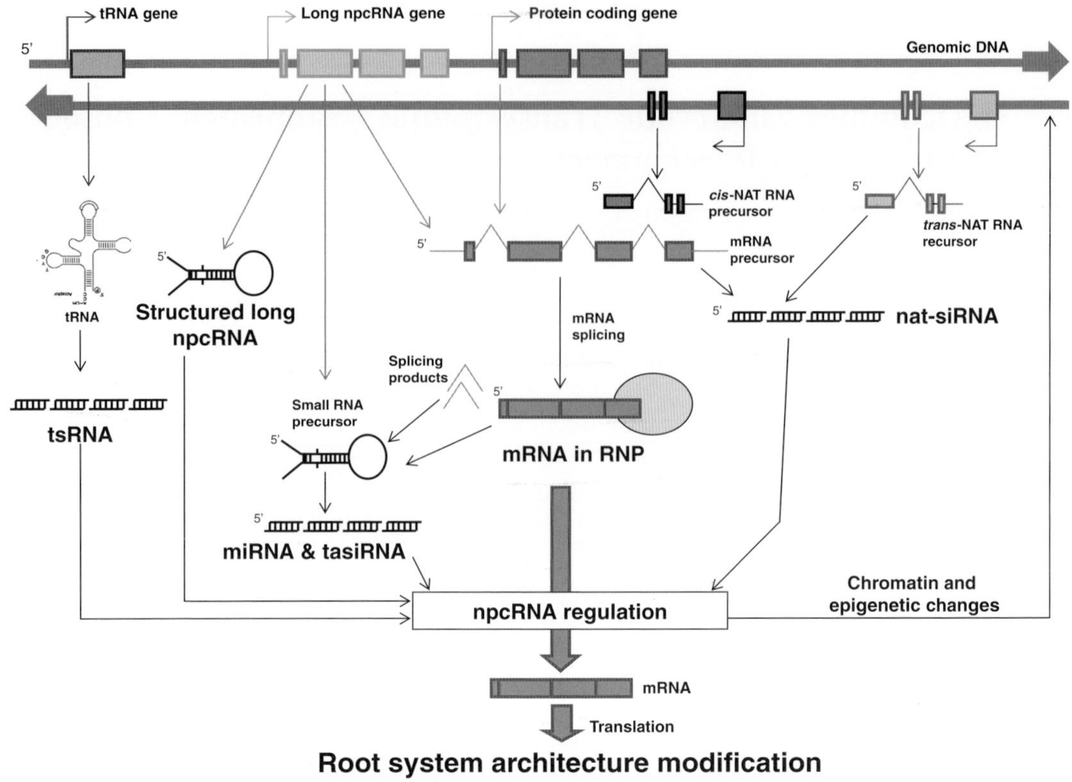

Figure 2.1 The nonprotein-coding RNA network in plants. Long npcRNAs can be precursors of small RNAs or *cis/trans*-antisense RNAs (NATs) of other transcripts. Once incorporated into the RISC effector complex, small RNAs from different pathways of RNA silencing (miRNA, siRNA, or nat-siRNA) can act directly on the target mRNA or other npcRNAs. These regulatory pathways can lead to epigenetic changes and DNA methylation (TGS). Hence, long and short npcRNAs form a network of ribonucleoproteins within the cell that may affect the expression patterns of coding mRNAs, affecting root development. (For a color version of this figure, see the color plate section.)

by specific RNA polymerases IV and V (Voinnet 2009) and then processed into smaller RNA molecules (see later). In fact, a transcriptome may be surprisingly complex with long npcRNAs often overlapping with or interspersed between coding transcripts. This dynamic molecular picture significantly changed the understanding of gene expression in eukaryotes, if we consider that a single DNA sequence can be transcribed in multiple sense and antisense transcripts, intronic npcRNAs, intergenic, or promoter-associated RNAs (Mercer et al. 2009). In *Arabidopsis thaliana*, whole-genome mapping based on the use of tiling arrays revealed that more than 30% of observed transcription was intergenic and that many antisense RNA transcripts actually exist (Yamada et al. 2003).

Long and small npcRNAs represent an emerging class of riboregulators, which either act directly in this long form or are processed to shorter molecules (Figure 2.1). Several long npcRNAs are processed into small RNAs due to their folding as double-stranded RNA (dsRNA) loops derived from endogenous loci (as the miRNAs) or due to the action of RNA-dependent RNA polymerases that generate long dsRNAs. These dsRNA structures are processed into siRNAs by member(s) of the Dicer family (DCL, Dicer like; Vaucheret 2006). Small si/miRNAs induce mRNA cleavage and translational inhibition in the so-called RISC complexes (RNA-induced silencing complex) through pairing with specific mRNA targets, mainly in the cytoplasm, or lead to transcriptional gene silencing, heterochromatin formation, and *de novo* DNA methylation in the nucleus (Vaucheret 2006; Jamalkandi and Masoudi-Nejad 2009; Verdel et al. 2009). Another class of endogenous siRNAs deriving from pairs of natural *cis*-antisense transcripts was discovered in plants. Natural antisense transcripts (NATs) can be defined as endogenous RNA molecules that are transcribed from the opposite DNA strands, resulting in partially or entirely complement transcripts. As in most species, the majority of *Arabidopsis* NATs pairs (72%) overlapped at their 3′ end (Wang and Metzlaff 2005) and, for 99% of them, the overlapping region included exon sequences. Both sense and antisense RNAs can encode proteins or be npcRNAs, although in most cases, antisense transcription consists of a protein-coding RNA overlapping with an npcRNA (Faghihi et al. 2010). NATs can give place to active siRNAs formed from their overlapped region (Borsani et al. 2005). Small RNAs have the fascinating property to move from cell to cell and act in recipient tissues. The recent discovery that microRNAs are also able to move between root tissues (Martienssen 2010) raises the intriguing possibility that different cells from a tissue signals other cells to coordinate growth. Although there are very few examples of miRNA movement, it is well established that siRNAs can move between plant tissues even at long distances (Dunoyer et al. 2010) and be biologically active in the recipient cells. Indeed, specific miRNAs (such as miR390) target long npcRNAs (e.g., TAS genes) to produce secondary tasiRNAs (for *trans*-acting siRNAs) that control mRNA expression patterns in a diffusible manner in meristems (Chitwood and Timmermans 2010) and during lateral root growth (Marin et al. 2010). These mobile siRNAs can repress gene expression at posttranscriptional level and even transmit new epigenetic patterns into root tissues (e.g., changes in DNA methylation) as elegantly shown using grafting experiments with different silencing mutants (Melnyk et al. 2011). If members of these classes of mobile signals (si/miRNAs) function in this manner in the root, their production and distribution in primary and lateral meristems during growth of the root system promises to be a key research goal for the coming years. Interestingly, recent data showed that plant miRNAs, like the rice miR168a, may be orally acquired by animals through food intake. Strikingly, miR168a was detected in the animal sera and tissues (Zhang et al. 2011). Moreover, it has been shown that miR168a could bind to the human/mouse low-density lipoprotein receptor adapter protein 1 (LDLRAP1) mRNA, inhibiting LDLRAP1 expression in liver, and consequently decreasing LDL removal from mouse plasma. These data indicate clearly an evidence of cross-kingdom regulation by miRNA.

Apart from the known described small RNAs (si/miRNAs), other small RNAs may play different roles in development. Deep sequencing analysis has recently revealed that under some growth or stress conditions, tRNA cleaved pathways could be activated in different organisms like human yeast or plants (e.g., during oxidative stress in *Arabidopsis* seedling, Thompson et al. 2008). Indeed, an abundant population of small RNAs of 30–40 nt aligns with tRNA sequences that can derive either from 5′ or 3′ end of mature tRNAs or from tRNA precursors by a cleavage near or in the anticodon loop (Pederson 2010). The two types of tRNA-derived small RNAs (tsRNAs) are nonrandom cleavage products: type I (5′-tsRNAs) and type II (3′-tsRNAs) (Haussecker et al. 2010). Both types are produced from tRNA precursors and, respectively, generated by Dicer (Babiarz et al. 2008) and endoribonuclease (Thompson and Parker 2009). In plants, it has been shown that phloem sap tsRNA interferes with wheat germ extract translation (Zhang et al. 2009). It has been shown that specific tRNA fragments may interfere with ribosomal activity, although these data were only obtained with artificially produced tRNA fragments in an *in vitro* experiment. However, the tsRNAs' ability to cause the assembly of stress granules suggests that they could affect the initiation of translation. More than 10 years ago, Asano et al. (2001) have shown that specific tsRNAs were able to displace eIF4F from capped mRNA. Moreover, pool-down analysis reveals that tsRNAs bind to a complex containing AGO2, which is involved in miRNAs/siRNAs silencing in animals. tsRNAs represent 34% of small RNA reads in phosphate-starved roots, and among this, the Gly-tRNATCC represented more than 80% of the population (Hsieh et al. 2009). Moreover, these tsRNAs were more accumulated in roots. The precise role of these small npcRNAs remains unknown, but recently, tsRNAs have been observed with a specific pattern in embryogenic and nonembryogenic rice calli (Chen et al. 2011). As certain aspects of the induction of calli formation may resemble an ectopic lateral root development program (Sugimoto et al. 2010), a role for tsRNAs in root development may also be considered.

Recently, bioinformatic approaches on full-length cDNA databases resulted in the identification of 76 *Arabidopsis* npcRNAs (Hirsch et al. 2006; Ben Amor et al. 2009), including several precursors of known miRNAs. Eleven of these npcRNAs turned out to be antisense to protein-coding mRNAs, suggesting that they act as *cis*-regulatory molecules, whereas five others are likely to be siRNA precursors. Expression analyses of the 76 npcRNAs in mutant plants of several RNA-binding proteins involved in npcRNA processing shed light on the molecular pathways in which some of them are involved. Abiotic stress, such as phosphate starvation, drought, or salt stress, altered the accumulation of 22 npcRNAs. In the same work, overexpression analyses in *Arabidopsis* identified two npcRNAs as regulators of root growth during salt stress and leaf morphology, respectively. These novel results supported the idea that together with small RNAs, long npcRNAs can also exert diverse roles during growth and differentiation.

Long npcRNAs have been shown to have specific localizations at tissular, cellular, and subcellular levels (Campalans et al. 2004; Zhan and Lukens 2010); however, their mechanisms of action remain largely unknown. It has been demonstrated in several organisms that long npcRNAs interact with histone modifiers (Swiezewski et al. 2009; Yu et al. 2009, 2010). Among the best-studied npcRNAs in animals, several were identified in the nucleus and act either as *cis*- or *trans*-acting epigenetic regulators of chromatin. This is the case for Xist that triggers X-chromosome inactivation (Clemson et al. 1996; Chaumeil et al. 2006), Kcnq1ot1 and Air that control genomic imprinting (Sleutels et al. 2002; Mancini-Dinardo et al. 2006; Nagano et al. 2008; Pandey et al. 2008; Terranova et al. 2008; Redrup et al. 2009), and *HOTAIR* that regulates homeobox (HOX) genes (Rinn et al. 2007). It has been shown that *HOTAIR* participates in epigenetic changes by recruiting chromatin-remodeling complexes to specific genomic loci (Rinn et al. 2007). In the same context, the long npcRNA MIRA activates the target genes *HOXa6* and *HOXa7* by recruiting epigenetic activators, implicating long npcRNAs in gene regulation during vertebrate cell-fate determination (Bertani et al. 2011). Recently, another long intergenic npcRNA called *HOTTIP* transcribed from the 5′ tip of the *HOXa* locus was shown to coordinate the activation of several *HOXa* genes *in vivo*. A chromosomal looping brings *HOTTIP* into close proximity to its target genes. *HOTTIP* RNA binds to adaptor proteins and target remodeling complexes across HOXA, driving histone H3 Lys 4 trimethylation and gene transcription (Wang et al. 2011). In *Arabidopsis*, the study of the regulation of the MADS box transcriptional regulator *FLOWERING LOCUS C* (*FLC*) revealed a set of antagonistic pathways comprising evolutionary conserved activities that link chromatin regulation, transcription level, and cotranscriptional RNA metabolism (Crevillén and Dean 2011). Cold triggers enrichment of trimethylated histone H3 Lys 27 at the chromatin of the floral repressor *FLC* and results in its epigenetically stable repression. It has been demonstrated that a long intronic npcRNA (termed as *COLDAIR*) physically associates with a component of polycomb repressive complex 2 (PRC2) and targets it to *FLC*, allowing the vernalization-mediated epigenetic repression of the locus (Heo and Sung 2011). In addition, a group of related antisense npcRNAs (termed *COOLAIR*) from *FLC* has been proposed to be involved in vernalization-mediated *FLC* repression (Swiezewski et al. 2009). The RNA-binding proteins (RBPs) FCA and FPA participate in the repression of *FLC*, but they may play a more global role in plant development and epigenetic control. Indeed, in addition to FLC, the double mutant *fca–fpa* misexpressed a large number of previously unannotated genomic segments corresponding to intergenic regions in which DNA methylation was altered. FCA and FPA would play important roles in the *Arabidopsis* genome in RNA 3′ processing and transcription termination, thus limiting intergenic transcription (Sonmez et al. 2011).

Long npcRNAs including antisense npcRNAs have to bypass several RNA quality control (RQC) mechanisms occurring in the cell that actively degrade "aberrant" mRNAs such as incompletely spliced products, transcripts without coding capacity, or pseudogenes (Lykke-Andersen et al. 2009). For example, a genome-wide analysis of exosome substrates in *Arabidopsis* revealed, in addition to mRNA and miRNA processing intermediates, hundreds of npcRNAs and antisense RNAs not previously described (Chekanova et al. 2007). The exosome is a macromolecular complex that mediates RNA processing and degradation and is generally essential for viability in eukaryotes. It was shown that the plant exosome exhibits a remarkable functional plasticity in comparison with yeast and metazoans. The npcRNAs detected only in exosome mutants include large numbers of antisense RNAs, as they are rapidly and actively degraded in wild-type plants (Chekanova et al. 2007). Similarly, the nonsense-mediated mRNA decay (or NMD) is an RQC mechanism related to cytoplasmic foci known as P-bodies, which recognizes premature nonsense or termination codons (PTC) within an mRNA (Conti and Izaurralde 2005). After recognition of an incorrectly positioned stop codon, the NMD system signals the elimination of the mRNAs through decapping, deadenylation, and exonucleolytic degradation. The UP-frameshift proteins are essential for NMD and 3 *UPF* genes exist in *Arabidopsis*. The analysis of these mutants revealed that, in addition to the expected NMD substrates, most npcRNAs including large numbers of antisense RNAs are degraded by this pathway, suggesting that one of the most important roles of NMD is the genome-wide suppression of aberrant or antisense RNAs (Kurihara et al. 2009). In *Saccharomyces cerevisiae*, regulatory npcRNAs are destabilized by the XRN1 $5' \rightarrow 3'$ RNA exonuclease, and the use of strand-specific RNA sequencing (RNA-seq) helped to identify a novel class of more than 1000 XRN1-sensitive unstable transcripts (XUTs) in which the majority are antisense to open reading frames (ORFS). These transcripts are polyadenylated and RNA pol II dependent. The accumulation of most of XUTs in response to stress suggests that they might have a role in adaptive responses to changes in growth conditions (van Dijk et al. 2011). Another major class of npcRNAs recently described consists of the cryptic unstable transcripts (CUTs), which are very short-lived RNA pol II transcripts only detectable when stabilized by mutations impairing the nuclear exosome (Neil and Jacquier 2011). On the other hand, the yeast stable uncharacterized transcripts (SUTs) are detectable also in wild-type cells. These transcripts are partially susceptible to the nuclear exosome and primarily degraded by the NMD (Marquardt et al. 2011). The role of various RNA-binding proteins in RQC mechanisms will be discussed later. However, it is clear that long npcRNAs, including small RNA precursors, regulating development must bypass these control mechanisms in order to reach the different RNP complexes where they act on their mRNA targets.

It is worth noting that the majority of the transcriptome studies usually measure the actual amount of transcripts present in a cell, resulting from the balance between transcription and degradation rates. This means that the complete set of sequences actually transcribed is significantly wider than the originally annotated genomic features. The steady state of npcRNAs should be considered in relation to their influence on gene expression and their own regulation may also affect the stability of their mRNA targets. Although several of these long npcRNAs have been experimentally identified in plants (Charon et al. 1999; Hirsch et al. 2006; Ma et al. 2008; Ben Amor et al. 2009; Franco-Zorrilla et al. 2009), their activity, subcellular localization, or molecular roles remain largely unknown. Globally, npcRNAs have been far less studied in plants than in animals, where diverse mechanisms involving npcRNAs in the regulation of gene expression have been discovered (Prasanth and Spector 2007; Voinnet 2009; Wilusz et al. 2009).

The Role of RNA-Binding Proteins in npcRNA Metabolism and Activity

The posttranscriptional regulation of gene expression mediated by npcRNAs as well as all aspects of RNA metabolism is globally determined by a variety of RBPs (Lorković 2009). Most RBPs contain one or more conserved domains, as the RNA-recognition motif (RRM) and the K homology (KH) motif. Based on the characteristics of those conserved domains, the analysis of the *A. thaliana* genome revealed that plants express a complex set of RBPs, with 196 RRM- and 26 KH-containing proteins, the majority of them being plant specific (Lorković and Barta 2002; Barta et al. 2008). Most of these proteins have not been characterized experimentally, and it is largely unclear how their action may control gene expression and development, primarily due to the difficulty in the identification of their RNA partners (Lorković 2009). Nevertheless, following forward and reverse genetic approaches, some studies begin to reveal the requirement of specific RBPs that have crucial roles in RNA metabolism during plant development.

Most RBPs are likely to have multiple RNA partners including mRNAs and npcRNAs (e.g., antisense RNAs, intergenic npcRNAs, or small RNAs) that may integrate different RNPs to generate RNA networks in which npcRNAs can determine their localization, interfere, or modulate their action (acting as competitors or activators against other substrates). Identification of the RBPs with which npcRNAs associate will contribute to understand their role in RNP networks in the cell. For example, several RBPs are involved in the biogenesis and action of small RNAs (e.g., DCLs, RDRs, or AGOs; Valencia-Sanchez et al. 2006). The RNAi pathways have been largely diversified and several steps occur in the nucleus and/or the cytoplasm. The different small RNA silencing pathways differ mainly in the way of generation of the small RNA (Vaucheret 2006). One RNA silencing mechanism is initiated by endogenous loci able to form double-stranded stem-loops that are processed by Dicer (dsRNAses) into small RNAs called miRNAs (miRNA pathway). In the siRNA pathway, a single-stranded RNA is targeted by RNA-dependent RNA polymerases to form long dsRNAs that then is cut by Dicers into small RNAs called siRNAs. In addition, long dsRNAs can be precursors of siRNAs when an npcRNA (a *trans*-acting or TAS long noncoding RNAs) is targeted by a specific miRNA and cleaved, becoming a substrate of RDR polymerases to form a long dsRNA qui sera processes in small siRNAs, called tasiRNAs (derived from TAS genes). Finally, two independent transcripts may form a complementary dsRNA (natural antisense genes) and this region of the dsRNA molecule is processed by Dicer into dsRNA small molecules or natsiRNAs. In all cases, one strand of the processed si/miRNAs duplex is incorporated into the RISC complex containing AGO proteins. The presence of this small RNA provides the RISC complex a sequence specificity to target a particular mRNA molecule. *G*ene *s*ilencing can occur at both *p*ost*t*ranscriptional (PTGS, through mRNA cleavage and/or translation inhibition) and/or *t*ranscriptional levels (TGS, through DNA methylation) (Vaucheret 2006).

In the cell, there exist several sources of long npcRNAs that may lack of coding capacity because they are aberrant RNAs, for example, mRNAs without either the 5′ cap or the 3′ poly(A) tail. Those aberrant RNAs can be converted in dsRNA by RDRs triggering silencing mechanisms through the action of DCLs or be eliminated by the normal degradation machinery, involving the exosome, XRN enzymes, or the NMD pathway (Conti and Izaurralde 2005; Chekanova et al. 2007; Kurihara et al.

2009). These survey mechanisms conform what we previously defined as RQC machinery and aim to ensure the correct expression of the genes (Houseley and Tollervey 2009).

As mentioned earlier, one of those control survey mechanisms is the NMD, for *n*onsense-*m*ediated mRNA *d*ecay, whose core RBPs are the UP-frameshift proteins (UPFs). This pathway recognizes PTC within an mRNA by the action of the exon-junction complex (EJC) that marks correctly fused exons (Pontes and Pikaard 2008). Such PTC-containing transcripts can arise as a result of genomic frameshifts, nonsense mutations, or inefficiently spliced pre-mRNAs, for example (Maquat 2004). In *Arabidopsis*, there are three UPF genes (UPF1, UPF2, and UPF3), and a genome-wide analysis using a tiling array of the mutants *upf1-1* and *upf3-1* revealed that in addition to the expected NMD substrates, that is coding mRNAs, most npcRNAs, including large numbers of antisense RNAs, are degraded by this pathway (the tiling-array data can be viewed at http://omicspace.riken.jp/gps/group/psca3). In these mutants, the percentage of upregulated messenger-like npcRNAs (mlnpcRNAs) to all expressed mlnpcRNAs was much higher than the percentage of upregulated mRNAs to all expressed mRNAs. This suggests that in fact, besides the recognition of nonsense mRNAs, another very important role of NMD is the genome-wide suppression of the mlnpcRNAs that are recognized as aberrant transcripts by this machinery (Kurihara et al. 2009). After recognition of an incorrectly positioned stop codon, the NMD system through the action of other RBPs, signals the elimination of this RNA by recruiting decapping enzymes followed by $5' \rightarrow 3'$ exonuclease activities (XRN enzymes), and/or deadenylation enzymes followed by $3' \rightarrow 5'$ exonucleolytic degradation (Conti and Izaurralde 2005; Lejeune and Maquat 2005; Amrani et al. 2006). This $3' \rightarrow 5'$ exonucleolytic degradation is known to take place in the exosome complex. Hence, npcRNAs may affect the stability of other aberrant transcripts by interfering with this pathway.

The exosome is an evolutionary conserved macromolecular complex that mediates numerous reactions of $3'$–$5'$ RNA processing and degradation, being essential for eukaryotic cell viability (Mitchell et al. 1997; Estévez et al. 2003). Some of its activities relay in the homeostatic mRNA turnover, decay of unstable mRNAs, nonsense-mediated mRNA decay, and degradation of the mRNA fragments derived from endonucleolytic cleavage mediated by mi/siRNAs in RISC complexes (Houseley et al. 2006). Combining genetic, proteomic, and whole-transcriptome analyses to investigate the function of the exosome complex, Chekanova et al. (2007) found that individual subunits of the exosome are functionally specialized, ranging from being dispensable for growth and development (CSL4) to being essential for the development of female gametophytes (RRP41) or embryogenesis (RRP4). These findings demonstrate an unexpected degree of functional plasticity in the plant exosome core. Concerning the exosome targets, they found multiple classes of stable structural RNAs, a select subset of mRNAs, primary miRNA (primiRNA) processing intermediates, tandem repeat-associated siRNA precursor species, as well as numerous long npcRNAs, as mentioned before, such as npcRNAs associated with heterochromatic regions in plants. In addition, the results revealed a novel layer of the transcriptome composed of intergenic npcRNAs that are tightly downregulated by constitutive exosome activity. This elegant work gave rise to a publicly available exosome-regulated transcriptome database (http://signal.salk.edu/cgi-bin/exosome) that will certainly help future work to elucidate regulatory mechanisms in complex eukaryotic transcriptomes.

Taking into account the roles of the NMD and the exosome, one could expect that the transcripts upregulated in the *upf* mutants ought to overlap with the exosome substrates, as aberrant transcripts recognized by the NMD complex should be degraded from the $3'$ end by deadenylation and subsequent $3' \rightarrow 5'$ exonuclease activity in the exosome (Lejeune et al. 2003; Mitchell and Tollervey 2003). However, it was not the case and only slight overlap was found between transcripts upregulated in the *upf* mutants and exosome substrates. In addition, in the population identified as exosome substrates, there were no natural antisense RNAs or NAT-RNAs. It is possible that these differences are probably due to the experimental variations, such as age of plants, difference in growth conditions, and statistical analysis (Kurihara et al. 2009) and certainly future work will need to address these discrepancies.

As mentioned earlier, after recognition of an abnormal RNA by one of the cellular survey mechanisms, the RNA molecules enter the decay mechanisms usually starting by deadenylation of the 3′ poly(A) tail and then by decapping of the 5′ cap structure, followed by degradation in the 3′→5′ and/or 5′→3′ directions by the exosome or XRN exoribonucleases, respectively (Chiba and Green 2009). In addition to aberrant transcripts derived from incorrect splicing or other cellular mechanisms, RNA substrates for this step also arrive from the silencing pathways after the cleavage of mRNA targets by si/miRNAs (Souret et al. 2004; Gy et al. 2007; Gregory et al. 2008; Rymarquis et al. 2011).

Despite that several components of the RQC machinery are required early on for plant growth and their mutations are lethal, inactivation of certain other components may lead to specific phenotypes. The *Arabidopsis* genome contains three XRN proteins (XRN2, XRN3, and XRN4) with different subcellular localizations and cellular functions. XRN2 and XRN3 are localized in the nucleus, are partially redundant, and are required for primary cleavage and processing of preribosomal RNAs (Zakrzewska-Placzek et al. 2010). On the other hand, XRN4, also called ethylene insensitive 5 (EIN5), is cytoplasmic and necessary for a correct ethylene signaling in the plant and for the degradation of the 3′ products resulting from the miRNA-mediated cleavage of target mRNAs (Roman et al. 1995; Olmedo et al. 2006; Gregory et al. 2008). Recent work pointed the association of XRN4 with transcripts encoding specific sequence motifs and select functional groups suggesting that these RNAs may be specifically targeted to the 5′→3′ decay pathway for degradation in *Arabidopsis*. The RNA sequence thus would play a major role in this targeting (Rymarquis et al. 2011). In addition to these functions on survey mechanisms, the three XRN proteins are involved in PTGS acting as endogenous RNA silencing suppressors. The exoribonuclease XRN4 was shown to act as a PTGS suppressor, possibly through the degradation of RNA-dependent RNA polymerase (RdRp) templates. In this way, mutations in XRN4 lead to the accumulation of aberrant, uncapped RNAs derived from transgenes, which could enhance PTGS (Gazzani et al. 2004). Likewise, in another work, XRN2 and XRN3, as well as FIERY1 (FRY1), were also shown to be suppressors of PTGS as the XRN activity is inhibited in a *fry1* background (Gy et al. 2007). FRY1, also known as SAL1 and HOS2, is a dual function 3′(2′),5′-bisphosphate nucleotidase/inositol polyphosphate 1-phosphatase orthologous to Hal2 and CysQ in yeast and *Escherichia coli*, respectively (Neuwald et al. 1992; Gläser et al. 1993). It was isolated in a genetic screen based on ABA- and stress-inducible gene transcription, and the *Arabidopsis* mutation *fry1* results in superinduction of ABA- and stress-responsive genes. Accordingly, *fry1* mutants are more sensitive to ABA or stress inhibition and present less tolerance to cold, drought, and salt stresses (Xiong et al. 2001). According to their activity as suppressors of PTGS, both *fry1* and the *xrn* double and triple mutants accumulate RNA intermediates of miRNA-directed posttranscriptional regulation, which are templates of XRN4 (Souret et al. 2004), and miRNA loops, which are templates of both XRN2 and XRN3 (Gy et al. 2007). To further analyze the role of XRN in the *fry1* phenotype, they generated an *xrn2 xrn3 xrn4* triple mutant that was fertile, unlike the sterile *xrn2 xrn3* double mutant. Although the mechanism for the partial phenotypic rescue is unclear, it suggests that *xrn4* mutations act to partially suppress the *xrn2 xrn3* phenotypic effects. Neither XRN2 nor XRN4 is critical for proper plant development. However, the embryo-lethality of null *xrn3* alleles and the developmental defects exhibited by hypomorphic *xrn3-3* mutants suggest an essential role of XRN3. The *xrn2 xrn3 xrn4* triple mutant displays the *fry1* lateral root and drought-tolerance phenotypes but does not affect primary root. Microscopical observations revealed that the altered root architecture in *fry1* mutants was due to reduced meristem activity in the primary root and to a lateral root initiation defect. Altogether, these results suggest that the pleiotropic phenotype of the *fry1* mutants results, at least in part, from a general perturbation in XRN activities.

Recently, in a genetic screen for *Arabidopsis* mutants deregulated in the expression of phosphate high-affinity transporter 1;4 (PHT1;4), a novel *fry1* allele was described. The authors identified a novel FRY1 function modulating the transcription of several Pi starvation markers in the root stele; however, this *fry1* mutant phenotype is independent of XRN activities. A transcriptomic analysis confirmed that the phenotype observed corresponded to a point mutation in the transcript corresponding to the *fry* gene.

The two periphery marks of the extreme borders of a eukaryotic mRNA are defined by the 5′ m7G-cap structure and the 3′ poly(A) tail, and in the nuclei of eukaryotic organisms, the 5′ cap is recognized by the cap-binding complex (CBC). In *Arabidopsis*, there are two single genes for both the large and small subunits, ABH1 (ABA *h*ypersensitive 1) and CBP20, respectively, that form the CBC heterodimer and play numerous roles in RNA metabolism (Hugouvieux et al. 2001, 2002; Aguilera 2005). Mutant plants for the ABH1 RBP present ABA hypersensitivity in seed germination, stomata closure, reduced wilting during drought, and ABA-induced guard cell calcium increases (Hugouvieux et al. 2001). *ABH1* activity is necessary for the correct expression level of a subset of genes in the *Arabidopsis* genome (Hugouvieux et al. 2001), suggesting a specific connection between mRNA metabolism and ABA signaling. Consistent with their intimately linked activities, inactivation of CBP20 causes a similar serrated leaf phenotype and increased drought resistance as seen in *abh1* mutants (Hugouvieux et al. 2001; Papp et al. 2004). Although the morphological and physiological effects of ABH1 and CBP20 have been quite well described (Hugouvieux et al. 2001; Bezerra et al. 2004; Papp et al. 2004), it is largely unclear how these proteins relate to the biochemical functions of the CBC. Nevertheless, through the analysis of the developmental defects on *xrn4-abh1* double mutant plants, Gregory et al. (2008) demonstrated surprising roles in RNA silencing pathways for these two proteins, XRN4 and ABH1, involved in general RNA metabolism. They found that the loss of ABH1 decreases the levels of mature miRNAs, suggesting that this protein functions in the miRNA-mediated RNA silencing pathway. Indeed the *Arabidopsis* CBC would be important for proper pri-miRNA processing eventually providing a platform for recruitment of miRNA maturation factors (Laubinger et al. 2008). On the other hand, XRN4 affected the abundance of a distinct class of mainly 21 nt small RNAs, processed from both sense and antisense strands of some endogenous transcripts. Apparently, those transcripts are converted to dsRNA and subsequently processed, and regularly accumulate in an uncapped form in *xrn4* mutant plants. Taken together, these results suggest that an additional fate for endogenous uncapped transcripts is shuttling into an RNA silencing pathway where they become small RNA-biogenesis substrates.

Accordingly, a very recent work showed that there is a link between the *Arabidopsis* cap-binding protein ABH1 and the suppression of silencing (Christie et al. 2011). Their results indicate that genes containing introns are less susceptible to PTGS and that this intron suppression of gene silencing requires an efficient splicing that is dependent on ABH1. At the end of the 1980s, it was already published that endogenous genes generate much higher levels of gene expression than their cDNA counterparts (Callis et al. 1987). As well, in transgene-expressed viral RNA genomes, the addition of introns has also been shown to significantly enhance their accumulation (Marillonnet et al. 2005). Finally, genome-wide mRNA decay rates show that transcripts from intronless genes are significantly more unstable than those from intron-containing genes (Narsai et al.

2007). Based on a GFP-transgenic reporter system and varying the number of introns, this recent study provides a molecular basis to elucidate those evidences. They show that efficiently spliced introns may reduce RDR6 activity along spliced GFP transcripts via a mechanism requiring the cap-binding protein ABH1. Indeed the ABH1 protein has been previously correlated to pre-mRNA splicing in plants (Kuhn et al. 2007; Gregory et al. 2008; Laubinger et al. 2008) and various studies in yeast and animals have shown that the CBC is functionally and physically coupled to pre-mRNA splicing (Izaurralde et al. 1994). To investigate this hypothesis, they transformed *abh1* mutants with the same GFP-transgenic reporter system and unexpectedly concluded that intron suppression of transgene silencing was lost in those mutants. They propose a model where an efficient intron splicing could decrease the quantity of aberrant RNA by-products produced by transcription (as uncapped or improperly terminated transcripts) or, alternatively, facilitate recruitment of enzymes that degrade aberrant RNA as they are formed. As discussed earlier, various RQC pathways exist in the cell and compete against endogenous RdR polymerases activities to prevent extensive amplification of silencing. In other words, there is a constant fight between RQC and the silencing pathways competing for the same aberrant or endogenous npcRNA substrates produced in the cell. Thus, it is possible that the large varieties of npcRNAs found in the cell may interfere with the RNA surveillance and degradation pathways and affect the processes of RNA maturation.

As mentioned earlier, the total number of protein-coding genes in diverse organisms varies much less than the number of different transcripts along evolutionary scales (Mattick and Makunin 2006; Yasuda and Hayashizaki 2008). This supports the idea that npcRNAs are essential to understand the huge complexity of multicellular organisms. Indeed, the large diversity of npcRNAs identified up to now in eukaryotes, and their increasing number, may reflect the importance of riboregulation, mediated by npcRNA–RBPs interactions, in the determination of differentiation and adaptability in eukaryotes. Concretely, plants display a notable flexibility in their architecture and growing patterns in response to external stimuli, characterized by a great developmental plasticity. This quality allows higher plants to adapt to different environmental conditions, with individuals with the same genotype giving rise to different phenotypes. Thus, future challenges lie in understanding the implication of the various RNP networks to determine growth and developmental outcomes under different environmental conditions.

Nonprotein-Coding RNAs in Root Development

In this section, we will discuss more specific npcRNAs linked to root development (Table 2.1). The root architecture of the plant constitutes an important model to study how developmental plasticity is translated into growth responses under stress conditions. Indeed, primary root development and the formation of *de novo* meristems to generate lateral roots (LRs) are conditioned by the soil environment (Osmont et al. 2007). The remarkable developmental plasticity called the attention of Charles Darwin and his son Francis. In their monograph on the *Power of Movements*, they referred to the behavior (i.e., gravitropism) of the growing root, postulating that the root tip acts like a plant brain: "We believe that there is no structure in plants more wonderful, as far as its functions are concerned, than the tip of the radicle . . . It is hardly an exaggeration to say that the tip of the radicle thus endowed, and having the power of directing the movements of the adjoining parts, acts like the brain of the lower animals" (cited by Kutschera and Niklas 2009).

Several of the best described riboregulators in plant biology, the miRNAs, have been linked to root development (Table 2.1), such as the control of primary root growth or the formation of organs from

Table 2.1 List of several miRNAs involved in root development, stress response, and hormone signaling of various plant species.

miR Name	Organism	Targets	Role	Reference
miR166	*Medicago truncatula*	HD-ZIP III	Root vascular tissues differentiation, lateral root and nodule formation	Boualem et al. 2008
miR165/166	*Arabidopsis thaliana*	PHB/PHV (HD-ZIP III)	Embryonic root development	Grigg et al. 2009
miR160	*A. thaliana*	ARF17	Adventicious root formation	Gutierrez et al. 2009
miR160	*A. thaliana*	ARF16	Lateral root initiation	Wang et al. 2005
miR160	*A. thaliana*	ARF10/ARF16	Auxin signaling and root growth and gravitropism	Wang et al. 2005
miR396	*A. thaliana*	GRF	Root growth	Rodriguez et al. 010
miR156	*Oedicago sativa*	SPL	Global growth	Jiao et al. 2010
miR156	*M. truncatula*	WD40	Root growth	Naya et al. 2010
miR164	*A. thaliana*	NAC1	Lateral root initiation and development	Guo et al. 2005
miR390	*O. sativa*	TAS3/ARF2/ARF3/ARF4	Potential role in auxin signaling and root development	Williams et al. 2005
miR390	*A. thaliana*	TAS3/ARF2/ARF3/ARF4	Auxin signaling and lateral root growth	Marin et al. 2010
miR167	*O. sativa*	ARF6/ARF8	Auxin signaling and adventitious root formation	Yang et al. 2006
miR167	*A. thaliana*	ARF6/ARF8	Auxin signaling and adventitious root formation	Gutierrez et al. 2009
miR169	*M. truncatula*	Mt-HAP2-1	Nodule differentiation	Combier et al. 2006
miR482	*Glycine max*	R gene receptor kinases	Nodule formation	Li et al. 2010
miR1511	*G. max*	Different	Nodule formation	Li et al. 2010
miR1512	*G. max*	Different	Nodule formation	Li et al. 2010
miR1515	*G. max*	DCL	Nodule formation	Li et al. 2010
miR167a	*A. thaliana*	ARF8	Root response to N and lateral root emergence	Gifford et al. 2008
miR393	*A. thaliana*	AFB1/AFB2/AFB3/TIR1	Root response to N, auxin and potential role in lateral root formation	Vidal et al. 2010
miR399	*O. sativa*	PHO2	Phosphate signaling, potential role in lateral root formation	Bari et al. 2006
miR399	*A. thaliana*	PHO2	Phosphate signaling and lateral root formation	Franco-Zorrilla et al. 2007
miR395	*A. thaliana*	SULTR2;1/APS1/APS3/APS4	Sulfate metabolism and potential role in laterale root formation	Liang et al. 2010
miR169	*A. thaliana*	NF-YA5	Drought stress, potential role in root development	Li et al. 2008

de novo meristems, such as lateral and adventitious roots, or the legume-specific nitrogen-fixing nodules, through symbiotic interactions with soil bacteria (Khan et al. 2011).

External cues influence plant growth by modulating hormone levels and signaling. Auxin is one of the main phytohormones regulating root growth and architecture. Auxin activity is mediated by the *AUXIN RESPONSE FACTOR* (*ARF*) genes, a plant-specific family of transcriptional regulators (Okushima et al. 2005). TAS3 is a *trans*-acting siRNA, whose biogenesis requires the initial miR390-mediated cleavage of the TAS3 precursor (Marin et al. 2010). The cleavage product is then converted to dsRNA through the RdR6 activity and sequential DCL4-mediated cleavage events (Peragine et al. 2004; Vazquez et al. 2004; Allen et al. 2005; Gasciolli et al. 2005; Xie et al. 2005; Yoshikawa et al. 2005; Adenot et al. 2006). Of the four tasiRNAs precursors identified (TAS1-4) in *Arabidopsis*, cleavage of TAS3 is unique, since it requires the specific action of the miR390/AGO7 complex for ta-siRNA production (Montgomery et al. 2008). These tasiRNAs inhibit ARF2, ARF3, and ARF4, thus releasing repression of lateral root growth (Marin et al. 2010). In addition, ARF2, ARF3, and ARF4 affect auxin-induced miR390 accumulation. Positive and negative feedback regulation of miR390 by ARF2, ARF3, and ARF4 thus ensures the proper definition of the miR390 expression pattern resulting in the adaptation of the root system architecture (RSA) by auxin. In *Arabidopsis*, it was shown that auxin-associated miRNAs tightly control adventitious root formation through a complex regulation that involved various ARFs (Gutierrez et al. 2009). Although miR160 positively regulates adventitious root formation by controlling ARF17, the auxin-related miR167, which targets ARF6 and 8, is a negative regulator of adventitious root development. In *Arabidopsis*, ARF8 and ARF17 play antagonist roles in auxin homeostasis (Tian et al. 2004; Sorin et al. 2005).

Other molecular mechanisms involved in root plasticity in response to the environment and endogenous signals are the natural antisense RNA (NATs), which are transcripts complementary to *cis-* or *trans*-mRNAs and exert a repressive activity on them. Considering that NATs may encode proteins, some of them can be classified as dual RNAs because of their double function: both NAT and mRNA (Bardou et al. 2011). NATs have been described in several organisms such as yeast, human, mice, and plants. We can distinguish two different classes of NATs: *cis*-NATs, which are generated by antisense transcription at the same genomic locus, and *trans*-NATs, which are generated from different loci. Interestingly, a large number of transcripts were predicted to have both *cis-* and *trans*-NATs, suggesting that antisense transcripts can form a complex regulatory network (Henz et al. 2007). In *Arabidopsis*, a *cis*-NAT pair encoding SRO5 and P5CDH was shown to repress translation and improve *Arabidopsis* salt tolerance (Borsani et al. 2005). High salt concentration is toxic for plants when up-taken from the soil by the root (Ariel et al. 2010), strongly affecting root metabolism and development. In response, plants have developed biological mechanisms that prevent NaCl accumulation or absorption (Munns et al. 2010). P5CDH is constitutively expressed and encodes the D1-pyrroline-5-carboxylate dehydrogenase, which prevents proline accumulation, whereas SRO5 is induced by salt stress and encodes an unknown protein. Under high levels of NaCl, both genes form a natural double-stranded pair of transcripts proved to be cleaved by DCL2 and DCL1 generating 24- and 21-nucleotide nat-siRNAs that will repress the constitutively expressed P5CDH mRNA and lead to increased salt tolerance (Borsani et al. 2005).

In many cases, the function or coding capacity of NAT-RNAs remain largely unknown. *PHO* genes participate in phosphate (Pi) transport in the cellular response to Pi starvation in plants. Phosphate is normally up-taken from the soil and is essential for plant development. In rice, three *PHO1* genes can form a *cis*-NAT pair with npcRNAs. Surprisingly, the *cis*-NAT associated with the OsPHO1-3 gene does not appear to be regulated either developmentally or in response to Pi deficiency, in contrast to the *cis*-NATs of OsPHO1-1 and OsPHO1-2, which are strongly upregulated by Pi starvation, while the expression of the complementary sense transcript remains relatively stable (Secco et al.

2010). However, considering that the analysis was performed using whole roots, a hypothesis is that the expression of the sense and antisense OsPHO1 transcripts may not occur in the same cells or tissues in all cases. In *Arabidopsis*, a search for npcRNAs (Ben Amor et al. 2009) allowed to identify 13 antisense npcRNAs complementary to protein-coding transcripts. One of these (npc536) forms a *cis*-NAT with AT1G67930, and its overexpression allowed plants to grow under salt stress without modifying AT1G67930 mRNA accumulation. Furthermore, *npc536* mutants do not show any misregulation of the antisense transcript. As this NAT contains a short ORF conserved in rice, npc536 may act through this encoded peptide. Alternatively, npc536 may regulate translation of the AT1G67930 mRNA or act as a *trans*-NAT, with an unidentified complementary target that plays a role in the salt stress response.

Apart from npcRNAs that lead to small RNAs such as the TAS or the NAT genes, only few npcRNAs have been implicated in root developmental processes (Charon et al. 2010). The *ENOD40* genes code for highly structured plant mRNAs that contain a series of short ORFs without any long ORF (Charon et al. 1999; Gultyaev and Roussis 2007) and are involved in legume-specific root nodule organogenesis. Root nodules are nitrogen-fixing symbiotic plant organs that result from the interaction of soil bacteria of the genus *Rhizobium* with the root cells of host legumes (Crespi and Frugier 2008; Oldroyd and Downie 2008). This process initiates with cell-specific division in the roots, where *ENOD40* is strongly expressed (Yang et al. 1993; Crespi et al. 1994). The *ENOD40* gene is characterized by specific conserved nucleotide sequences that can also be found in some nonleguminous plants (Gultyaev and Roussis 2007). Furthermore, transgenic lines with increased or decreased levels of *ENOD40* exhibit accelerated or reduced nodulation, respectively (Charon et al. 1999). It was first proposed that *ENOD40* was an npcRNA due to its highly stable RNA secondary structure, a characteristic of known npcRNAs (Crespi et al. 1994; Hofacker et al. 2002); however, other authors proposed that this transcript may encode a small primary oligopeptide of around 13 amino acids (Charon et al. 2010). Translational analysis identified two short ORFs (sORF I and II; 13 and 27 amino acids long, respectively) that could be translated from this transcript in *Medicago truncatula* (Sousa et al. 2001) and sORF I contains a conserved nucleotide region across legumes, but not other plants, in contrast to the highly conserved stem-loops of the *ENOD40* RNA throughout all known plant species (Girard et al. 2003). A cell-specific assay for the action of *ENOD40* in *Medicago sativa* was developed using a biolistic process and suggested that translation of these sORF may be biologically relevant. Interestingly, mutations in the predicted structured RNA region also strongly inhibited this biological activity (Sousa et al. 2001). These results confirm the importance of both the sORF peptides and the RNA secondary structure of *ENOD40* in its activity and suggest that *ENOD40* encodes a bifunctional or dual RNA. To gain further insight into the action of *ENOD40*, molecules that interact with the peptides or RNA were identified. A novel RNA-binding protein MtRBP1 (for *M. truncatula* RBP 1), which interacts with the *ENOD40* RNA, was identified using a yeast three-hybrid screen. Immunolocalization studies and the use of an MtRBP1-DsRed2 fusion construct showed that MtRBP1 localized to nuclear "speckles," which are nuclear RNP complexes known to house the splicing machinery in plant cells (Cioce and Lamond 2005; Handwerger and Gall 2006; Li et al. 2006). These nuclear speckles (or interchromatin granule clusters) are spotted shapeless structures containing elevated concentrations of splicing snRNPs and other splicing-related proteins that participate in the cotranscriptional splicing of mRNAs at the chromosomes (Shaw and Brown 2004). Interestingly, MtRBP1 was located in the cytoplasm of *ENOD40* expressing cells in *M. truncatula* nodules. The direct involvement of the *ENOD40* RNA in MtRBP1 relocalization into cytoplasmic granules was confirmed using a transient expression assay and an MS2 bacteriophage system to tag the *ENOD40* RNA (Campalans et al. 2004). This *in vivo* approach to monitor RNA–protein interactions demonstrated that the cytoplasmic relocalization of MtRBP1 was mediated by *ENOD40* and suggested that the relocalization of nuclear RBPs during specific developmental processes could be a new function mediated by npcRNAs (Campalans et al. 2004). On the other hand, the *ENOD40* peptides expressed in soybeans were shown to bind to sucrose synthase (SUC1), suggesting a potential role of these peptides in the regulation of sucrose utilization in the nodules (Rohrig et al. 2002). These results further highlight that npcRNA genes may act as bifunctional RNAs in plants as many genes contain potentially active sORF-encoded peptides. In *Arabidopsis*, more than 3000 sORFs are transcribed, suggesting that huge numbers of sORF-encoded peptides are still hidden in genomic regions that have not been annotated yet (Hanada et al. 2007). For example, the *POLARIS* (*PLS*) gene in *Arabidopsis* was identified experimentally using a promoter-trap approach and shown to have a root-specific expression pattern (Topping et al. 1997). *pls* mutant plants have short roots with radially expanded cells and reduced leaf vascularization (Casson et al. 2002). The *PLS* gene is transcribed as a relatively short, 500-nucleotide mRNA, which contains three short ORFs that encode putative peptides of 8, 9, and 36 amino acids. Overexpression of the ORF encoding the 36-amino acid peptide partially rescues the short-root phenotype. Although the function of *PLS* has not been fully elucidated, a role in hormonal homeostasis, including ethylene signaling and auxin transport, and in the regulation of microtubule cytoskeletal dynamics was proposed (Chilley et al. 2006). These dual RNA further highlights the large variety of mechanisms that npcRNAs may trigger in root tissues.

Future Perspectives

In recent years, growing evidence has demonstrated that RNAs can not only encode proteins but can also exert a wide range of molecular functions, including the modulation of mRNA expression and RNA processing or localization, the regulation of protein activity and structure, and can act as precursors to small RNAs or sORF-encoded peptides. The transcriptome complexity and the remarkable portion of npcRNAs indicate that many of the roles played by these RNAs remain largely unknown. Plant developmental plasticity in response to environmental changes depends entirely on molecular flexibility in gene expression. Therefore, future work will help to unravel the crucial role of npcRNAs in root development.

Acknowledgments

Our recent studies cited in this review were supported by grants from ANR (RNAPATHS) and from the program Saclay Plant Sciences (SPS, ANR-10-LABX-40). F. Ariel was supported by an EMBO Long-Term Fellowship and F. Bardou was supported by a grant from the Ministère Français de l'Enseignement Supérieur et de la Recherche. We apologize to our colleagues whose work we were unable to cite.

References

Adenot, X. et al. (2006) DRB4-dependent TAS3 trans-acting siRNAs control leaf morphology through AGO7. *Current Biology: CB*, 16(9), 927–932.

Aguilera, A. (2005) Cotranscriptional mRNP assembly: from the DNA to the nuclear pore. *Current Opinion in Cell Biology*, 17(3), 242–250.

Allen, E. et al. (2005) microRNA-directed phasing during trans-acting siRNA biogenesis in plants. *Cell*, 121(2), 207–221.

Amrani, N. et al. (2006) Early nonsense: mRNA decay solves a translational problem. *Nature Reviews Molecular Cell Biology*, 7(6), 415–425.

Ariel, F. et al. (2010) Environmental regulation of lateral root emergence in *Medicago truncatula* requires the HD-Zip I transcription factor HB1. *The Plant Cell*, 22(7), 2171–2183.

Asano, K. et al. (2001) A multifactor complex of eIF1, eIF2, eIF3, eIF5, and $tRNA_i^{Met}$ promotes initiation complex assembly and couples GTP hydrolysis to AUG recognition. *Cold Spring Harbor Symposia on Quantitative Biology*, 66, 403–416.

Babiarz, J.E. et al. (2008) Mouse ES cells express endogenous shRNAs, siRNAs, and other microprocessor-independent, Dicer-dependent small RNAs. *Genes & Development*, 22(20), 2773–2785.

Bardou, F. et al. (2011) Dual RNAs in plants. *Biochimie*, 93, 1950–1954.

Bari, R. et al. (2006) Phosphate-signaling pathway in plants. *Plant Physiology*, 141, 988–999.

Barta, A. et al. (2008) Plant SR proteins and their functions. *Current Topics in Microbiology and Immunology*, 326, 83–102.

Ben Amor, B. et al. (2009) Novel long non-protein coding RNAs involved in Arabidopsis differentiation and stress responses. *Genome Research*, 19(1), 57–69.

Bertani, S. et al. (2011) The noncoding RNA Mistral activates Hoxa6 and Hoxa7 expression and stem cell differentiation by recruiting MLL1 to chromatin. *Molecular Cell*, 43(6), 1040–1046.

Bezerra, I.C. et al. (2004) Lesions in the mRNA cap-binding gene ABA HYPERSENSITIVE 1 suppress FRIGIDA-mediated delayed flowering in Arabidopsis. *The Plant Journal*, 40(1), 112–119.

Borsani, O. et al. (2005) Endogenous siRNAs derived from a pair of natural *cis*-antisense transcripts regulate salt tolerance in Arabidopsis. *Cell*, 123(7), 1279–1291.

Boualem, A. et al. (2008) MicroRNA166 controls root and nodule development in *Medicago truncatula*. *The Plant Journal*, 54(5), 876–887.

Brown, J.W.S. et al. (2008) Intronic noncoding RNAs and splicing. *Trends in Plant Science*, 13(7), 335–342.

Callis, J. et al. (1987) Introns increase gene expression in cultured maize cells. *Genes & Development*, 1(10), 1183–1200.

Campalans, A. et al. (2004) Enod40, a short open reading frame-containing mRNA, induces cytoplasmic localization of a nuclear RNA binding protein in *Medicago truncatula*. *The Plant Cell*, 16, 1047–1059.

Carninci, P. (2010) RNA dust: where are the genes? *DNA Research*, 17(2), 51–59.

Casson, S.A. et al. (2002) The POLARIS gene of Arabidopsis encodes a predicted peptide required for correct root growth and leaf vascular patterning. *The Plant Cell*, 14(8), 1705.

Charon, C. et al. (1999) Alteration of enod40 expression modifies *Medicago truncatula* root nodule development induced by *Sinorhizobium meliloti*. *The Plant Cell*, 11(10), 1953–1966.

Charon, C. et al. (2010) Non-protein-coding RNAs and their interacting RNA-binding proteins in the plant cell nucleus. *Molecular Plant*, 3(4), 729–739.

Chaumeil, J. et al. (2006) A novel role for Xist RNA in the formation of a repressive nuclear compartment into which genes are recruited when silenced. *Genes & Development*, 20(16), 2223–2237.

Chekanova, J.A. et al. (2007) Genome-wide high-resolution mapping of exosome substrates reveals hidden features in the Arabidopsis transcriptome. *Cell*, 131(7), 1340–1353.

Chen, C.J. et al. (2011) Genome-wide discovery and analysis of microRNAs and other small RNAs from rice embryogenic callus. *RNA Biology*, 8(3), 538–547.

Chiba, Y. and Green, P.J., (2009) mRNA degradation machinery in plants. *Journal of Plant Biology*, 52(2), 114–124.

Chilley, P.M. et al. (2006) The POLARIS peptide of Arabidopsis regulates auxin transport and root growth via effects on ethylene signaling. *The Plant Cell*, 18(11), 3058–3072.

Chitwood, D.H. and Timmermans, M.C.P. (2010) Small RNAs are on the move. *Nature*, 467(7314), 415–419.

Christie, M. et al. (2011) Intron splicing suppresses RNA silencing in Arabidopsis. *The Plant Journal*, 68(1), 4676–4676.

Cioce, M. and Lamond, A.I. (2005) Cajal bodies: a long history of discovery. *Annual Review of Cell and Developmental Biology*, 21, 105–131.

Clemson, C.M. et al. (1996) XIST RNA paints the inactive X chromosome at interphase: evidence for a novel RNA involved in nuclear/chromosome structure. *The Journal of Cell Biology*, 132(3), 259–275.

Combier, J.P. et al. (2006) MtHAP2-1 is a key transcriptional regulator of symbiotic nodule development regulated by microRNA169 in *Medicago truncatula*. *Genes & Development*, 20(22), 3084–3088.

Conti, E. and Izaurralde, E. (2005) Nonsense-mediated mRNA decay: molecular insights and mechanistic variations across species. *Current Opinion in Cell Biology*, 17(3), 316–325.

Crespi, M. and Frugier, F. (2008) De novo organ formation from differentiated cells: root nodule organogenesis. *Science Signaling*, 1(49), re11.

Crespi, M.D. et al. (1994) Expressed during nodule codes for a non-translatable RNA involved in plant growth. *EMBO Journal*, 13(21), 5099–5112.

Crevillén, P. and Dean, C. (2011) Regulation of the floral repressor gene FLC: the complexity of transcription in a chromatin context. *Current Opinion in Plant Biology*, 14(1), 38–44.

Djebali, S. et al. (2008) Efficient targeted transcript discovery via array-based normalization of RACE libraries. *Nature Methods*, 5(7), 629–635.

Dunoyer, P. et al. (2010) Small RNA duplexes function as mobile silencing signals between plant cells. *Science*, 328(5980), 912–916.

Estévez, A.M. et al. (2003) The roles of intersubunit interactions in exosome stability. *The Journal of Biological Chemistry*, 278(37), 34943–34951.

Faghihi, M.A. et al. (2010) Evidence for natural antisense transcript-mediated inhibition of microRNA function. *Genome Biology*, 11(5), R56.

Franco-Zorrilla, J.M. et al. (2007) Target mimicry provides a new mechanism for regulation of microRNA activity. *Nature Genetics*, 39(8), 1033–1037.

Franco-Zorrilla, J.M. et al. (2009) Genome-wide identification of small RNA targets based on target enrichment and microarray hybridizations. *The Plant Journal*, 59(5), 840–850.

Jiao, Y. et al. (2010) Regulation of OsSPL14 by OsmiR156 defines ideal plant architecture in rice. *Nature Genetics*, 42(6), 541–544.

Gasciolli, V. et al. (2005) Partially redundant functions of Arabidopsis DICER-like enzymes and a role for DCL4 in producing trans-acting siRNAs. *Current Biology*, 15(16), 1494–500.

Gazzani, S. et al. (2004) A link between mRNA turnover and RNA interference in Arabidopsis. *Science*, 306(5698), 1046–1048.

Gifford, M.L. et al. (2008) Cell-specific nitrogen responses mediate developmental plasticity. *Proceedings of the National Academy of Sciences USA*, 105(2), 803–808.

Girard, G. et al. (2003) Structural motifs in the RNA encoded by the early nodulation gene enod40 of soybean. *Nucleic Acids Research*, 31(17), 5003–5015.

Gläser, H.U. et al. (1993) Salt tolerance and methionine biosynthesis in *Saccharomyces cerevisiae* involve a putative phosphatase gene. *The EMBO Journal*, 12(8), 3105–3110.

Gregory, B.D. et al. (2008) A link between RNA metabolism and silencing affecting Arabidopsis development. *Developmental Cell*, 14(6), 854–866.

Grigg, S.P. et al. (2009) Repression of apical homeobox genes is required for embryonic root development in Arabidopsis. *Current Biology*, 19(17), 1485–1490.

Gultyaev, A.P. and Roussis, A. (2007) Identification of conserved secondary structures and expansion segments in enod40 RNAs reveals new enod40 homologues in plants. *Nucleic Acids Research*, 35(9), 3144–3152.

Guo, H.-S. et al. (2005) MicroRNA directs mRNA cleavage of the transcription factor NAC1 to downregulate auxin signals for arabidopsis lateral root development. *Development*, 17, 1376–1386.

Gutierrez, L. et al. (2009) Phenotypic plasticity of adventitious rooting in Arabidopsis is controlled by complex regulation of AUXIN RESPONSE FACTOR transcripts and microRNA abundance. *The Plant Cell*, 21(10), 3119–3132.

Gy, I. et al. (2007) Arabidopsis FIERY1, XRN2, and XRN3 are endogenous RNA silencing suppressors. *The Plant Cell*, 19(11), 3451–3461.

Hanada, K. et al. (2007) A large number of novel coding small open reading frames in the intergenic regions of the *Arabidopsis thaliana* genome are transcribed and/or under purifying selection. *Genome Research*, 17(5), 632–640.

Handwerger, K.E. and Gall, J.G. (2006) Subnuclear organelles: new insights into form and function. *Trends in Cell Biology*, 16(1), 19–26.

Haussecker, D. et al. (2010) Human tRNA-derived small RNAs in the global regulation of RNA silencing. *RNA*, 16(4), 673–695.

Henz, S.R. et al. (2007) Distinct expression patterns of natural antisense transcripts in Arabidopsis. *Plant Physiology*, 144(3), 1247–1255.

Heo, J.B. and Sung, S. (2011) Vernalization-mediated epigenetic silencing by a long intronic noncoding RNA. *Science*, 331(6013), 76–79.

Hirsch, J. et al. (2006) Characterization of 43 non-protein-coding mRNA genes in Arabidopsis, including the *MIR162a*-derived transcripts. *Plant Physiology*, 140, 1192–1204.

Hofacker, I.L. et al. (2002) Secondary structure prediction for aligned RNA sequences. *Journal of Molecular Biology*, 319(5), 1059–1066.

Houseley, J. and Tollervey, D. (2009) The many pathways of RNA degradation. *Cell*, 136(4), 763–776.

Houseley, J. et al. (2006) RNA-quality control by the exosome. *Nature reviews. Molecular Cell Biology*, 7(7), 529–539.

Hsieh, L.C. et al. (2009) Uncovering small RNA-mediated responses to phosphate deficiency in Arabidopsis by deep sequencing. *Plant Physiology*, 151(4), 2120–2132.

Hugouvieux, V. et al. (2001) An mRNA cap binding protein, ABH1, modulates early abscisic acid signal transduction in Arabidopsis. *Cell*, 106(4), 477–487.

Hugouvieux, V. et al. (2002) Localization, ion channel regulation, and genetic interactions during abscisic acid signaling of the nuclear mRNA cap-binding protein, ABH1 1. *Plant Physiology*, 130, 1276–1287.

Izaurralde, E. et al. (1994) A nuclear cap binding protein complex involved in pre-mRNA splicing. *Cell*, 78(4), 657–668.

Jamalkandi, S.A. and Masoudi-Nejad, A. (2009) Reconstruction of *Arabidopsis thaliana* fully integrated small RNA pathway. *Functional & Integrative Genomics*, 9(4), 419–432.

Jay, F. et al. (2010) Biotic stress-associated microRNAs: identification, detection, regulation, and functional analysis. *Methods Molecular Biology*, 592, 183–202.

Khan, G.A. et al. (2011) MicroRNAs as regulators of root development and architecture. *Plant Molecular Biology*, 77(1–2), 47–58.

Kuhn, J.M. et al. (2007) mRNA metabolism of flowering-time regulators in wild-type Arabidopsis revealed by a nuclear cap binding protein mutant, abh1. *The Plant Journal*, 50(6), 1049–1062.

Kurihara, Y. et al. (2009) Genome-wide suppression of aberrant mRNA-like noncoding RNAs by NMD in Arabidopsis. *Proceedings of the National Academy of Sciences USA*, 106(7), 2453–2458.

Kutschera, U. and Niklas, K.J. (2009) Evolutionary plant physiology: Charles Darwin's forgotten synthesis. *Die Naturwissenschaften*, 96(11), 1339–1354.

Laubinger, S. et al. (2008) Dual roles of the nuclear cap-binding complex and SERRATE in pre-mRNA splicing and microRNA processing in *Arabidopsis thaliana*. *Proceedings of the National Academy of Sciences USA*, 105(25), 8795–8800.

Lejeune, F. and Maquat, L.E. (2005) Mechanistic links between nonsense-mediated mRNA decay and pre-mRNA splicing in mammalian cells. *Current Opinion in Cell Biology*, 17(3), 309–315.

Lejeune, F. et al. (2003) Nonsense-mediated mRNA decay in mammalian cells involves decapping, deadenylating, and exonucleolytic activities. *Molecular Cell*, 12(3), 675–687.

Li, C.F. et al. (2006) An ARGONAUTE4-containing nuclear processing center colocalized with Cajal bodies in *Arabidopsis thaliana*. *Cell*, 126(1), 93–106.

Li, H. et al. (2010) Misexpression of miR482, miR1512, and miR1515 increases soybean nodulation. *Plant Physiology*, 153(4), 1759–1770.

Li, W.-X. et al. (2008) The Arabidopsis NFYA5 transcription factor is regulated transcriptionally and posttranscriptionally to promote drought resistance. *The Plant Cell*, 20(8), 2238–2251.

Liang, G. et al. (2010) MicroRNA395 mediates regulation of sulfate accumulation and allocation in *Arabidopsis thaliana*. *The Plant Journal*, 62(6), 1046–1057.

Lorković, Z.J. (2009) Role of plant RNA-binding proteins in development, stress response and genome organization. *Trends in Plant Science*, 14(4), 229–236

Lorković, Z.J. and Barta, A. (2002) Genome analysis: RNA recognition motif (RRM) and K homology (KH) domain RNA-binding proteins from the flowering plant *Arabidopsis thaliana*. *Nucleic Acids Research*, 30(3), 623–635.

Lykke-Andersen, S. et al. (2009) Origins and activities of the eukaryotic exosome. *Journal of Cell Science*, 122(Pt 10), 1487–1494.

Ma, L. et al. (2008) A novel RNA-binding protein associated with cell plate formation. *Plant Physiology*, 148(1), 223–234.

Mancini-Dinardo, D. et al. (2006) Elongation of the Kcnq1ot1 transcript is required for genomic imprinting of neighboring genes. *Genes & Development*, 20(10), 1268–1282.

Maquat, L.E. (2004) Nonsense-mediated mRNA decay: splicing, translation and mRNP dynamics. *Nature Reviews. Molecular Cell Biology*, 5(2), 89–99.

Marillonnet, S. et al. (2005) Systemic *Agrobacterium tumefaciens*-mediated transfection of viral replicons for efficient transient expression in plants. *Nature Biotechnology*, 23(6), 718–723.

Marin, E. et al. (2010) miR390, Arabidopsis TAS3 tasiRNAs, and their AUXIN RESPONSE FACTOR targets define an autoregulatory network quantitatively regulating lateral root growth. *The Plant Cell*, 22(4), 1104–1117.

Marquardt, S. et al. (2011) Distinct RNA degradation pathways and 3′ extensions of yeast non-coding RNA species. *Transcription*, 2(3), 145–154.

Martienssen, R. (2010) Small RNA makes its move. *Science*, 328(5980), 834–835.

Mattick, J.S. and Makunin, I.V. (2006) Non-coding RNA. *Human Molecular Genetics*, 15(1), 17–29.

Melnyk, C.W. et al. (2011) Intercellular and systemic movement of RNA silencing signals. *The EMBO Journal*, 30(17), 3553–3563.

Mercer, T.R. et al. (2009) Long non-coding RNAs: insights into functions. *Nature Reviews Genetics*, 10, 155–159.

Mitchell, P. et al. (1997) The exosome: a conserved eukaryotic RNA processing complex containing multiple 3′-5′ exoribonucleases. *Cell*, 91(4), 457–466.

Mitchell, P. and Tollervey, D. (2003) An NMD pathway in yeast short article involving accelerated deadenylation and exosome-mediated 3 → 5 degradation. *Yeast*, 11, 1405–1413.

Montgomery, T.A. et al. (2008) Specificity of ARGONAUTE7-miR390 Interaction and dual functionality in TAS3 trans-acting siRNA formation. *Cell*, 133(1), 128–141.

Munns, R. et al. (2010) New phenotyping methods for screening wheat and barley for beneficial responses to water deficit. *Journal of Experimental Botany*, 61(13), 3499–3507.

Nagano, T. et al. (2008) The air noncoding RNA epigenetically silences transcription by targeting G9a to chromatin. *Science*, 322, 1717–1720.

Narsai, R. et al. (2007) Genome-wide analysis of mRNA decay rates and their determinants in *Arabidopsis thaliana*. *The Plant Cell*, 19(11), 3418–3436.

Naya, L. et al. (2010) Cleavage of a non-conserved target by a specific miR156 isoform in root apexes of *Medicago truncatula*. *Plant Signaling Behavior*, 5(3), 328–331.

Neil, H. and Jacquier, A., (2011) Enrichment of unstable non-coding RNAs and their genome-wide identification. *Methods in Molecular Biology*, 759(Pt 2), 87–106.

Neuwald, A.F. et al. (1992) cysQ, a gene needed for cysteine synthesis in *Escherichia coli* K-12 only during aerobic growth. *Journal of Bacteriology*, 174(2), 415–425.

Okushima, Y. et al. (2005) Functional genomic analysis of the AUXIN RESPONSE FACTOR gene family members in *Arabidopsis thaliana*: unique and overlapping functions of ARF7 and ARF19. *The Plant Cell*, 17, 444–463.

Oldroyd, G.E.D. and Downie, J.A. (2008) Coordinating nodule morphogenesis with rhizobial infection in legumes. *Annual Review of Plant Biology*, 59, 519–546.

Olmedo, G. et al. (2006) *ETHYLENE-INSENSITIVE5* encodes a 5′-3′ exoribonuclease required for regulation of the EIN3-targeting F-box proteins EBF1/2. *Proceedings of the National Academy of Sciences USA*, 103(36), 13286–13293.

Osmont, K.S. et al. (2007) Hidden branches: developments in root system architecture. *Annual Review of Plant Biology*, 58, 93–113.

Pandey, R.R. et al. (2008) Kcnq1ot1 antisense noncoding RNA mediates lineage-specific transcriptional silencing through chromatin-level regulation. *Molecular Cell*, 32(2), 232–246.

Papp, I. et al. (2004) A mutation in the cap binding protein 20 gene confers drought tolerance to Arabidopsis. *Plant Molecular Biology*, 55(5), 679–686.

Pederson, T. (2010) Regulatory RNAs derived from transfer RNA? Regulatory RNAs derived from transfer RNA? *RNA*, 16, 1865–1869.

Peragine, A. et al. (2004) *SGS3* and *SGS2/SDE1/RDR6* are required for juvenile development and the production of *trans*-acting siRNAs in *Arabidopsis*. *Genes & Development*, 18, 2368–2379.

Pontes, O. and Pikaard, C.S. (2008) siRNA and miRNA processing: new functions for Cajal bodies. *Current Opinion in Genetics & Development*, 18(2), 197–203.

Prasanth, K.V. and Spector, D.L. (2007) Eukaryotic regulatory RNAs: an answer to the "genome complexity" conundrum. *Genes & Development*, 21(1), 11–42.

Redrup, L. et al. (2009) The long noncoding RNA Kcnq1ot1 organises a lineage-specific nuclear domain for epigenetic gene silencing. *Development (Cambridge, England)*, 136(4), 525–530.

Rinn, J.L. et al. (2007) Functional demarcation of active and silent chromatin domains in human HOX loci by noncoding RNAs. *Cell*, 129(7), 1311–1323.

Rodriguez, R. et al. (2010) Control of cell proliferation in *Arabidopsis thaliana* by microRNA miR396. *Development*, 137, 103–112.

Rohrig, H. et al. (2002) Soybean ENOD40 encodes two peptides that bind to sucrose synthase. *Proceedings of the National Academy of Sciences USA*, 99(4), 1915–1920.

Roman, G. et al. (1995) Genetic analysis of ethylene signal transduction in *Arabidqsis thaliana*: five novel mutant loci integrated into a stress response pathway. *Genetics*, 139, 1393–1409.

Rymarquis, L.A. et al. (2011) Evidence that XRN4, an Arabidopsis homolog of exoribonuclease XRN1, preferentially impacts transcripts with certain sequences or in particular functional categories. *RNA*, 17(3), 501–11.

Secco, D. et al. (2010) Characterization of the rice PHO1 gene family reveals a key role for OsPHO1;2 in phosphate homeostasis and the evolution of a distinct clade in dicotyledons. *Plant Physiology*, 152(3), 1693–1704.

Shaw, P.J. and Brown, J.W.S. (2004) Plant nuclear bodies. *Current Opinion in Plant Biology*, 7(6), 614–20.

Sleutels, F. et al. (2002) The non-coding air RNA is required for silencing autosomal imprinted genes. *Nature*, 415, 810–813

Sonmez, C. et al. (2011) RNA 3′ processing functions of Arabidopsis FCA and FPA limit intergenic transcription. *Proceedings of the National Academy of Sciences USA*, 108(20), 8508–8513.

Sorin, C. et al. (2005) Auxin and light control of adventitious rooting in Arabidopsis require ARGONAUTE1. *The Plant Cell*, 17, 1343–1359.

Souret, F.F. et al. (2004) AtXRN4 degrades mRNA in Arabidopsis and its substrates include selected miRNA targets. *Molecular Cell*, 15(2), 173–183.

Sousa, C. et al. (2001) Translational and structural requirements of the early nodulin gene enod40, a short-open reading frame-containing RNA, for elicitation of a cell-specific growth response in the alfalfa root cortex. *Molecular and Cellular Biology*, 21(1), 354–366.

Sugimoto, K. et al. (2010) Article Arabidopsis regeneration from multiple tissues occurs via a root development pathway. *Developmental Cell*, 18(3), 463–471.

Sunkar, R. (2010) MicroRNAs with macro-effects on plant stress responses. *Seminars in Cell and Developmental Biology*, 21(8), 805–811.

Swiezewski, S. et al. (2009) Cold-induced silencing by long antisense transcripts of an Arabidopsis Polycomb target. *Nature*, 462(7274), 799–802.

Terranova, R. et al. (2008) Polycomb group proteins Ezh2 and Rnf2 direct genomic contraction and imprinted repression in early mouse embryos. *Developmental Cell*, 15(5), 668–679.

Thompson, D.M. and Parker, R. (2009) Stressing out over tRNA cleavage. *Cell*, 138(2), 215–219.

Thompson, D.M. et al. (2008) tRNA cleavage is a conserved response to oxidative stress in eukaryotes tRNA cleavage is a conserved response to oxidative stress in eukaryotes. *RNA*, 14(10), 2095–2103.

Tian, C. et al. (2004) Disruption and overexpression of auxin response factor 8 gene of Arabidopsis affect hypocotyl elongation and root growth habit, indicating its possible involvement in auxin homeostasis in light condition. *The Plant Journal*, 40(3), 333–343.

Topping, J.F. et al. (1997) Mutations in the HYDRA1 gene of Arabidopsis perturb cell shape and disrupt embryonic and seedling morphogenesis. *Development*, 124(21), 4415–4424.

Valencia-Sanchez, M.A. et al. (2006) Control of translation and mRNA degradation by miRNAs and siRNAs. *Genes & Development*, 20(5), 515–524.

van Dijk, E.L. et al. (2011) XUTs are a class of Xrn1-sensitive antisense regulatory non-coding RNA in yeast. *Nature*, 475(7354), 114–117.

Vaucheret, H. (2006) Post-transcriptional small RNA pathways in plants: mechanisms and regulations. *Genes & Development*, 759–771.

Vazquez, F. et al. (2004) Endogenous trans-acting siRNAs regulate the accumulation of Arabidopsis mRNAs. *Molecular Cell*, 16, 69–79.

Verdel, A. et al. (2009) Common themes in siRNA-mediated epigenetic silencing pathways. *International Journal of Development Biology*, 53, 245–257.

Vidal, E.A. et al. (2010) Nitrate-responsive miR393/AFB3 regulatory module controls root system architecture in *Arabidopsis thaliana*. *Proceedings of the National Academy of Sciences USA*, 107(9), 4477–4482.

Voinnet, O. (2009) Origin, biogenesis, and activity of plant microRNAs. *Cell*, 136(4), 669–687.

Wang, J. et al. (2005) Control of root cap formation by microRNA-targeted auxin response factors in *Arabidopsis*. *Plant Cell*, 17, 2204–2216.

Wang, K.C. et al. (2011) A long noncoding RNA maintains active chromatin to coordinate homeotic gene expression. *Nature*, 472(7341), 120–124.

Wang, M.-B. and Metzlaff, M. (2005) RNA silencing and antiviral defense in plants. *Current Opinion in Plant Biology*, 8(2), 216–222.

Williams, L. et al. (2005) A database analysis method identifies an endogenous trans-acting short-interfering RNA that targets the Arabidopsis ARF2, ARF3, and ARF4 genes. *Proceedings of the National Academy of Sciences USA*, 102(27), 9703–9708.

Wilusz, J.E. et al. (2009) Long noncoding RNAs: functional surprises from the RNA world. *Genes & Development*, 23(13), 1494–1504.

Xie, Z. et al. (2005) DICER-LIKE 4 functions in trans-acting small interfering RNA biogenesis and vegetative phase change in *Arabidopsis thaliana*. *Proceedings of the National Academy of Sciences*, 102, 12985–12989.

Xiong, L. et al. (2001) FIERY1 encoding an inositol polyphosphate 1-phosphatase is a negative regulator of abscisic acid and stress signaling in Arabidopsis. *Genes & Development*, 15(15), 1971–84.

Yamada, K. et al. (2003) Empirical analysis of transcriptional activity in the Arabidopsis genome. *Science*, 302(5646), 842–846.

Yang, W.C. et al. (1993) Characterization of *GmENOD40*, a gene showing novel patterns of cell-specific expression during soybean nodule development. *The Plant Journal*, 3(4), 573–585.

Yang, J.H. et al. (2006) Evidence of an auxin signal pathway, microRNA167-ARF8-GH3, and its response to exogenous auxin in cultured rice cells. *Nucleic Acids Research*, 34(6), 1892–1899.

Yang, Z. et al. (2011) Ends of vaccinia virus early mRNAs delineates regulatory sequences of annotated and anomalous transcripts genome-wide analysis of the 5′ and 3′ ends of vaccinia virus early mRNAs delineates regulatory sequences of annotated and anomalous transcripts. *Journal of Virology*, 85(12), 5896–5908.

Yap, K.L. et al. (2010) Molecular interplay of the noncoding RNA ANRIL and methylated histone H3 lysine 27 by polycomb CBX7 in transcriptional silencing of INK4a. *Molecular Cell*, 38(5), 662–674.

Yasuda, J. and Hayashizaki, Y. (2008) The RNA continent. *Advances in Cancer Research*, 99(07), 77–112.

Yoshikawa, M. et al. (2005) A pathway for the biogenesis of trans-acting siRNAs in Arabidopsis A pathway for the biogenesis of trans-acting siRNAs in Arabidopsis. *Genes & Development*, 2164–2175.

Yu, Q. et al. (2009) *Saccharomyces cerevisiae* linker histone Hho1p functionally interacts with core histone H4 and negatively regulates the establishment of transcriptionally silent chromatin. *Journal of Biological Chemistry*, 284, 740–750.

Zakrzewska-Placzek, M. et al. (2010) *Arabidopsis thaliana* XRN2 is required for primary cleavage in the pre-ribosomal RNA. *Nucleic Acids Research*, 38(13), 4487–4502.

Zhan, S. and Lukens, L. (2010) Identification of novel miRNAs and miRNA dependent developmental shifts of gene expression in *Arabidopsis thaliana*. *PloS One*, 5(4), e10157.

Zhang, S. et al. (2009) The phloem-delivered RNA pool contains small noncoding RNAs and interferes with translation. *Plant Physiology*, 150, 378–387.

Zhang, L. et al. (2011) Exogenous plant MIR168a specifically targets mammalian LDLRAP1: evidence of cross-kingdom regulation by microRNA. *Cell Research*, 1–20.

3 Genomics of Auxin Action in Roots

Elisabeth L. Williams and Ive De Smet

Introduction

Roots are vital for plant survival and contribute toward many fundamental processes required for normal plant development. Onset of root development occurs during embryogenesis, and as maturation occurs, the root develops into a highly organized structure, where files of cells are formed from the root apical meristem (RAM) and move upward toward the elongation zone. The cell layers in the root from outer layer to the internal vascular bundle consist of the epidermis, cortex, endodermis, and pericycle (Figure 3.1; Dolan et al. 1993; Petricka and Benfey 2008). Cells within some of these concentrical layers will give rise to other structures, namely epidermal cells will form root hairs and pericycle cells will form lateral roots (LRs; Beeckman et al. 2001; Schiefelbein 2003).

The phytohormone auxin influences almost all aspects of development throughout the plant, including the above aspects of root development, through influencing processes such as cell division, cell elongation and differentiation, and organ formation (Vanneste and Friml 2009). Auxin structures and patterns the root system via its biosynthesis, active transport, and primary and secondary response, each of which will be discussed in more detail in the first part of this chapter. In a second part, we will describe the auxin-mediated changes on the level of transcriptome and proteome during several root developmental processes in various plant species.

The Basis of Auxin Biology

Auxin biology has been widely investigated in plants, and different auxin-related aspects of root development have been reported, including cell differentiation, signaling pathways, transport, biosynthesis, response genes, gravitropism, and LR development (Palme et al. 2006; De Smet and Jürgens 2007; Guilfoyle and Hagen 2007; Petricka and Benfrey 2008; Overvoorde et al. 2010; Perrot-Rechenmann 2010; Benková and Bielach 2010; Zhao 2010; Stablowski 2011). Over the years, various forward genetics approaches revealed mutants that displayed striking root defects, which turned out to be associated with auxin biology (reviewed in De Smet et al. 2006; Lau et al. 2008; Péret et al. 2009).

Root Genomics and Soil Interactions, First Edition. Edited by Martin Crespi.

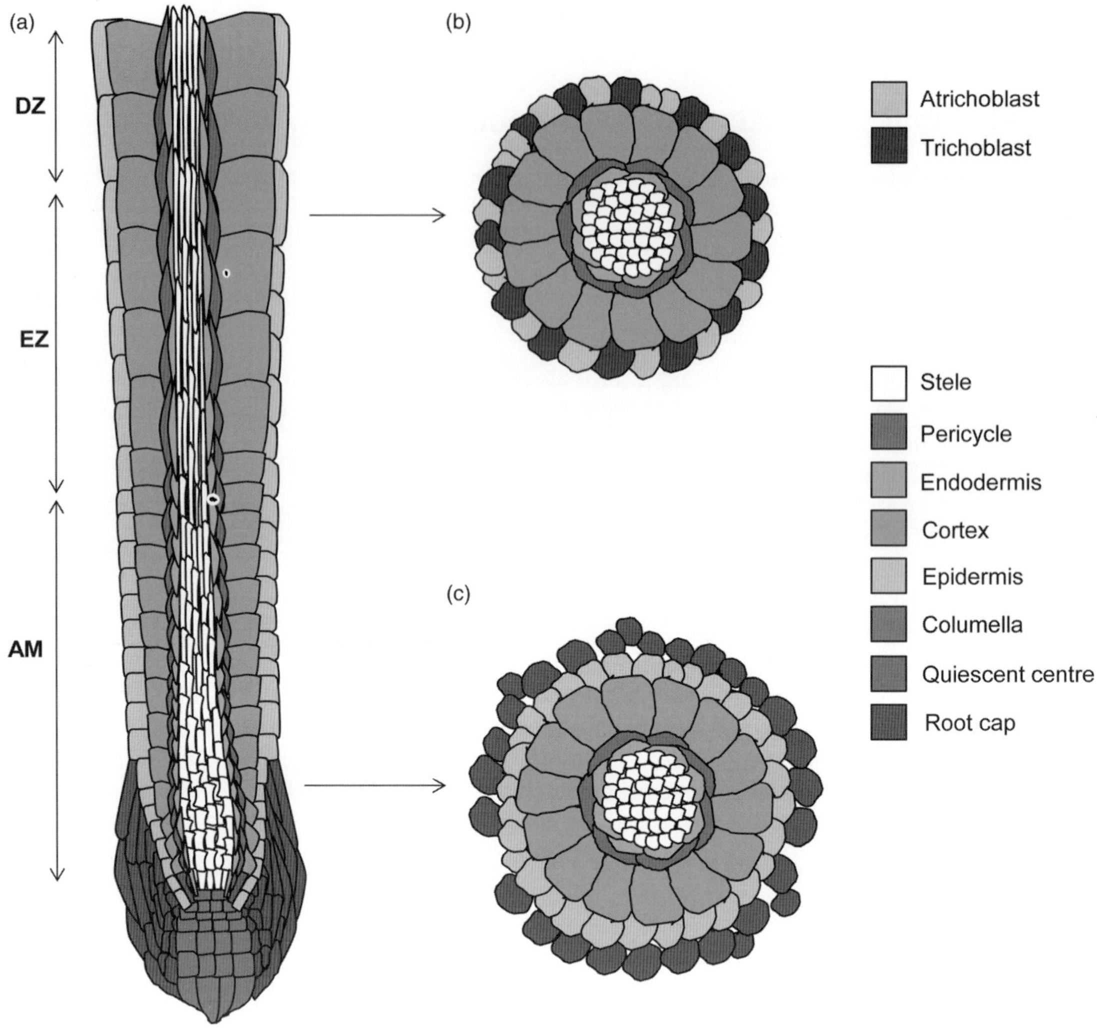

Figure 3.1 Cellular organization of the *Arabidopsis* root. (a) Root tissue develops from cell division and differentiation of different types of daughter cells, situated adjacently to the quiescent center. The resulting cell divisions give rise to the proximal distal organization of the root, with the root cap/epidermal initials forming the root cap and the epidermal cells, the columella initials establish the columella cells, central to the root cap, cortex/endodermis initials give rise to the formation of ground tissue, and the vascular initials produce the vascular and pericycle cells. The apical meristem (AM), elongation zone (EZ), and differentiation zone (DZ) of the root are indicated on the left of the figure. (b) Cross section of the elongation zone highlighting the patterning of the trichoblasts and the atrichoblasts responsible for hair cell fate. (c) Radial organization of the immature root is shown in a cross section. (For a color version of this figure, see the color plate section.)

Auxin Synthesis and Transport

Indole-3-acetic acid (IAA) has been shown to be the most biologically active and naturally occurring member from the class of auxins (Bonner and Banderski 1952). Auxin is synthesized from tryptophan (TRP) and indole predominantly in the developing seeds, leaf primordia, and young leaves, but auxin synthesis also occurs in the root (Muller et al. 1998; Ljung et al. 2001, 2005; Ikeda et al. 2009; Yamada et al. 2009). Therefore, auxin availability in the root is controlled via two pathways: transport from the aerial tissues via the vascular tissue throughout the root and synthesis in the root (Vanneste and Friml 2009).

Multiple pathways have been shown to be responsible for auxin biosynthesis in plants, and interfering with auxin biosynthesis results in dramatic phenotypes on the level of primary root growth and root branching (Boerjan et al. 1995; Celenza et al. 1995; King et al. 1995; Delarue et al. 1998; Barlier et al. 2000; Stepanova et al. 2008). The two main pathways known to synthesize IAA are known as the TRP-dependent and -independent pathways. Compared with transport and response of auxin, relatively little is known about auxin biosynthesis; however, recent discoveries provided novel insights in the biosynthesis field. Because of genetic redundancy between biosynthetic genes and multiple pathways capable of auxin synthesis, auxin-deficient mutants were not detected in genetic screens. Three steps in the TRP-dependent pathway have now been characterized, firstly *superroot1 (sur1)/rooty (rty)/aberrant lateral root formation 1 (alf1)* and *superroot2 (sur2)* were shown to be auxin overproduction mutants responsible for conversion of indole-3-aceldoxime (IAOx) to indolic glucosinolates (Boerjan et al. 1995; Celenza et al. 1995; King et al. 1995; Delarue et al. 1998; Barlier et al. 2000). Another auxin overproduction mutant, which overexpresses P450 *CYP79B2*, has been shown to convert TRP to IAOx (Hull et al. 2000; Zhao et al. 2002). More mutants in the auxin biosynthetic pathway have been identified from an activation-tagging screen, which showed that *(YUC)* genes are involved in auxin homeostasis (Zhao et al. 2001). YUC1 has been shown to be capable of the catalysis of tryptamine to *N*-hydroxyl trypamine *in vitro,* and acts as a flavin monooxygenase-like enzyme (Zhao et al. 2001). A series of recent studies suggested that *TRP AMINO-TRANSFERASE OF ARABIDOPSIS (TAA)* and related genes encode for a pyridoxal-phosphate-dependent enzyme that converts TRP to indole-3-pyruvate, and that aminotransferase enzymes are involved in auxin biosynthesis (Stepanova et al. 2008; Tao et al. 2008; Yamada et al. 2009).

In addition to auxin production in the young primordia and cotyledons of plants, many of the genes known to be responsible for auxin biosynthesis are expressed in root tissue, which suggests that plant roots have the capacity to produce auxin (Muller et al. 1998; Ljung et al. 2001, 2005; Stepanova et al. 2005, 2008; Yamada et al. 2009; Ikeda et al. 2009; Petersson et al. 2009). After synthesis, auxin is transported throughout the plant to create the unique differential patterns that are characteristic of auxin-mediated development in plants. Long distance transport of auxin from "source" tissue to "sink" tissue occurs via the phloem channels in the vascular bundle. Where auxin is concerned, source tissues are parts of the plant that produce relatively high levels of auxin, such as young leaves and cotyledons, and sink refers to tissues that contain relatively low levels of auxin, such as the root (Ljung et al. 2001). Long distance transport of auxin is performed in the same way as carbohydrates and other hormones (ABA and cytokinins) along with mRNA and proteins (Robert and Friml 2009). The initial flow of auxin through the vascular bundle forms a positive feedback loop, regulated by auxin itself, and controls the directional throughput of auxin flow. The measured flow of auxin results in development of auxin-transporting cell files that dictate the localization of additional auxin-transporting vascular strands, this process is known as the auxin canalization model (Sauer et al. 2006; Robert and Friml 2009).

Along with long distance transport, active cell-to-cell auxin transport has been shown to have a profound effect on developmental processes, including root development and branching (Overvoorde et al. 2010). Short distance transport of auxin is required to produce an auxin gradient, which is controlled by PIN-FORMED (PIN) efflux carriers (Vieten et al. 2005), AUXIN1 (AUX1)/LIKE AUX1 (LAX) influx carriers (Bennett et al. 1996; Swarup et al. 2001), and ATP-BINDING CASSETTE subfamily P-GLYCOPROTEIN (PGP)/MULTIDRUG RESISTANCE (MDR) transporters

(Geisler and Murphy 2006; Petrasek et al. 2006), referred to as polar auxin transport. The resulting areas of differential auxin concentration have been shown to greatly contribute to the developmental processes, including cell differentiation and elongation. The predicted auxin maximum is found in the quiescent centre of the root and creates a central point for cell differentiation via the stem cells either side (Van den Berg et al. 1997; Sabatini et al. 1999; Petersson et al. 2009).

Taken together, auxin biosynthesis and transport generate and maintain auxin gradients in the root, controlling its architecture and patterning.

Auxin Response-Mediated Transcription

Several key regulators of auxin response have been identified through forward genetics screens, identifying mutants with striking root phenotypes (De Smet et al. 2006; Lau et al. 2008; Péret et al. 2009). However, root-specific genome-wide transcriptome and proteome studies concerning auxin response are limited, and those studies that have been performed concentrate mainly on auxin response of LR development (Sawa et al. 2002; Himanen et al. 2004; Vanneste et al. 2005; Laskowski et al. 2006; De Smet et al. 2008). Overall, many gene families have been shown to have involvement in various auxin responses in the root, and we will discuss the most prominent ones below.

Canonical Auxin Response Machinery

Canonical auxin response occurs via the recognized auxin response elements (AuxREs), which have been shown to be responsible for binding of members of the AUXIN RESPONSE FACTOR (ARF) protein family (Ulmasov et al. 1997). Next to the ARFs, the major components involved in canonical auxin response are the AUX/IAA, TRANSPORT INHIBITOR RESPONSE 1/AUXIN F-BOX (TIR1/AFB), and TOPLESS (TPL) proteins, which are all members of gene families (Reed 2001; Dharmasiri et al. 2005b; Lau et al. 2008; Szemenyei et al. 2008).

The auxin signaling pathway was the source of many research groups' studies for many years, and the characterization of the auxin receptor was a major step forward (Kepinski and Leyser 2004; Dharmasiri et al. 2005a; Tan et al. 2007). The TIR1 auxin receptor has been shown to encode for an F-box protein (Ruegger et al. 1998). Auxin binds to the TIR1 receptor and acts as a molecular glue between the SCF complex and members of the AUX/IAA family, which is shown in Figure 3.2. This interaction results in ubiquitination of the AUX/IAA by the E3 ubiquitin ligase activity of the SCF complex and subsequent degradation via the 26S proteasome (Gray et al. 1999, 2001; Kepinski and Leyser 2004; Dharmasiri et al. 2005a). The removal of AUX/IAA repressor results in ARF-dependent transcription of auxin-regulated genes (Gray et al. 2001; Figure 3.2).

The *AUX/IAA* gene family is a relatively large group of 29 members, these genes encode for short-lived proteins, which consist of four conserved domains (I–IV) (Abel et al. 1994). Domain I confers transcription repression activity, domain II encodes a degron, which has been shown to be required for degradation by the 26S proteasome. Domains III and IV have regions that are responsible for binding to other AUX/IAA and ARF proteins. For example, the lack of a primary root and the altered LR density and positioning in the gain-of-function *bodenlos* (*bdl*)/*iaa12* mutant suggest a profound role for BDL/IAA12 in various aspects of root development (Hamann et al. 1999; Weijers et al. 2006; De Smet et al. 2010). Another AUX/IAA, which appears to affect LR formation, is SOLITARY ROOT (SLR)/IAA14. A gain-of-function *slr* mutation leads to a lack of LRs, therefore including another member of the *AUX/IAA* gene family in LR formation (Fukaki et al. 2005).

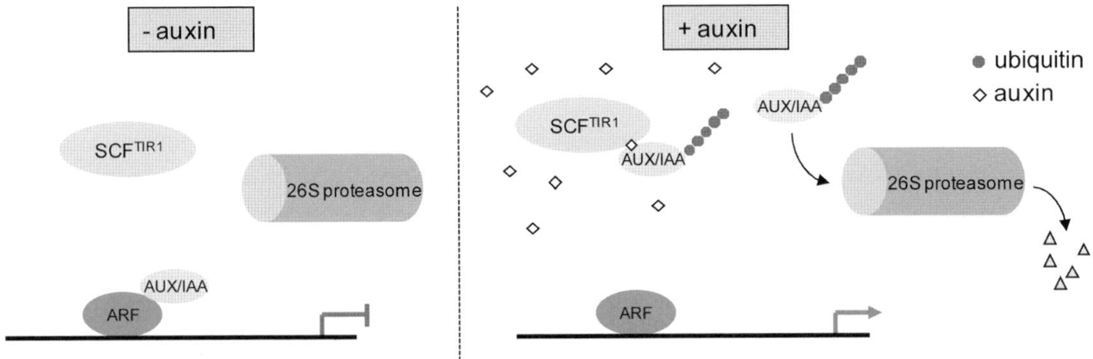

Figure 3.2 Basics and evolution of auxin signaling and response. Simplified scheme of auxin-responsive gene regulation through an activating ARF in the absence or presence of auxin. ARFs, binding to the consensus TGTCTC auxin-responsive promoter element, are the main transcriptional regulators of auxin-regulated gene expression and can be grouped into activators and repressors. AUX–IAAs are repressors of the (activating) ARFs and are degraded in an auxin-dependent manner. In the absence of auxin, AUX–IAAs repress ARF activity (left). Auxin, functioning as a molecular glue, enhances the interaction between AUX–IAAs and the auxin receptors of the TIR1–AFB family, which are an integral component of the S-PHASE KINASE-ASSOCIATED PROTEIN1–CULLIN–F-BOX PROTEIN (SCF)$^{TIR1–AFB}$ complex. This targets the bound and ubiquitinated AUX–IAA for degradation by the 26S proteasome (right). The (activating) ARFs, released from repression, induce gene expression. (Reprinted from Trends in Plant Science, Lau et al, Auxin signaling in algal lineages: fact or myth? 14: 182–188, Copyright 2009, with permission from Elsevier.)

The family of ARFs is also a reasonably large family of proteins made up of 23 members. The *ARF* genes encode transcription factor proteins and therefore contain a conserved DNA-binding domain; ARF proteins also contain a dimerization domain (Ulmasov et al. 1997; Remington et al. 2004). The middle region of the ARF protein demonstrates a more divergent region; however, if this region consists of a high proportion of Glu residues, then it has been shown that these proteins act as transcriptional activators, whereas if the region lacks the Glu residues, then it acts as a transcriptional repressor (Ulmasov et al. 1999).

As mentioned earlier, many of the proteins involved in the auxin signaling complex are members of multigene families, which suggest genetic redundancy between these family members (Remington et al. 2004; Dharmasiri et al. 2005a; Okushima et al. 2005a; Overvoorde et al. 2005). Indeed, genetic analysis of the *arf* mutants has demonstrated that most single mutants display very subtle phenotypes (Okushima et al. 2005a). Higher order mutants display strong phenotypes, for example, *arf7arf19* double mutants have no LRs which suggests that both these members of the *ARF* family are involved in LR initiation (Okushima et al. 2005a). However, the multitude of proteins capable of performing multiple parts of the pathway may lead to subtle but highly specific variations for auxin signaling. For example, expression of a certain gene is much higher in specific tissues and therefore may play a role in a specific auxin development pathway, such as LR formation, whereas a different group of genes from the same family may be responsible for a different process, such as leaf expansion. In this respect, ARF7 and ARF19 have been shown to influence (lateral) root development, whereas ARF1 and ARF2 are involved in control of senescence and floral organ abscission (Ellis et al. 2005; Okushima et al. 2005a; De Rybel et al. 2010; De Smet et al. 2010). The recent catalog of *ARF* gene expression, where several *ARF* genes show visible expression in different layers of the root cell files, will provide valuable insights (Rademacher et al. 2011).

One can distinguish between primary and secondary auxin response. Primary auxin response is fast, requires no de novo protein synthesis and depends on AUX/IAA degradation (see earlier).

Secondary auxin response is downstream in the cascade and requires de novo protein synthesis. Several transcript profiling studies have been performed to identify auxin-responsive genes, and some of these focused on the root. Nevertheless, only few of these were designed to identify direct ARF targets. Most prominent is the analysis of MONOPTEROS (MP)/ARF5 targets, which lead to the identification of the mobile transcription factor TARGET OF MP 7 (TMO7) involved in embryonic root development (Schlereth et al. 2010). In addition, MP/ARF5 is capable of regulating its own expression and also the expression of its inhibitor, the AUX/IAA BDL. In this pathway, auxin behaves as a threshold trigger and promotes the degradation of BDL (Lau et al. 2011). Also, the *LATERAL ORGAN BOUNDARIES DOMAIN (LBD)* gene family of transcriptional regulators has recently been discovered to have implications on auxin responsiveness (Iwakawa et al. 2002; Liu et al. 2006; Okushima et al. 2007; Paponov et al. 2008). Specifically, *LBD16* and *LBD29* are direct targets of ARF7 and ARF19 in the control of LR formation. These *LBD* genes were identified in a microarray screen for candidate genes that control LR development in the *arf7arf19* double mutant. Overexpression of both *LBD16* and *LBD29* was shown to partially rescue LR formation in the *arf7/arf19* double mutant and therefore suggested that the *LBDs* promote the formation of LRs (Okushima et al. 2007).

Genome-Wide Analyses of Transcriptional Changes

Many microarray studies have recently been performed to show that many gene families are regulated in response to auxin (Sawa et al. 2002; Pufky et al. 2003; Zhao et al. 2003; Armstrong et al. 2004; Himanen et al. 2004; Nemhauser et al. 2004; Redman et al. 2004; Nagpal et al. 2005; Okushima et al. 2005a, 2005b; Overvoorde et al. 2005; Vanneste et al. 2005; Wirta et al. 2005; Nemhauser et al. 2006; Paponov et al. 2008). Experimental data has been obtained from various different tissues of the plant, including whole seedlings, flowers, suspension cultures, and also roots. Analyses of the root tissue showed that after 2 and 6 h of naphthalene acetic acid (NAA) treatment, 688 and 1645 genes were upregulated, respectively, whereas 238 and 1314 were downregulated, respectively (Paponov et al. 2008). Among the differentially regulated genes in the root, several gene families were present, which include *AUX/IAAs*, *ARFs*, *SAURs*, *GH3s*, *AUX1/LAXs*, *LBDs,* and IAA biosynthesis genes (Paponov et al. 2008).

Most of the important genes involved in auxin response and transport are known to be auxin-responsive themselves and are part of complex regulatory networks (Vieten et al. 2005; Lau et al. 2011). Members of the *PIN* family (such as *PIN1*, *PIN3,* and *PIN7),* which encode auxin efflux carriers, were upregulated by auxin throughout the plant and also specifically in root tissue (Vieten et al. 2007; Paponov et al. 2008). The control on tissue-specific expression of *PIN* genes in an AUX/IAA-dependent manner suggests that the auxin feedback loop on *PIN* expression leads to stabilization of auxin gradients in the plant (Vieten et al. 2005). Auxin feedback actions described here allow for greater control of its own transport and response. Also members of the *AUX1/LAX* group of genes encoding auxin influx carriers showed fast auxin-mediated upregulation (Paponov et al. 2008; Swarup et al. 2008).

Previous reports have stated that out of the *AUX/IAA* gene family, the majority of members from a variety of different species are upregulated (Remington et al. 2004). In the root tissue, it is known that *IAA7, IAA19,* and *IAA14* have implications in root development (Okushima et al. 2005a). Microarray analysis demonstrates that levels of both *IAA7* and *IAA19* are upregulated in root tissue, whereas *IAA14* appears to be downregulated after 2 and 6 h of auxin treatment (Vanneste et al. 2005). Microarray analysis has highlighted three members that are upregulated in roots after auxin treatment: *ARF4*, *ARF16,* and *ARF19*. All three *ARF* genes were upregulated after both 2 and 6 h (Vanneste et al. 2005; Paponov et al. 2008).

The *SMALL AUXIN-UP RNA (SAUR)* family of genes is a very large group, which contains more than 70 members; within the family, two subgroups have been grouped together in subclades using phylogenetics (Paponov et al. 2008). Subclade I has 6 members and subclade II has 26 members. The members of these two subclades appear to have increased responsiveness to auxin when compared with the members that could not be grouped by phylogenetics (Paponov et al. 2008). *SAUR* genes encode highly unstable mRNAs (McClure and Guilfoyle 1989; Franco et al. 1990), and it has been suggested that these *SAUR* genes may be involved in cell elongation (Knauss et al. 2003). However, little is known about this group of genes. The *SAUR* gene family showed regulation with different members throughout the plant; however, in the roots, mainly downregulation of members was observed (Paponov et al. 2008). *SAUR31*, *SAUR36*, *SAUR59*, and *SAUR72* were all downregulated in response to auxin, whereas *SAUR34* and *SAUR45* were upregulated (Paponov et al. 2008).

IAA biosynthesis genes showed only downregulation after auxin treatment. After 2 h of treatment, only *SUPERROOT2 (SUR2)* and *CYP79B3* showed differential expression patterns; however, after a 6 h of treatment, *AMI1*, *SUR1*, *SUR2*, *CYP79B2*, and *CYP79B3*, all displayed downregulated gene expression (Paponov et al. 2008). Additionally, *GRETCHEN HAGEN3 (GH3)* genes encode for a group of conjugation enzymes induced by auxin, of which there are 19 members in *Arabidopsis* split into three groups (Staswick et al. 2005). The GH3 proteins are acyl-acid-amido synthases that are responsible for catalyzing amino acid conjugates, in an ATP-dependent manor, to alter active plant hormone levels, including auxin (Staswick et al. 2005). The *GH3* gene family was mainly upregulated in response to auxin treatment. One member, *GH3.14,* exhibited upregulation only in root tissue (Paponov et al. 2008). *GH3.14* is a member of group III; however, the function for this subgroup of *GH3s* is unknown. Regulation of this member only in the root suggests that it may perform a specific auxin-responsive role in the root.

Finally, the *LBD* family of genes has recently been shown to have implications in root development (Iwakawa et al. 2002), and microarray analysis of the *LBD* family in roots after auxin treatment has confirmed this involvement (Paponov et al. 2008). After 6 h of auxin treatment on *Arabidopsis*, differential expression levels of the *LBD* family was observed, mainly upregulation (*LBD16*, *LBD17*, *LBD18*, *LBD29*, *LBD33*, and *LBD41*), but three genes were also downregulated (*LBD4*, *LBD25*, and *LBD40*). The remaining members of the family did not show any differential expression in response to auxin (Paponov et al. 2008).

Auxin Genomics in Root Development

The specific role of auxin in the root has been mainly addressed in *Arabidopsis*, which we will primarily report on here, but analyses of some other species also revealed aspects of auxin biology in the root. Below, we will highlight some important developmental processes where auxin plays a role, but as will become clear, very few genome-wide data on the impact of auxin on these processes are available.

Nevertheless, auxin-related genome-wide analyses have been done in various species, and some even focused on the root and even occasionally on specific processes and tissues. For instance, in *Medicago truncatula*, the auxin-mediated regulation of the root proteome was investigated (van Noorden et al. 2007). In cereal crops, large-scale analyses of transcriptome and proteome revealed cell- and tissue-specific changes, mainly using laser microdissection. For example, transcript profiling of the rice root system (Takehisa et al. 2011) and of the maize root tip (Jiang et al. 2006) revealed differential regulation of auxin-related genes. Also in maize, the role of auxin has been further explored on the transcriptome and proteome level using a mutant with stabilized auxin response, namely *rootless with undetectable meristems 1* (*rum1*) on the level of LR initiation (Woll et al. 2005; Liu et al. 2010) and in the primary root (Liu et al. 2006).

Auxin Plays a Role in Primary Root Development

The embryonic root develops via asymmetric division of the hypophysis, which forms a small cell that produces the quiescent center in the RAM and a larger basal cell that leads to formation of the columella cell layers below the RAM. Without proper hypophysis division, the plant is unable to form a root meristem, which ultimately results in rootless seedlings (Hamann et al. 1999; Weijers et al. 2006; Lau et al. 2011). The transcription factor ARF5/MP plays an important role in hypophysis specification by inducing transport of auxin to the hypophysis precursor from the embryo. More recently, it was demonstrated that the transcriptional target genes TMO5 and TMO7 are also required for MP-dependent root initiation (Schlereth et al. 2010). It has been suggested that recruitment of the adjacent cells from the apical cell lineage, via signaling from the quiescent center, develop into an upper layer of stem cells for development of the root tissue (Van den Berg et al. 1997).

The RAM is a highly structured organ, which undergoes specific specification of cells during its development; the radial patterning of the root meristem is initiated at the embryonic stage of cell development; and maintenance is controlled via the stem cell niche. Stem cells control development of the root by asymmetric division, where a smaller daughter cell is produced, which retains stem cell identity, whereas the larger cell undergoes differentiation events and becomes part of the root tissue. During longitudinal patterning of the root meristem, defined areas are formed in the root tip, consisting of the stem cell niche, columella and lateral root cap, and the proximal meristem. The PIN-dependent formation of auxin gradients in this region is essential for root patterning and particularly PIN1, PIN3, PIN4, and PIN7 play an important role (Sabatini et al. 1999; Blilou et al. 2005). Consequently, auxin-dependent gene expression plays a major role in positioning the stem cell niche, even if this has to be formed de novo (Sena et al. 2009).

In the proximal meristem, the cells are rapidly dividing, after which they enter the elongation zone, leaving the meristem region and undergoing rapid growth in the form of elongation. As cells enter the basal meristem, the interface between the apical meristem and the elongation zone, differentiation slows and cell expansion begins (Ishikawa and Evans 1995; Figure 3.1). While it is known that auxin plays a crucial role (Dello Ioio et al. 2008), no large-scale genome-wide data have been obtained on these processes.

Auxin and Root Hair Development

Root hair development from the epidermal layer of root cells on the outer surface has, for many years, been used as a model system for the analysis of cell patterning (Leavitt 1904; Cormack 1935; Sinnot and Bloch 1939). Epidermal cell development is limited, as only two types exist: trichoblasts, which produce root hairs, and atrichoblasts, which do not produce root hairs. Figure 3.1 shows the distribution of trichoblasts/atrichoblasts in the epidermal cell layer. *Arabidopsis* root hair cell fate is decided in a mechanistic manner; the positioning of the underlying cortex cells dictates the position of the cells that produce root hairs. The epidermal cells overlapping a junction between two cortex cells underneath produces root hairs (the "H" cell position), and the epidermal cells overlaying just one cortex cell do not develop root hairs (the "N" cell position) (Cormack 1935; Dolan et al. 1994; Galway et al. 1994). This mechanistic patterning is not universal among all plant species.

For example, in some grass species, root hair patterning is decided by asymmetric division from an epidermal cell precursor, which results in the smaller cell developing the ability to produce root hairs while the larger cell does not (Berger et al. 1998; Clowes 2000). While the involvement of auxin here is apparent (Jones et al. 2009), no genome-wide data is available on the role of auxin.

Auxin in Gravitropism

Gravitropism in the root is due to differential auxin concentrations on either side of the root, and root bending occurs when auxin levels are higher on the outside of the bend where cells are elongating, whereas cells on the inside of the bend contain low auxin levels. The transport of auxin from cell to cell creates differential concentrations on either side of the bending root, and proper auxin distribution is required for a normal gravitropic response (Friml et al. 2002; Vieten et al. 2007; Petrášek and Friml 2009). For example, PIN3 shows specific functions in the gravitropic response (Friml et al. 2002). PIN3 accumulates in the lateral cell surface in the gravity-sensing tissues of the root. Symmetric positioning of the PIN3 protein at the plasma membrane exists until gravitropic stimulation, where rapid relocalization of PIN3 occurs. Relocalization of PIN3 upon gravitropic stimulation creates a redirection of auxin to promote asymmetric growth (Friml et al. 2002). And, as supported by modeling, AUX1 and PIN2 are also essential for gravitropic response (Paciorek et al. 2005; Swarup et al. 2005). Reorientation of plants to produce a gravitropic response is known to be perceived by the sedimentation of starch granules present in the root columella cells (Kiss et al. 1989, 1996; Blancaflor et al. 1998). Again, no genome-wide data are available on the role of auxin in these processes.

Auxin in LR Initiation

LRs are initiated from a subset of pericycle cells that develop the identity to be founder cells (Dubrovsky et al. 2000; Dubrovsky et al. 2008). To form a LR, the founder cells divide asymmetrically, resulting in small daughter cells flanked by larger cells, which have the capability to divide further anticlinally. Subsequent cell division and differentiation gives rise to the lateral root primordium (LRP). The LRP carries on growing and emerges through the files of root cells; after emergence of the primary root, the apical meristem develops to assume control of mature LR growth (Malamy and Benfey 1997; De Smet et al. 2007; Swarup et al. 2008). Mature LRs are found at the upper levels of the root, whereas young LRs and emerging primordia are found lower down the root. All aspects of LR development strongly rely on auxin (De Smet et al. 2006; Péret et al. 2009). For example, auxin signaling in the basal meristem (De Smet et al. 2007; Moreno-Risueno et al. 2010) controls LR initiation. A large number of genes oscillate in phase or antiphase with the auxin reporter DR5, and this network is essential for LR initiation and positioning (Moreno-Risueno et al. 2010). Subsequently, various auxin response modules—such as IAA14/SLR-ARF7-ARF19 and IAA12/BDL-ARF5/MP—play a key role in establishing LR primordia (Fukaki et al. 2002; Okushima et al. 2005b; De Smet et al. 2010). Following these events, through subtractive analyses of microarray data, it was shown that the auxin influx carrier LAX3 is an important factor strengthening the auxin induction of cell wall remodeling enzymes responsible for cell separation in preparation for the emergence of LRs (Swarup et al. 2008).

Many studies have been performed to elucidate the genomic regulation of LR initiation in response to auxin. Some have even used cell sorting on plant roots expressing specific cell markers to identify the genes that are differentially regulated in different cell types (De Smet et al. 2008). Other studies deduced more broad transcriptional regulation during LR initiation (Himanen et al. 2002, 2004; Vanneste et al. 2005), using a LR-inducible system. In this system, successive treatment with an auxin transport inhibitor (*N*-1-naphthylphthalamic acid, NPA) and exogenously applied auxin (NAA) inhibits the first formative cell division and subsequently activates the entire pericycle, respectively (Himanen et al. 2002).

Conclusion

Auxin has a profound effect on all aspects of plant development. For root development in particular, we have highlighted the importance of auxin to promote cellular patterning and gravitropism of the main root, development of root hairs and LRs, along with negative feedback loops involved with auxin biosynthesis and transport. Most analyses have been performed in *Arabidopsis*; however, more data is becoming available in crop species such as maize, which we have highlighted here. While the influence of auxin on plant root development has been extensively studied, more detailed spatial and temporal transcriptome and proteome data will be needed before we fully understand the role of auxin in all aspects of root development.

Acknowledgments

This work was supported by a BBSRC David Phillips Fellowship (BB_BB/H022457/1), a Marie Curie European Reintegration Grant (PERG06-GA-2009-256354), the Research Foundation Flanders (FWO09/PDO/064 A 4/5 SDS), and a University of Nottingham Bridging the Gaps Feasibility Award. We apologize to those colleagues whose work could not be incorporated due to space restrictions.

References

Abel, S. et al. (1994) Early auxin-induced genes encode short-lived nuclear proteins. *Proceedings of the National Academy of Science of the United States of America*, 91, 326–330.

Armstrong, J.I. et al. (2004) Identification of inhibitors of auxin transcriptional activation by means of chemical genetics in Arabidopsis. *Proceedings of the National Academy of Science of the United States of America*, 101, 14978–14983.

Barlier, I. et al. (2000) The *SUR2* gene of *Arabidopsis thaliana* encodes the cytochrome P450 CYP83B1, a modulator of auxin homeostasis. *Proceedings of the National Academy of Science of the United States of America*, 97, 14819–14824.

Beeckman, T. et al. (2001) The peri-cell-cycle in Arabidopsis. *Journal of Experimental Botany*, 52, 403–411.

Benková, E. and Bielach, A. (2010) Lateral root organogensis—from cell to organ. *Current Opinion in Plant Biology*, 13, 677–683.

Bennett, M.J. et al. (1996) Arabidopsis *AUX1* gene: a permease-like regulator of root gravitropism. *Science*, 273, 948–950.

Berger, F.et al. (1998) Positional information in root epidermis is defined during embryogenesis and acts in domains with strict boundaries. *Current Biology*, 8, 421–430.

Blancaflor, E.B. et al. (1998) Mapping the functional roles of cap cells in the response of Arabidopsis primary roots to gravity. *Plant Physiology*, 116, 213–222.

Blilou, I. et al. (2005) The PIN auxin efflux facilitator network controls growth and patterning in Arabidopsis roots. *Nature*, 433, 39–44.

Boerjan, W. et al. (1995) Superroot, a recessive mutation in Arabidopsis, confers auxin overproduction. *Plant Cell*, 7, 1405–1419.

Bonner, J. and Bandurski, R.S. (1952) Studies of the physiology, pharmacology, and biochemistry of the auxins. *Annual Reviews in Plant Physiology*, 3, 59–86.

Celenza, J.L. Jr. et al. (1995) A pathway for lateral root formation in *Arabidopsis thaliana*. *Genes and Development*, 9, 2131–2142.

Clowes, F.A.L. (2000) Pattern in root meristem development in angiosperms. *New Phytologist*, 146, 83–94.

Cormack, R.G.H. (1935) Investigations on the development of root hairs. *New Phytologist*, 34, 30–54.

Delarue, M. et al. (1998) *Sur2* mutations of *Arabidopsis thaliana* define a new locus involved in the control of auxin homeostasis. *Plant Journal*, 14, 603–611.

Dello Ioio, R. et al. (2008) A genetic framework for the control of cell division and differentiation in the root meristem. *Science*, 322, 1380–1384.

De Smet, I. et al. (2006) Lateral root initiation or the birth of a new meristem. *Plant Molecular Biology*, 60, 871–887.

De Smet, I. et al. (2007) Auxin-dependent regulation of lateral root positioning in the basal meristem of Arabidopsis. *Development*, 134, 681–690.

De Smet, I. and Jürgens, G. (2007) Patterning the axis in plants—auxin in control. *Current Opinion in Genetics and Development*, 17, 337–343.

De Smet, I. et al. (2008) Receptor-like kinase ACR4 restricts formative cell divisions in the Arabidopsis root. *Science*, 322, 594–597.

De Smet, I. et al. (2010) Bimodular auxin response controls organogenesis in Arabidopsis. *Proceedings of the National Academy of Science of the United States of America*, 107, 2705–2710.

De Rybel, B. et al. (2010) A novel aux/IAA28 signaling cascade activates GATA23-dependent specification of lateral root founder cell identity. *Current Biology*, 20, 1697–1706.

Dharmasiri, N., Dharmasiri, S. and Estelle, M. (2005a) The F-box protein TIR1 is an auxin receptor. *Nature*, 435, 441–445.

Dharmasiri, N. et al. (2005b) Plant development is regulated by a family of auxin receptor F box proteins. *Developmental Cell*, 9, 109–119.

Dolan, L. et al. (1993) Cellular organisation of the *Arabidopsis thaliana* root. *Development*, 119, 71–84.

Dolan, L. et al. (1994) Clonal relations and patterning in the root epidermis of Arabidopsis. *Development*, 120, 2465–2474.

Dubrovsky, J.G. et al. (2000) Pericycle cell proliferation and lateral root initiation in Arabidopsis. *Plant Physiol*, 124, 1648–1657.

Dubrovsky, J.G. et al. (2008) Auxin acts as a local morphogenetic trigger to specify lateral root founder cells. *Proceedings of the National Academy of Science USA*, 105, 8790–8794.

Ellis, C.M. et al. (2005) AUXIN RESPONSE FACTOR1 and AUXIN RESPONSE FACTOR2 regulate senescence and floral organ abscission in *Arabidopsis thaliana*. *Development*, 132, 4563–4574.

Franco, A.R. et al. (1990) Induction and superinduction of auxin-responsive mRNAs with auxin and protein synthesis inhibitors. *Journal of Biological Chemistry*, 265, 15845–15849.

Friml, J. et al. (2002) Lateral relocation of auxin efflux regulator PIN3 mediates tropism in Arabidopsis. *Nature*, 415, 806–809.

Fukaki, H. et al. (2002) Lateral root formation is blocked by a gain-of-function mutation in the *SOLITARY-ROOT/IAA14* gene of Arabidopsis. *Plant Journal*, 29, 153–168.

Fukaki, H. et al. (2005) Tissue-specific expression of stabilized SOLITARY-ROOT/IAA14 alters lateral root development in Arabidopsis. *Plant Journal*, 44, 382–395.

Galway, M.E. et al. (1994) The TTG gene is required to specify epidermal fate and cell patterning in the Arabidopsis root. *Development Biology*, 166, 740–754.

Geisler, M. and Murphy, A.S. (2006) The ABC of auxin transport: the role of p-glycoproteins in plant development. *FEBS Letters*, 580, 1094–10102.

Gray, W.M. et al. (1999) Identification of an SCF ubiquitin–ligase complex required for auxin response in *Arabidopsis thaliana*. *Genes and Development*, 13, 1678–1691.

Gray, W.M. et al. (2001) Auxin regulates SCF(TIR1)-dependent degradation of AUX/IAA proteins. *Nature*, 414, 271–276.

Guilfoyle, T.J. and Hagen, G. (2007) Auxin response factors. *Current Opinion in Plant Biology*, 10, 453–460.

Hamann, T. et al. (1999) The auxin-insensitive bodenlos mutation affects primary root formation and apical–basal patterning in the Arabidopsis embryo. *Development*, 126, 1387–1395.

Hull, A.K. et al. (2000) Arabidopsis cytochrome P450s that catalyze the first step of tryptophan-dependent indole-3-acetic acid biosynthesis. *Proceedings of the National Academy of Science USA*, 97, 2379–2384.

Himanen, K. et al. (2002) Auxin-mediated cell cycle activation during early lateral root initiation. *Plant Cell*, 14, 2339–2351.

Himanen, K. et al. (2004) Transcript profiling of early lateral root initiation. *Proceedings of the National Academy of Science of the United States of America*, 101, 5146–5151.

Ikeda, Y. et al. (2009) Local auxin biosynthesis modulates gradient-directed planar polarity in Arabidopsis. *Nature Cell Biology*, 11, 731–738.

Ishikawa, H. and Evans, M.L. (1995) Specialized zones of development in roots. *Plant Physiology*, 109, 725–727.

Iwakawa, H. et al. (2002) The *ASYMMETRIC LEAVES2* gene of *Arabidopsis thaliana*, required for formation of a symmetric flat leaf lamina, encodes a member of a novel family of proteins characterized by cysteine repeats and a leucine zipper. *Plant Cell Physiology*, 43, 467–478.

Jiang, K. et al. (2006) Transcription profile analyses identify genes and pathways central to rootcap functions in maize. *Plant Molecular Biology*, 60, 343–363.

Jones, A.R. et al. (2009) Auxin transport through non-hair cells sustains root-hair development. *Nature Cell Biology*, 11, 78–84.

Kepinski, S. and Leyser, O. (2004) Auxin-induced SCFTIR1–Aux/IAA interaction involves stable modification of the SCFTIR1 complex. *Proceedings of the National Academy of Science of the United States of America*, 101, 12381–21386.

King, J.J. et al. (1995) A mutation altering auxin homeostasis and plant morphology in Arabidopsis. *Plant Cell*, 7, 2023–2037.

Kiss, J.Z. and Sack, F.D. (1989) Reduced gravitropic sensitivity in roots of a starch-deficient mutant of *Nicotiana sylvestris*. *Planta*, 180, 123–130.

Kiss, J.Z. et al. (1996) Gravitropism in roots of intermediate-starch mutants of Arabidopsis. *Physiologia Plantarum*, 97, 237–244.

Knauss, S. et al. (2003) The auxin-induced maize gene *ZmSAUR2* encodes a short-lived nuclear protein expressed in elongating tissues. *Journal of Biological Chemistry*, 278, 23936–23943.

Laskowski, M. et al. (2006) Expression profiling of auxin-treated Arabidopsis roots: toward a molecular analysis of lateral root emergence. *Plant Cell Physiology*, 47, 788–792.

Lau, S. et al. (2008) The evolving complexity of the auxin pathway. *Plant Cell*, 20, 1738–1746.

Lau, S. et al. (2011) Auxin triggers a genetic switch. *Nature Cell Biology,* 13, 611–615.

Leavitt, R.G. (1904) Trichomes of the root in vascular cryptograms and angiosperms. *Proceedings of the Boston Society Natural History*, 31, 273–313.

Liu, Y. et al. (2006) Comparative proteome analyses of maize (*Zea mays* L.) primary roots prior to lateral root initiation reveal differential protein expression in the lateral root initiation mutant rum1. *Proteomics*, 6, 4300–4308.

Liu, Y. et al. (2010) Regulation of the pericycle proteome in maize (*Zea mays* L.) primary roots by RUM1 which is required for lateral root initiation. *European Journal of Cell Biology*, 89, 236–241.

Ljung, K. et al. (2001) Sites and homeostatic control of auxin biosynthesis in Arabidopsis during vegetative growth. *Plant Journal*, 28, 465–474.

Ljung, K. et al. (2005) Sites and regulation of auxin biosynthesis in Arabidopsis roots. *Plant Cell*, 17, 1090–10104.

Malamy, J.E. and Benfey, P.N. (1997) Organisation and cell differentiation in lateral roots of *Arabidopsis thaliana*. *Development*, 124, 33–44.

McClure, B.A. and Guilfoyle, T. (1989) Rapid redistribution of auxin-regulated RNAs during gravitropism. *Science*, 243, 91–93.

Moreno-Risueno, M.A. et al. (2010) Oscillating gene expression determines competence for periodic Arabidopsis root branching. *Science*, 329, 1306–1311.

Muller, A. et al. (1998) Indole-3-acetic acid is synthesized from L-tryptophan in roots of *Arabidopsis thaliana*. *Planta*, 206, 362–369.

Nagpal, P. et al. (2005) Auxin response factors ARF6 and ARF8 promote jasmonic acid production and flower maturation. *Development*, 132, 4107–4118.

Nemhauser, J.L. et al. (2004) Interdependency of brassinosteroid and auxin signaling in Arabidopsis. *PLoS Biology*, 2, E258.

Nemhauser, J.L. et al. (2006) Different plant hormones regulate similar processes through largely nonoverlapping transcriptional responses. *Cell*, 126, 467–475.

van Noorden, G.E. et al. (2007) Overlap of proteome changes in *Medicago truncatula* in response to auxin and *Sinorhizobium meliloti*. *Plant Physiology*, 144, 1115–1131.

Okushima, Y. et al. (2005a) Functional genomic analysis of the *AUXIN RESPONSE FACTOR* gene family members in *Arabidopsis thaliana*: unique and overlapping functions of ARF7 and ARF19. *Plant Cell*, 17, 444–463.

Okushima, Y. et al. (2005b) AUXIN RESPONSE FACTOR 2 (ARF2): a pleiotropic developmental regulator. *Plant Journal*, 43, 29–46.

Okushima, Y. et al. (2007) ARF7 and ARF19 regulate lateral root formation via direct activation of *LBD/ASL* genes in Arabidopsis. *Plant Cell*, 19, 118–130.

Overvoorde, P.J. et al. (2005) Functional genomic analysis of the *AUXIN/INDOLE-3-ACETIC ACID* gene family members in *Arabidopsis thaliana*. *Plant Cell*, 17, 3282–3300.

Overvoorde, P. et al. (2010) Auxin control of root development. *Cold Spring Harbour Perspectives in Biology*, 2, a001537.

Paciorek, T. et al. (2005) Auxin inhibits endocytosis and promotes its own efflux from cells. *Nature*, 435, 1251–1256.

Palme, K. et al. (2006) Auxin transport and gravitational research: perspectives. *Protoplasma*, 229, 175–181.

Paponov, I.A. et al. (2008) Comprehensive transcriptome analysis of auxin responses in Arabidopsis. *Molecular Plant*, 1, 321–337.

Péret, B. et al. (2009) Arabidopsis lateral root development: an emerging story. *Trends in Plant Science*, 14, 399–408.

Perrot-Rechenmann, C. (2010) Cellular responses to auxin: division versus expansion. *Cold Spring Harbour perspectives in Biology*, 2, a001446.

Petersson, S.V. et al. (2009) An auxin gradient and maximum in the Arabidopsis root apex shown by high-resolution cell-specific analysis of IAA distribution and synthesis. *Plant Cell*, 21, 1659–1668.

Petrášek, J. et al. (2006) PIN proteins perform a rate-limiting function in cellular auxin efflux. *Science*, 312, 914–918.

Petrášek, J. and Friml, J. (2009) Auxin transport routes in plant development. *Development*, 136, 2675–2688.

Petricka, J.J. and Benfey, P.N. (2008) Root layers: complex regulation of developmental patterning. *Current Opinion in Genetics and Devevopment*, 18, 354–361.

Pufky, J. et al. (2003) The auxin-induced transcriptome for etiolated Arabidopsis seedlings using a structure/function approach. *Functional and Integrative Genomics*, 3, 135–143.

Rademacher, E.H. et al. (2011) A cellular expression map of the Arabidopsis *AUXIN RESPONSE FACTOR* gene family. *Plant Journal*, 68, 597–606.

Redman, J.C. et al. (2004) Development and evaluation of an Arabidopsis whole genome Affymetrix probe array. *Plant Journal*, 38, 545–561.

Reed, J.W. (2001) Roles and activities of Aux/IAA proteins in Arabidopsis. *Trends in Plant Science*, 6, 420–425.

Remington, D.L. et al. (2004) Contrasting modes of diversification in the *Aux/IAA* and *ARF* gene families. *Plant Physiology*, 135, 1738–1752.

Ruegger, M. et al. (1998) The TIR1 protein of Arabidopsis functions in auxin response and is related to human SKP2 and yeast grr1p. *Genes and Development*, 12, 198–207.

Robert, H.S. and Friml, J. (2009) Auxin and other signals on the move in plants. *Nature Chemical Biology*, 5, 325–332.

Sabatini, S. et al. (1999) An auxin-dependent distal organizer of pattern and polarity in the Arabidopsis root. *Cell*, 99, 463–472.

Sablowski, R. (2011) Plant stem cell niches: from signalling to execution. *Current Opinion in Plant Biology*, 14, 4–9.

Sauer, M. et al. (2006) Canalization of auxin flow by Aux/IAA-ARF-dependent feedback regulation of PIN polarity. *Genes and Development*, 20, 2902–2911.

Sawa, S. et al. (2002) The *HAT2* gene, a member of the *HD-Zip* gene family, isolated as an auxin inducible gene by DNA microarray screening, affects auxin response in Arabidopsis. *Plant Journal*, 32, 1011–1022.

Schiefelbein, J. (2003) Cell-fate specification in the epidermis: a common patterning mechanism in the root and shoot. *Current Opinion in Plant Biology*, 6, 74–78.

Stablowski, R. (2011) Plant stem cell niches: from signalling to execution. *Currrent Opinion in Plant Biology*, 14, 4–9.

Schlereth, A. et al. (2010) MONOPTEROS controls embryonic root initiation by regulating a mobile transcription factor. *Nature*, 454, 913–916.

Sena, G. et al. (2009) Organ regeneration does not require a functional stem cell niche in plants. *Nature*, 457, 1150–1153

Sinnot, E.W. and Bloch, R. (1939) Cell polarity and the differentiation of root hairs. *Proceedings of the National Academy of Sciences of the United States of America*, 25, 248–252.

Staswick, P.E. et al. (2005) Characterisation of an Arabidopsis enzyme family that conjugates amino acids to indole-3-acetic acid. *Plant Cell*, 17, 616–627.

Stepanova, A.N. et al. (2005) A link between ethylene and auxin uncovered by the characterization of two root-specific ethylene-insensitive mutants in Arabidopsis. *Plant Cell*, 17, 2230–2242.

Stepanova, A.N. et al. (2008) TAA1-mediated auxin biosynthesis is essential for hormone crosstalk and plant development. *Cell*, 133, 177–191.

Swarup, R. et al. (2001) Localization of the auxin permease AUX1 suggests two functionally distinct hormone transport pathways operate in the Arabidopsis root apex. *Genes and Development*, 15, 2648–2653.

Swarup, R. et al. (2005) Root gravitropism requires lateral root cap and epidermal cells for transport and response to a mobile auxin signal. *Nature Cell Biology*, 7, 1057–1065.

Swarup, K. et al. (2008) The auxin influx carrier LAX3 promotes lateral root emergence. *Nature Cell Biology*, 10, 946–954.

Szemenyei, H. et al. (2008) TOPLESS mediates auxin-dependent transcriptional repression during Arabidopsis embryogenesis. *Science*, 319, 1384–1386.

Takehisa, H. et al. (2011) Genome-wide transcriptome dissection of the rice root system: implications for developmental and physiological functions. *Plant Journal*, 69, 126–140.

Tan, X. et al. (2007) Mechanism of auxin perception by the TIR1 ubiquitin ligase. *Nature*, 446, 640–645.

Tao, Y. et al. (2008) Rapid synthesis of auxin via a new tryptophan-dependent pathway is required for shade avoidance in plants. *Cell*, 133, 164–176.

Ulmasov, T. et al. (1997) ARF1, a transcription factor that binds to auxin response elements. *Science*, 276, 1865–1868.

Ulmasov, T. et al. (1999) Activation and repression of transcription by auxin-response factors. *Proceedings of the National Academy of Science of the United States of America*, 96, 5844–5849.

Van den Berg, C. et al. (1997) Short-range control of cell differentiation in the *Arabidopsis* root meristem. *Nature*, 390, 287–289.

Vanneste, S. et al. (2005) Cell cycle progression in the pericycle is not sufficient for SOLITARY ROOT/IAA14-mediated lateral root initiation in *Arabidopsis thaliana*. *Plant Cell*, 17, 3035–3050.

Vanneste, S. and Friml, J. (2009) Auxin: a trigger for change in plant development. *Cell*, 136, 1005–1016.

Vieten, A. et al. (2005) Functional redundancy of PIN proteins is accompanied by auxin-dependent cross-regulation of PIN expression. *Development*, 132, 4521–4231.

Vieten, A. et al. (2007) Molecular and cellular aspects of auxin-transport-mediated development. *Trends in Plant Science*, 12, 160–168.

Weijers, D. et al. (2006) Auxin triggers transient local signaling for cell specification in Arabidopsis embryogenesis. *Developmental Cell*, 10, 265–270.

Wirta, V. et al. (2005) Assembly of a gene sequence tag microarray by reversible biotin-streptavidin capture for transcript analysis of *Arabidopsis thaliana*. *BMC Biotechnology*, 5, 5.

Woll, K. et al. (2005) Isolation, characterization, and pericycle-specific transcriptome analyses of the novel maize lateral and seminal root initiation mutant rum1. *Plant Physiology*, 139, 1255–1267.

Yamada, M. et al. (2009) The *TRANSPORT INHIBITOR RESPONSE2* gene is required for auxin synthesis and diverse aspects of plant development. *Plant Physiology*, 151, 168–179.

Zhao, Y. et al. (2001) A role for flavin monooxygenase-like enzymes in auxin biosynthesis. *Science*, 291, 306–309.

Zhao, Y. et al. (2002) Trp-dependent auxin biosynthesis in Arabidopsis: involvement of cytochrome P450s CYP79B2 and CYP79B3. *Genes Development*, 16, 3100–3112.

Zhao, D. et al. (2003) Members of the *Arabidopsis-SKP1*-like gene family exhibit a variety of expression patterns and may play diverse roles in Arabidopsis. *Plant Physiology*, 133, 203–217.

Zhao, Y. (2010) Auxin biosynthesis and its role in plant development. *Annual Reviews in Plant Biology*, 61, 49–64.

4 Cell-Type Resolution Analysis of Root Development and Environmental Responses

José R. Dinneny

Introduction

The advent of multicellularity was a key innovation in evolution that did not just occur once. Plants, animals, fungi, and microbes all display varying levels of organization among cells, which facilitates survival (Meyerowitz 2002). One key feature of a multicellular body is the specification of cells with different identities and the organization of these cell types into tissue layers and organs with coordinated functions. A major goal of developmental biology has been to understand the mechanisms that control the specification of cell types and the communication that must occur between these cells to properly organize them in the organism. In plants and animals, these studies have identified an important role for transcription factors in controlling these events. As a consequence, a significant emphasis has been placed on understanding the primary outcome of transcriptional regulation, gene expression.

At the single gene level, gene expression patterns can be visualized at the cell-type-specific level using common methods such as *in situ* hybridization (Brewer et al. 2006). Performing such experiments at the genomic scale is technically challenging. Genome-scale technologies such as microarrays and high-throughput sequencing enable information on gene expression levels to be readily gathered (Mockler et al. 2005; Galbraith 2006; Schmitz and Zhang 2011), but until recently, few options were available to plant scientists interested in performing a cell-type-specific analysis. Today, a wide variety of techniques have been developed, which allow cell-type transcriptomes (Birnbaum et al. 2005), translatomes (Mustroph et al. 2009b), epigenomes (Deal and Henikoff 2010), and such to be generated (Table 4.1). These techniques can be combined with various methods for quantitating the output (RNA or chromatin) and have led to a large collection of data sets. The goal of this chapter is to describe in detail the various methods developed for cell-type-specific analysis, present the biological insight the associated studies have generated, and critically compare the utility of these methods for future investigations. In the process, we will highlight the important advances in understanding developmental regulation these data have enabled.

An important extension of recent studies into cell-type-specific gene expression is in the area of plant–environment interactions (Dinneny 2010; Wee and Dinneny 2010). Rarely do developmental biologists delve into the area of physiology; however, recently cell-type-specific profiling techniques have allowed a developmental perspective to be taken in understanding classical physiological stimuli. The ultimate goal of these studies, whether stated or not, is to develop a functional understanding of cell identity. Just as structural features of a cell can be used to classify and characterize the outcome

Root Genomics and Soil Interactions, First Edition. Edited by Martin Crespi.

Table 4.1 Summary and comparison of methods for cell-type-specific analysis.

Method	Description of Method	Biological Information Obtained	Strengths of Method	Potential Complications/Caveats
Fluorescence-activated cell sorting (FACS) of GFP-labeled cells	Birnbaum et al. (2005)	Gene expression from total RNA. Hormone quantitation.	Many available GFP reporter lines available. Many published data sets.	Effect of protoplasting and cell sorting hard to define. Technical and logistical difficulties associated with use of a FACS machine.
Laser capture microdissection (LCM)		Gene expression	No need for transgenic resources.	Specialized equipment needed. Quality of RNA obtained marginal.
Translatome	Mustroph et al. 2009	Ribosome-associated transcripts	Data potentially reflects protein composition of cell better. Cryopreserved tissue used. Large collection of FLAG-RPL18 lines available.	No defined methods for validation. Changes in transcript accumulation can be the result of many layers of regulation.
INTACT	Deal and Henikoff 2011	Nuclear RNA, chromatin	Likely the best method for cell-type-specific ChIP studies. Cryopreserved tissue used.	Few available lines expressing NTF available. Nuclear RNA characterization requires methods new to many researchers.

of cell-type specification pathways, the differential response of a cell to a change in environment is a direct outcome of these developmental mechanisms, as well. Cell-identity regulators will control the expression of receptors and signaling pathways, which can influence how environmental signals are processed. Thus, an integrated view of development and physiology is ultimately necessary to understand plants, organisms whose biology is a superb collaboration between the genome and the environment.

Tools for Cell-Type Resolution Analysis

In the Beginning . . .

An interest in developing cell-type-specific data did not begin with the recent development of the cell-sorting methodology for *Arabidopsis* roots. In fact, biologists have been using various ingenious methods to isolate enriched populations of particular cell types. Three approaches are highlighted here.

Hairs are important structures that are thought to increase the absorptive surface area of the root (Grierson 2002). Hairs initiate from prespecified cells of the epidermis, and this process is under tight regulation by developmental and environmental pathways. Hairs are also critical for the interaction of plants with the biotic environment (Murray 2011). For example, rhizobium infection is initiated at the tip of a developing root hair. Thus, understanding the regulatory mechanisms controlling root hair development may lead to the design of plants with improvements in nutrient and water absorption. To understand the biology occurring specifically in hairs, biologists have developed simple methods for generating hair-cell-enriched fractions (Lauter et al. 1996). To do this, roots are frozen in liquid nitrogen and lightly agitated. Since root hairs are thinner than the primary root, the agitation of the frozen hairs causes them to break off. By filtering this mixture through a strainer, a hair-cell-enriched fraction can be isolated. Lauter et al. developed this method to identify nitrate transporters that might act specifically in root hairs. A tomato cDNA library was created from RNA isolated from the root-hair-enriched fraction and screened using *Arabidopsis* clones for two known nitrogen transporters. These studies identified three genes that were confirmed to be involved in ammonium or nitrate transport in root hair cells. More recently, the group of Gary Stacey has taken a similar approach to isolate root hairs from soybeans for studies at the genomic and proteomic level (Libault et al. 2010).

An alternative approach to obtaining large quantities of a particular cell type is to engineer the plant to produce more of the cell type of interest. Dominique Bergmann used this technique in *Arabidopsis* to obtain a guard-cell-enriched sample (Bergmann et al. 2004). Stomata are pores in the epidermis of the shoot that allow for gas exchange. Two guard cells make up the stoma and the patterning of these cells in the epidermis has become an important model for understanding cell-type specification processes and stem-cell regulation in plants. While guard cells are of great importance, they constitute a small portion of the cells in the shoot. To circumvent this problem, Bergmann utilized a mutant strain, *yoda* (yda), which leads to the massive overaccumulation of guard cells in the plant. She also used a gain-of-function version of YDA, to create a plant without guard cells, for comparison. Gene expression profiles were then generated using microarrays for seedlings of these different genetic backgrounds and analyzed. On the basis of these studies, a set of bHLH transcription factors that control the different steps of stomatal patterning were identified.

One method that has proven difficult to use widely in plants is the *in vitro* differentiation of cells. Culturing techniques have been developed in animals for maintaining or inducing the identity of highly pure populations of cells with specific identities. However in plants, isolation of cells during culturing leads to dedifferentiation. One important exception to this rule is the mesophyll cell of *Zinnia elegans*. The Fukuda lab has pioneered culturing methods for transdifferentiating this cell type into xylem tracheary elements (Karkonen et al. 2011). Differentiation is induced in liquid culture by treating isolated cells with auxin and cytokinin. Their studies have lead to insight into the process of regulated cell death, apoptosis, and the transcriptional networks involved in secondary cell wall biosynthesis (Fukuda 2000; Ohashi-Ito and Fukuda 2003; Endo et al. 2009). These studies have identified genes that play similar roles in *Arabidopsis*, including members of the class III HD-ZIP family, which controls vascular patterning and organ polarity (McConnell et al. 2001; Emery et al. 2003).

Fluorescence-Activated Cell Sorting of Green Fluorescent Protein-Marked Cells

Fluorescence-activated cell sorting (FACS) technology (Figure 4.1) has proven useful in the study of cell-type-specific biology. FACS is primarily used in the animal field, particularly in immunology, where the rapid characterization of different populations of cells in the blood is a daily exercise. FACS requires that the sample of cells used be dissociated or individualized. For blood, this requirement is easily achieved. Other animal tissues can be used simply by incubating the sample in an enzyme mixture to digest the extracellular matrix, which holds the cells together. In plants, the use of similar methods required the careful optimization of protoplasting conditions to enable efficient digestion of the cell wall (Birnbaum et al. 2005). Currently, protoplasting works most efficiently on roots or mature leaves and shoot meristems (Harkins et al. 1990; Yadav et al. 2009).

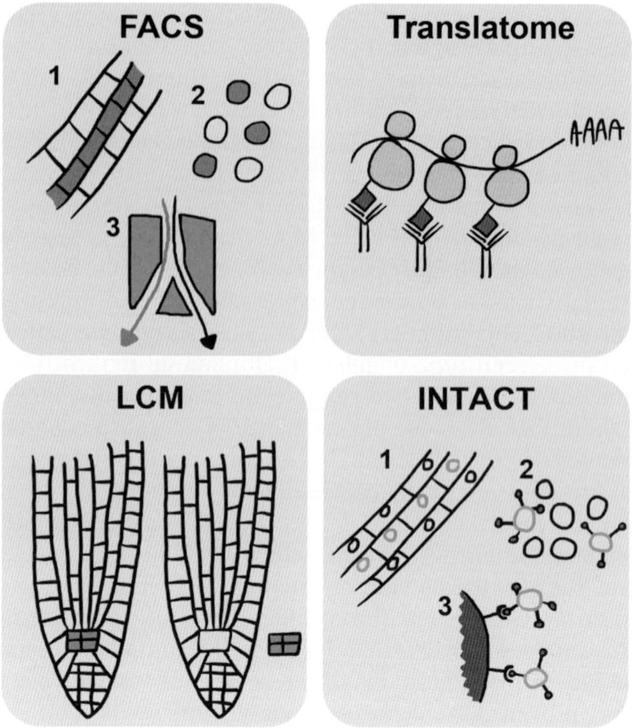

Figure 4.1 Methods for cell-type-specific analysis. Fluorescence-activated cell sorting (FACS) utilizes a plant expressing GFP in a cell type of interest (1). Harvested tissue is protoplasted into individual cells (2). These cells are feed into a FACS machine to isolate GFP-positive or negative enriched cell populations. Translatome:immunoprecipitation of ribosome-transcript complex. A ribosomal subunit such as RPL18 is fused to a tag, such as FLAG, and expressed in the cell using a tissue-specific promoter. Cryopreserved tissue is used and transcripts are immunoprecipitated for later analysis. LCM, laser-capture microdissection. Fixed and embedded sections of tissue are used. A special microscope fitted with a laser is used to cut out and isolate cells of interest. Isolation of nuclei tagged in specific cell types (INTACT) utilizes a nuclear membrane targeted protein, NTF, which is expressed in a cell-type-specific manner (1). Cryopreserved tissue is disrupted to obtain a crude nuclear fraction (2). Affinity purification of nuclei uses the biotin–streptavidin interaction (3).

In order to isolate a specific cell type from a heterogeneous population, FACS requires that some optical property of the cell of interest distinguish it from other cells. Typically, a fluorophore is used to mark the cell. This can be done through immunohistochemistry or through the expression of a fluorescent protein such as green fluorescent protein (GFP). Initial experiments in plants utilizing FACS took advantage of the autofluorescence produced by chloroplasts to separate leaf cells into photosynthetic and nonphotosynthetic subpopulations (Harkins et al. 1990). These experiments revealed that transgenes expressed in green tissues of intact plants maintained their cell-type specificity even in protoplasted cells. This discovery indicated that it would be possible to gain information on cell-type-specific transcriptional programs using FACS.

While chloroplasts are useful endogenous markers, additional reporter lines are necessary to distinguish between the multitude of other cell types composing the plant. GFP enhancer trap transgenic lines developed by the Haseloff lab helped to fill this gap (Kiegle et al. 2000). The GAL4-UAS system, developed in Drosophila to identify tissue-specific enhancers, was adapted for *Arabidopsis* and led to the creation of hundreds of different GFP-reporter lines with reproducible and cell-type-specific expression patterns. In parallel, the identification of genes such as *WEREWOLF* (Lee and Schiefelbein 1999), *SCARECROW* (Di Laurenzio et al. 1996), and *WOODENLEG* (Mahonen et al. 2000), which regulate the specification of tissue types in the root, led to the development of additional cell-type-specific GFP reporter lines.

In short, the FACS procedure begins by treating thousands of harvested roots with cell-wall-degrading enzymes to release the protoplasts. Protoplasted root cells are then fed into a FACS machine to isolate cells expressing GFP. RNA is then isolated, processed, and hybridized on Affymetrix microarrays containing probe-sets for nearly 24,000 genes. This method, developed in Birnbaum et al., was initially used to profile the expression of five different populations of cells in the root (Birnbaum et al. 2003; Birnbaum et al. 2005). Analysis of these data led to the development of several additional GFP reporter lines, which were utilized subsequently to profile 11 additional cell populations in the root.

The digestion of the cell wall takes time and can lead to changes in the expression of genes. This protoplasting effect was characterized in Birnbaum et al. by comparing whole roots to protoplasted cells (Birnbaum et al. 2003). It was found that protoplasting affects several hundred transcripts, which are typically removed from later analysis. However, this control experiment is unlikely to identify those genes that respond to protoplasting in a cell-type-specific manner. Additional controls are possible. For example, if roots are treated with a stimulus, the response to that stimulus can be analyzed by comparing the transcriptional profile of that cell type before and after treatment (Dinneny et al. 2008). This analysis controls for the cell-type-specific sorting effect, but not for the effect of cell sorting on the cell-type-specific stimulus response. In the end, it may be extremely difficult to control for all the effects of protoplasting and sorting, thus it is always highly recommended that data obtained by FACS be validated using independent methods such as reporter lines or *in situ* hybridization.

Laser Capture Microdissection

In situ hybridization is a tried and true method for the cell-type-specific expression analysis of an individual transcript. Development of laser capture microdissection (LCM) has updated this technique to allow for genome-scale information to be gathered (Jiao et al. 2009; Figure 4.1). Similar to *in situ* hybridization, tissue is fixed and embedded in a solid matrix of wax. Tissue sections are made and affixed to a slide. A specialized microscope fitted with a laser enables the user to cut out cells of interest by various methods. Cells are collected and RNA extracted. Because of the small amount of material typically collected, several rounds of RNA amplification are often required to obtain enough samples for microarray hybridization. Because of sample fixation, RNA quality is often relatively low, thus strong 3′ bias in the transcripts profiled is usually observed. Another concern is that the user-guided collection of cell samples may result in data that is difficult to reproduce. Nevertheless, LCM is a highly useful method, especially for those species where there are few transgenic resources or where a cell-type-specific marker does not exist.

Immunoprecipitation of Ribosome-Associated Transcripts

While transcriptional regulation is certainly important in determining the biological state of a cell, it is not the only form of regulation. In particular, environmental stress has been shown to strongly affect the rate of translation for some transcripts more than others. This was shown to be particularly true during hypoxia, where transcriptional changes induced by this stress are minimal, but translational changes are dramatic (Branco-Price et al. 2008). Transcripts that are actively being translated are often associated with multiple ribosomes in polysome complexes. By tagging a ribosomal subunit with an affinity tag such as FLAG, the Bailey-Serres lab has developed a technique in which the ribosome-transcript complex can be immunoprecipitated (Mustroph et al. 2009a, 2009b; Figure 4.1). The method starts with cryopreserved tissue and does not require protoplasting or preprocessing. Immunoprecipitation of transcripts before and after hypoxia has led to an impression as to which proteins are being synthesized under these two conditions. Expression of the ribosome-FLAG fusion using cell-type-specific promoters enables tissue-specific information to be gathered (Mustroph et al. 2009b). Mustroph et al. used this method to profile the "translatome" of tissues from the shoot and root of *Arabidopsis*. The information gathered with this approach is likely to provide the best reflection of the actual proteins present in a cell. Transcriptional regulation, however, will be more difficult to discern from these data as additional layers of regulation separate the measurements of RNA quantity from the actual transcriptional events. Furthermore, there are currently no straightforward methods to validate the association of ribosomes with a specific transcript of interest in live tissue, making a spatiotemporal understanding of the process difficult.

Affinity Purification of Nuclei Using the INTACT Method

The ability to work with cryopreserved tissue and the elimination of FACS and protoplasting are important steps in the development of a method that is easy and reliable to implement. While the FACS methodology has been utilized by several labs worldwide, there are many steps that need to be optimized. Furthermore, the equipment needed is expensive, and trained technical expertise can often be difficult to find. Recently, a method was developed by Deal and Henikoff, which circumvents these issues (Deal and Henikoff 2010, 2011; Figure 4.1). Their method, termed INTACT (isolation of nuclei tagged in specific cell types), utilizes a nuclear targeting fusion protein (NTF) to tag nuclei in specific cell types. The NTF is a fusion of a subdomain of the Ran-GTPase-activating protein1 (RanGAP1), GFP and a biotin-ligase recognition peptide (BLRP). RanGAP1 localizes the NTF to the nuclear envelope, while the BLRP enables the chimeric protein to be biotinylated in the presence of biotin ligase, which must also be expressed in the cell for the system to work. The biotin–streptavidin interaction is one of the strongest noncovalent interactions known and enables affinity purification of NTF-labeled nuclei from a lysate made from frozen tissue. The isolation of nuclei can be confirmed by fluorescence microscopy. The NTF can be expressed in any tissue of interest. Currently, there are few NTF lines available, but this is likely to change.

Deal et al. performed an initial analysis using hair and nonhair cell promoters to drive expression of the NTF. Microarray-based expression analysis of RNA isolated from purified nuclei identified 946 genes showing a hair-cell bias in expression, while 118 genes showed a nonhair cell bias. Comparative analysis of hair-cell and nonhair-cell enriched genes was performed with a validated list of 24 hair-cell specific genes. Of these genes, 19 were correctly identified as hair-cell enriched in the INTACT data set. Deal and Henikoff also performed a similar comparison using the Brady et al. list of hair-cell enriched genes. Here, 11 of the 24 were correctly identified. Deal and Henikoff suggest the differences may indicate a lower correspondence between the data obtained by FACS and actual cell-type-specific transcriptional programs. However, this interpretation may be tenuous, especially when considering the vastly different analyses that led to the identification of hair-cell-enriched gene lists in each study. Deal and Henikoff provide data that the FACS method provides a lower purity of GFP-positive elements (50% purity of cells for FACS compared to 93–95% purity of nuclei for INTACT). This result may also be in some doubt as other studies (Birnbaum et al. 2005), in addition to our own unpublished results, confirm a purity of GFP-positive cells in the range of 90% when FACS settings are properly established.

The isolation of nuclei by INTACT presents some advantages and disadvantages depending on the biological process of interest. Unlike FACS, where total RNA is isolated, nuclei contain unprocessed or partially processed transcripts. In addition, miRNA regulation, which occurs in the cytoplasm, will not affect transcript abundance. Microarray resources such as the Affymetrix ATH1 array are designed specifically to hybridize to cRNA amplified from processed transcripts, and probe synthesis protocols often utilize an oligo-dT primer to generate cDNA. Thus, other methods of transcript quantitation are likely necessary, such as tiling arrays or high-throughput RNA-sequencing. While this alters the typical workflow many labs are used to, the data obtained may provide information on transcript abundance that better reflects the status of transcriptional activity at the promoter. A more thorough analysis needs to ultimately be performed to understand the difference between nuclear, cytoplasmic, and total RNA populations. A clear advantage of the INTACT system is that chromatin can also be characterized. This is technically also possible for FACS sorted cells; however, this is not recommended as changes in chromatin features are likely to be very rapidly altered during protoplasting.

Analysis of Spatiotemporal Expression Patterns in the Arabidopsis Root

The study by Birnbaum et al. provided the first opportunity to explore spatial expression patterns in an organ at the genomic scale (Birnbaum et al. 2003). In this study, five different subpopulations of cells or radial zones were profiled using FACS. Most of these radial zones were composed of 1–2 tissue layers of the root due to the lack of available cell-type-specific GFP reporter lines at the time. In addition, gene expression was profiled along the longitudinal axis by microdissecting the root tip into three sections (columella/stem cells, meristematic, and elongation zones). Thus, gene expression was assayed along two dimensions: radial and longitudinal. For visualization purposes and the identification of dominant gene expression patterns, data from the two axes were integrated. An apportioning method was used to calculate the expression values for 15 subzones of the root comprising five radial zones for each of three longitudinal zones. In essence, the method scaled the radial zone data based on the relative strength of expression calculated for a particular longitudinal zone. The apportioned data was then used in various clustering methods to identify eight dominant expression patterns that were well represented in the data set. Of the expression patterns identified, several were enriched for genes regulating hormonal pathways, suggesting that hormone response centers may be present in the root. Importantly, hormones such as indole-3-acetic acid (IAA), jasmonic acid, and gibberellin were represented in different clusters indicating that root development may be under the control of a complex spatial and temporal pattern of hormone action.

This initial analysis provided an important testing ground for ideas pertaining to the analysis of complex expression data sets. However, several gaps in understanding remained. The most important being that many cell types had not been profiled and the majority of GFP reporters used did not exhibit true cell-type specificity. Furthermore, gene expression along the longitudinal axis was profiled at a relatively crude level where higher resolution data may be needed to identify highly dynamic expression patterns.

To extend the root map data set, Brady et al. utilized several additional FACS-based experiments that had been published in the intervening time period (Nawy et al. 2005; Lee et al. 2006; Levesque et al. 2006; Brady et al. 2007b). In addition, 11 new cell populations were profiled, resulting in a total of 19 subpopulations of cells in the root. Together, these data assayed the expression of 14 of the 15 cell types present in the root and was the highest resolution data set available at the date of publication. In addition, expression along the longitudinal axis was examined by microdissection of a single root into 13 sections and microarray profiling of each of these sections individually. As a comparison, a second root was also analyzed in the same way.

The advanced root map data set identified 51 radial and 40 longitudinal dominant expression patterns. Many expression patterns were cell-type-specific; however, a large portion spanned several cell types. This transcriptional similarity observed between cell types did not always correspond to positional similarity. For example, xylem cells and root hairs shared a transcriptional module enriched for cell-wall biosynthesis genes. These data suggest shared regulatory pathways may span obvious developmental boundaries. In Birnbaum et al., the localization of hormone response centers was determined. With the additional data, hormonal pathways could be described at much higher resolution. For auxin, biosynthesis (quiescent center, lateral roots, and pericycle), transport (root hairs, columella root cap and cortex), and homeostasis (columella root cap) could be spatially defined to reveal complex regulation only partly envisioned at the single gene level before.

Of particular novelty, expression profiling along the longitudinal axis identified 17 out of 40 clusters that showed oscillatory expression patterns. Genes in these clusters exhibited multiple peaks of expression, suggesting that their transcriptional programs were not simply activated during a particular developmental window. Analysis of these fluctuating expression patterns in the second root prepared confirmed the presence of these patterns and argued that they did not simply represent transcriptional noise. Interestingly, a few clusters exhibited nonuniformity in their expression pattern in the second root. The authors suggest that this may be an indication of a shifting wave of transcriptional activity captured at different times in the two roots. This hypothesis is consistent with a recent report revealing highly dynamic waves of gene expression in the root tip (Moreno-Risueno et al. 2010). Using a DR5::LUCIFERASE reporter, which allows highly dynamic transcriptional events to be tracked, Moreno-Risueno et al. found that reporter expression oscillated at a ~5 h period. This study led to the identification of hundreds of other genes exhibiting similar oscillatory behavior. Thus, a single cell may activate the same transcriptional program multiple times. It is hypothesized that the coincidence of other positional cues may provide a context for interpreting these fluctuating signals during lateral root founder cell specification.

Analysis of Cell-Type-Specific Expression Patterns in the Rice Root

The creation of a cell-type-specific gene expression map as published in Brady et al. requires a large collection of tissue-specific GFP-reporter lines. In other species with fewer transgenic resources than *Arabidopsis*, other methods, such as LCM enable similar resolution data sets to be generated. Jiao et al. took such an approach to profile 40 different cell types in the root, shoot, and seed (Jiao et al. 2009; Petersson et al. 2009). Many tissues were profiled at multiple developmental stages. The utility of performing cell-type-specific analysis of gene expression was confirmed in this study. They found that 87% of genes detected as expressed in shoot cell types were not detected as expressed when whole shoots were profiled. The expression data in rice was compared to the *Arabidopsis* root map and, for 112 orthologs identified, 77 showed a similar pattern of expression. Hormonal response pathways were also inspected for bias in the rice gene expression map. Here, more differences than similarities were identified between *Arabidopsis* and rice. For example, auxin response is centered in the rice endodermis, while in *Arabidopsis*, it is centered in the stele. This may reflect true differences in the distribution of auxin in the rice root or may indicate that the location of the response center is highly dependent upon the specific genes used to estimate its location.

Cell-Type-Specific Analysis of Auxin

One of the most intensely studied areas of plant biology is the role of auxin in development (Vanneste and Friml 2009). The chemical nature of auxin as a weak acid enables it to easily pass into cells in its protonated form, while inside the cell it is deprotonated and unable to leave. PIN-FORMED transporters, which are localized to discrete domains within each cell, provide a route to exit the cell. The gene expression pattern of PINs, the localization pattern of the protein in the cell, and regulation of protein stability all contribute to the directional transportation of auxin within an organ. In the root, auxin transport is directed toward the stem-cell organizer, the quiescent center, and subsequently redirected shoot-ward through the lateral root cap and epidermis (Blilou et al. 2005; Grieneisen et al. 2007). Reporters such as DR5, which track the activity of the canonical transcriptional regulatory pathway, suggest that auxin may accumulate in the QC. Indeed, auxin response in the QC is thought to be necessary for its specification (Sabatini et al. 1999). Transcriptome analysis of the *Arabidopsis* and rice root also suggest that auxin response may be localized in the root tip (Brady et al. 2007a; Jiao et al. 2009).

To obtain direct evidence for the distribution of auxin within the root meristem, Petersson et al. utilized gas chromatography-selected reaction mode-mass spectrometry (GC-SRM-MS) on FACS isolated cells to determine the concentration of the natural auxin (IAA) in each of the tissue layers of the root (Petersson et al. 2009). Control experiments revealed that protoplasted cells do not "leak" auxin, even after prolonged incubation, and indicate that the amount of auxin in the cell should be stable throughout the FACS procedure. IAA concentration was normalized by comparing the concentration of IAA in GFP-positive cells to all other cells in the sample. In addition, cell diameter in protoplasts and in cells of intact roots was determined and used to convert mass measurements to concentrations in each tissue. Consistent with expectations, the QC was found to contain the highest levels of IAA in the root. This was followed by the cortex, endodermis, and stele, while the columella root cap and epidermis had IAA levels that were below background. Measurements of auxin biosynthesis were also performed, but these data did not reveal significant differences between cells. Thus, IAA transport appears to play the lead role in defining a concentration gradient in the root centered at the QC.

Cell-Type-Specific Analyses of Chromatin

In addition to the analysis of gene expression, the INTACT method enables chromatin features to be characterized at cell-type resolution (Deal and Henikoff 2010). Deal and Henikoff utilized hair-cell-specific and nonhair-specific NTF lines to profile the state of histone modifications in these two cell types. Histones are important structural and regulatory proteins that affect the rate of gene transcription (Pfluger and Wagner 2007). Residues in the N-terminal tail of histone H3 are modified by various enzymes, and these marks correlate with the transcriptional activity of the locus. Trimethylation of H3 at the fourth lysine residue (H3K4me3) is associated with loci that are actively being transcribed while trimethylation of H3 at the 27th lysine is associated with Polycomb-complex-dependent silencing. Deal and Henikoff used antibodies raised against these two histone modifications to immunoprecipitate chromatin associated with the marks. As a control for variation in histone occupancy across loci, an antibody that recognizes all forms of H3 was also used. Immunoprecipitated chromatin was amplified, labeled, and hybridized to tiling microarrays.

Analysis of the ChIP-chip data revealed that H3K4me3 is associated with strongly expressed genes and localized at the 5′ start of transcription. H3K27me3, on the other hand, is depleted from strongly expressed genes and was enriched primarily in the promoters of moderately expressed genes and throughout the gene body in genes with low expression. An analysis of genes expressed in a cell-type-specific manner found that the H3K4me3 mark was generally associated with genes showing a higher level of expression in that cell type, as expected. However, this trend was not always followed and may indicate that other mechanisms besides histone modification mediate cell-type-specific expression patterns.

Cell-Type-Specific Analyses of Responses to Environmental Change

Roots are intimately associated with their environment and exquisitely sensitive to minor variation in a number of diverse stimuli. Charles Darwin and his son, Francis, famously studied several of these responses in a book titled *The Power of Movement in Plants* (Darwin and Darwin 1881). In this work, they were able to observe that different parts of the root were involved in sensing and responding to stimuli such as touch, heat, moisture and gravity. Through this work, it was realized that many growth decisions are made at the root tip. Considering the important role environmental stimuli have on root development, it is surprising to find that little emphasis is placed on taking a developmental approach when studying environmental stimuli. However, this is beginning to change. The advent of cell-type-specific approaches has enabled the creation of data sets that reveal high spatial complexity in environmental responses (Figure 4.2). These data provide opportunities for future studies aimed

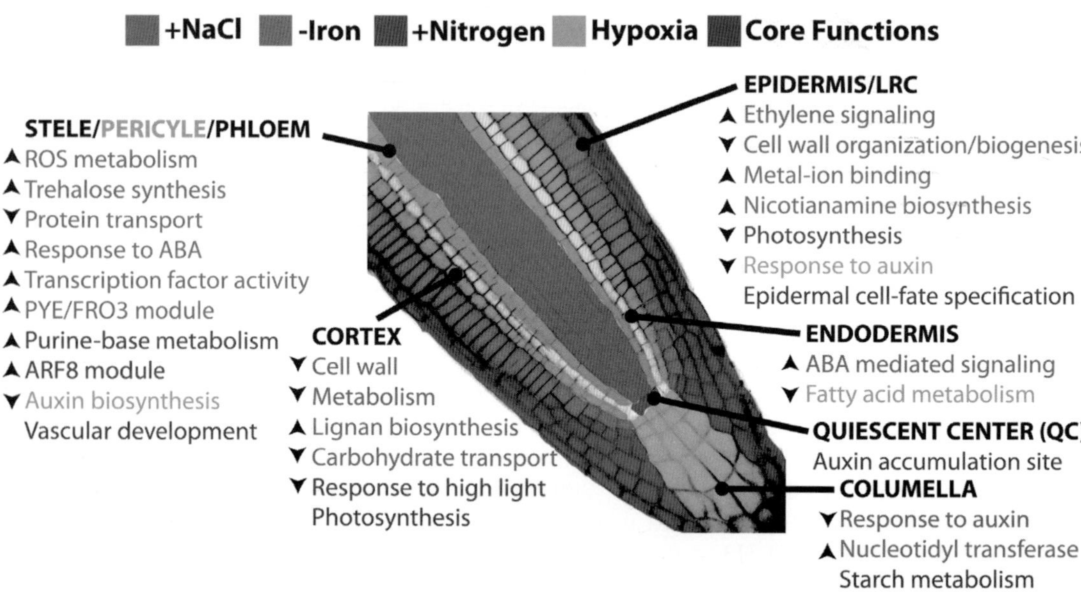

Figure 4.2 Summary of cell-type-specific data sets. False-colored root showing a summary of cell-type-specific regulation uncovered through FACS and translatome data sets.

at understanding the mechanisms that guide tissue-specificity in environmental responses and new avenues for the design of stress-resistant plants.

Nitrate

One of the most important macronutrients that is limiting to plant growth is nitrogen. Nitrogen is a vital building block for nucleic and amino acids and is absorbed predominately as nitrate. To understand what role individual cell types play in the response, Gifford et al. used FACS to isolate five different tissue layers under nitrogen-limiting conditions and as well as after 3.5 h treatment with nitrate (Gifford et al. 2008). Analysis of these data revealed 5396 differentially expressed genes. Spatial information governed the majority of transcriptional responses with 87% of genes responding in at least one cell type, but not all. Instances of mixed regulation were also observed with expression being activated in one cell layer and repressed in another. Genes identified as differentially expressed in a whole root microarray experiment tended to be regulated in 3.5 out of 5 cell layers, on average, while genes identified in the FACS data set were regulated in 2.4 out of 5 cell layers. These data indicate that cell sorting is able to identify a larger number of differentially expressed genes, likely due to less heterogeneity in the FACS-sorted cell population compared to the whole organ.

Gifford et al. utilized their cell-type-specific response map to characterize a transcriptional module involved in regulating root branching (lateral root development). Lateral roots develop from founder cells specified in the pericycle cell layer of the stele. Examination of genes regulated by nitrogen treatment in this cell layer identified the auxin response transcription factor, ARF8, as strongly upregulated. Auxin signaling promotes lateral root development, and ARF-type transcription factors are typically involved in mediating this response (Ulmasov et al. 1997). *ARF8* is particularly interesting as it is repressed at the posttranscriptional level by the *miR167* microRNA (Wu et al. 2006). Since *ARF8* is activated by nitrate, Gifford et al. hypothesized that this effect is mediated by nitrate suppression of *miR167*. This turned out to be correct, and plants that overexpress the miR167 or were mutant for *ARF8* were insensitive to nitrogen treatment. Thus, nitrogen regulation of *miR167* activates lateral root development through *ARF8*. With nitrogen treatment, it was found that the initiation stage was primarily induced, while outgrowth was largely suppressed. These data suggest that nitrate treatment activates a foraging strategy, where lateral root primordia are initiated, but kept in stasis until the time at which nitrogen levels become limiting again.

Salt Stress

High salinity is an important stress found in the natural and agricultural context (Xiong 2002). Plants can be roughly categorized as halophytes or glycophytes depending on whether they are resistant or susceptible to high levels of salt in their soil environment, respectively. This distinction is arbitrary and undervalues the importance of salt tolerance mechanisms that do exist in glycophytes, which enable them to survive under mild stress conditions. *Arabidopsis*, for example, is considered a glycophyte; however, it has proven quite useful for identifying genes that are necessary for salt tolerance. AtHKT1;1, for instance, promotes the removal of sodium ions from the vascular stream (Maser et al. 2002). Recently, work analyzing natural variation in salt tolerance identified AtHKT1;1 as an important loci associated with higher levels of tolerance (Baxter et al. 2010). Other studies examining natural variation in salt tolerance have identified additional loci that contribute to differences between ecotypes (Ren et al. 2010). Such work strengthens the hypothesis that evolution is acting on salt tolerance pathways in *Arabidopsis*.

Salt stress causes a wide range of physiological and developmental changes in the root, many of which may have a cell-type-specific basis in regulation (Dinneny 2010). Root hair development, for instance, is dynamically regulated by salt treatment, with hair initiation initially being suppressed for a period of several hours (Dinneny et al. 2008). Subsequently, hair initiation resumes, which is suggestive of some level of acclimation being attained. The cortex and epidermis also show marked changes in cell shape with radial cell expansion and reductions in cell length being most obvious (Burssens et al. 2000). The columella root cap, which is necessary for gravity sensing and signaling, has been reported to show changes in the accumulation of starch in the amyloplasts (Sun et al. 2007). These modified organelles are necessary for sensing the direction of gravity, which is suppressed upon salt-stress treatment as part of a presumed avoidance mechanism (Sun et al. 2007; Dinneny et al. 2008).

To understand the mechanisms that drive these important changes in the root, we utilized FACS and microdissection to develop a high-resolution transcriptional map of the salt-stress response (Dinneny et al. 2008). This data set consisted of a time-course analysis of salt response in whole roots, an analysis of the response of four sections along the longitudinal axis and an analysis of the cell-type response using FACS to isolate six different subpopulations of cells. Similar to the results of Gifford et al., we observed in our data sets a strong bias toward zone-specific responses. Along the longitudinal axis, the elongation zone exhibited the highest number of activated and repressed genes. This zone is of particular interest as decisions related to growth rate, root hair development, and lateral root patterning are made here. Among the different tissue layers profiled, the cortex exhibited the strongest response. These transcriptional changes occurred in genes associated with structural components of the cell (cell wall and tubulin) as well as metabolic pathways (fatty acid biosynthesis).

To understand the regulatory pathways controlling the salt response, an analysis of *cis*-regulatory element enrichment was performed. Present in the regulatory regions of genes are short DNA sequences termed *cis*-regulatory elements (CREs). These sequences recruit specific DNA-binding transcription factors to promoters, which then determine the rate of transcription by interacting with basal transcriptional machinery or through epigenetic mechanisms. Genes expressed with similar patterns tend to share CREs, thus bioinformatics analysis of microarray data sets can identify known or novel sequences involved in the transcriptional changes under investigation. On the basis of our cell-type-specific data, we found that genes regulated in multiple cell types by salt could be distinguished from genes regulated in a cell-type-specific manner based on the CREs present in their promoters. For example, the ABA response element (Mundy et al. 1990) and the drought response element (Yamaguchi-Shinozaki and Shinozaki 1994) exhibited strongest statistical enrichment in genes activated by salt in at least three cell layers. These data suggest that cell-type-specific transcriptional regulation by salt stress may not work by the simple combination of stress CREs with developmental CREs (modular design). Few known CREs were associated with cell-type-specific responses, suggesting that their regulation may be under the control of unknown CREs.

To understand the role of classical developmental pathways in the salt response, mutants that alter the specification of the two cell types in the epidermis (hairs and nonhairs) were used. Mutants altering the ratio of these two cell types had a significant affect on the salt response, both within the epidermis and in internal tissue layers. These data reveal that an important downstream function of developmental pathways is in the control of environmental responses.

Iron Deprivation

Similar to our studies of the salt-stress response, time course, longitudinal, and cell-type-specific data sets were generated to understand the iron deprivation response. Iron is a vital nutrient in the human diet; however, plants are often poor sources for such nutrition. To develop plants that accumulate higher levels of this nutrient, it is necessary to understand how plants perceive differences in iron content in their environment and regulate uptake (Kim and Guerinot 2007). Similar to the high degree of cell-type-specific responses observed under salt stress, iron deprivation regulated a majority of spatially restricted responses (Dinneny et al. 2008). For example, in the epidermis, genes involved in chelating iron or iron transportation were upregulated. Interestingly, the stele represents the strongest response center in the root. This is the site for iron storage in the embryo and iron transportation in seedlings (Kim et al. 2006). To follow up on these results, Long et al. performed an additional analysis on the response of the pericycle to iron deprivation (Long et al. 2010). In this analysis, the pericycle (which is a cell type of the stele) exhibited the highest responsiveness compared to all other cell layers. Transcription factor activity was strongly enriched in the pericycle response module, which included the gene POPEYE (PYE). PYE encodes a bHLH transcription factor that directly regulates the expression of the iron transporter, FRO3. Mutations in *PYE* lead to reduced growth under iron-deprivation conditions, yet genes such as *FRO3* exhibit higher iron-inducibility. Interestingly, iron stores are also higher in *pye* mutants. These data suggest that the lack of tolerance of *pye* to low iron conditions may be caused by an inability to mobilize these stores.

Meta-Analysis of Stress Responses

The utilization of a standard set of methods for sample collection, RNA processing, microarray hybridization, and data processing enabled us to perform a meta-analysis of the FACS-based data sets generated in Dinneny et al. Essentially, gene expression had been assayed under three environmental conditions: standard, salt stress, and iron deprivation (Dinneny et al. 2008; Licausi 2011). First, we asked whether there was a shared stress response. Interestingly, a significant portion of the salt and iron response measured by FACS overlapped. Most of these genes exhibited a high degree of cell-type specificity in the two responses. This may make sense as the biological functions that define a cell under standard conditions are likely to be important targets under environmental change. Perhaps most interesting was the observation that biological functions, which define particular cell types under standard conditions, frequently change under stress conditions. Only 15% of the Gene Ontology categories enriched in a cell-type-specific manner are conserved under both salt stress and iron deprivation conditions. Thus, as a consequence of environmental change, new biological functions define each cell layer, which results in a massive reprogramming of root biology. An important pathway that resists environmental change is development. Genes involved in cell-type specification tended to exhibit robust cell-type enriched expression, which is consistent with the overall conservation of tissue-layer organization even upon stress treatment.

Hypoxia

The ability of plants to generate and use metabolic energy is dependent upon gas exchange. In addition to carbon dioxide, which is necessary for plants to fix carbon into sugar, oxygen is also necessary for respiration during the night. In roots, where little to no photosynthesis occurs, limitations in oxygen availability can have lethal effects. Hypoxia is the environmental condition where the availability of oxygen is severely limited, which can occur upon flooding (Licausi 2011). In rice paddies, flooding is a common occurrence and the development of cultivars better able to grow under these conditions is a major goal of applied research.

To better understand the stresses caused by hypoxia and the responses that occur subsequently, the Bailey-Serres lab has developed approaches for translatome characterization in *Arabidopsis*. On the basis of initial work in whole plants, hypoxia most strongly affects the translational rather than the transcriptional process in the cell (Branco-Price et al. 2008). To gain a cell-type-specific understanding, Mustroph extended their translatome studies to survey hypoxia-treated seedlings, as well (Mustroph et al. 2009b). Thirteen different transgenic lines expressing a FLAG-tagged ribosomal protein, RPL18, were used for immunoprecipitation experiments after two hours of hypoxia. Analysis of these data identified 6416 genes differentially regulated by hypoxia. Transcripts that exhibited ubiquitous increases in ribosome association were involved in substrate-level ATP production and fermentation, consistent with the known effect of hypoxia on energy-dependent processes. Cell-type-specific changes in the translatome were less frequent than that observed in the previous data sets described here. Nevertheless, a pronounced feature of the hypoxia response was a suppression of translation for cell-type-specific transcripts. For example, auxin biosynthesis in the stele, fatty acid metabolism in the endodermis, and the response to auxin in the epidermis were all strongly suppressed. The authors suggest that hypoxia may induce similar pathways in all cells of the organism as a consequence of the universal importance of energy production.

Future Prospects

The new generation of cell-type-specific techniques and data sets represent a boon for biologists interested in the interaction between plants and the environment or for developmental biologists hoping to understand the functional significance of their pathways. However, these data sets are far from useful unless they inspire functional studies. An important next step is to understand the regulatory mechanisms that control the complex spatial and temporal transcriptional programs that exist in the root. Large-scale data sets, however, require high-throughput methods downstream as well. The development of phenotyping platforms to evaluate growth and reporter gene expression changes in mutants or different ecotypes will be important in this process (Miller et al. 2007; Iyer-Pascuzzi et al. 2010; Wee and Dinneny 2010).

Another future goal for all of these studies will be to understand how cell-type-specific transcriptional programs function in the natural environment. All studies described here have utilized some form of tissue-culture environment. These conditions include the use of nutrients, buffering agents, and sterility. Comparison of these standard conditions with stress conditions reveals enormous differences and it is also highly likely that these transcriptional programs will differ significantly in a soil environment. But which soil environment is most appropriate? Soil conditions differ substantially between geographic locations and the ability to replicate experiments requires some form of standardization. It represents a great challenge in root biology to design future growth conditions to balance the need for greater relevance to the natural environment while also enabling high-resolution and high-throughput observation and experimentation.

Acknowledgments

Funding for work on root biology in the Dinneny lab is from the Temasek Lifesciences Laboratory, National Research Foundation, Singapore, the National Science Foundation, U.S.A. and the Carnegie Institution for Science Endowment.

References

Baxter, I. et al. (2010) A coastal cline in sodium accumulation in *Arabidopsis thaliana* is driven by natural variation of the sodium transporter AtHKT1;1. *PLoS Genetics*, 6(11), e1001193.

Bergmann, D.C. et al. (2004) Stomatal development and pattern controlled by a MAPKK kinase. *Science*, 304(5676), 1494–1497.

Birnbaum, K. et al. (2005) Cell type-specific expression profiling in plants via cell sorting of protoplasts from fluorescent reporter lines. *Nature Methods*, 2(8), 615–619.

Birnbaum, K. et al. (2003) A gene expression map of the Arabidopsis root. *Science*, 302(5652), 1956–1960.

Blilou, I. et al. (2005) The PIN auxin efflux facilitator network controls growth and patterning in Arabidopsis roots. *Nature*, 433(7021), 39–44.

Brady, S.M. et al. (2007a) A high-resolution root spatiotemporal map reveals dominant expression patterns. *Science*, 318(5851), 801–806.

Brady, S.M. et al. (2007b) Combining expression and comparative evolutionary analysis. The COBRA gene family. *Plant Physiology*, 143(1), 172–187.

Branco-Price, C. et al. (2008) Selective mRNA translation coordinates energetic and metabolic adjustments to cellular oxygen deprivation and reoxygenation in *Arabidopsis thaliana*. *The Plant Journal*, 56(5), 743–755.

Brewer, P.B. et al. (2006) *In situ* hybridization for mRNA detection in Arabidopsis tissue sections. *Nature Protocols*, 1(3), 1462–1467.

Burssens, S. et al. (2000) Expression of cell cycle regulatory genes and morphological alterations in response to salt stress in *Arabidopsis thaliana*. *Planta*, 211(5), 632–640.

Darwin, C. and Darwin, F. (1881) *The Power of Movement in Plants*. D. Appleton and Company, New York.

Deal, R.B. and Henikoff, S. (2010) The INTACT method for cell type-specific gene expression and chromatin profiling in *Arabidopsis thaliana*. *Nature Protocol*, 6(1), 56–68.

Deal, R.B. and Henikoff, S. (2011) The INTACT method for cell type-specific gene expression and chromatin profiling in *Arabidopsis thaliana*. *Nature Protocol*, 6(1), 56–68.

Di Laurenzio, L. et al. (1996) The *SCARECROW* gene regulates an asymmetric cell division that is essential for generating the radial organization of the *Arabidopsis* root. *Cell*, 86(3), 423–433.

Dinneny, J.R. (2010) Analysis of the salt-stress response at cell-type resolution. *Plant, Cell & Environment*, 33(4), 543–551.

Dinneny, J.R. et al. (2008) Cell identity mediates the response of Arabidopsis roots to abiotic stress. *Science*, 320(5878), 942–945.

Emery, J.F. et al. (2003) Radial patterning of *Arabidopsis* shoots by class III HD-ZIP and KANADI genes. *Current Biology*, 13(20), 1768–1774.

Endo, S. et al. (2009) Identifying new components participating in the secondary cell wall formation of vessel elements in zinnia and Arabidopsis. *The Plant Cell*, 21(4), 1155–1165.

Fukuda, H. (2000) Programmed cell death of tracheary elements as a paradigm in plants. *Plant Molecular Biology*, 44(3), 245–253.

Galbraith, D.W. (2006) DNA microarray analyses in higher plants. *Omics*, 10(4), 455–473.

Gifford, M.L. et al. (2008) Cell-specific nitrogen responses mediate developmental plasticity. *Proceedings of the National Academy of Sciences of the United States of America*, 105(2), 803–808.

Grieneisen, V.A. et al. (2007) Auxin transport is sufficient to generate a maximum and gradient guiding root growth. *Nature*, 449(7165), 1008–1013.

Grierson, C. and Schiefelbein, J. (2002) Root hairs. In: C.R. Somerville and E.M. Meyerowitz (eds) *The Arabidopsis Book*. The American Society of Plant Biologists, Rockville, MD.

Harkins, K.R. et al. (1990) Expression of photosynthesis-related gene fusions is restricted by cell type in transgenic plants and in transfected protoplasts. *Proceedings of the National Academy of Sciences of the United States of America*, 87(2), 816–820.

Iyer-Pascuzzi, A.S. et al. (2010) Imaging and analysis platform for automatic phenotyping and trait ranking of plant root systems. *Plant Physiology*, 152(3), 1148–1157.

Jiao, Y. et al. (2009) A transcriptome atlas of rice cell types uncovers cellular, functional and developmental hierarchies. *Nature Genetics*, 41(2), 258–263.

Karkonen, A. et al. (2011) Plant tissue cultures. *Methods in Molecular Biology*, 715, 1–20.

Kiegle, E. et al. (2000) Cell-type-specific calcium responses to drought, salt and cold in the Arabidopsis root. *Plant Journal*, 23(2), 267–278.

Kim, S.A. and Guerinot, M.L. (2007) Mining iron: iron uptake and transport in plants. *FEBS Letters*, 581(12), 2273–2280.

Kim, S.A. et al. (2006) Localization of iron in Arabidopsis seed requires the vacuolar membrane transporter VIT1. *Science*, 314(5803), 1295–1298.

Lauter, F.R. et al. (1996) Preferential expression of an ammonium transporter and of two putative nitrate transporters in root hairs of tomato. *Proceedings of the National Academy of Sciences of the United States of America*, 93(15), 8139–8144.

Lee, J.Y. et al. (2006) Transcriptional and posttranscriptional regulation of transcription factor expression in Arabidopsis roots. *Proceedings of the National Academy of Sciences of the United States of America*, 103(15), 6055–6060.

Lee, M.M. and Schiefelbein, J. (1999) WEREWOLF, a MYB-related protein in Arabidopsis, is a position-dependent regulator of epidermal cell patterning. *Cell*, 99(5), 473–483.

Levesque, M.P. et al. (2006) Whole-genome analysis of the SHORT-ROOT developmental pathway in Arabidopsis. *PLoS Biol*, 4(5), e143.

Libault, M. et al. (2010) Root hair systems biology. *Trends in Plant Science*, 15(11), 641–650.

Licausi, F. (2011) Regulation of the molecular response to oxygen limitations in plants. *New Phytology*, 190(3), 550–555.

Long, T.A. et al. (2010) The bHLH transcription factor POPEYE regulates response to iron deficiency in Arabidopsis roots. *The Plant Cell*, 22(7), 2219–2236.

Mahonen, A.P. et al. (2000) A novel two-component hybrid molecule regulates vascular morphogenesis of the Arabidopsis root. *Genes & Development*, 14(23), 2938–2943.

Maser, P. et al. (2002) Altered shoot/root Na^+ distribution and bifurcating salt sensitivity in Arabidopsis by genetic disruption of the Na^+ transporter AtHKT1. *FEBS Letters*, 531(2), 157–161.

McConnell, J.R. et al. (2001) Role of *PHABULOSA* and *PHAVOLUTA* in determining radial patterning in shoots. *Nature*, 411(6838), 709–713.

Meyerowitz, E.M. (2002) Plants compared to animals: the broadest comparative study of development. *Science*, 295(5559), 1482–1485.

Miller, N.D. et al. (2007) Computer-vision analysis of seedling responses to light and gravity. *Plant Journal*, 52(2), 374–381.

Mockler, T.C. et al. (2005) Applications of DNA tiling arrays for whole-genome analysis. *Genomics*, 85(1), 1–15.

Moreno-Risueno, M.A. et al. (2010) Oscillating gene expression determines competence for periodic Arabidopsis root branching. *Science*, 329(5997), 1306–1311.

Mundy, J. et al. (1990) Nuclear proteins bind conserved elements in the abscisic acid-responsive promoter of a rice rab gene. *Proceedings of the National Academy of Sciences of the United States of America*, 87(4), 1406–1410.

Murray, J.D. (2011) Invasion by invitation: rhizobial infection in legumes. *Molecular Plant–Microbe Interactions*, 24(6), 631–639.

Mustroph, A. et al. (2009a) Isolation of plant polysomal mRNA by differential centrifugation and ribosome immunopurification methods. *Methods in Molecular Biology*, 553, 109–126.

Mustroph, A. et al. (2009b) Profiling translatomes of discrete cell populations resolves altered cellular priorities during hypoxia in Arabidopsis. *Proceedings of the National Academy of Sciences of the United States of America*, 106(44), 18843–18848.

Nawy, T. et al. (2005) Transcriptional profile of the Arabidopsis root quiescent center. *Plant Cell*, 17(7), 1908–1925.

Ohashi-Ito, K. and Fukuda, H. (2003) HD-zip III homeobox genes that include a novel member, ZeHB-13 (Zinnia)/ATHB-15 (Arabidopsis), are involved in procambium and xylem cell differentiation. *Plant and Cell Physiology*, 44(12), 1350–1358.

Petersson, S.V. et al. (2009) An auxin gradient and maximum in the Arabidopsis root apex shown by high-resolution cell-specific analysis of IAA distribution and synthesis. *The Plant Cell*, 21(6), 1659–1668.

Pfluger, J. and Wagner, D. (2007) Histone modifications and dynamic regulation of genome accessibility in plants. *Current Opinion in Plant Biology*, 10(6), 645–652.

Ren, Z. et al. (2010) RAS1, a quantitative trait locus for salt tolerance and ABA sensitivity in Arabidopsis. *Proceedings of the National Academy of Sciences of the United States of America*, 107(12), 5669–5674.

Sabatini, S. et al. (1999) An auxin-dependent distal organizer of pattern and polarity in the Arabidopsis root. *Cell*, 99(5), 463–472.

Schmitz, R.J. and Zhang, X. (2011) High-throughput approaches for plant epigenomic studies. *Current Opinion in Plant Biology* 14(2), 130–136.

Sun, F. et al. (2007) Salt modulates gravity signaling pathway to regulate growth direction of primary roots in *Arabidopsis thaliana*. *Plant Physiology*, 146(1), 178–188.

Ulmasov, T. et al. (1997) ARF1, a transcription factor that binds to auxin response elements. *Science*, 276(5320), 1865–1868.

Vanneste, S. and Friml, J. (2009) Auxin: a trigger for change in plant development. *Cell*, 136(6), 1005–1016.

Wee, C.W. and Dinneny, J.R. (2010) Tools for high-spatial and temporal-resolution analysis of environmental responses in plants. *Biotechnology Letters*, 32(10), 1361–1371.

Wu, M.F. et al. (2006) Arabidopsis microRNA167 controls patterns of ARF6 and ARF8 expression, and regulates both female and male reproduction. *Development*, 133(21), 4211–4218.

Xiong, L. and Zhu, J. (2002) *Salt Tolerance*, American Society of Plant Biologists, Rockville, MD.

Yadav, R.K. et al. (2009) Gene expression map of the Arabidopsis shoot apical meristem stem cell niche. *Proceedings of the National Academy of Sciences of the United States of America*, 106(12), 4941–4946.

Yamaguchi-Shinozaki, K. and Shinozaki, K. (1994) A novel *cis*-acting element in an Arabidopsis gene is involved in responsiveness to drought, low-temperature, or high-salt stress. *Plant Cell*, 6(2), 251–264.

5 Toward a Virtual Root: Interaction of Genomics and Modeling to Develop Predictive Biology Approaches

Julien Lavenus, Leah Band, Alistair Middleton, Michael Wilson,
Mikael Lucas, Laurent Laplaze and Malcolm Bennett

Introduction

Our understanding of root biology has advanced dramatically in the last twenty years, thanks largely to the use of molecular genetic approaches, the completion of the *Arabidopsis* genome and the experimental tools and resources that have resulted. Genetic studies in *Arabidopsis* have, and will continue to play, a vital role in the molecular dissection of root biology. However, the field has recently moved beyond studying the individual gene products, focusing instead on determining the functional relationships between multiple components of regulatory pathways that control root biology. Mathematical modelling approaches are set to become crucial as networks become increasingly complex and their behaviour and outputs less intuitive. In this Chapter, we first review how genetics and genomics were used to build small gene regulatory networks (GRN), and then describe how mathematical and computational models can probe the biological significance of their well-characterised regulatory relationships. Next, we discuss how computational approaches can also be used to infer new GRNs. Finally, we review examples describing how small GRNs can be studied in the more realistic context of root tissue(s) using multi-scale models that span network, cellular, tissue and organ scales.

Assembling Root Gene Regulatory Pathways Using Genomics

Before the millennium, root gene regulatory networks (GRN) were elucidated primarily employing genetic and molecular biology techniques. During the last decade, genomic approaches such as transcriptomics have come to the fore. The main technology used to profile global gene expression levels has been oligonucleotide-based gene arrays (Affymetrix Chips; www.nasc.org). Arrays have been very useful for monitoring root responses to various stimuli on a whole organ basis (http://bar.utoronto.ca/efp/cgi-bin/efpWeb.cgi).

The integration of large-scale gene expression datasets with existing genetic information has been used to assemble several important regulatory pathways that govern root development. For example, Okushima et al. (2005) identified two key members of the *AUXIN RESPONSE FACTOR (ARF)* gene family, *ARF7* and *ARF19*, that were necessary for auxin-regulated root gravitropism and lateral root development. Although the *arf7* single mutant showed only a weak root phenotype, the double

Root Genomics and Soil Interactions, First Edition. Edited by Martin Crespi.

knockout *arf7 arf19* developed almost no lateral roots and was agravitropic. In accordance with this observation, a transcriptomic analysis performed on the single and the double mutants subsequently showed that a significant proportion of auxin-regulated genes were redundantly targeted by both ARF7 and ARF19 in roots. Several *LATERAL ORGAN BOUNDARY (LOB/LBD)* transcription factors encoding genes were identified as targets of *ARF7* and *ARF19*. The same authors went on to show that ARF7 and ARF19 directly controlled the expression of *LBD16* and *LBD29* in roots to regulate lateral root formation (Okushima et al. 2007).

Expression analysis of specific cell types and developmental stages is also possible in *Arabidopsis* roots (Figure 5.1). Several techniques, such as laser capture microdissection, have been employed (Kerk et al. 2003). The most widely used technique relies on sorting marked cells (usually with GFP) using a fluorescence-activated cell sorter (FACS). Transgenic lines expressing GFP in a

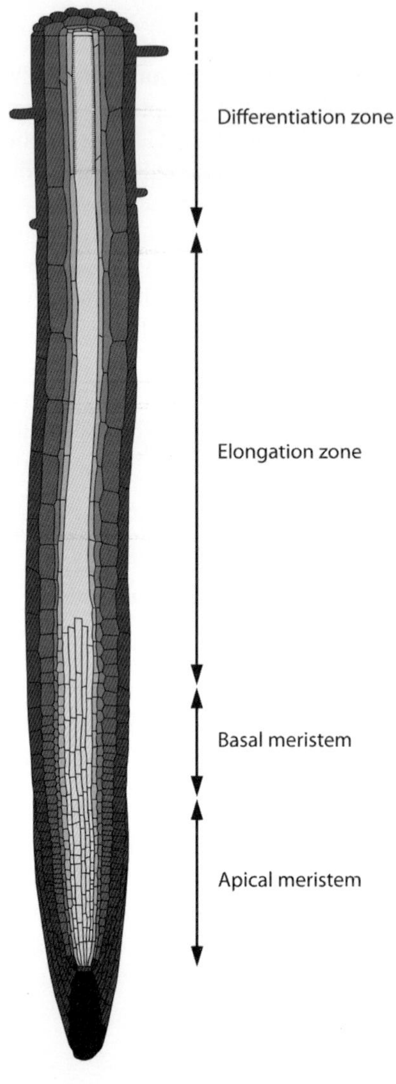

Figure 5.1 Schematic of tissue organization and developmental zones within the *Arabidopsis* root apex.

tissue-specific manner are used to prepare protoplasts for FACS sorting, then total RNA is isolated (Birnbaum et al. 2003). A third method involves expressing a tagged ribosomal protein behind a tissue-specific promoter. Immunoprecipitation of ribosomes from such transgenic plants should yield mRNA that was in the process of being translated (Zanetti et al. 2005) and hence has the advantage of profiling only the actively translated transcripts. These high spatial resolution datasets have proved invaluable to define a GRN for the root stele (Brady et al. 2011; Figure 5.1).

One of the most promising strategies for assembling GRNs is to perturb a network by engineering a loss of function, gain of function, or mis-expression of one or more genes and then combine the sets of transcriptomic results to infer other network components. Such an approach has been used to elucidate the pathway regulated by the SHORT ROOT (SHR) transcription factor (Levesque et al. 2006). In this study, the authors identified targets of SHR by combining transcriptomic datasets generated by profiling null mutants and lines expressing an inducible SHR fusion protein (fused to the steroid-regulated glucocorticoid receptor). These targets were then validated as direct targets of SHR and included several components of hormonal and cell cycle pathways (Levesque et al. 2006; Sozzani et al. 2010).

One of the key challenges to building GRNs lies in defining and confirming the *cis*-acting regulatory elements that exist in gene promoters of transcription factor targets. Recent genome-aided investigations of the developmental pathways have begun to provide insights into how useful this can be (Won et al. 2009). This approach is highly dependent on the knowledge of the consensus binding motifs for the transcription factors, which are currently lacking in *Arabidopsis*. High-throughput genomic approaches such as ChIP-seq are currently filling in this knowledge gap, but the large number of known and putative transcription factors suggests that in silico scans with subsequent validation remains the approach generally employed in this field.

Modeling Well-Characterized Small Root Gene Regulatory Networks

The advent of molecular genetics and genomics has enabled many of key regulators of root development to be identified (Péret et al. 2009; Stahl and Simon 2010; Sablowski 2011). These components will necessarily form part of a larger network of interactions. In this section, we will describe how mathematical and computational models are being employed to probe the biological significance of these interactions in increasingly complex GRNs.

Abscisic acid (ABA) is a key hormone regulating root growth and responses to biotic and abiotic stresses. ABA signals by binding to a family of receptors termed PYR/PYL/RCAR proteins that promote the formation of a complex with several protein phosphatase type 2C (PP2C), which activates ABA responses. PYR/PYL/RCAR family members either bind ABA as a monomer or dimer. Dupeux et al. (2011) recently developed a computational model capturing this relatively simple network of interactions (Figure 5.2a) to probe the response to ABA when both monomeric and dimeric PYR/PYL/RCAR receptor are competing for ABA and P2C molecules. Intriguingly, the study revealed that monomeric receptors have a competitive advantage for binding, particularly at lower ABA concentrations. Hence, a tissues receptor composition and their oligomerization properties are likely to impact ABA responsiveness.

The phytohormone auxin regulates the key cellular processes of division, elongation and differentiation in roots (Péret et al. 2009; Stahl and Simon 2010; Sablowski 2011). Middleton et al. (2010) recently developed a mathematical model of the auxin response network employing ordinary differential equations (ODEs). The model captures all key auxin response network components. These include auxin, its receptor protein TIR1; ARF transcription factors; and interacting AUX/IAA

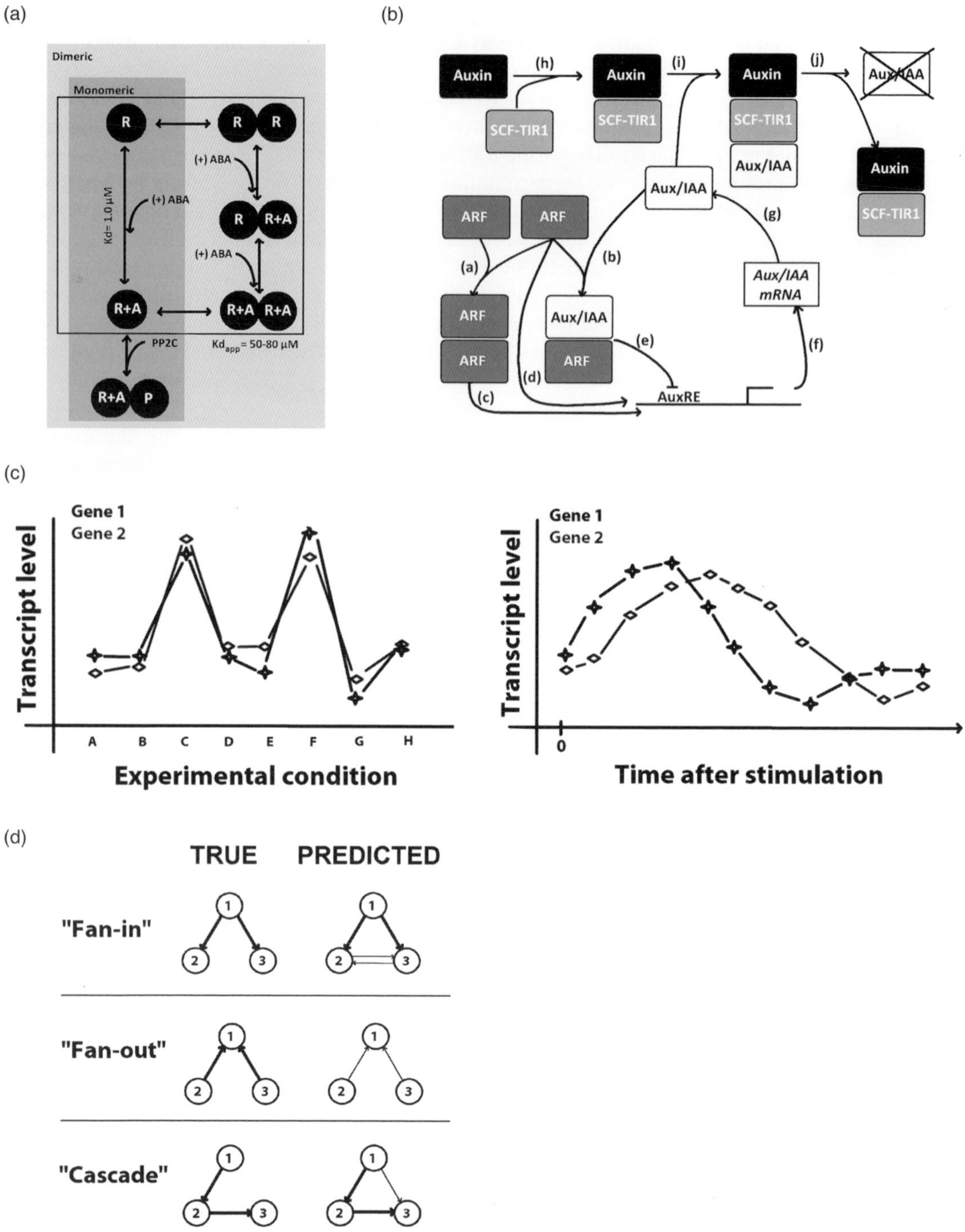
(a)
Dimeric
Monomeric
R
R R
(+) ABA
Kd= 1.0 μM
(+) ABA
R R+A
(+) ABA
R+A
R+A R+A
PP2C
Kd$_{app}$= 50-80 μM
R+A P
(b)
Auxin
(h)
Auxin
(i)
Auxin
(j)
Aux/IAA
SCF-TIR1
SCF-TIR1
SCF-TIR1
Aux/IAA
Auxin
SCF-TIR1
Aux/IAA
ARF
ARF
(a)
(b)
(g)
Aux/IAA mRNA
ARF
ARF
Aux/IAA
ARF
(e)
(f)
(d)
(c)
AuxRE
(c)
Gene 1
Gene 2
Transcript level
A B C D E F G H
Experimental condition
Gene 1
Gene 2
Transcript level
0
Time after stimulation
(d)
TRUE
PREDICTED
"Fan-in"
"Fan-out"
"Cascade"
1
2
3

repressor proteins, together with their known regulatory relationships (Figure 5.2b). Auxin functions in part by mediating the degradation of the Aux/IAA family of proteins. Aux/IAA proteins regulate various downstream genes that define the auxin responsive cellular processes described earlier. Aux/IAA proteins also repress auxin-inducible *Aux/IAA* gene expression by binding to ARFs and so the associated pathway comprises multiple negative feedback loops that are disrupted by auxin mediating the turnover of Aux/IAA proteins. This constitutes a nonlinear GRN. While one can revert to direct experiments to study the behavior of such a system, it is not always clear whether the known interactions are sufficient to explain a particular set of observations. For example, the model developed by Middleton et al. (2010) revealed some interesting discrepancies between its behavior and the experimental data. In experiments, fold changes in some of the mRNAs can be up to 30-fold (Abel and Theologis 1996). In the model, to achieve this, it requires an extremely large (and possibly unrealistic) fold change in auxin levels. This points to the existence of additional regulatory mechanisms that might amplify the response (as measured by *Aux/IAA* mRNA levels) made by cells to changes in auxin. A number of ARF-encoding genes are themselves auxin responsive (Okushima et al. 2005). It has recently been demonstrated by Lau et al. (2011) that ARF family member MONOPTEROS (MP) upregulates both its own expression and the expression of the gene encoding Aux/IAA family member BODENLOS (BDL). Thus, the network consists of an ARF-based positive feedback loop coupled to the Aux/IAA negative feedback.

The presence of a positive feedback loop can lead to bistability in a model. This corresponds to a system that can rest in any of two stable states, rather than just one. In this case, increasing levels of exogenous auxin can switch the system from a low-MP/BDL concentration steady-state to a high one. This switch-like behavior can be viewed as a form of signal amplification. Another hallmark of a bistable system is hysteresis. In the case of MP-BDL, we can imagine the following scenario. As auxin levels increase, the *MP* gene is upregulated and the system approaches its high-MP/BDL steady state. However, as auxin levels decrease, MP and BDL levels will decrease much more slowly (i.e., the MP positive feedback loop is self-sustaining) and in effect remain upregulated for several hours. Moreover, the system can generate switch-like responses to changes in auxin concentration, so that only over a certain threshold could an MP-dependent developmental program be initiated.

Figure 5.2 Gene regulatory networks: modeling, inference, and errors. (**a**) Model for ABA signaling through the family of PYR/PYL receptors. The receptors (R) can be found either in a monomeric or dimeric form. The capacity of the receptors to form dimers varies among the family members. In both forms, the receptors can bind ABA (R+A). When bound to ABA, a monomeric receptor can bind and inhibit certain PP2C phosphatases (P), which triggers the ABA response. The competition for ABA binding between the monomeric and dimeric forms of the receptors generates various degrees of ABA sensitivity depending on the expression level of each member of the PYR/PYL family. (**b**) Model for auxin signaling through Aux/IAA proteins degradation. The transcription factors of the auxin response factors (ARF) family may form dimers with themselves (a) or with Aux/IAA repressor proteins (b). ARFs have the ability to bind auxin response element (AuxRE) in the promoter of primary auxin responsive genes. Monomeric (c) or dimeric (d) ARFs activates transcriptions, whereas ARF-Aux/IAA complexes (e) repress transcription (f). Following translation (g), the accumulation level of Aux/IAA proteins and therefore the transcription level of primary auxin responsive genes is regulated through an auxin-dependent protein degradation mechanism. Auxin can bind the SCF-TIR1 complex (h) and act as a "molecular glue" between the SCF-TIR1 complex and Aux/IAA proteins (i). This interaction causes Aux/IAA to be polyubiquitinated by SCF-TIR1, which signals them for degradation by the 26S-proteasome (j). (**c**) Principle of GRN inference. Genes showing similar expression profiles may be regulating each other or be coregulated. Interactions can be inferred from steady-state expression data (left) or time-series expression data (right). Time delay during time course help determine the directionality of the interactions. Though no such information is available for steady-state expression data, it is still possible in some cases to infer the directionality of the interactions. (**d**) Systematic errors made by inference algorithm (Inspired from Marbach et al. 2011). On the left, the true structure that inference algorithm seek to infer. On the right, the structure of the network as predicted by the inference algorithm. The confidence of the predictions is indicated by the width of the edges.

Once the switch has been activated (and the system is in a MP upregulated state), the network's behavior will be robust to large changes in the level of auxin. Hence, modeling has helped reveal how auxin gradients can be read to induce binary developmental outputs.

Another increasingly important aspect of plant developmental biology is the study of cross-regulation between two or more hormones. Perhaps the most notable example of this is the interplay between auxin and cytokinin, which control the size of the root meristem in an antagonistic manner (Ioio et al. 2008). Local accumulation of auxin promotes root cell division, whereas high cytokinin concentrations promote differentiation. To study the interplay between auxin and cytokinin signaling, Muraro et al. (2011) developed an ODE-based mathematical model of the known interactions. In addition to the auxin-mediated upregulation of *Aux/IAA* genes (notably *SHY2*), auxin is also assumed to upregulate the transcription of genes encoding ARR-A that are inhibitors of cytokinin response. In contrast, cytokinin upregulates the expression of an ARR-B transcription factor, which in turn can activate transcription of *SHY2*. In silico perturbations to the model were found to predict experimental changes in cell division observed in both the gain-of-function and the loss-of-function *shy2* mutants, suggesting that the network had accurately captured the regulatory mechanisms controlling meristem size.

In the preceding examples, the models developed are typically found to be consistent with the available data. However, this need not always be the case. Discrepancies between the model and the data can indeed point to the existence of interactions or components that are missing from the network. This can in fact be the most valuable scenario to help an experimentalist pinpoint missing components from their GRNs, provided a suitable candidate mechanism can be identified and tested for.

Building New Large-Scale Root Gene Regulatory Network

To date, GRNs have been characterized mainly using genetic and genomic approaches. Although these approaches have enabled the discovery of many important genes in the GRN underlying various biological processes of interest, they are not suitable for large-scale network identification. Instead, these approaches explore GRN in a gene-by-gene manner.

Two distinct methods have been employed to elucidate large-scale root GRN. The first one is based on a direct approach involving a systematic experimental screen of pairwise interactions among a carefully selected subset of potentially interacting molecular species. The second one employs an indirect approach that exploits the recent availability of large transcriptomic datasets to infer GRN by running novel statistical analysis tools and by identifying regulatory relationships between tens to many thousands of genes. In this next section, we review both approaches using recent examples.

Building Large-Scale Root GRN Using Systematic Screens for Interactions

Brady et al. (2011) recently unraveled the GRN controlling the identity of *Arabidopsis* stele tissues, the innermost tissues of the root. The authors made use of a high-resolution spatiotemporal root gene expression map (Brady et al. 2007) to select 167 stele-enriched TF-encoding genes. Subsequently, they first carried out a yeast-1-hybrid (Y1H) screen using these 167 TF as preys and the promoters of 65 out of the 167 genes as bait. They also included in their screen the promoters of 28 miRNA that target some of the selected TF. Next, they completed their network with a yeast-2-hybrid (Y2H) screen to identify protein–protein interactions between the 167 TF.

The two screens allowed them to build a network showing 66 protein–DNA interactions and 25 protein–protein interactions. About 60% of the protein–DNA interaction could be validated *in planta* by ChIP-qPCR. In order to determine whether the protein–DNA interactions they detected by Y1H and ChIP-PCR resulted in positive or negative effects on transcription, they next analyzed the change in the transcript accumulation level of the predicted targets in lines knockdowned or overexpressing the putative regulator. This approach allowed them to quantify the strength of the interactions and identify the strongest regulators for each target.

By adopting this systematic screening approach, the authors successfully recovered the core of the GRN controlling stele tissues identity and quantitatively measured the strength of the interactions in their network. While this method for GRN identification is very time-consuming and yields a partial GRN, it provides an excellent experimentally validated framework to undertake later more detailed analysis of network topology through further lab-based and *in silico* modeling studies.

Building Large-Scale Root GRN Using Statistical Inference Algorithms

Besides the direct approach adopted by Brady et al. (2011), indirect approaches based on statistical analysis of transcriptomic data are also being employed. Several algorithms based on various types of mathematical formalisms have been developed over the last decade to infer GRN topology from transcriptomic data (for a detailed review and classification of the algorithms, see De Smet and Marchal 2010; for a comparison of their performances on in silico data, see Marbach et al. 2010).

Inferred GRN can be represented in the form of graphs composed of nodes featuring the genes and edges representing the regulatory interactions between the genes. Directed graphs are composed of edges whose directionality is determined, that is, it is known which of the two interacting genes is upstream of the other. On the contrary, graphs lacking this directionality information are called undirected graphs. Similarly, cyclic graphs contain feedback loops in contrast to acyclic graphs.

Two types of transcriptomic data can be used to infer GRN. Steady-state expression data (e.g., mutant vs. wild type or drug treated vs. mock treated) or time-series expression data following experimental induction of the system (e.g., datasets for cell cycle, organ formation, response to hormone treatment; Figure 5.2c).

Clustering Analyses

The first analyses to be done on huge transcriptomic datasets are very often clustering analyses (Eisen et al. 1998; Bansal et al. 2007). Hierarchical clustering uses a pairwise measure of distance between gene expression profiles to compute distance trees (dendrograms) that delimit clusters of coexpressed genes. The most commonly used measures of distance are Pearson's correlation (Eisen et al. 1998) and the Euclidian distance (Bergkvist et al. 2010). Various methods to construct dendrograms from distance matrices exist.

These analyses do not predict the topology of the GRN sensu stricto but enable one to group genes into clusters of genes showing similar expression profiles. The idea behind the clustering approach is that genes showing a similar expression profiles are likely to be coregulated. Though it can help identify regulatory modules, the clustering methods do not yield any information about the regulatory relationship between the genes inside a cluster. If two genes cluster together, they could either be both targets of another third gene or one of them could be directly or indirectly regulating the other.

Correlation-Based Inference Methods

Unlike clustering approaches, correlation network reconstruction algorithms aim to infer the regulatory relationships between correlated genes. The simplest way of building a correlation network consists in first calculating all pairwise Pearson's correlation between expression profiles and second determining the edges of the network through a thresholding procedure. However, these simple correlation networks analyses have exactly the same drawbacks as clustering analyses and are not suitable to properly infer GRN. To infer *causal* networks using correlation analyses, several methods have been proposed, which are all based on the computation of partial correlations (de la Fuente et al. 2004; Opgen-Rhein and Strimmer 2007a, 2007b; Ma et al. 2007; Yuan et al. 2011). The partial correlation can be regarded as the correlation that remains between two variables if the effect of other variables has been regressed away.

Information Theory-Based Inference Methods

Like correlation based methods, information theory (IT) based methods need a measure of distance (or similarity) between gene expression profiles to be computed. However, while the correlation based methods use Pearson's correlation as measurement of similarity, IT based methods rather use mutual information. (MI; Steuer et al. 2002; Bansal et al. 2007). MI quantifies the level of dependence of pairs of variables from the estimated joint probability density. Unlike Pearson's correlation, MI is able to detect nonlinear relationships between variables. The method, which was initially developed to solve inference problems from steady-state expression data (ARACNE; Margolin et al. 2006) was later adapted to infer GRN from time-series data (Time-Delay ARACNE, Zoppoli et al. 2010). ARACNE only produces undirected graphs. However, TDARACNE is able to assign directionality to the edges by taking into account time delays between expression profiles.

Bayesian Statistics-Based Inference Methods

Generally speaking, a Bayesian network is a probabilistic graphical model representing a set of random variables and their conditional dependencies in the form of an acyclic directed graph (Friedman et al. 2000; Bansal et al. 2007; Needham et al. 2009). When using Bayesian approaches to solve an inference problem, one therefore implicitly assumes that the network to be inferred is deprived of any feedback loops.

In a Bayesian network, the nodes directly upstream of another are called its parents, whereas the nodes directly downstream are called its children. Parenthood cannot be interpreted primarily as direct causality. Instead, it should be regarded in terms of conditional dependence. For example, if both genes X and Y are parents of gene Z, it only means that X and Y expression levels, Z expression level becomes statistically independent of every other variables in the network. Therefore, learning the structure of a Bayesian network basically consists in determining for each variable X_i, the minimum set of parents $\{X_j, \ldots ,X_{j+p}\}$ that "shields" it from every other variable (Friedman et al. 2000). In order to learn the structure of the network that best explains the data, the various possible graphs are evaluated using a scoring function. This is achieved using Bayes' formula, which may incorporate prior knowledge. Given the huge size of the space of possible graphs, it is computationally impossible to fully explore it. Therefore, a heuristic approach is used to search for high-scoring graphs. As a consequence, there is no guarantee that the result of the search is optimal. Very often several high-scoring graphs may be found to be equally able to explain the data. In this case, the graphs can be averaged. This generates a partially directed acyclic graph whenever the various models do not agree about the directionality of an edge.

Bayesian approaches were initially developed to infer GRN from steady-state expression data (Friedman et al. 2000). They were later extended to solve inference problems on time-series expression data (Dynamic Bayesian Network, Perrin et al. 2003; Yu et al. 2004; Morrissey et al. 2010, 2011). Unlike Bayesian Networks, Dynamic Bayesian Networks are not necessarily acyclic, that is, they may contain feedback loops (Yu et al. 2004; Bansal et al. 2007; Morrissey et al. 2010, 2011).

Ordinary Differential Equations-Based Inference Methods
Several GRN inference algorithms are based on systems of linear ODEs. In these algorithms, a system of ODEs is used to model the evolution of the gene transcripts concentrations over time as a function of themselves. The various coefficients of the equations give quantitative estimation of the effect of each of the genes on the expression of the others and can therefore be used to reconstruct the network. ODEs-based inference algorithms are used on both steady-state data (Gardner et al. 2003; di Bernardo et al. 2005) and time-series data (Bansal et al. 2006).

Systematic Errors Made by the Inference Techniques and Other Limitations
One of the problems encountered when using inference approaches is the assessment of the performance of the algorithm. Indeed, our current knowledge on GRN is very fragmented and therefore cannot be of much help to assess the validity of the predictions. An alternative way to test the performance of inference algorithms consists in running them on *in silico* data generated from biologically plausible random network structure. For example, Marbach et al. (2010) used *in silico* generated data (Marbach et al. 2009) to compare the performance of 29 algorithms submitted to them for the DREAM3 challenge. One of the interesting results brought by this comparative study is that no particular formalism seems to generally solve better the inference problem than the others. This suggests that the way the algorithm is implemented is more important than the formalism chosen. Sadly, the authors also noticed that a surprisingly high number of methods did not perform better in average than random guessing, which they imputed to the lack of experience in the field of network inference of many competitors. The algorithm that was found to be the best one combines methods exploiting both steady-state and time-series expression data (Yip et al. 2010), underlying the importance of combining different type of data to properly infer GRN. Noticeably, Marbach et al. showed that "community predictions" combining the results of all the algorithms are very often more reliable and more robust than individual predictions. However, whatever the method chosen, systematic prediction errors are made (to various extents, depending on the method; Figure 5.2d). The "fan-in" error corresponds to the tendency not to find combinatorial regulation ("Genes A and B → Gene C"). The "fan-out" error consists in misinterpreting coregulations as interactions ("Gene A → B and C" yields "B →C" or "C →B"). Finally, the "cascade" error consists of introducing shortcuts in regulatory cascades ("A → B → C" yields "A→C").

Many methods for inferring regulatory relationships between genes assume that these relationships are approximately linear, which may not always be the case in particular concerning combinatorial relationships. A loss of linearity in combinatorial relationships could at least partly explain the systematic "fan-in" errors highlighted by Marbach et al. (2010). Up to now, inference techniques have mainly been tested on datasets generated from unicellular organisms such as *Escherichia coli*, the yeast *Saccharomyces cerevisiae*, or from mammalian cells. However recently, a few studies have been published, which aimed at reconstructing the *Arabidopsis* circadian clock network from microarray data using various methods (Ma et al. 2007; Needham et al. 2009; Morrissey et al. 2010, 2011). Network inference studies are currently being performed on *Arabidopsis* root time-course datasets at the Centre for Plant Integrative Biology (CPIB; www.cpib.ac.uk).

Most of the algorithms described above perform well on relatively small datasets (up to 100 genes). Interestingly, Bonneau et al. (Bonneau et al. 2006) managed to reconstruct a GRN at the genome scale level in the archaeon *Halobacterium*. Though network predictions by inference algorithms need to be validated using experimental approaches (Brady et al. 2011), they promise to provide a very useful tool to help biologists initially identify biologically relevant interactions with limited recourse to time-consuming and often expensive lab-based experiments.

Multi-Scale Modeling Approaches to Study Root Growth and Development

Although analyses of GRNs have provided insights into the control of root development, a full understanding requires one to consider the interplay between the subcellular processes and many other processes that occur over multiple spatial scales. One must consider the feedback between these signaling networks and hormone transport, biomechanics, and growth on the cell scale; how these cell-scale processes create tissue-scale variations in hormone concentrations and growth rates; and how these tissue-scale distributions determine root phenotype. The interplay between these processes is extremely complicated and an integrative systems approach is essential to understand the underlying biological mechanisms and determine how the system behaves as a whole.

Many multi-scale models of root development have focused on the role of the plant hormone auxin. Auxin moves through the root in a polar manner, facilitated by the spatial distribution of influx and efflux carriers (Figure 5.3), which increase the auxin flux across specific cell membranes

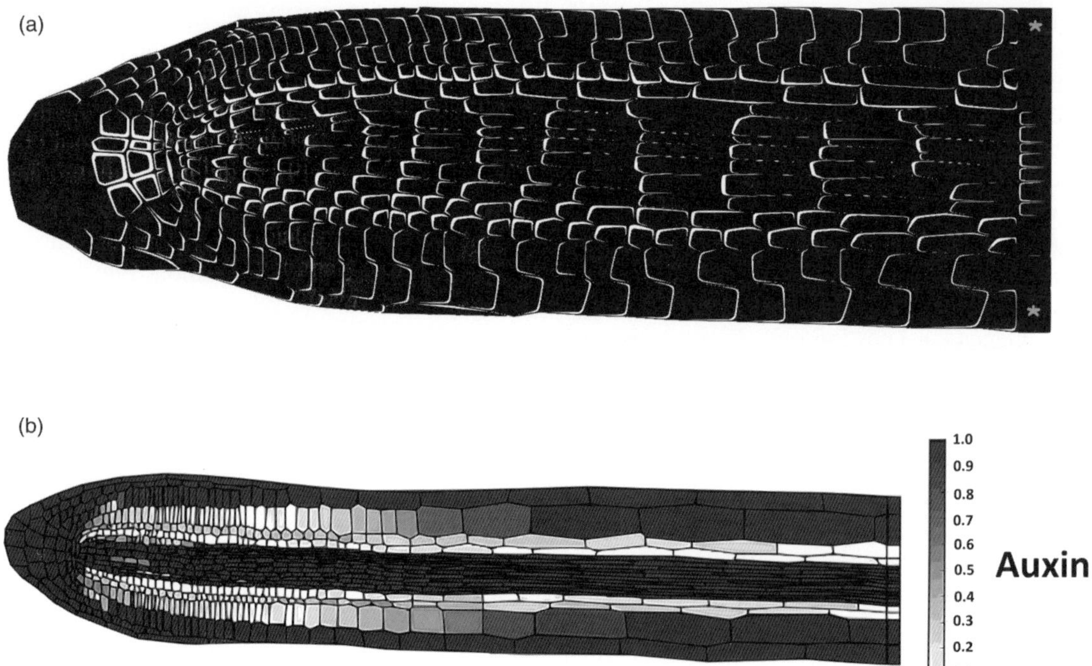

Figure 5.3 Simulation of auxin distribution within *Arabidopsis* root apical cells. (a) Optical slices of *Arabidopsis* roots imaged in confocal microscopy were digitized and used as a vertex-based graph to simulate the dynamic of auxin fluxes within the root tissue. Auxin influx and efflux carriers (represented with the lighter color) were distributed within the virtual root tissues according to previous studies (Stoma et al. 2008). (b) The gray scale indicates the relative amount of auxin within each root cell after running the simulation.

(Kramer and Bennett 2006). A series of modeling studies have investigated how these cell-scale fluxes produce a distinctive tissue-scale auxin distribution, and we shall discuss a few key examples here.

Swarup et al. (2005) described the first multi-scale model of root growth focused on auxin's role during gravitropism. By developing a cell-based model of auxin transport in the outer three layers of the root with the pattern of influx and efflux carriers (as shown in Figure 5.3a), Swarup et al. (2005) showed that the presence of influx carriers in the epidermis created high auxin concentrations in this tissue, resulting in the epidermal efflux carriers creating a shootward flux through these layers. This shootward auxin flux is essential to mobilize a lateral auxin gradient, generated in gravity-sensing columella cells at the root tip, to the elongation zone, where the root bends. The model provided insight into the distribution of auxin in root tissues and helped direct experimental studies that demonstrated that the epidermis represents the elongation zone tissue that was critical for auxin-dependent differential root growth.

Another key development in multi-scale modeling of auxin transport in roots was the study by Grieneisen et al. (2007), which featured a cell-based model in which all the tissue layers in the root elongation zone and meristem were considered and auxin-dependent cell division and cell growth were incorporated. The model demonstrated a spatial gradient of auxin concentration, which the authors suggest provides cells with information about their location, enabling controlled passage between the division phase in the meristem and rapid growth in the elongation zone. The model explained the maintenance of a high auxin concentration at the quiescent center, which is thought to specify these cells.

As the studies described earlier demonstrate, multi-scale modeling can be an excellent tool for understanding the underlying biological mechanisms governing root growth and development. However, there are many future challenges. To date, most studies have exploited the regular multicellular geometry of the root to approximate cell geometries as regular rectangles (Swarup et al. 2005; Grieneisen et al. 2007; Laskowski et al. 2008; Mironova et al. 2010; Twycross et al. 2010). While this is a reasonable approximation in the root elongation zone, cell shapes in the meristem and emerging lateral roots are irregular and accurately representing these shapes will improve predictions of auxin distribution in these regions. The use of modeling frameworks like OpenAlea (Pradal et al. 2008) greatly assists a researcher's ability to capture realistic root tissue geometries and program cells to contain key signaling components or transport properties (Figure 5.3a). Recent auxin-transport models of the primary root (Stoma et al. 2008) and Frankia-infected root nodules (Perrine-Walker et al. 2010) have shown the advantages of representing root tissues with realistic cell geometries (Figure 5.3b). Another criticism of many current auxin-transport models is the use of static cell geometries. Although some studies of the primary root suggest that this does not affect model predictions (Grieneisen et al. 2007; Band and King 2011), this is an area that requires more attention, especially in complex evolving geometries such as in emerging lateral roots.

Conclusions and Future Challenges

Our understanding of root biology has dramatically advanced in the last 20 years, thanks largely to genetic approaches in the model plant *Arabidopsis*. During the last several years, the field has moved beyond studying the functions of individual gene products, focusing instead on the relationships between multiple components that compose macromolecular complexes or GRNs that control root biology. High-throughput genomic technologies generating datasets such as high-resolution spatiotemporal transcript maps of individual cell types and developmental zones in the root (Birnbaum et al. 2003; Brady et al. 2007) have proved critical, as demonstrated by Brady et al. (2011).

Computer and mathematical approaches to model root regulatory pathways are set to become much more important in the near future. Modeling approaches are being used to build GRNs (reviewed in Section "Building New Large-Scale Root Gene Regulatory Network"). Nevertheless, network inference remains a challenging task, in part because the number of interactions to infer is typically far higher than the number of available independent samples. Several ways exist to circumvent this problem (De Smet and Marchal 2010). The first consists in restricting the inference problem to a subset of selected genes. Another method consists in identifying modules of coregulated genes (Reiss et al. 2006) and working on the relationships between these modules (Bonneau et al. 2006) and not the individual gene they contain or studying each module individually (Segal et al. 2003). Inference approaches are also limited by the fact that transcriptomic data only offer a narrow view of the state of the system. Many relevant processes controlling gene expression remain hidden, such as post-transcriptional, translational, post-translational, and chromatin-based controls. In addition, many of the GRN algorithms developed so far use linear models, which might not be well suited to describe gene interactions particularly in the case of combinatorial regulations. Therefore, one of the current challenges of system biology independently of inference methods development consists of precisely understanding how combinatorial regulation impacts gene expression.

Computer and mathematical models are also ideally suited to study the behavior of root regulatory pathways as these networks become increasingly complex and their outputs less intuitive (see examples in Section "Modeling Well-Characterized Small Root Gene Regulatory Networks"). For example, the auxin response pathway contains a number of feedback loops (Figure 5.2b), making it difficult for a researcher to predict response behavior. Models also promise to help test hypotheses that, when simulations disagree with experimental data. This can helped identify important missing components or regulatory relationships within a network (Lau et al. 2011). Knowledge of the GRN structure along with quantitative data about the interactions it contains could enable systems biologists on the one hand to predict the system response to untested conditions and on the other hand to predict modifications in the GRN that could enhance a given biologically meaningful trait, such as root architecture and tolerance to abiotic or biotic stresses. Ultimately, system biologists aim to integrate individual models for these processes into a multi-scale model for root development in the form of a "virtual root" (www.cpib.ac.uk).

To create a "virtual root," it will also be essential to embed GRN models in a multicellular context. Regulatory pathways do not often function within just a single cell type but operate across several tissues to control root development (Swarup et al. 2008). Hence, models of transcellular GRN should be studied in a realistic spatial context, factoring in the geometry of the cells and tissues in which they operate while capturing the spatial distribution of key response and output gene products.

In addition to capturing spatiotemporal information, modeling root development in the future will necessitate that the biomechanics of cell division and tissue growth are also factored in. Much more work is needed to understand the mechanisms of growth regulation. We lack crucial knowledge of how the established signaling networks regulate the cell biomechanics and how such changes manifest on the organ scale. Plant biomechanics is an active area of research (see Baskin 2005; Winship et al. 2010 for reviews). It is generally thought that reconstruction of signaling networks will show hormones to affect the levels of remodeling enzymes (e.g., XTH, expansins), which will loosen the cell wall, enabling cells to grow. The work of Chavarrıa-Krauser et al. (2005) used this idea to model the root as a growing cell file, incorporating phenomenological relationships between auxin, cell-wall remodeling enzymes, and growth. However, accurately describing the biomechanical processes and incorporating them into multi-scale models remains a challenge for the future.

References

Abel, S. and Theologis, A. (1996) Early genes and auxin action. *Plant Physiology*, 111, 9.

Band, L.R. and King, J.R. (2011) Multiscale modeling of auxin transport in the plant-root elongation zone. *Journal of Mathematical Biology* (in press).

Bansal, M. et al. (2007) How to infer gene networks from expression profiles. *Molecular Systems Biology*, 3, 78.

Bansal, M. et al. (2006) Inference of gene regulatory networks and compound mode of action from time course gene expression profiles. *Bioinformatics*, 22, 815.

Baskin, T.I. (2005) Anisotropic expansion of the plant cell wall. *Annual Review of Cell and Developmental Biology*, 21, 203–222.

Bergkvist, A. et al. (2010) Gene expression profiling—clusters of possibilities. *Methods*, 50, 323–335.

di Bernardo, D. et al. (2005) Chemogenomic profiling on a genome-wide scale using reverse-engineered gene networks. *Nature Biotechnology*, 23, 377–383.

Birnbaum, K. et al. (2003) A gene expression map of the Arabidopsis root. *Science*, 302, 1956–1960.

Bonneau, R. et al. (2006) The Inferelator: an algorithm for learning parsimonious regulatory networks from systems-biology data sets de novo. *Genome Biology*, 7, R36.

Brady, S.M. et al. (2011) A stele-enriched gene regulatory network in the Arabidopsis root. *Molecular Systems Biology*, 7, 459.

Brady, S.M. et al. (2007) A high-resolution root spatiotemporal map reveals dominant expression patterns. *Science*, 318, 801–806.

Chavarría-Krauser, A. et al. (2005) Primary root growth: a biophysical model of auxin-related control. *Functional Plant Biology*, 32, 849–862.

Dupeux, F. et al. (2011) A thermodynamic switch modulates abscisic acid receptor sensitivity. *The EMBO Journal*, 30, 4171–4184.

Eisen, M.B. et al. (1998) Cluster analysis and display of genome-wide expression patterns. *Proceedings of the National Academy of Sciences*, 95, 14863–14868.

Friedman, N. et al. (2000) Using Bayesian networks to analyze expression data. *Journal of Computational Biology*, 7, 601–620.

de la Fuente, A. et al. (2004) Discovery of meaningful associations in genomic data using partial correlation coefficients. *Bioinformatics*, 20, 3565–3574.

Gardner, T.S. et al. (2003) Inferring genetic networks and identifying compound mode of action via expression profiling. *Science*, 301, 102.

Grieneisen, V. et al. (2007) Auxin transport is sufficient to generate a maximum and gradient guiding root growth. *Nature*, 449, 1008–1013.

Ioio, R.D. et al. (2008) A genetic framework for the control of cell division and differentiation in the root meristem. *Science*, 322, 1380.

Kerk, N.M. et al. (2003) Laser capture microdissection of cells from plant tissues. *Plant Physiology*, 132, 27.

Kramer, E.M. and Bennett, M.J. (2006) Auxin transport: a field in flux. *Trends in Plant Science*, 11, 382–386.

Laskowski, M. et al. (2008) Root system architecture from coupling cell shape to auxin transport. *PLoS Biology*, 6, e307.

Lau, S. et al. (2011) Auxin triggers a genetic switch. *Nature Cell Biology*, 13, 611–615.

Levesque, M.P. et al. (2006) Whole-genome analysis of the SHORT-ROOT developmental pathway in Arabidopsis. *PLoS Biology*, 4, e143.

Lucas, M. et al. (2011) Plant systems biology: network matters. *Plant, Cell & Environment*, 34, 535–553.

Marbach, D. et al. (2010) Revealing strengths and weaknesses of methods for gene network inference. *Proceedings of the National Academy of Sciences*, 107, 6286.

Marbach, D. et al. (2009) Generating realistic in silico gene networks for performance assessment of reverse engineering methods. *Journal of Computational Biology*, 16, 229–239.

Margolin, A. et al. (2006) ARACNE: an algorithm for the reconstruction of gene regulatory networks in a mammalian cellular context. *BMC Bioinformatics*, 7, S7.

Middleton, A.M. et al. (2010) Mathematical modelling of the Aux/IAA negative feedback loop. *Bulletin of Mathematical Biology*, 72, 1383–1407.

Mironova, V. et al. (2010) A plausible mechanism for auxin patterning along the developing root. *BMC Systems Biology*, 4, 98.

Morrissey, E.R. et al. (2010) On reverse engineering of gene interaction networks using time course data with repeated measurements. *Bioinformatics*, 26, 2305–2312.

Morrissey, E.R. et al. (2011) Inferring the time-invariant topology of a nonlinear sparse gene regulatory network using fully Bayesian spline autoregression. *Biostatistics*, 12, 682–694.

Ma, S. et al. (2007) An Arabidopsis gene network based on the graphical Gaussian model. *Genome Research*, 17, 1614.

Muraro, D. et al. (2011) The influence of cytokinin–auxin cross-regulation on cell-fate determination in *Arabidopsis thaliana* root development. *Journal of Theoretical Biology*, 283, 152–167.

Needham, C.J. et al. (2009) From gene expression to gene regulatory networks in *Arabidopsis thaliana*. *BMC Systems Biology*, 3, 85.

Okushima, Y. et al. (2007) ARF7 and ARF19 regulate lateral root formation via direct activation of LBD/ASL genes in Arabidopsis. *Plant Cell*, 19, 118–130.

Okushima, Y. et al. (2005) Functional genomic analysis of the AUXIN RESPONSE FACTOR gene family members in *Arabidopsis thaliana*: unique and overlapping functions of ARF7 and ARF19. *Plant Cell*, 17, 444–463.

Opgen-Rhein, R. and Strimmer, K. (2007a). From correlation to causation networks: a simple approximate learning algorithm and its application to high-dimensional plant gene expression data. *BMC Systems Biology*, 1, 37.

Opgen-Rhein, R. and Strimmer, K. (2007b). Learning causal networks from systems biology time course data: an effective model selection procedure for the vector autoregressive process. *BMC Bioinformatics*, 8, S3.

Perrin, B.-E. et al. (2003) Gene networks inference using dynamic Bayesian networks. *Bioinformatics*, 19, ii138–ii148.

Perrine-Walker, F. et al. (2010) Auxin carriers localization drives auxin accumulation in plant cells infected by Frankia in *Casuarina glauca* actinorhizal nodules. *Plant Physiology*, 154, 1372–1380.

Pradal, C. et al. (2008) OpenAlea: a visual programming and component-based software platform for plant modelling. *Functional Plant Biology*, 35, 751–760.

Péret, B. et al. (2009) Arabidopsis lateral root development: an emerging story. *Trends in Plant Science*, 14, 399–408.

Reiss, D. et al. (2006) Integrated biclustering of heterogeneous genome-wide datasets for the inference of global regulatory networks. *BMC Bioinformatics*, 7, 280.

Sablowski, R. (2011) Plant stem cell niches: from signalling to execution. *Current Opinion in Plant Biology*, 14, 4–9.

Segal, E. et al. (2003) Module networks: discovering regulatory modules and their condition specific regulators from gene expression data. *Nature Genetics*, 34, 166–176.

De Smet, R. and Marchal, K. (2010) Advantages and limitations of current network inference methods. *Nature Reviews Microbiology*, 8, 717–729.

Sozzani, R. et al. (2010) Spatiotemporal regulation of cell-cycle genes by SHORTROOT links patterning and growth. *Nature*, 466, 128–132.

Stahl, Y. and Simon, R. (2010) Plant primary meristems: shared functions and regulatory mechanisms. *Current Opinion in Plant Biology*, 13, 53–58.

Steuer, R. et al. (2002) The mutual information: detecting and evaluating dependencies between variables. *Bioinformatics*, 18, S231.

Stoma, S. et al. (2008) Flux-based transport enhancement as a plausible unifying mechanism for auxin transport in meristem development. *PLoS Computational Biology*, 4, e1000207.

Swarup, K. et al. (2008) The auxin influx carrier LAX3 promotes lateral root emergence. *Nature Cell Biology*, 10, 946–954.

Swarup, R. et al. (2005) Root gravitropism requires lateral root cap and epidermal cells for transport and response to a mobile auxin signal. *Nature Cell Biology*, 7, 1057.

Twycross, J. et al. (2010) Stochastic and deterministic multiscale models for systems biology: an auxin-transport case study. *BMC Systems Biology*, 4, 34.

Winship, L.J. et al. (2010) Under pressure, cell walls set the pace. *Trends in Plant Science*, 15, 363–369.

Won, S.K. et al. (2009) *Cis*-element-and transcriptome-based screening of root hair-specific genes and their functional characterization in Arabidopsis. *Plant Physiology*, 150, 1459.

Yip, K.Y. et al. (2010) Improved reconstruction of in silico gene regulatory networks by integrating knockout and perturbation data. *PloS One*, 5, e8121.

Yu, J. et al. (2004) Advances to Bayesian network inference for generating causal networks from observational biological data. *Bioinformatics*, 20, 3594–3603.

Yuan, Y. et al. (2011) Directed partial correlation: inferring large-scale gene regulatory network through induced topology disruptions. *PloS One*, 6, e16835.

Zanetti, M.E. et al. (2005) Immunopurification of polyribosomal complexes of Arabidopsis for global analysis of gene expression. *Plant Physiology*, 138, 624.

Zoppoli, P. et al. (2010) TimeDelay-ARACNE: reverse engineering of gene networks from time-course data by an information theoretic approach. *BMC Bioinformatics*, 11, 154.

6 Genomics of Root Hairs

Hyung-Taeg Cho

Genomics with Single Cells

Plants consist of diverse cell types that are specialized for optimal function. Therefore, knowledge about the roles of individual cell types is essential to understanding whole-plant development. Different cell types are defined by different morphology and biochemical function, and these in turn are determined by differential gene expression. Although organ- or tissue-specific gene expression analyses have long been conducted, the development of modern technologies has greatly increased the resolution and precision of gene expression analysis at the single-cell level. Laser capture microdissection and fluorescence-activated cell sorting (FACS) are currently the most efficient methodologies to acquire homogeneous single-cell types. These isolated and purified single cells are then subjected to high-throughput omics analyses to characterize the regulatory relationships and networks between cellular components. Among the many types of plant cells, root hairs, trichomes, and pollens have been given the most attention due to their ease of isolation, the availability of large amounts of cells, and distinctive morphogenesis and cell growth processes (tip growth). These properties enable those cells to serve as models for understanding the mechanisms underlying general plant cell differentiation at the genome level. This chapter reviews studies of root hair genomics and highlights the genes that are involved in root hair development.

Root hairs form the interface between the plant root and soil, where the absorption of water and nutrients, anchorage, and the interaction between plants and soil microbes take place. The primary functions of root hairs stem from their expansion of the root surface area. Root hairs approximately double the root surface area, compared with naked roots, in seven studied plant species (Föhse et al. 1991). Root hairs operate diverse nutrient transporters as well as water channels (Libault et al. 2010c). The increase in root surface area conferred by root hairs greatly enhances nutrient uptake by the root. For example, in low-phosphorus soil, 90% of the phosphorus is absorbed from the soil through the root hairs (Föhse et al. 1991). Conversely, the secretion of heavy metal ions through root hairs provides the plant with a mechanism of resistance to metal toxicity (Kim et al. 2007a). Root hairs are also the primary contact point for symbiotic soil bacteria in legume plants (Limpens and Bisseling 2008). These root hair functions have prompted attempts to elucidate the mechanisms underlying root hair development and response to environmental factors. In particular, high-throughput approaches have identified many genes, proteins, and metabolites affecting root hair functions and have provided a means to characterize their interactomes.

Root Genomics and Soil Interactions, First Edition. Edited by Martin Crespi.

Root Hair Development

Root hairs are tubular protrusions from the root epidermal cells of vascular plants. The epidermis is made up of hair-forming cells and nonhair cells, and the distribution of hair cells includes three basic types (Dolan 1996; Schiefelbein 2000). In type 1, root hairs emerge from any kind of cell, and so can be considered a random type. In type 2, root hairs emerge from a population of root epidermal cells composed of longitudinally alternating short and long cells; the hairs emerge from the shorter cells. In type 3, root hairs emerge from cells localized between two underlying cortical cells, with nonhair cells contacting only a single cortical cell. Development into a type 2 cell seems to stem from the acquisition of certain determinants for root hair cell fate after asymmetrical cell division. On the other hand, development into a type 3 cell is determined by position-dependent signaling processes (Grierson and Schiefelbein 2008). According to a survey of root hair distribution patterns by Clowes (2000), type 3 development only occurs in eudicots and type 2 development only in monocots, lower dicots, ferns, and lycopodiophytes, suggesting that type 2 is more ancient and type 3 emerged with the eudicots (Kim et al. 2006). Because type 1 is seen throughout the vascular plant lineage, and even within the same species in which type 2 or type 3 are observed (Clowes 2000), their origin could be related to the disruption of the fate-determining machinery for type 2 or type 3.

The components and network for type 3 fate determination have been well characterized in *Arabidopsis* (for a recent review, see Grierson and Schiefelbein 2008). In the *Arabidopsis* nonhair cell position, WEREWOLF (WER, a MYB transcription factor), together with GLABRA3/ENHANCER OF GLABRA3 (GL3/EGL3, bHLH transcription factors) and TRANSPARENT TEST GLABRA (TTG, a WD40 protein), activates the expression of GLABRA2 (GL2, a homeodomain transcription factor), which in turn suppresses the expression of downstream target genes for root hair morphogenesis. In the hair cell position, a Leu-rich repeat receptor-like protein kinase (LRR-RLK) known as SCRAMBLED (SCM) seems to receive external signals from the inner tissues and, in response to these, suppresses the expression of WER and thus of GL2, resulting in the activation of root hair morphogenesis. The external ligand for SCM, which probably works as a positional cue, remains to be identified. Several recent studies have also demonstrated that chromatin modification is implicated in the positional regulation of type 3 fate determinants. The chromatin structure of the *GL2* locus is activated in the nonhair cell position, whereas it is inactivated in the hair cell position (Costa and Shaw 2006). The expression of *GL2* and *CAPRICE* is also modulated by histone modification (Xu et al. 2005; Caro et al. 2007). These root hair fate determinants or their paralogs also play roles in trichome development, suggesting that a common regulatory network has been adopted for upstream fate determination steps in both root hairs and trichomes.

The hair-forming morphogenetic step can be divided into three stages: initiation, bulge formation, and tip growth. These actions are achieved via morphogenetic gene products for cell wall organization, vesicle secretion, signal transduction, and many other cellular processes (Grierson and Schiefelbein 2008). ROOT HAIR DEFECTIVE6 (RHD6, a bHLH transcription factor) seems to be the prime positive modulator for root hair initiation in the root hair cell and is negatively regulated by GL2 in the nonhair cell (Masucci and Schiefelbein 1996; Menand et al. 2007). Hormones such as auxin and ethylene are involved in hair morphogenesis and they act downstream of RHD6, and root hair defects in *rhd6* are rescued by the exogenous application of these hormones (Masucci and Schiefelbein 1996; Cho and Cosgrove 2002). RHD6 is expressed in root hair cells in the meristem and elongation zones of the *Arabidopsis* root (Menand et al. 2007). RHD6-LIKE4 (RSL4, a bHLH transcription factor) is a major downstream target of RHD6, and RSL4 regulates the major root hair morphogenetic genes (Yi et al. 2009). A plethora of gene products are involved in root hair morphogenesis (Grierson and Schiefelbein 2008), many of which have general morphogenetic functions for other cell types as well as root hair cells. A recent high-throughput approach has identified many root hair-specific morphogenetic genes (*ROOT HAIR SPECIFIC*, *RHS*) whose proximal promoter regions include the common root hair-specific *cis*-element (RHE; Won et al. 2009). In contrast to the fate determinants, the root hair morphogenetic genes have not been implicated in trichome development. This suggests that the downstream morphogenetic genes between two hair cell types have adopted independent upstream regulatory machinery after GL2.

High-Throughput Approaches for the Characterization of Root Hairs

High-throughput approaches so far performed for the characterization of root hairs include classical genetic screening to find root hair mutants; omics approaches such as transcriptomics, proteomics, and metabolomics; and genome-wide in silico approaches to find root hair-specific promoters and genes. Root hair genetic screening relies on forward genetics: root hair-defective mutants are screened and the corresponding genes and their functions are determined. Omics approaches adopt a genome-wide comparative analysis between root hair cells and nonhair cell tissues, or between the wild type and root hair-defective mutant lines. To obtain root hair-specific omics resources (mRNA, protein, and metabolite profiles), root hair cell-specific cytoplasms are isolated by methods such as root hair stripping, single-cell aspiration, or FACS. For root hair omics analyses without these root hair purification steps, the omics data from whole root tissues of the wild type and mutant lines are simply compared in order to extract root hair-specific omics profiles.

Genetic Screening for Root Hair Mutants and Genes

Many genetic screening studies have been performed to obtain root hair mutants and identify the corresponding genes, mostly in *Arabidopsis*. Many targeted screening efforts have found the mutants and genes responsible for specific root hair developmental steps; in addition, several screening studies have added a massive resource of root hair mutants and genes, which has prompted much follow-up research.

Systematic screening of root hair mutants was pioneered by Schiefelbein and Somerville (1990). They screened root hair-defective (*rhd*) mutants from ethyl methane sulfonate (EMS)-mutagenized *Arabidopsis* M2 seeds. More than 40 mutants defective in root hair morphogenesis were obtained, and four loci (*RHD1–4*) were further analyzed for their genetic relationship and roles in root hair morphogenesis. These mutants did not show any trichome phenotypes, indicating that these gene functions were specific to root hair morphogenetic steps rather than to fate-determination steps.

A mutant screening by Schneider et al. (1997) was designed to find root hair fate determinants so that root hairless (*rhl*) or ectopic root hair (*erh*) mutants would be selected. Three *RHL* and another three *ERH* loci were isolated from EMS-mutagenized *Arabidopsis* seeds. Most of these mutants exhibited pleiotropic phenotypes, including trichome phenotypes, implying that they play roles in upstream fate-determining steps for both root hairs and trichomes.

In a search for root hair morphogenetic mutants in an EMS-mutagenized *Arabidopsis* population, Parker et al. (2000) isolated eight new loci, *SHAVEN*, *CENTIPEDE*, *BRISTLED1*, and *SUPER-CENTIPEDE1*, and conducted intensive genetic analysis with these new loci and previously known root hair mutants. These new mutants did not change root hair density or distribution patterns, indicating that they do not function in hair/nonhair cell fate determination; they only function in root hair morphogenesis. After a massive double mutant analysis investigating 79 gene pairs, this study proposed a genetic model allocating those genes' sequential contribution to root hair morphogenesis from hair initiation to tip growth.

Diet et al. (2004), in an effort to find mutants enhancing the root hair-defective *lrx* mutant phenotype, acquired four root hair morphogenetic *ENL* loci from the screening of an EMS-mutagenized pool of *lrx1* mutants (Baumberger et al. 2001). Although these mutants were screened in the *lrx1* background so as to enhance the *lrx1* phenotypes (short and frequently burst root hairs), three *enl* single mutants showed their own root hair phenotypes, suggesting that these *ENL* gene products function in root hair morphogenesis independently of LRX1 (LEUCINE-RICH EXTENSIN1).

Ringli et al. (2005) also performed root hair morphogenetic mutant screening in *Arabidopsis* and identified nine *der* (*deformed root hairs*) mutants, of which five represented new loci for root hair morphogenesis. This study, which performed double mutant analysis with *der* and other known root hair mutants, also assigned the function of DERs to each step of root hair morphogenesis from bulge formation to late tip growth.

Transcriptomics and Root Hairs

To construct a root hair-specific transcriptome, it is most critical to obtain pure root hair cells. Although the FACS method, in which fluorescent proteins expressed by root hair-specific promoters are used to purify the marked root hair cells, is the best option to obtain pure root hair cells, early studies took advantage of the morphology of a root hair cell. The extended outgrowths of root hair cells enable them to be easily stripped off after freezing in liquid nitrogen and then purified by filtration (Röhm and Werner 1987). mRNAs from root hairs and stripped roots have often been used to screen out root hair-specific genes using differential expression analysis (Bucher et al. 1997).

Using the root hair stripping method, Covitz et al. (1998), in an effort to elucidate mechanisms underlying plant–bacterial symbiosis in legumes, obtained the first root hair transcriptome data. They constructed a root hair cDNA library of *Medicago truncatula* and obtained 899 root hair expressed sequence tags (ESTs) representing genes underlying diverse cellular processes. Similarly, 5468 ESTs were determined in this way from sorghum root hairs (Baerson et al. 2008). This root hair stripping method is an efficient way to easily obtain large amounts of root hair samples for RNA isolation (up to 60 μg total RNA from 800 mg root hairs). However, the root hairs isolated by this method are likely to be mostly mature root hairs, providing expression profiles corresponding to that region but excluding genes responsible for root hair initiation and bulge formation.

This limitation can be overcome by single-cell RNA preparation. Jones and Grierson (2003) constructed a root hair-specific cDNA library using the cell aspiration technique. The cytoplasms of single root hair cells were individually aspirated through a glass micropipette, and a cDNA library was constructed from the small amounts of root hair-specific RNA isolated from these cytoplasms. *RHD3* and *Rop2* (Rho of plants 2) were able to be amplified by PCR using this cDNA as the template. This single-cell aspiration method, aided by PCR amplification of the cDNAs, could provide a reasonable tool to generate gene expression profiles of root hair or nonhair cells that are in specific developmental stages along the longitudinal root axis.

A genome-scale comparison of transcriptomes between wild-type and root hair-defective mutant plants was conducted for the first time by Jones et al. (2006). Total RNA from the differentiation zone of wild-type and root hair-defective *rhd2-1* mutant *Arabidopsis* plants (Schiefelbein and Somerville 1990; Foreman et al. 2003) was isolated, and these transcriptomes were compared using *Arabidopsis* microarray chip analysis. A total of 606 genes (the so-called root hair morphogenesis transcriptome) showing higher expression in wild-type plants than in the root hair-defective mutants were selected. Of these, 303 genes showed a more than twofold higher expression in the wild-type plants than in the mutant plants. Among these 303 genes, screening using the SALK T-DNA insertion database revealed 159 genes that had T-DNA insertions in their 5′-untranslated regions or exons. From this T-DNA mutant collection, seven mutant lines showing consistent root hair phenotypes were finally selected. Out of these seven genes, one had been previously characterized; the other six (*MRH1*∼*6*) were further analyzed in this study. These six *MRH* genes were found to encode proteins for cellular signaling (LRR-RLK, glycerophosphoryl diester phosphodiesterase-like protein, and IP3 phosphatase-like protein), intracellular trafficking (kinesin-related), and morphogenesis (COBRA-like protein). Accordingly, their mutants showed diverse phenotypes in root hair length and morphology. This was the first study to screen root hair morphogenetic genes using genome-wide root hair transcriptome analysis and to characterize the biological functions of these genes during root hair morphogenesis. An intriguing finding from this study was that the expression of many morphogenetic genes requires normal RHD2 function. *RHD2* encodes an NADPH oxidase, which produces reactive oxygen species (ROS) (Foreman et al. 2003), suggesting that RHD2-mediated ROS production is linked to certain transcription-regulatory modules for root hair morphogenetic genes.

Recently, Kwasniewski et al. (2010) conducted a genome-wide comparative transcriptome analysis with wild-type and root hairless mutant (*rhl1.a*) barley plants. Ten genes showing a more than 14-fold downregulation in the hairless mutant were screened in this comparison. Most of these gene products, including peroxidase, xyloglucan endotransglycosylase, arabinogalactan protein, extensin, LRR protein, RhoGTPase GDP dissociation inhibitor (a AtSCN1 ortholog), and phosphatidylinositol phosphatidylcholine transfer protein (a AtCOW1 ortholog), seemed to have cell wall- or membrane-related functions. These same genes were expressed in another root hair-defective barley mutant (*rhp1.b*) that forms root hair initials but is defective in hair elongation, suggesting that the genes are required for root hair initiation rather than for elongation. Although this study did not characterize the function of these genes, their *Arabidopsis* orthologs are known to play roles in root hair morphogenesis, indicating that common genes are operational for root hair morphogenesis in both monocots and dicots.

In an effort to characterize the early nodulation mechanism between soybean and the symbiotic bacterium *Bradyrhizobium japonicum*, global root hair transcriptome analyses were conducted (Libault et al. 2010a, 2010b). Limpense and Bisseling (2008) had previously shown that legume root hairs determine the specificity of interaction between the plant and the symbiotic bacteria by recognizing the Nod factor from the bacteria and by providing the route of bacterial entrance into the plant body. Using several high-throughput transcriptome analysis tools such as soybean microarray hybridization, Illumina cDNA sequencing, and quantitative RT-PCR, Libault et al. (2010a) concluded that approximately 2000 genes are regulated in the root hair in response to *B. japonicum* inoculation. Soybean seedling roots were inoculated with *B. japonicum* or mock-treated, and their transcriptomes were obtained from isolated root hairs by stripping or from the stripped roots for a comparative analysis. This study showed that similar numbers of genes were up- or downregulated in response to the inoculation and that the expression of 41 genes was specifically induced by the inoculation. Using a similar approach, Libault et al. (2010b) demonstrated that among the ∼2000 genes comprising this collection, 451 were preferentially expressed in the root hair. Of these, 69 genes were root hair-specific and three were highly root hair-specific.

Brady et al. (2007a), using the FACS method, provided genome-wide root hair cell transcriptomes as well as transcriptomes of other root cell types. They purified 14 *Arabidopsis* root cell types (out of 15 whole cell types) that were marked by each cell type-specific promoter::GFP reporter. Thirteen developmental sections, from the columella root cap to the hair-differentiation zone, were added to these cell type categories, allowing the assessment of transcriptome changes along the spatiotemporal developmental stages of each root cell type. For isolation of root hair cells, the *COBRA-LIKE9* (*COBL9*) promoter::GFP line was used, which shows the trichoblast-specific expression pattern from the elongation to the hair differentiation zones of the root (Brady et al. 2007a, 2007b). The graphical illustration of these root cell type-specific expression profiles is available in http://bar.utoronto.ca/efp/cgi-bin/efpWeb.cgi?dataSource = Root.

Proteomics and Root Hairs

From root hairs prepared by the stripping method after the inoculation of soybean roots with *B. japonicum*, Wan et al. (2005) identified 37 2D-PAGE spots that were unique, absent, or upregulated in root hairs and their protein sequences were determined by tandem mass spectrometry (MS/MS). This comparative proteome collection included not only the proteins (peroxidase and phenylalanine-ammonia lyase) known to respond to rhizobial inoculation, but also novel proteins (phospholipase D and phosphoglucomutase), which had not been identified as having roles in symbiosis. A proteomics tool using comparative analysis of the proteomes of wild-type and root hair-defective lines was also used to determine root hair-specific protein profiles in maize (Wen et al. 2005).

Brechenmacher et al. (2009), using 2D-PAGE and shotgun proteomics (1D-PAGE-liquid chromatography [LC] and multidimensional protein identification technology), provided a root hair protein profile for soybean. They identified a total of 1492 different proteins in soybean root hairs. These proteins represent functions not only for basic cell metabolism but also for root hair morphogenesis and physiology, such as vesicle trafficking, cell wall synthesis and reassembly, water and nutrient uptake, secondary metabolism, and polarity-related functions.

The monocot root hair proteome profile was obtained from stripped maize root hairs (Nestler et al. 2011). Using 1D-PAGE-nanoLC-MS/MS technology, this study identified 2573 maize root hair proteins. The most abundant (~33%) were proteins related to the metabolism and localization of proteins and RNAs. Proteins related to stress, signaling, and amino acid metabolism accounted for 4~5% each. Cell wall-related proteins made up 1.7% of the proteome. This maize root hair protein collection included many homologs (252 proteins) with known functions in dicots related to root hair development. These homologs included GEM (GL2 EXPRESSION MODULATOR, a repressor of GL2 by histone modification), K^+ transporters such as ARABIDOPSIS K^+ TRANSPORTER1, and TYNY ROOT HAIR1, which are involved in hair elongation, and RHD2, RHD3, MICROTUBULE ORGANIZATION1, ERH3, and many Ca^{2+}-, exocytosis- and cytoskeleton-related proteins. Comparison of this maize root hair proteome with the soybean root hair proteome (Brechenmacher et al. 2009) indicated that two representative species for monocots and dicots not only have conserved common root hair morphogenetic machinery but also that this machinery is related to specific physiological functions. For example, while most cell wall-related proteins were shared by both species, proteins representing secondary metabolism showed more divergence between maize and soybean. One representative example of the latter are the proteins responsible for the synthesis of chalcones and flavonoids, which are specific to soybean and are most likely related to nodulation signaling.

The advantage of proteomics over transcriptomics is that the former can provide information about co- or posttranslational protein modification that is unavailable from the DNA sequence data. Protein modifications can be easily detected by 2D-PAGE or by MS analysis. So far, this type of proteomic analysis has not been conducted with root hairs. Future studies of this type may provide insight into understanding specific molecular mechanisms related to polar root hair morphogenesis.

Metabolomics and Root Hairs

Root hair chemistry has been a major area of interest in studies of nodulation because host root hairs play important roles in communication with symbiotic rhizobia as well as providing the means for the rhizobia to enter the plant. Nodule formation in legume plants takes place through serial chemical communications between the host and the rhizobia (Limpense and Bisseling 2008). Flavonoids secreted from the host root determine symbiont specificity and induce the expression of nodulation genes in the symbiont (Peters et al. 1986; Subramanian et al. 2007). In turn, rhizobial nodulation factors (lipochitooligosaccharides), which were produced by the nodulation gene products, cause root hair deformation and nodule formation (Limpense and Bisseling 2008). In addition to these representative signaling chemicals, many other chemical factors are involved in the modulation of nodulation (Gibson et al. 2008). In this context, the root hair chemical profile (metabolome) of legume plants can help to identify as-yet-unknown signaling chemicals and major metabolites produced by host root hairs during nodulation processes. Although root hair transcriptomic and proteomic data can be used to infer such metabolites, the direct analysis of root hair metabolites can definitively profile these chemical candidates.

Although metabolomic approaches have been used in studies of plant–pathogen interactions and late nodulation processes, the metabolites in the early stages of nodulation have not been characterized. A recent study, using gas chromatography–MS and ultraperformance LC–quadrupole time of flight–MS, analyzed root and root hair metabolites during *B. japonicum*-mediated nodulation in soybean (Brechenmacher et al. 2010). This study identified a total of 2610 metabolites from stripped root hairs after up to 48 h inoculation with the rhizobia. Among these, 166 metabolites were regulated by *B. japonicum* inoculation: 114 were increased, and 50 were decreased in response to the inoculation. These rhizobium-regulated root hair metabolites originated from both the host plant and the bacteria. These metabolites include diverse chemicals, including (iso)flavonoids, amino acids, carboxylic acids, fatty acids, and carbohydrates. The rhizobia induced the synthesis of flavonoid precursors as soon as 12 h after inoculation. Amino acids such as Glu (the precursor of gamma-amino butyric acid [GABA]) and GABA were decreased in response to the inoculation; this was interpreted as functioning to suppress plant innate immunity since GABA has been implicated in the plant defense response (Shelp et al. 1999). The increase in fatty acids in the root hair in response to the bacterial inoculation can be attributed to the release of fatty acids from membrane lipids in order to increase membrane fluidity for easy bacterial penetration. This chemical profile also includes metabolites with unidentified functions during early nodulation processes, providing an opportunity for future research to assess these novel roles in nodule formation.

In Silico Approaches for Screening Root Hair-specific Genes

Root hair mutant screening can reveal the function of specific genes during hair morphogenesis and has greatly contributed to the determination of hair morphogenetic processes. However, the function of these deduced genes may not be root hair cell-specific but may instead be involved in the general differentiation processes of many cell types. Root hair transcriptome analysis allows us to test the functions of many genes expressed in root hair cells, but the transcriptome necessarily includes genes that are generally expressed and function in various cell types. A recent study took advantage of in silico and transcriptomic approaches to find root hair cell-specific genes that are implicated in hair morphogenesis.

Genes expressed in the same tissue are likely to share common *cis*-regulatory elements in their promoter regions (Jareborg et al. 1999; Kaplinsky et al. 2002; Guo and Moose 2003; Inada et al. 2003; Haberer et al. 2004; Gachon et al. 2005; Lisso et al. 2005; Persson et al. 2005; Zhang et al. 2005; Jones et al. 2006; Kim et al. 2006). Conversely, a defined cell type-specific *cis*-element can be used to identify new cell type-specific genes from whole genome data. Won et al. (2009), using this approach, identified 19 *RHS*s that had not been reported previously.

The root hair-specific *cis*-regulatory region was first identified by promoter analysis of two root hair-specific *Arabidopsis EXPA* (*EXPA7* and *18*) genes (Cho and Cosgrove 2002). Further detailed promoter analyses of *EXPA7* orthologs/paralogs and other previously known root hair-specific genes deciphered the 16∼17 bp-long RHE, which has been conserved in the angiosperm lineage (Kim et al. 2006). The consensus RHE sequence is WHHDTGNNN(N)KCACGWH, where W = A/T, H = A/T/C, D = G/T/A, K = G/T, and N = A/T/C/G. RHEs from monocots and dicots are reciprocally functional (i.e., directing root hair expression of the reporter gene) in both angiosperm lineages, suggesting that the root hair-specific gene regulatory module of RHE and its transcription factor has been conserved despite the divergence of upstream fate-determining processes in angiosperms. Thus far, this RHE is the only identified *cis*-element directing root hair-specific gene expression.

This conservation of RHE led Won et al. to attempt to screen new RHE-containing root hair-specific genes in the *Arabidopsis* genome. They first screened 904 RHE-containing genes from a total of 33,282 *Arabidopsis* genes. The Patmatch program from the TAIR website was used to screen gene promoters located within 1000 bp from the start codon of RHEs. In the second step, 904 genes were further filtered through the Gene Atlas microarray database (www.genevestigator.com; Hruz et al. 2008) so as to select only root-specific genes, resulting in 81 genes. This second filtering process necessarily excluded many root hair-expressing genes that were also expressed in shoot tissues. These 81 root-specific and RHE-containing genes were further screened through triple root hair-mutant microarray filters. The root hair-mutant microarray filters represent genes that were suppressed greater than twofold in root hair-defective mutant (*rhd6*) and transgenic (*EXPA7 promoter::axr2-1* or *EXPA7 promoter::GL2*) lines. Root hair-specific genes involved in hair morphogenesis are likely to be suppressed in these root hairless lines, since the root hair defects occurred only in the root hair cells in these lines. This third filtration step produced 37 genes that were RHE-containing, root-specific, and tentatively root hair-specific. Thirty-four out of these 37 genes were tested for their root hair specificity using promoter::reporter analysis in *Arabidopsis* plants, resulting in 24 genes with clear root hair specificity and two genes with stronger expression in root hair cells than in nonhair cells. Among the 24 genes, 5 were previously known to be root hair-specific and 19 had not been reported previously. The molecular roles of these novel RHE-containing *RHS*s were assigned into several major root hair morphogenesis and physiological functions, such as cell wall-related (six genes), protein kinases or kinase-related (five genes), and transporters (two genes) (Table 6.1). Among these 19 *RHS*s, 8 T-DNA insertion mutants were analyzed, and 4 *RHS* mutants showed phenotypes related to root hair growth. In the overexpression analysis with the *EXPA7* promoter, seven *RHS*s showed apparent phenotypes related to root hair morphology and growth.

While five positions in the 16∼17 bp-long RHE motif were strictly conserved, other positions were functionally quite flexible (Kim et al. 2006). However, this in silico screening used a stringent RHE consensus, as mentioned earlier. This indicates that there must be many more root hair-specific genes with RHE-like *cis*-elements. For example, five out of six *MRH* genes that were downregulated in the *rhd2* mutant had RHE-like *cis*-elements in their proximal promoter regions, suggesting the possibility of root hair-specific expression.

Table 6.1 Root hair-specific genes and their functions for root hair morphogenesis.[a]

Functional Category	Gene	RHE (Location from ATG)	Protein Function	Root Hair Phenotype of Mutants	Reference
Transcription regulation	*RHD6* (At1G66470)	No	bHLH transcription factor	Few hairs	Masucci and Schiefelbein 1994; Menand et al. 2007
	RSL4 (At1G27740)	−249, −297, −447	bHLH transcription factor	Fewer, short hairs	Yi et al. 2009
Signaling	*RHD2* (At5G51060)	−346	NADPH oxidase	Short hairs	Schiefelbein and Somerville 1990; Foreman et al. 2003
	RHS1 (At1 G 05990)	−38, −739	Calcium binding	Long hair (ox,[b] short hairs)	Won et al. 2009
	RHS5 (At1G34760)	−52, −559	14-3-3 protein	n.d.[c]	Won et al. 2009
Kinase-related	*RHS3* (At1G16440)	−257, −454, −547	AGC-type protein kinase	Ox, spiral, bent, branched hairs	Won et al. 2009
	RHS6 (At1G51880)	−214, −373	LRR-RLK	n.d.	Won et al. 2009
	RHS10 (At1G70460)	−447, −636, −767	RLK with extensin	Longer hairs (ox, no, or short hairs)	Won et al. 2009
	RHS16 (At4G29180)	−155	LRR-RLK	Ox, branched hairs	Won et al. 2009
	RHS11/RopGEF4 (At2G45890)	−96, −220; −706, −877	Kinase partner protein (KPP)	Short hair (ox, multiple-branched short hairs)	Won et al. 2009
	RopGEF10 (At5G19560)	−402, −551	KPP	n.d.	Won et al. 2009
Cytoskeletone/ trafficking	*ADF8* (At4G00680)	−179	Actin-depolymerizing factor	n.d.	Kandasamy et al. 2007
	SYP123 (At4G03330)	−282	Syntaxin	n.d.	Enami et al. 2009
	ZWI (At5 G 65930)	−238	Kinesin-related	n.d.	Reddy and Reddy 2004
Cell wall	*KJK/CSLD3* (At3G03050)	−518, −575	Cellulose synthase	Burst, swollen hairs	Favery et al. 2001
	OsCSLD1 (Os10G0578200)	−789	Cellulose synthase	Short, kinked hairs	Kim et al. 2007b
	COBL9/MRH4/ DER9/SHV2 (At5G49270)	−130, −183, −236, −807	Glycosylphosphatidylinositol (GPI)-anchored protein	Short, wide burst hairs	Parker et al. 2000; Ringli et al. 2005; Jones et al. 2006; Brady et al. 2007b
	RHS8 (At1G63450)	−73, −301	Exostosin	n.d.	Won et al. 2009
	EXPA7 (At1g12560)	−127, −299, −357	Expansin	Short hairs (ox, rescue *osexpa17* mutant)	Kim et al. 2006; Lin et al. 2011; Yu et al. 2011
	EXPA18 (At1G62980)	−161, −206, −417	Expansin	n.d.	Kim et al. 2006
	OsEXPA30 (OsJNBb0060I05)	−201, −277	Expansin	Ox, rescue *osexpa17* mutant	Kim et al. 2006; Yu et al. 2011
	OsEXPA17 (AP000616)	−117, −155, −796	Expansin	Short hairs	Kim et al. 2006; Yu et al. 2011

(Continued)

Table 6.1 *(Continued)*

Functional Category	Gene	RHE (Location from ATG)	Protein Function	Root Hair Phenotype of Mutants	Reference
	BoEXPA7 (DQ899785)	−129, −304	Expansin	n.d.	Kim et al. 2006
	MtEXPA7 (DQ899790)	−87, −140, −350	Expansin	n.d.	Kim et al. 2006
	IbEXPA7 (DQ899786)	−517, −105	Expansin	n.d.	Kim et al. 2006
	TaEXPA7 (DQ899793)	−243, −193	Expansin	n.d.	Kim et al. 2006
	ZmEXPA7 (DQ899789)	−231, −188	Expansin	n.d.	Kim et al. 2006
	OsEXPB5 (AF261273)	−134, −451	Expansin	n.d.	Won et al. 2010
	HvEXPB1 (AY351785)	−188, −215, −365, −434	Expansin	n.d.	Won et al. 2010
	RHS12 (At3G10710)	−166, −1288	Pectinesterase	n.d.	Won et al. 2009
	RHS14 (At4G22080)	−110, −348, −597	Pectate lyase family protein	n.d.	Won et al. 2009
	RHS18 (At5G22410)	−129, −485, −535	Peroxidase	Ox, short hairs	Won et al. 2009
	RHS19 (At5G67400)	−199, −538	Peroxidase	n.d.	Won et al. 2009
	LRX1 (At1G12040)	−370, −417	LRR extensin	Short, swollen, burst hairs	Baumberger et al. 2001, 2003; Kim et al. 2006
	LRX2 (At1G62440)	−241	LRR extensin	Severe *lrx1* phenotype in *lrx1 lrx2*	Baumberger et al. 2001, 2003; Kim et al. 2006
	RHS13 (At4G02270)	−200, −215, −253	Extensin family protein	n.d.	Won et al. 2009
	RHS7/PRP1 (At1G54970)	−212	Pro-rich protein	n.d.	Won et al. 2009
	PRP3 (At3G62680)	−207	Pro-rich protein	n.d.	Bernhardt and Tierney 2000; Kim et al. 2006
Membrane transport	*RHS15* (At4G25220)	−510, −879, −1159	Putative transporter	n.d.	Won et al. 2009
	RHS2 (At1G12950)	−536, −566	MATE transporter	Short hairs (ox, long hairs)	Won et al. 2009
Others	*RHS4* (At1G30850)	−111, −306	Hypothetical protein	n.d.	Won et al. 2009
	RHS9 (At1G69240)	−91, −202	Hydrolase	n.d.	Won et al. 2009
	RHS17 (At4G38390)	−591	Expressed protein	n.d.	Won et al. 2009
	PLP1 (At4G37070)	−661, −771, −924	Patatin-like protein	n.d.	Won et al. 2009

[a]This list represents the genes whose root-hair specificity was confirmed by promoter::reporter analyses at least in the root tissues.
[b]The root hair phenotype of overexpressors.
[c]Not determined.

The RHE filter used in the approach of Won et al. greatly enriched root hair-specific genes, even though it served as just one of the three filters used to screen root hair-specific genes. While the use of the other two filters (root-specific and root hair mutant) without the RHE filter resulted in the identification of 247 genes, the use of the RHE filter with either of those filter narrowed this down to 60 (with the root hair mutant filter) or 81 (with the root-specific filter) genes. Once the cell type-specific *cis*-element has been defined, this in silico approach can be applied to other cell types.

Functions of Root Hair-Specific Genes

Although many essential morphogenetic genes without cell type specificity should be operational in root hair morphogenesis, it is likely that root hair-specific gene products direct general morphogenetic genes to function specifically in hair morphogenesis. So far, expression of at least 44 genes has been confirmed specifically (or predominantly) in root hair cells or trichoblast through the use of promoter::reporter transgenic analysis (Table 6.1). Here, "specifically" does not necessarily mean that the gene is expressed solely in the root hair. These root hair cell- or trichoblast-specific genes represent genes that are expressed predominantly in root hair cells, or at least in the root epidermis, owing to the RHE or other root hair cell-specific *cis*-elements. Here, the putative functions of these root hair-specific genes are reviewed according to their functional categories. Among the 44 root hair-specific genes, root hair mutant phenotypes of 18 genes have been demonstrated. The functions of the other genes in root hair development can be inferred from their functions in other cell types.

Transcription Regulation

Two root hair cell-specific bHLH transcription factors (RHD6 and RSL4) are implicated in transcription regulation related to root hair development. These two bHLH transcription factors show slightly different expression patterns along the root hair cell file. While *RHD6* is expressed in the meristem and elongation zones (Menand et al. 2007), *RSL4* is expressed in the elongation and differentiation zones (Yi et al. 2009). The expression domains of these two genes overlap approximately in the elongation zone, indicating that temporally early expressed RHD6 regulates the late *RSL4* gene. *RSL4* appears to be a direct transcriptional target of RHD6 since RHD6 induces *RSL4* expression regardless of cycloheximide (a translation inhibitor) treatment (Yi et al. 2009). Loss of RHD6 almost completely blocks root hair initiation (Masucci and Schiefelbein 1994), whereas the *rsl4* mutant shows a less severe phenotype, continuing to grow some short hairs (Yi et al. 2009), suggesting that RHD6 regulates root hair initiation through multiple pathways, including RSL4. Functional analysis showed that RHD6 and its closest paralog, RSL1, work together to regulate *RSL4* and its close paralog, *RSL2*, the mutation of which also causes short hair phenotypes (Yi et al. 2009). RSL4 affects the transcription of many root hair morphogenetic genes, including most RHS genes (Yi et al. 2009). The hormone pathway seems to work through RSL4, since the auxin-mediated restoration of root hairs in *rhd6* requires RSL4, and *RSL4* expression is induced by auxin (Yi et al. 2009). These results are consistent with the finding that *RHS* genes are significantly suppressed both in the *rhd6* mutant and in the root hair-specific *axr2-1* (the nondegradable form of AXR2/IAA7) overexpression line (Won et al. 2009).

Signaling and Protein Phosphorylation

Tip-focused cytosolic Ca^{2+}-gradient formation is important for directing and maintaining the tip growth of root hairs and pollen tubes, which is accomplished through cytoskeleton-mediated vesicle trafficking (Gilroy and Jones 2000; Grierson and Schiefelbein 2002; Cole and Fowler 2006). Although Ca^{2+}-binding proteins and targets are generally well defined, those for tip growth are not well characterized. Calmodulins (CaMs) are the best studied Ca^{2+} sensors. Plants have CaM-like proteins (CMLs) that contain CaM-like EF hands and are potential Ca^{2+}-sensing proteins (McCormack et al. 2005). There are 50 CMLs in *Arabidopsis* but their functions remain largely unknown.

RHS1/CML7 is the only reported root hair-specific CML that is thought to have a unique function in Ca^{2+} sensing during root hair tip growth (Won et al. 2009; Table 6.1). While only positive roles for Ca^{2+} during tip growth have been reported so far, RHS1 shows a negative function; its knockout mutant has longer hairs and its overexpressors grow shorter hairs than wild-type cells (Won et al. 2009). A protein microarray analysis in *Arabidopsis* showed that CaMs and CMLs primarily target diverse protein kinases and transcription factors, indicating their involvement in cellular signaling and transcriptional regulation (Popescu et al. 2007). Further study of the downstream targets of RHS1 should reveal how it negatively transduces a Ca^{2+} signal to inhibit hair tip growth.

RHD2 (which encodes an NADPH oxidase) is expressed in the elongation and differentiation zones of the root hair cell file (Foreman et al. 2003). It carries one RHE in its proximal promoter region (Table 6.1), suggesting that *RHD2* can also be regulated by the RHE-binding transcription factor (RHF). In the *Arabidopsis* root hair, RHD2-mediated production of ROS has been implicated in cellular Ca^{2+} influx (Foreman et al. 2003). Furthermore, a recent study showed that Ca^{2+} positively modulates RHD2, most likely by directly binding to its Ca^{2+}-binding EF hand motifs (Takeda et al. 2008). This positive feedback mechanism for Ca^{2+} influx probably facilitates Ca^{2+}-mediated polar vesicle trafficking for hair tip growth. In the *rhd2* mutant, root hair elongation is impaired (Schiefelbein and Somerville 1990) and many root hair morphogenetic genes, including *RHS*s, are suppressed in comparison with wild-type cells (Jones et al. 2006; Won et al. 2009), indicating that RHD2-mediated ROS production not only regulates polar vesicle trafficking but also elicits expression of RHE-containing *RHS* genes. In this context, the expression of *RHD2* in the root hair cell file via an RHE (Table 6.1) suggests that RHD2 could be amplified by positive feedback through its own downstream signaling.

RHS5 is a root hair-specific 14-3-3 protein gene (Won et al. 2009). 14-3-3 Proteins are phosphoserine-binding proteins with diverse targets and have been implicated in signaling, transcription, metabolism, membrane transport, and vesicle trafficking (Roberts 2003). Protein–protein interaction analyses have shown that plant 14-3-3 proteins interact with signaling components, transcription factors, and metabolic enzymes (Roberts 2003; Schoonheim et al. 2007). However, the function of 14-3-3 protein during root hair development remains to be studied.

Six root hair-specific kinase- or protein phosphorylation-related genes that may relay signaling cues or modulate cellular components have been identified. Among these six were three RLKs, an AGC (cAMP-dependent protein kinase A, cGMP-dependent protein kinase G, and phospholipid-dependent protein kinase C)-type protein kinase, and two guanine nucleotide exchange factor (RopGEF)/kinase partner proteins (KPP) (Table 6.1). Reversible protein phosphorylation and its relay to other proteins are pivotal biochemical reactions that modulate protein molecular functions and cellular development. More than 1000 kinases are recognized in *Arabidopsis*, including ~600 RLKs, 39 AGC kinases, and 14 RopGEF/KPPs (Bögre et al. 2003; www.arabidopsis.org), indicating the critical roles of kinases in diverse plant developmental processes.

Two (RHS6 and RHS16) of the three RLK-type RHSs belong to the LRR-RLK family. *RHS6* is one of the closest of eight tandem paralogs in *Arabidopsis*. RHS6 shows some homology to RHS16, sharing similarities in amino acid sequence, size, and domain order of the extracellular, transmembrane, and kinase regions. The putative extracellular domains of RHS6 and RHS16 include two consecutive LRR motifs, LxxLxxLxxLxLxxNxLxGxIPxx (Jones and Jones 1997). The loss of *RHS6* did not lead to any obvious root hair phenotype under optimized growth conditions, possibly due to the overlapping function of its four closest paralogs or to its unique function mediating specific external, rather than internal, cues for root hair development (Won et al. 2009). Conversely, root hair-specific *RHS16* overexpression caused severe morphological changes, such as branched, shortened, and kinked root hairs (Won et al. 2009). Thus, root hair-specific LRR-RLKs may be instrumental for internal root hair developmental processes, or their activity (or expression) may be upregulated by an external stimulus (or stimuli) to modulate root hair morphogenesis.

The RHS10 RLK has a Pro-rich N-terminal region and is described as a member of the PERK (Pro-rich extensin-like receptor kinase) family (Humphrey et al. 2007). As PERKs have extensin-like motifs, the extracellular domain of these RLKs might integrate with the cell wall to mediate cell wall-related signals. The loss of *RHS10* enhanced root hair elongation and its overexpression inhibited hair growth, indicating that RHS10 functions as a pacemaker in root hair elongation (Won et al. 2009). Together with the RHS1 function, this RHS10 function is in contrast to most of the root hair genes, the defects of which largely cause hair tip growth inhibition. *Arabidopsis* THESEUS1 (THE1), a receptor-like protein kinase belonging to the CrRLK1 subfamily, inhibits cell growth upon alteration of cell wall composition (Hematy et al. 2007). Among the proteins encoded by THE1-dependent, upregulated genes are an apoplastic peroxidase and an extensin that can stiffen cell walls by cross-linking the wall polymers (Held et al. 2004). Thus, RHS10 may also negatively modulate root hair elongation by activating components that inhibit cell wall extension. There exist at least two root hair-specific cell wall peroxidase genes (*RHS18* and *RHS19*) (Table 6.1), and root hair-specific overexpression of these genes resulted in reduced hair growth (Won et al. 2009). Plant cell growth accompanies cyclic events in the cell wall, such as cell wall loosening, the integration of new wall materials into existing wall polymers, and wall polymer cross-linking (Carpita and Gibeaut 1993). Cell wall peroxidases and extensins, probably together with turgor regulators, could therefore be major factors that control the pace of cell elongation.

The only RLK whose role in root hair development is known is SCM, an LRR-RLK (Kwak et al. 2005). However, SCM works at the upstream end of the root hair development pathway where hair/nonhair cell fate is determined. SCM seems to perceive extracellular cues that will determine hair or nonhair cell fate for root epidermal cells, depending on their position relative to the underlying cortical cells (Grierson and Schiefelbein 2008).The root hair-specific RLKs should not possess SCM-like functions because they are far-downstream components that engage at a point in the pathway where morphogenetic processes, rather than fate determination, occur. RHS3/AGC1-6 is a non-RLK, AGC-type protein kinase (Bögre et al. 2003). Several AGC-family protein kinases are involved in root hair development, including AGC2-1/OXIDATIVE SIGNAL-INDUCIBLE 1 (OXI1; Rentel et al. 2004) and INCOMPLETE ROOT HAIR ELONGATION (Oyama et al. 2002). The *oxi1* mutants develop shorter root hairs. Root hair-specific expression of *PINOID* inhibits root hair elongation by facilitating auxin efflux, resulting in insufficient cellular auxin levels for root hair elongation (Lee and Cho 2006). Root hair-specific *RHS3* overexpression leads to spiral and branched hair morphologies (Won et al. 2009), suggesting that the molecular function of RHS3 might be linked to cytoskeleton-mediated cell polarity.

There are two root hair-specific RopGEFs (Table 6.1). RhoGEFs switch GDP-bound inactive Rho to GTP-bound active Rho. At least two plant-specific RhoGEFs can interact with Rop

(Rho of plants), and RopGEF1 exerts guanine exchange activity on Rop (Berken et al. 2005; Gu et al. 2006). Upon overexpression, RopGEF1 can cause the depolarization of tobacco pollen tube growth, similar to that caused by Rop hyperactivity, indicating an *in vivo* role for RopGEFs in activating Rops (Gu et al. 2006). Tip-growing pollen tube cells and root hair cells require the massive consumption of cell wall materials, delivered by actin-mediated polar vesicle trafficking, at the growing tip. Rop GTPases may modulate tip growth by coordinating actin organization and membrane trafficking (Kost 2008; Yalovsky et al. 2008). Hyperactivity of *Arabidopsis* Rop2, Rop4, or Rop6 depolarizes root hair tip growth (Molendijk et al. 2001; Jones et al. 2002). The root hair-specific RopGEF4 (RHS11) might be the major activator of these Rop GTPases in the root hair, as GEF overexpression causes severely depolarized root hair phenotypes such as multiple hair initiation and branched hairs (Won et al. 2009). A tomato RopGEF homolog is named KPP since it interacts with a pollen LRR-RLK (LePRK1) and is phosphorylated *in vivo* (Kaothien et al. 2005). This suggests a signaling pathway in which the order of components is RLK, RopGEF, Rop, and tip growth.

Cytoskeleton and Trafficking

Actin deficiencies in the *act2-1* mutant cause short and stunted root hairs, suggesting that proper organization of the actin cytoskeleton is necessary for root hair tip growth (Ringli et al. 2002). Actin-depolymerizing factor (ADF) severs F-actin in a pH-dependent manner (Ketelaar and Emons 2008). *Arabidopsis* ADF8 is an RHE-containing root hair-specific ADF (Kandasamy et al. 2007; Table 6.1). Although its function in root hair development has yet to be characterized, ADF8 may target actins in the root hair cell to modulate the dynamics of the actin cytoskeleton. Ectopic expression of ACT1 led to dwarfism of the *Arabidopsis* shoot, but coexpression of ADF8 with ACT1 suppressed the dwarfism, suggesting that ADF8 regulates actin dynamics *in vivo* (Kandasamy et al. 2007).

Kinesins are motor proteins that move directionally along microtubules. *ZWICHEL* (*ZWI*) encodes a Ca^{2+}/CaM-regulated kinesin with an RHE located in its proximal promoter region. *ZWI* is expressed in the root hair cell, though a longer version of the promoter directs expression in other tissues as well (Reddy and Reddy 2004). Although its function in trichome morphogenesis is known (Oppenheimer et al. 1997), the role of ZWI in root hair morphogenesis has not yet been determined. Kinesins deliver diverse cargos to certain points of the cell (Goldstein and Philip 1999), an essential process for the secretion of large amounts of cell wall materials to the hair tip region. Though root hair-specific expression has not been shown, a kinesin-related protein has been implicated in root hair morphogenesis. Constitutively active (CA) forms of Rop GTPases cause defects in root hair polarity, most likely through the disruption of cytoskeleton organization. An enhancer of this CA-Rop effect turned out to be the mutation in a kinesin-related protein, CAE1 (CA-rop2 enhancer, Yang et al. 2007) or MRH2 (Jones et al. 2006). Mutations in CAE1 or MRH2 for its own led to wavy or branched root hair phenotypes, reminiscent of root hair disruptions induced by microtubule antagonists, and caused disruption in microtubule organization, suggesting that MRH2/CAE1 kinesin-related proteins are implicated in Rop-mediated microtubule organization during root hair tip growth (Jones et al. 2006; Yang et al. 2007).

SYP123 is an RHE-containing root hair-specific SNARE (soluble *N*-ethylmaleimide-sensitive factor attachment protein receptor) protein (Enami et al. 2009; Table 6.1). SNAREs are essential eukaryotic proteins for membrane fusion during vesicle trafficking. Nine out of over 60 *Arabidopsis* SNAREs are localized to the plasma membrane (Uemura et al. 2004). The plasma membrane SYP123 predominantly localizes to the growing root hair tip (Enami et al. 2009). Although its mutant phenotype has yet to be analyzed, the subcellular localization of SYP123 indicates a function related to polar secretion during hair tip growth.

Cell Wall Dynamics

Plant cell shapes are primarily determined by the cell wall. Root hair morphogenesis and growth accompany many cell wall synthesis and assembly events (Galway 2006). Cell wall-related root hair-specific gene products include cellulose synthases, COBRA protein, hydroxyproline-rich glycoproteins (HRGPs), pectin-metabolic enzymes, peroxidases, and expansins (EXP).

Arabidopsis has 29 cellulose synthase-related genes. *CSLD3* (encoding cellulose-synthase-like D3) shows stronger expression in the root hair cell, although it is expressed in many other tissues as well (Favery et al. 2001). The stronger expression of *CSLD3* in root hairs could be due to the presence of two RHEs in its promoter (Table 6.1). On the other hand, the rice ortholog *OsCSLD1*, which has a single RHE, is expressed predominantly in root hair cells (Kim et al. 2007b). Root hair growth and morphology are impaired in *Arabidopsis csld3* and rice *csld1* mutants due to defects in cell wall components (Favery et al. 2001; Kim et al. 2007b). Thus far, it is not clear exactly which cell wall polysaccharides are synthesized by CSLD3, though it is known to synthesize (1,4)-β-glucans. (1,4)-β-glucans can form the backbone of either cellulose or hemicellulose. They may affect the synthesis of both wall polysaccharides since CSLD3 activity has been shown to be linked with the alteration of both cellulose and xyloglucan organization (Galway et al. 2011). The promiscuous subcellular localization of CSLD3 in both internal compartments (Golgi and ER), as well as the plasma membrane, is indicative of its dual role in the biosynthesis of cell wall polysaccharides (Park et al. 2011).

Root hair-specific *COBL9* encodes a GPI-anchored protein with two cellulose-binding sites that is thought to affect cellulose biosynthesis in the plasma membrane (Roudier et al. 2002; Jones et al. 2006). Mutations in *COBL9* cause the inhibition and bursting of root hairs, most likely due to weakened cell walls caused by the impairment of cellulose synthesis. *COBL9* mutants have been repeatedly screened in several studies (Parker et al. 2000; Ringli et al. 2005; Jones et al. 2006). The *COBL9* promoter includes four RHE motifs within a promoter region 1 kb from the start codon (Table 6.1) and this may direct its expression in root hairs. However, its spatial expression pattern is more extended toward the early elongation zone of the root, which is different from the patterns of most *RHS* genes (Brady et al. 2007a). This suggests that the *COBL9* promoter may interact with additional *cis*-elements to bring about this extended expression and that the gene product might also function during the elongation of the root hair epidermal cell.

The *RHS8* gene product includes an exostosin (xyloglucan galactosyltransferase)-like domain that probably mediates cell wall component biosynthesis. The N-terminal cytosolic domain of a close *Arabidopsis* homolog (KATAMARI1/MURUS3, KAM1/MUR3) of RHS8 interacts with actin filaments to organize endomembranes, regardless of its C-terminal exostosin-like domain (Tamura et al. 2005). Because RHS8 is predicted to have a similar transmembrane topology to KAM1/MUR3 (a short N-terminal cytosolic stretch and transmembrane helix followed by a long C-terminal exostosin domain), it can be speculated that RHS8 might also be involved in membrane trafficking during root hair tip growth, which is accompanied by intense vesicle trafficking toward the tip area.

The major HRGP groups are arabinogalactan proteins, extensins, and Pro-rich proteins (PRPs) (Kieliszewski and Lamport 1994; Estévez et al. 2006). HRGPs are conserved in plant lineages, including algae, and most likely function in cell wall integrity (Cassab 1998). HRGP-related RHSs include two LRXs (LRX1 and LRX2), two extensin motif-containing PRPs (RHS7/PRP1 and PRP3), and a PRP-like protein (RHS13). Extensins contain (hydroxy)proline-rich repeats (Ser[Pro]$_4$) and contribute to cell wall assembly by forming isodityrosine cross-links between extensin monomers (Held et al. 2004). LRX1 and LRX2 are required for root hair tip growth and likely mediate cell wall assembly since their defects often cause swelling and rupture of the root hair tip (Baumberger et al. 2001, 2003). PRP3 (Bernhardt and Tierney 2000) and RHS7/PRP1, the two closest *Arabidopsis* root hair-specific PRP paralogs, belong to one of two PRP subclasses characterized by Pro-rich N-terminal and non-PRP C-terminal domains (Fowler et al. 1999).

PRPs carry PPPXYK repeats (X = any amino acid) and contribute to cell wall structure via peroxidase-mediated cross-linking, similar to the extensins (Kleis-San Francisco and Tierney 1990; Bradley et al. 1992). *RHS13* encodes a predicted cell wall protein of unknown function that is homologous to the C-terminal non-PRP domain of RHS7/PRP1 and PRP3. Since RHS13 lacks PRP motifs, the protein may not function as a typical cell wall integrator through cross-linking, as do canonical HRGPs. The root hair-specific HRGPs identified here and previously could also mediate interactions between microbes and root hairs. *Ext26G*, a *Vigna unguiculata* extensin gene with two RHE-like motifs in its promoter region, is expressed in root hairs and is suppressed by rhizobia inoculation (Arsenijevic-Maksimovic et al. 1997). This suppression coincides with the start of root hair deformation, an early-stage hallmark of rhizobium–plant interaction. This result implies that root hair-specific extensins normally protect the root from hostile microbe invasions, possibly by forming a physical barrier, but are downregulated in order to allow the entrance of symbiotic microbes.

The pectin metabolism-related RHSs include a putative pectin methyl esterase (RHS12) and a putative pectate lyase (RHS14). Pectins are essential matrix polysaccharides in primary cell walls; they mediate the elongation of pollen tubes, whose apical regions are particularly enriched with pectins (Bosch et al. 2005). Pectin methyl esterase (PME) demethylesterifies pectins, exposing carboxyl groups that cross-link with calcium ions to increase cell wall rigidity (Willats et al. 2001). RHS12 is a type I PME since it has a PME inhibitor-like N-terminal region (Micheli 2001) that negatively modulates PME activity in the Golgi (Bosch et al. 2005). PMEs have been implicated in diverse plant developmental processes; in particular, the introduction of exogenous PMEs or the loss of endogenous PMEs retard pollen tube growth, most likely by altering the balanced cell wall plasticity at the tip (Bosch et al. 2005; Jiang et al. 2005; Bosch and Hepler 2006). The pectin-degrading enzyme pectate lyase is also highly expressed in pollen (Pina et al. 2005) and is thought to mediate oscillatory pollen tube growth (Bosch et al. 2005).

While not enriched with as much pectin as pollen tube cell walls, immunohistological approaches using antibodies against specific wall epitopes have revealed that root hairs do contain pectins (Galway 2006). Although direct evidence for the role of pectin metabolism in the root hair has been in paucity so far, the identification and characterization of root hair-specific PME and pectate lyase may help to elucidate the importance of pectin dynamics during root hair growth. Tip growth is confined to the very end of the growing root hair or pollen tube, and the growing apical cell wall requires both flexibility (for the incorporation of new wall materials) and rigidity (to bear cellular turgor) (Bosch et al. 2005). Root hair-specific PME (a rigidifying factor) and pectate lyase (a loosening factor) may contribute to wall-rheological tuning processes during root hair tip growth.

Pectins were recently implicated in root hair growth. The *rol* (*repressor of lrx1*) mutant showed a defect in root hair elongation, and the corresponding locus encoded the enzyme RHAMNOSE BIOSYNTHESIS1 (RHM1), which biosynthesizes rhamnose, a pectin monosaccharide unit (Diet et al. 2006). Comparison of the *rol1-2 lrx1* double and the *lrx1* single mutant transcriptomes revealed the induction of many genes for cell wall proteins and cell wall-modifying enzymes, including extensins, PRPs, peroxidases, xyloglucan endotransglycosylases, and pectin-modifying enzymes. Among these *rol1*-induced cell wall genes were four *RHS*s: *RHS14* (pectate lyase), *RHS18* (peroxidase), *RHS19* (peroxidase), and *RHS7* (*PRP1*). The fact that impaired pectin biosynthesis leads to phenotypic effects on root hair growth and transcriptional alteration of cell wall-related genes suggests that pectins may mediate root hair development.

RHS18 and RHS19 are class III peroxidases that are thought to act on the cell wall (Bakalovic et al. 2006). Class III peroxidases have peroxidative and hydroxylic activities (Passardi et al. 2004). In the peroxidative cycle, these enzymes catalyze cross-linking between ferulates, extensins, and lignin monomers, ultimately resulting in cell wall rigidification. In contrast, the hydroxylic cycle produces ROS that can induce wall loosening by nonenzymatically breaking the cell wall polymers. Thus, a balance between these two antagonistic activities should optimize cell wall plasticity and cell expansion. Class III peroxidases also modulate cell wall loosening/stiffening by controlling the level of H_2O_2, which is required for cross-linking between phenolic compounds (Passardi et al. 2004). Root hair-specific overexpression of RHS18 slightly decreased root hair elongation compared with the control (Won et al. 2009). RHS18 overexpression might shift the balance of peroxidase-modulated cell wall dynamics toward wall stiffening in the root hair. For example, if cross-linking substrates such as extensins are more prominent in the root hair cell wall, increased peroxidase levels may favor wall stiffening.

EXP loosen the cell wall in acidic conditions via a nonhydrolytic mechanism (Cosgrove 1999; Choi et al. 2006). Root hair bulge formation seems to require local wall loosening at the bulge-forming site. A priori local wall acidification occurs at the hair-initiating point, which would be expected to activate EXP to loosen cell walls at the hair-bulging point (Bibikova et al. 1998). Exogenous application of expansin proteins causes swelling (at low doses) or bursting (at high doses) of the root hair tip (Cosgrove et al. 2002). Thus, root hair-specific EXP may mediate hair bulge formation by loosening cell walls in response to acidic apoplastic conditions and may mediate hair elongation by maintaining wall plasticity at the hair tip. There are two subgroups of EXPs, EXPA, and EXPB, which are characterized by different biochemical properties (Cosgrove 1999). *Arabidopsis EXPA7* and *EXPA18* are the first known root hair-specific expansin genes with expression in the late elongation zone of the root that is highly correlated with root hair initiation (Cho and Cosgrove 2002). The RHE has been described in detail for these two *Arabidopsis EXPA*s and their orthologous *EXPA*s from angiosperms, as previously described (Kim et al. 2006; Table 6.1).

Recently, Won et al. (2010) identified two root hair-specific monocot *EXPB*s in rice (*OsEXPB5*) and barley (*HvEXPB1*) and showed that RHEs act to regulate root hair-specific expression of these *EXPB*s. Interestingly, these root hair-specific *EXPB* orthologs are seen only in grasses (Won et al. 2010), whereas root hair-specific *EXPA* orthologs occur both in grasses and dicots (Kim et al. 2006). This suggests that root hair-specific EXPBs have evolved for grass-specific cell wall modification during root hair morphogenesis. While xyloglucans (XGs) are the primary hemicelluloses of gymnosperms, dicots, and nongraminaceous monocots (so-called type 1 cell walls), graminaceous hemicelluloses include glucuronoarabinoxylans (GAXs) and (1,3)(1,4)-β-mixed glucans, in addition to some XGs (so-called type 2 cell walls) (Carpita and Gibeaut 1993). EXPA and EXPB show substrate preferences for type 1 and type 2 cell walls, respectively (Cosgrove 1999). Therefore, it is conceivable that EXPA modifies XG-linked cell wall structures in both type 1 and 2 cell walls, whereas EXPB only works on GAX- or mixed glucan-linked wall organization in type 2 cell walls. This might be indicative of why dicots do not have root hair-specific EXPBs.

Two recent genetic studies have demonstrated that EXPAs are necessary for root hair elongation in *Arabidopsis* and rice (Lin et al. 2011; Yu et al. 2011). RNA interference of *Arabidopsis EXPA7* significantly inhibited the accumulation of *EXPA7* transcripts, which inhibited root hair growth (Lin et al. 2011). Mutations of *OsEXPA17*, the *AtEXPA7* ortholog in rice, resulted in very short root hair phenotypes in rice roots, and this root hair defect could be rescued not only by *OsEXPA17* but also by *OsEXPA30* (a paralog of *OsEXPA17*) and even by *AtEXPA7* (Yu et al. 2011). This result suggested that these root hair-specific EXPA orthologs and paralogs have been functionally conserved for root hair growth regardless of the long evolutionary divergence between monocots and dicots.

Transporters

The root hair-specific morphogenetic gene collection includes two putative membrane transporter genes: *RHS2* and *RHS15*. The *RHS15* gene product appears to be a member of the major facilitator superfamily (MFS), which typically has 12 transmembrane helices. MFS transporters are found in all organisms from bacteria to higher eukaryotes and function as secondary active transporters for small solutes when energized by chemiosmotic ion gradients (Pao et al. 1998). Plant MFS transporters can transport sugars, inorganic ions, and metal ions (Pao et al. 1998; Haydon and Cobbett 2007). RHS2 belongs to the MATE (multidrug and toxin efflux) transporter family, which contains 12–14 transmembrane domains (Brown et al. 1999). Plant MATE transporters have been implicated in the transport of diverse substrates, such as flavonoids, alkaloids, salicylic acid, and citrate (Durrett et al. 2007). However, the functions of MATE and MFS transporters in root hairs have not yet been studied. Root hairs are likely to act as major interfaces for exchanging compounds between the root and soil, and root hair-specific transporters may participate in this process. The root hair-specific MATE transporter RHS2 positively mediates root hair elongation, as indicated by loss-of-function and overexpression studies: loss-of-function decreased and overexpression increased hair length (Won et al. 2009). RHS2 might contribute to tip growth by transporting certain solutes or ions that direct tip polarity and/or sustain tip growth.

The Regulatory Pathway for Root Hair-Specific Genes

Root hair cells originate from epidermal stem cells of the root meristem. They then develop through three regions, the meristem, elongation, and differentiation zones, which represent developmental ages (in sequential order). Won et al. (2009) used the term "H-cell-specific genes" (H genes) to denote cells that are specifically expressed along the hair cell file regardless of the developmental zone and proposed three types of H genes, classified according to the spatial expression pattern along the hair cell file. The first type of H gene is expressed in the meristem and elongation zones (e.g., *RHD6*, Menand et al. 2007). The second type is expressed in the elongation and differentiation zones (e.g., *RHD2*, Foreman et al. 2003; *COBL9*, Brady et al. 2007b; *RSL4*, Yi et al. 2009). The third type is expressed in the differentiation zone only (*RHS*s; Won et al. 2009]). This spatially (and also temporally) sequential and partially overlapping expression pattern of H genes may indicate that the early genes sequentially regulate the late genes downstream. Namely, the type 1 genes (meristem and elongation zones) regulate the type 2 genes (elongation and differentiation zones), and the type 2 genes in turn regulate the type 3 genes (morphogenetic worker genes expressed in the differentiation zone).

Table 6.1 shows that the promoter of *RHD6* (a type 1 gene) does not carry an RHE, but that the promoters of type 2 (*RHD2*, *COBL9*, and *RSL4*) and type 3 (*RHS*s) genes do include RHEs. This suggests that RHD6 is regulated independently of the RHF. Conversely, RHE-containing type 2 and 3 genes could be directly regulated by the RHF. *RSL4* has been shown to be a direct target of RHD6 (Yi et al. 2009). Likewise, RHD6 might also directly target other type 2 genes such as *RHD2* and *COBL9*. In turn, these type 2 genes could directly or indirectly modulate the type 3 *RHS*s. RSL4, a bHLH transcription factor, upregulates most of the *RHS*s, as shown by transcriptome analysis of *rsl4* comparing RSL4-overexpressing and wild-type *Arabidopsis* plants (Yi et al. 2009). RSL4 might directly upregulate RHE-containing RHS genes by acting as an RHF or indirectly by way of additional transcription cascades. Another transcriptome comparison showed that RHS genes are also downregulated in the *rhd2* mutant (Jones et al. 2006), suggesting that NADPH oxidase-mediated ROS signaling is implicated in transcriptional regulation of the morphogenetic RHS genes. However, this ROS-mediated regulatory pathway for *RHS*s remains to be determined.

Perspective

Root hairs have been an attractive biological model system due to their distinctive fate-determining mechanisms and morphogenetic processes as well as their roles in root physiology. Recently developed high-throughput approaches using intensive genetic screening, omics tools, and bioinformatics have yielded an enormous amount of data. In order to enable deeper understanding of this single cell model and of fundamental cell differentiation mechanisms in general, the next challenge will be to use this data to characterize the systematic regulatory networks underlying these processes.

Regarding gene regulatory networks, in addition to genetic screening and transcriptomic and proteomic profiling, epigenomic profiling of root hair cells is needed. In particular, high-throughput epigenomic analysis, combined with transcriptomic and proteomic analyses, of the root hair cell files along the root developmental zones may help to elucidate the details of the gene-regulatory network underlying spatiotemporal root hair development.

Currently, only limited root hair metabolomic data, which is needed to understand plant–soil interactions, is available. In addition to symbiotic processes in legume plants, root hair metabolomics underlying other biotic and abiotic interactions should reveal a much wider functional spectrum of root hair metabolites, as well as the functions and interactions of their biosynthetic genes, which so far might have been simply treated as functional orphans embedded in the genomic data.

General cell-morphogenetic genes could work together with root hair-specific morphogenetic genes to construct root hairs. The root hair-specific genes, modulated by corresponding root hair-specific upstream cues, may play decisive roles in root hair morphogenesis according to hair cell position. Several dozen root hair-specific genes have been identified, but many of these genes still remain to be functionally characterized. The characterization of the regulatory relationship between general and root hair-specific morphogenetic genes presents an intriguing research goal. The understanding of this relationship will help to explain why cell type-specific genes are required, as well as the nature of their specific roles.

Acknowledgments

This work was supported by grants from the Mid-career Researcher Program (2011-0017242, NRF, MEST) and the Next-Generation BioGreen 21 programs (TAGC PJ00820701 and SSAC PJ00814102) of Rural Development Administration.

References

Arseninevic-Maksimovic, I. et al. (1997) Rhizobia modulate root-hair-specific expression of extensin genes. *Molecular Plant–Microbe Interaction*, 10, 95–101.

Baerson, S.R. et al. (2008) A functional genomics investigation of allelochemical biosynthesis in *Sorghum bicolor* root hairs. *Journal of Biological Chemistry*, 283, 3231–3247.

Bakalovic, N. et al. (2006) PeroxiBase: a class III plant peroxidase database. *Phytochemistry*, 67, 534–539.

Baumberger, N. et al. (2001) The chimeric leucine-rich repeat/extensin cell wall protein LRX1 is required for root hair morphogenesis in *Arabidopsis thaliana*. *Genes & Development*, 15, 1128–1139.

Baumberger, N. et al. (2003) Synergistic interaction of the two paralogous *Arabidopsis* genes *LRX1* and *LRX2* in cell wall formation during root hair development. *Plant Journal*, 35, 71–81.

Berken, A. et al. (2005) A new family of RhoGEFs activates the Rop molecular switch in plants. *Nature*, 436, 1176–1180.

Bernhardt, C. and Tierney, M.L. (2000) Expression of AtPRP3, a proline-rich structural cell wall protein from Arabidopsis, is regulated by cell-type-specific developmental pathways involved in root hair formation. *Plant Physiology*, 122, 705–714.

Bibikova, T.N. et al. (1998) Localized changes in apoplastic and cytoplasmic pH are associated with root hair development in *Arabidopsis thaliana*. *Development*, 125, 2925–2934.

Bögre, L. et al. (2003) Growth signalling pathways in *Arabidopsis* and the AGC protein kinases. *Trends in Plant Science*, 8, 424–431.

Bosch, M. et al. (2005) Pectin methylesterase, a regulator of pollen tube growth. *Plant Physiology*, 138, 1334–1346.

Bosch, M. and Hepler, P.K. (2006) Silencing of the tobacco pollen pectin methylesterase NtPPME1 results in retarded in vivo pollen tube growth. *Planta*, 223, 736–745.

Bradley, D.J. et al. (1992) Elicitor- and wound-induced oxidative cross-linking of a proline-rich plant cell wall protein: a novel, rapid defense response. *Cell*, 70, 21–30.

Brady, S.M. et al. (2007a) A high-resolution root spatiotemporal map reveals dominant expression patterns. *Science*, 318, 801–806.

Brady, S.M. et al. (2007b) Combining expression and comparative evolutionary analysis. The COBRA gene family. *Plant Physiology*, 143, 172–187.

Brechenmacher, L. et al. (2009) Establishment of a protein reference map for soybean root hair cells. *Plant Physiology*, 149, 670–682.

Brechenmacher, L. et al. (2010) Soybean metabolites regulated in root hairs in response to the symbiotic bacterium *Bradyrhizobium japonicum*. *Plant Physiology*, 153, 1808–1822.

Brown, M. et al. (1999) The multidrug efflux protein NorM is a prototype of a new family of transporters. *Molecular Microbiology*, 31, 393–395.

Bucher, M. et al. (1997) Two genes encoding extensin-like proteins are predominantly expressed in tomato root-hair cells. *Plant Molecular Biology*, 35, 497–508.

Caro, E. et al. (2007) A chromatin link that couples cell division to root epidermis patterning in *Arabidopsis*. *Nature*, 447, 213–217.

Carpita, N.C. and Gibeaut, D.M. (1993) Structural models of primary cell walls in flowering plants: consistency of molecular structure with the physical properties of the walls during growth. *Plant Journal*, 3, 1–30.

Cassab, G.I. (1998) Plant cell wall proteins. *Annual Review of Plant Physiology and Plant Molecular Biology*, 49, 281–309.

Cho, H.-T. and Cosgrove, D.J. (2002) Regulation of root hair initiation and expansin gene expression in *Arabidopsis*. *Plant Cell*, 14, 3237–3253.

Choi, D. et al. (2006) Expansins: expanding importance in plant growth and development. *Physiologia Plantarum*, 126, 511–518.

Clowes, F.A.L. (2000) Pattern in root meristem development in angiosperms. *New Phytologist*, 146, 83–94.

Cole, R.A. and Fowler, J.E. (2006) Polarized growth: maintaining focus on the tip. *Current Opinion in Plant Biology*, 9, 1–10.

Cosgrove, D.J. (1999) Enzymes and other agents that enhance cell wall extensibility. *Annual Review of Plant Physiology and Plant Molecular Biology*, 50, 391–417.

Cosgrove, D.J. et al. (2002) The growing world of expansins. *Plant and Cell Physiology*, 43, 1436–1444.

Costa, S. and Shaw, P. (2006) Chromatin organization and cell fate switch respond to positional information in Arabidopsis. *Nature*, 439, 493–496.

Covitz, P.A. et al. (1998) Expressed sequence tags from a root-hair-enriched *Medicago trunculata* cDNA library. *Plant Physiology*, 117, 1325–1332.

Diet, A. et al. (2004) The *enl* mutants enhance the *lrx* root hair mutant phenotype of *Arabidopsis thaliana*. *Plant and Cell Physiology*, 45, 734–741.

Diet, A. et al. (2006) The Arabidopsis root hair cell wall formation mutant lrx1 is suppressed by mutations in the RHM1 gene encoding a UDP-L-rhamnose synthase. *Plant Cell*, 18, 1630–1641.

Dolan, L. (1996) Pattern in the root epidermis: an interplay of diffusible signals and cellular geometry. *Annals of Botany*, 77, 547–553.

Durrett, T.P. et al. (2007) The FRD3-mediated efflux of citrate into the root vasculature is necessary for efficient iron translocation. *Plant Physiology*, 144, 197–205.

Enami, K. et al. (2009) Differential expression control and polarized distribution of plasma membrane-resident SYP1 SNAREs in *Arabidopsis thaliana*. *Plant and Cell Physiology*, 50, 280–289.

Estévez, J.M. et al. (2006) Characterization of synthetic hydroxyproline-rich proteoglycans with arabinogalactan protein and extensin motifs in Arabidopsis. *Plant Physiology*, 142, 458–470.

Favery, B. et al. (2001) KOJAK encodes a cellulose synthase-like protein required for root hair cell morphogenesis in Arabidopsis. *Genes Dev*, 15, 79–89.

Föhse, D. et al. (1991) Phosphorus efficiency of plants. 2. Significance of root radius, root hairs and catioin-anion balance for phosphorus influx in seven plant species. *Plant Soil*, 132, 261–272.

Foreman, J. et al. (2003) Reactive oxygen species produced by NADPH oxidase regulate plant cell growth. *Nature*, 422, 442–446.

Fowler, T.J. et al. (1999) Characterization and expression of four proline-rich cell wall protein genes in Arabidopsis encoding two distinct subsets of multiple domain proteins. *Plant Physiology*, 121, 1081–1091.

Gachon, C.M., Langlois-Meurinne, M., Henry, Y., and Saindrenan, P. (2005) Transcriptional co-regulation of secondary metabolism enzymes in Arabidopsis: functional and evolutionary implications. *Plant Molecular Biology*, 58, 229–245.

Galway, M.E. (2006) Root hair cell walls: filling in the framework. *Canadian Journal of Botany*, 84, 613–621.

Galway, M.E. et al. (2011) Root hair-specific disruption of cellulose and xyloglucan in AtCSLD3 mutants, and factors affecting the post-rupture resumption of mutant root hair growth. *Planta*, 233, 985–999.

Gibson, K.E. et al. (2008) Molecular determinants of a symbiotic chronic infection. *Annual Review of Genetics*, 42, 413–441.

Gilroy, S. and Jones, D.L. (2000) Through form to function: root hair development and nutrient uptake. *Trends in Plant Science*, 5, 56–60

Goldstein, L.S.B. and Philip, A.V. (1999) The road less traveled: emerging principles of kinesin motor utilization. *Annual Review of Cell and Developmental Biology*, 15, 141–183.

Grierson, C. and Schiefelbein, J. (2002) Root hairs. In: C.R. Somerville and E.M. Meyerowitz (eds) *The Arabidopsis Book*. American Society of Plant Biologists, Rockville, MD.

Grieson, G. and Schiefelbein, J. (2008) Genetics of root hair formation. In: A.M.C. Emons and T. Ketelaar (eds) *Root Hairs*. Springer-Verlag, Heidelberg, Germany, pp. 1–25.

Gu, Y. et al. (2006) Members of a novel class of Arabidopsis Rho guanine nucleotide exchange factors control Rho GTPase-dependent polar growth. *Plant Cell*, 18, 366–381.

Guo, H. and Moose, S.P. (2003) Conserved noncoding sequences among cultivated cereal genomes identify candidate regulatory sequence elements and patterns of promoter evolution. *Plant Cell*, 15, 1143–1158.

Haberer, G., Hindemitt, T., Meyers, B.C., and Mayer, K.F.X. (2004) Transcriptional similarities, dissimilarities, and conservation of cis-elements in duplicated genes of Arabidopsis. *Plant Physiology*, 136, 3009–3022.

Haydon, M.J. and Cobbett, C.S. (2007) A novel major facilitator superfamily protein at the tonoplast influences zinc tolerance and accumulation in Arabidopsis. *Plant Physiology*, 143, 1705–1719.

Held, M.A. et al. (2004) Diisodityrosine is the intermolecular cross-link of isodityrosinerich extensin analogs cross-linked in vitro. *Journal of Biological Chemistry*, 279, 55474–55482.

Hematy, K. et al. (2007) A receptor-like kinase mediates the response of Arabidopsis cells to the inhibition of cellulose synthesis. *Current Biology*, 17, 922–931.

Hruz, T. et al. (2008) Genevestigator V3: a reference expression database for the meta-analysis of transcriptomes. *Advanced Bioinformatics*, 2008, 420747.

Humphrey, T.V. et al. (2007) Sentinels at the wall: cell wall receptors and sensors. *New Phytologist*, 176, 7–21.

Inada, D.C. et al. (2003) Conserved noncoding sequences in the grasses. *Genome Research*, 13, 2030–2041.

Jareborg, N. et al. (1999) Comparative analysis of noncoding regions of 77 orthologous mouse and human gene pairs. *Genome Research*, 9, 815–824.

Jiang, L. et al. (2005) VANGUARD1 encodes a pectin methylesterase that enhances pollen tube growth in the Arabidopsis style and transmitting tract. *Plant Cell*, 17, 584–596.

Jones, M.A. and Grierson, C.S. (2003) A simple method for obtaining cell-specific cDNA from small numbers of growing root-hair cells in *Arabidopsis thaliana*. *Journal of Experimental Botany*, 54, 1373–1378.

Jones, D.A. and Jones, J.D.G. (1997) The role of leucine-rich repeat proteins in plant defenses. *Advances in Botanical Research*, 24, 89–167.

Jones, M.A. et al. (2006) Analysis of the root-hair morphogenesis transcriptome reveals the molecular identity of six genes with roles in root-hair development in Arabidopsis. *Plant Journal*, 45, 83–100.

Jones, M.A. et al. (2002) The Arabidopsis Rop2 GTPase is a positive regulator of both root hair initiation and tip growth. *Plant Cell*, 14, 763–776.

Kandasamy, M.K. et al. (2007) Class-specific interaction of profilin and ADF isovariants with actin in the regulation of plant development. *Plant Cell*, 19, 3111–3126.

Kaothien, P. et al. (2005) Kinase partner protein interacts with the LePRK1 and LePRK2 receptor kinases and plays a role in polarized pollen tube growth. *Plant Journal*, 42, 492–503.

Kaplinsky, N.J. et al. (2002) Utility and distribution of conserved noncoding sequences in the grasses. *Proceedings of the National Academy of Science of the United States of America*, 99, 6147–6151.

Ketelaar, T. and Emons, A.M. (2008) The actin cytoskeleton in root hairs: a cell elongation device. In: A.M.C. Emons and T. Ketelaar (eds) *Root Hairs*. Springer-Verlag, Heidelberg, Germany, pp. 211–232.

Kieliszewski, M.J. and Lamport, D.T.A. (1994) Extensin: repetitive motifs, functional sites, post-translational codes, and phylogeny. *Plant Journal*, 5, 157–172.

Kim, D.W. et al. (2006) Functional conservation of a root hair cell-specific *cis*-element in angiosperms with different root hair distribution patterns. *Plant Cell*, 18, 2958–2970.

Kim, D.Y. et al. (2007a) The ABC transporter AtPDR8 is a cadmium extrusion pump conferring heavy metal resistance. *Plant Journal*, 50, 207–218.

Kim, C.M. et al. (2007b) OsCSLD1, a cellulose synthase-like D1 gene, is required for root hair morphogenesis in rice. *Plant Physiology*, 143, 1220–1230.

Kleis-San Francisco, S.M. and Tierney, M.L. (1990) Isolation and characterization of a proline-rich cell wall protein from soybean seedlings. *Plant Physiology*, 94, 1897–1902.

Kost, B. (2008) Spatial control of Rho (Rac-Rop) signaling in tip-growing plant cells. *Trends in Cell Biology*, 18, 119–127.

Kwak, S.H. et al. (2005) Positional signaling mediated by a receptor-like kinase in Arabidopsis. *Science*, 307, 1111–1113.

Kwasniewski, M. et al. (2010) Global analysis of the root hair morphogenesis transcriptome reveals new candidate genes involved in root hair formation in barley. *Journal of Plant Physiology*, 167, 1076–1083.

Lee, S.H. and Cho, H.-T. (2006) PINOID positively regulates auxin efflux in Arabidopsis root hair cells and tobacco cells. *Plant Cell*, 18, 1604–1616.

Libault, M. et al. (2010a). Root hair systems biology. *Trends in Plant Science*, 15, 641–650.

Libault, M. et al. (2010b). Complete transcriptome of soybean root hair cell, a single cell model, and its alteration in response to *Bradyrhizobium japonicum* infection. *Plant Physiology*, 152, 541–552.

Libault, M. et al. (2010c) An integrated transcriptome atlas of the crop model Glycine max and its use in comparative analyses in plants. *Plant Journal*, 63, 86–99.

Limpens, E. and Bisseling, T. (2008) Nod factor signal transduction in the rhizobium–legume symbiosis. In: A.M.C. Emons and T. Ketelaar (eds) *Root Hairs*. Springer-Verlag, Heidelberg, Germany, pp. 249–276.

Lin, C. et al. (2011) Root hair-specific EXPANSIN A7 is required for root hair elongation in Arabidopsis. *Molecular Cells*, 31, 393–397.

Lisso, J., Steinhauser, D., Altmann, T., Kopka, J., and Mussig, C. (2005) Identification of brassinosteroid-related genes by means of transcript co-response analyses. *Nucleic Acids Research*, 33, 2685–2696.

Masucci, J. and Schiefelbein, J.W. (1994) The rhd6 mutation of *Arabidopsis thaliana* alters root-hair initiation through an auxin and ethylene-associated process. *Plant Physiology*, 106, 1335–1346.

Masucci, J.D. and Schiefelbein, J.W. (1996) Hormones act downstream of TTG and GL2 to promote root hair outgrowth during epidermis development in the Arabidopsis root. *Plant Cell*, 8, 1505–1517.

McCormack, E. et al. (2005) Handling calcium signaling: Arabidopsis CaMs and CMLs. *Trends in Plant Science*, 10, 383–389.

Menand, B. et al. (2007) An ancient mechanism controls the development of cells with a rooting function in land plants. *Science*, 316, 1477–1480.

Micheli, F. (2001) Pectin methylesterases: cell wall enzymes with important roles in plant physiology. *Trends in Plant Science*, 6, 414–419.

Molendijk, A.J. et al. (2001) *Arabiodpsis thaliana* Rop GTPases are localized to tips of root hairs and control polar growth. *EMBO Journal*, 20, 2779–2788.

Nestler, J. et al. (2011) Conserved and unique features of the maize (*Zea mays* L.) root hair proteome. *Journal of Proteome Research*, 10, 2525–2537.

Oppenheimer, D.G. et al. (1997) Essential role of a kinesin-like protein in Arabidopsis trichome morphogenesis. *Proceedings of the National Academy of Science of the United States America*, 94, 6261–6266.

Oyama, T. et al. (2002) The IRE gene encodes a protein kinase homologue and modulates root hair growth in Arabidopsis. *Plant Journal*, 30, 289–299

Pao, S.S. et al. (1998) Major facilitator superfamily. *Microbiology and Molecular Biology Review*, 62, 1–34.

Park, S. et al. (2011) A role for CSLD3 during cell-wall synthesis in apical plasma membranes of tip-growing root-hair cells. *Nature Cell Biology*, 13, 973–980.

Parker, J.S. et al. (2000) Genetic interactions during root hair morphogenesis in Arabidopsis. *Plant Cell*, 12, 1961–1974.

Passardi, F. et al. (2004) Performing the paradoxical: how plant peroxidases modify the cell wall. *Trends in Plant Science*, 11, 534–540.

Persson, S., Wei, H., Milne, J., Page, G.P., and Somerville, C.R. (2005) Identification of genes required for cellulose synthesis by regression analysis of public microarray data sets. *The Proceedings of the National Academy of Sciences USA*, 102, 8633–8638.

Peters, N.K. et al. (1986) A plant flavone, luteolin, induces expression of *Rhizobium meliloti* nodulation genes. *Science*, 233, 977–980.

Pina, C. et al. (2005) Gene family analysis of the Arabidopsis pollen transcriptome reveals biological implications for cell growth, division control and gene expression regulation. *Plant Physiology*, 138, 744–756.

Popescu, S.C. et al. (2007) Differential binding of calmodulin-related proteins to their targets revealed through high-density Arabidopsis protein microarrays. *Proceedings of the National Academy of Science of the United States of America*, 104, 4730–4735.

Reddy, V.S. and Reddy, A.S. (2004) Developmental and cell-specific expression of ZWICHEL is regulated by the intron and exon sequences of its gene. *Plant Molecular Biology*, 54, 273–293.

Rentel, M.C. et al. (2004) OXI1 kinase is necessary for oxidative burst-mediated signaling in Arabidopsis. *Nature*, 427, 858–861.

Ringli, C. et al. (2002) ACTIN2 is essential for bulge site selection and tip growth during root hair development of Arabidopsis. *Plant Physiology*, 129, 1464–1472.

Ringli, C. et al. (2005) The Arabidopsis root hair mutants der2-der9 are affected at different stages of root hair development. *Plant Cell Physiology*, 46, 1046–1053.

Roberts, M.R. (2003) 14-3-3 Proteins find new partners in plant cell signaling. *Trends in Plant Science*, 8, 218–223.

Röhm, M. and Werner, D. (1987) Isolation of root hairs from seedlings of *Pisum sativum*. Identification of root hair-specific proteins by in situ labeling. *Physiologia Plantarum*, 69, 129–136.

Roudier, F. et al. (2002) The COBRA family of putative GPI-anchored proteins in Arabidopsis. A new fellowship in expansion. *Plant Physiology*, 130, 538–548.

Schiefelbein, J. (2000) Specification of root hair cells. In: R.W. Ridge and A.M.C. Emons (eds) *Root Hairs*. Springer, Heidelberg, Germany, pp. 197–209.

Schiefelbein, J. and Somerville, C. (1990) Genetic control of root hair development in *Arabidopsis thaliana*. *Plant Cell*, 2, 235–243.

Schneider, K. et al. (1997) Structural and genetic analysis of epidermal cell differentiation in Arabidopsis primary roots. *Development*, 124, 1789–1798.

Schoonheim, P.J. et al. (2007) A comprehensive analysis of the 14-3-3 interactome in barley leaves using a complementary proteomics and two-hybrid approach. *Plant Physiology*, 143, 670–683.

Shelp, B.J. et al. (1999) Metabolism and functions of gamma-aminobutyric acid. *Trends in Plant Science*, 4, 446–452.

Subramanian, S., Stacey, G., and Yu, O. (2007) Distinct, crucial roles of flavonoids during legume nodulation. *Trends in Plant Science*, 12, 282–285.

Takeda, S. et al. (2008) Local positive feedback regulation determines cell shape in root hair cell. *Science*, 319, 1241–1244.

Tamura, K. et al. (2005) KATAMARI1/MURUS3 is a novel Golgi membrane protein that is required for endomembrane organization in Arabidopsis. *Plant Cell*, 17, 1764–1776.

Uemura, T. et al. (2004) Systematic analysis of SNARE molecules in Arabidopsis: dissection of the post-Golgi network in plant cells. *Cell Structure and Function*, 29, 49–65.

Wan, J. et al. (2005) Proteomic analysis of soybean root hairs after infection by *Bradyrhizobium japonicum*. *Molecular Plant–Microbe Interactions*, 18, 458–467.

Wen, T.J. et al. (2005) The *roothairless1* gene of maize encodes a homolog of sec3, which is involved in polar exocytosis. *Plant Physiology*, 138, 1637–1643.

Willats, W.G. et al. (2001) Modulation of the degree and pattern of methyl-esterification of pectic homogalacturonan in plant cell walls. Implications for pectin methyl esterase action, matrix properties, and cell adhesion. *Journal of Biological Chemistry*, 276, 19404–19413.

Won, S.-K. et al. (2010) Root hair-specific *EXPANSIN B* genes have been selected for Graminaceae root hairs. *Molecular Cells*, 30, 369–376.

Won, S.-K. et al. (2009) *Cis*-element- and transcriptome-based screening of root hair-specific genes and their functional characterization in Arabidopsis. *Plant Physiology*, 150, 1459–1473.

Yalovsky, S. et al. (2008) Regulation of membrane trafficking, cytoskeleton dynamics, and cell polarity by ROP/RAC GTPases. *Plant Physiology*, 147, 1527–1543.

Yang, G. et al. (2007) A mutation in MRH2 kinesin enhances the root hair tip growth defect caused by constitutively activated ROP2 small GTPase in Arabidopsis. *PLos One*, 10, e1074.

Yi, K. et al. (2009) A basic helix-loop-helix transcription factor controls cell growth and size in root hairs. *Nature Genetics*, 42, 264–267.

Yu, Z. et al. (2011) Root hair-specific expansins modulate root hair elongation in rice. *Plant Journal*, 66, 725–734.

Xu, C.R. et al. (2005) Histone acetylation affects expression of cellular patterning genes in the Arabidopsis root epidermis. *Proceedings of the National Academy of Science of the United States of America*, 102, 14469–14474.

Zhang, W. et al. (2005) *Cis*-regulatory element-based targeted gene finding: genome-wide identification of abscisic acid- and abiotic stress-responsive genes in *Arabidopsis thaliana*. *Bioinformatics*, 21, 3074–3081.

7 The Effects of Moisture Extremes on Plant Roots and Their Connections with Other Abiotic Stresses

Laura M. Vaughn and Henry T. Nguyen

Introduction

There is a large amount of crop loss on a yearly basis due to various abiotic stresses. Billions of dollars in yield losses are accrued every growing season because of drought, flooding, and other abiotic stressors. Climate change is expected to exacerbate the problem on both micro- and macroscales. Regions currently affected by adverse environmental conditions may expand or shift into areas that were previously unaffected. Because of the challenges of crop production today and in the future, it is important that we develop an arsenal of crops that can maintain yield and quality in a variety of soil and climatic conditions. Ultimately, plants bred to withstand and thrive under any number of stress types will be crucial to allowing us to feed humanity in a changeable world.

Attempts to breed for or genetically engineer abiotic stress tolerance into important crop plants have been largely unsuccessful (Sinclair 2011). This is in part due to the complex nature of stress tolerance. It is not always clear what phenotype should be measured to qualify a plant as tolerant. Also, many possible combinations and severity levels of abiotic stress may be faced by plants, and in turn, the plants have a number of biochemical pathways that are altered to combat the changes in the environment. This has led breeders and scientists to genomic methods of analysis to try and understand what kinds of responses work in concert to lead to tolerance. These methods often utilize the natural range of tolerance to abiotic stresses available in germplasm stocks. Natural variation can be exploited to carry out quantitative trait loci (QTL) mapping and genome-wide association studies to locate genomic regions that contribute to tolerance. Additionally, by observing RNA, protein, metabolite, and ion levels in specific genotypes under control and stress conditions, we can begin to understand the changes a plant makes when faced with stress on a global level.

In the quest for stress-adapted plants, the root system has been underexploited as a resource for the breeding of tolerance (Lynch 2008). A strong root system is imperative for the plant to efficiently utilize available water and nutrients, as well as excluding compounds causing toxicity. Part of the reason the roots are overlooked is due to the difficulty of phenotyping them. Plants must usually be sacrificed to observe their intact root system, causing challenges for observing root dynamics over time. Even upon sacrifice, it is difficult to extract an intact root system from the growth medium and labor-intensive to quantify lengths, diameters, and orders of branching. Specialized, high-throughput root phenotyping facilities are now appearing and various observation methods are being applied (Zhu et al. 2011), but more refined technologies need to be developed for better imaging and analysis in the greenhouse and in the field. In addition to the problem of

Root Genomics and Soil Interactions, First Edition. Edited by Martin Crespi.

root phenotyping, when analysis requires root tissue from specific regions or even cell types, a large number of biological replicates are needed to gather enough tissue. This is often the case for transcriptome, proteome, and metabolome studies.

Even though studies of plant root systems face challenges, it is increasingly clear that roots play important roles in response to abiotic stress. Therefore, the complex interactions between roots and stresses must be tackled by breeders and geneticists to arrive at tolerant plants with acceptable yields (Gewin 2010). In particular, unfavorable environmental conditions involving too much or too little water cause a great deal of yield loss every year. Both of these stresses bring about changes in the appearance and biology of plant root systems. In this chapter, we explore how root growth and morphology change under these two stresses and what biochemical alterations accompany them. The combination of phenomics and genomics draws an emerging picture of differences and commonalities between stresses, and this may be the key to the development of multi-stress resistant crops.

Low Water Availability—Drought

The Anatomical and Physiological Effects of Low Water Stress and Drought on Plants

Today, drought is a significant cause of crop loss globally (Cattivelli et al. 2008). Both predicted changes in global precipitation and a looming scarcity of ground water sources for irrigation threaten to make drought an even larger problem in the future (Foley et al. 2011). So, proper land and water management are crucial, but there is also hope that crops able to thrive with less water input can be bred to keep up with global food demands. About 70% of freshwater used globally is for irrigation, and yields would drop an estimated 20% without it (Siebert and Döll 2010; Foley et al. 2011). There are many possible approaches to breeding crops that require less water and perform better in times of low water stress. These include modifying the mechanisms plants use for drought escape, drought avoidance, and those they employ for bonafide drought tolerance (Ludlow 1989). Much effort to date has been placed into altering the shoot for traits such as increased water use efficiency, altered stomatal conductance, slower wilting, improved osmotic adjustment, and increased scavenging of reactive oxygen species (ROS) (Valliyodan and Nguyen 2006; Reynolds and Tuberosa 2008; Manavalan et al. 2009). However, it has been observed over many species that upon low water stress, biomass accumulation is directed toward the root system, causing an increased root : shoot ratio (Sharp and Davies 1989). This indicates that control of the amount and type of root growth is an important mechanism for drought response.

During low water stress, there is an increase in tap root length and increased numbers and spans of other root types (reviewed in Sharp et al. 2004). Before any stress occurs, though, it is beneficial for the plant to have a well-developed root system in place. Varieties with a combination of good constitutive root growth prior to water deficit and continued elongation ability when water is scarce are often better performers under drought (Azhiri-Sigari et al. 2000; Kamoshita et al. 2000). Therefore, in screening crops for varieties with improved drought tolerance, evaluations of root growth under favorable conditions may help in the selection of lines to assess under poor water availability.

In studies of root growth in relation to drought tolerance, the main adaptive behavior focused on is the ability to extend at least some branches of the root system deeper into the soil in search of water reservoirs. Generally, this has meant measuring the maximum length a root system can achieve under various levels of water stress. However, some investigations have focused on location-specific elongation of roots when faced with water deficit. In maize seedlings, the 3 mm nearest to the root tip of water-stressed plants continues to elongate at a rate similar to that of control roots, although elongation is repressed further from the tip (Sharp et al. 1988). A similar trend was noted for water-stressed soybean seedling root tips (Yamaguchi et al. 2010). So, root extension can continue during times of low water, but it is restricted to the apical portion of the organ in these cases.

Many other root traits are also thought to have a role in adapting plants to low water conditions. Thick roots can be important for drought survival by increasing root volume for water uptake and, in some areas, helping the plant to penetrate hard soils (Hemamalini et al. 2000). Using the model plant *Arabidopsis thaliana*, it has been observed that lateral root development is inhibited under drought-simulating conditions, and the hormone abscisic acid (ABA) is involved (Xiong et al. 2006). So, root branching can be altered when plants are faced with water shortages. Other studies using *A. thaliana* have found that root hairs become misshapen, shortened, or disappear under drought-like conditions (Schnall and Quatrano 1992; Vartanian et al. 1994). However, these observations are not always repeatable (Xiong et al. 2006), possibly because root hair development is very sensitive to slight changes in the environment. Both branching and root hairs increase root system surface area and so may be critical areas of focus to increase water uptake under drought. Another root trait that was recently linked to drought tolerance is the formation of a high amount of root cortical aerenchyma (Zhu et al. 2010). Cortical aerenchymas are air spaces opened up in plant tissues via programmed cell death (PCD). The authors show evidence that the reduction of living tissue in the core of maize roots frees up metabolic resources to continue root growth and therefore soil exploration for water resources.

For all plant tissues, it is important to note that desiccation, low water stresses, and drought are related but distinct stresses. All three types are often used to evaluate plant response to water-limiting conditions. Desiccation involves the removal of a plant from its soil medium and a period of air drying, which does remove roots from their water source but also creates other stresses due to the sudden exposure of the below-ground organs to air. Low water stresses can be imposed at varying severities, but, especially in laboratory settings, the rest of the plant's environment is ideal. Drought in the agricultural sense means a prolonged period of time with no moisture input for crops, but also very often involves a coinciding period of high heat. The combination of water and heat stress hinders plant development and causes yield losses to an even greater extent than low water stress alone (Mittler 2006). For many areas prone to drought involving heat and low precipitation, it will be important that crop improvement efforts can provide varieties with vigor under both.

Studies of Low Water Stress and Rice Root Response

In the search for root characteristics that lead to better tolerance and yield under low water conditions, rice (*Oryza sativa*) is a crop that has been studied extensively. In particular, upland rice varieties are mostly dependent on rainfall and therefore subjected to drought on occasion. However, even lowland varieties that usually start the growing season with an excess of water sometimes face a deficit later on in rain-dependent agricultural systems (Kamoshita et al. 2002). The timing of the drought is also important, as rice is more vulnerable to the stress when it is entering into the reproductive stage.

QTL studies have been carried out to locate genetic regions contributing to various characteristics associated with drought tolerance, and some consider root traits. Measured traits include maximum root length, total root length, total root number, root volume, root thickness, total root dry weight, and root–shoot ratio (Hemamalini et al. 2000; Price et al. 2002; Venuprasad et al. 2002; Zheng et al. 2003; Macmillan et al. 2006). However, even more QTL mapping studies have focused on rice root traits without application of stress to find genomic areas leading to a robust root system (Champoux et al. 1995; Yadav et al. 1997; Kamoshita et al. 2002). The mapped loci should contribute to the establishment of a strong root system and may also be important in maintaining favorable root dynamics during low water stress. These basic morphological traits are studied in concert with the mechanical trait known as root penetration index in some QTL studies (Zheng et al. 2000). Artificial-environment experiments for root penetration expose roots to a layer of wax and petroleum and count how many and how far the roots can proceed through the matrix (Yu et al. 1995). This trait is a focus of root drought studies because of the hardpan soil conditions that can arise during times of water deficit. However, it is important to be mindful that soil conditions vary in drought-prone regions and microenvironments can occur even in a single field. Also, the ability to penetrate the wax system may not translate to better rooting in field settings (Clark et al. 2000; Gowda et al. 2011).

Through QTL analysis, many genetic regions contributing a large percentage of the predicted phenotypic variance for root traits have been mapped. Under water stress conditions, loci contributing to over 20% of the phenotypic variance were found for root volume, root thickness, and root dry weight (Hemamalini et al. 2000; Venuprasad et al. 2002). These studies used a mapping population created from the same parental lines, but they were grown in different environments. Robust QTL for root dry weight and root volume on chromosome 3 as well as root thickness on chromosome 2 were shared between both studies. The same regions on chromosomes 2 and 3 were found to contribute to root traits under nonstress conditions in other QTL experiments, and some used different mapping populations (Champoux et al. 1995; Yadav et al. 1997; Zheng et al. 2000).

Also under nonstress conditions, a QTL for root length, deep rooting length, and/or root thickness is consistently found in IR64 × Azucena populations on chromosome 9 (Yadav et al. 1997; Hemamalini et al. 2000; Zheng et al. 2000; Venuprasad et al. 2002). In other mapping populations using Azucena as a parent, QTLs for root traits under water deficit have been mapped to the same region as well (Price et al. 2002; Zheng et al. 2003; MacMillan et al. 2006). A QTL for root thickness in nonstress conditions mapped from CO39 × Moroberekan recombinant inbred lines (RILs) is also located in this area (Champoux et al. 1995) as well as one in an IR58821 × IR52561 population (Kamoshita et al. 2002). In another double haploid population, CT9993 × IR62266, a QTL for penetrated root dry mass under water stress conditions mapped to this same region of chromosome 9 (Zhang et al. 2001; Nguyen et al. 2004).

In order to make the data from various QTL studies for these traits more useful to breeders, meta-QTL experiments have been performed to locate genetic regions important for drought tolerance over many genetic backgrounds and environmental settings. A meta-QTL analysis for a Bala × Azucena mapping population concluded that there was not much overlap between QTLs for root traits and QTLs for drought avoidance parameters (Khowaja et al. 2009). This does not mean root traits could not contribute to drought tolerance in this population. Environmental conditions such as water source depth and soil matrix penetrability may be important to the manifestation of root morphology for drought tolerance, and this has not been addressed in many experiments (Khowaja et al. 2009). The study did find a meta-QTL for root traits, including length, mass, and thickness, on the same region of chromosome 9 as found in the studies listed earlier. This region of chromosome 9 was taken from the variety Azucena and placed in Kalinga III, a variety used in the uplands of India, and gave a longer primary root under both control and water stress conditions (Steele et al. 2006). Courtois et al. (2009) performed a meta-QTL analysis of rice root traits over several rice mapping populations that narrowed the chromosome 9 QTL interval down to 20 kb. Neither of the genes in the interval stands out as a good candidate for rooting architecture, but a gene 40 kb away is an auxin efflux carrier. Auxin's role in root elongation and tropic responses is well established, and further studies may prove this locus is a major player in this hormonal regulation of root growth.

Another meta-QTL analysis looked specifically for regions affecting grain yield under drought conditions (Swamy et al. 2011). Defined, repeatable QTL contributing to higher yield during low water stress will be very important to include in drought tolerance breeding programs as yield is the most important factor for the farmer. In this meta-analysis, 14 meta-QTL for grain yield were mapped using data from 15 independent QTL trials. They conclude that a meta-QTL on chromosome 12 is the most important trait to incorporate into elite lines for yield under drought tolerance since it is present in a high percentage of tolerant lines and is predicted to be responsible for 28% of the phenotypic variance for this trait. The meta-QTL on chromosome 12 does not map to regions with known root QTL for drought stress tolerant traits, but many overlap with other drought yield meta-QTL regions. These include $MQTL_{1.2}$ (Yadav et al 1997; Zheng et al. 2000; Price et al. 2002; Kamoshita et al. 2002; Khowaja et al 2009), $MQTL_{1.3}$ (Hemamalini et al. 2000; Zheng et al. 2000, 2003, 2008; Price et al. 2002; Khowaja et al. 2009), $MQTL_{1.4}$ (Hemamalini et al. 2000; Zheng et al. 2000; Kamoshita et al. 2002; Price et al. 2002; Nguyen et al. 2004), $MQTL_{3.1}$ (Champoux et al. 1995; Yadav et al. 1997; Hemamalini et al. 2000; Zheng et al. 2000, 2003; Venuprasad et al. 2002; Khowaja et al. 2009), $MQTL_{4.1}$ (Kamoshita et al. 2002; Price et al. 2002), $MQTL_{8.1}$ (Yadav et al. 1997; Hemamalini et al. 2000; Kamoshita et al. 2002), and $MQTL_{8.2}$ (Champoux et al. 1995; Khowaja et al. 2009). Of these, $MQTL_{1.3}$ and $MQTL_{1.4}$ are also ranked as high-priority regions for breeding programs (Khojawa et al. 2009; Swamy et al. 2011). It is not clear at this time if any of the root traits are responsible for the yield increase under drought associated with these QTL. Therefore, further study will be important to establish how closely they are linked, and ultimately what the causal loci are. Also, it must be remembered that there may be a negative correlation between yield and increased root length, mass, and/or thickness at these loci (Cattivelli et al. 2008). In some cases, grain yield may be negatively associated with root traits, like length, under nonstress conditions and positively associated with it under stress conditions (Venuprasad et al. 2002). These correlations may be difficult to reconcile, but overall it is likely that the best strategy is to plant lines with a good root system in place to buffer yield against any water stress that arises.

Further genomic studies to find genes contributing to drought stress response include transcriptome, proteome, and metabolome profiling. However, for rice root tissue specifically under low water stress conditions, not many experiments of this nature exist. Yang et al. (2004), Rabello et al. (2008), and Wang et al. (2011a) looked for gene expression differences in rice roots under simulated drought conditions. Using an Affymetrix rice genome array chip, Wang et al. (2011a) profiled control and stressed roots at each of two developmental stages and found 1154 and 1114 differentially expressed genes (DEGs) at tillering and panicle elongation stages, respectively. Root-specific upregulated genes included many transcription factors as well as genes involved in metabolism, cell wall expansion, and response to phytohormones. This study also looked at DEGs in leaves and the panicle under drought. They found a group of transcripts commonly induced in all three tissues studied, including late-embryogenesis abundant (LEA) proteins, dehydrin family, protein phosphatase 2C family, and abiotic stress response proteins. Yang et al. (2004) found 121 transcripts affected by water deficit in various sections of the rice root system, and Rabello et al. (2008) found 84 transcripts expressed exclusively in their drought-tolerant genotype. For both studies, there is much overlap with the functional categories found in the genome-wide profiling. The highest number of transcripts found by Yang et al. (2004) are in the category of cell organization and cell wall biogenesis. This group includes expansins, several of which were also DEGs in Wang et al. (2011a) and have been considered as candidate genes for rice root QTLs (Vinod et al. 2006). A critical finding of the Wang et al. (2011a) study was that there were several genes regulated in opposing manners depending on tissue type and/or stage. This finding highlights the importance of temporal tissue and even cell-specific studies instead of making generalizations about abiotic stress response based on a single tissue type at one time point.

Currently, there are only a few examples where altering transcript levels of specific root-related rice genes have led to increased drought tolerance. The overexpression of a metallothionein expressed highly in roots, *OsMT1a*, allowed plants to grow better under osmotic stress and dehydration (Yang et al. 2009). The authors believe that the increased *OsMT1a* may activate antioxidant enzymes and zinc finger transcription factors, leading to enhanced ROS scavenging and, therefore, drought tolerance. Both Rabello et al. (2008) and Wang et al. (2011a) showed changes in transcript levels of metallothionein-family proteins under drought stress. Other overexpression experiments involved NAC (NAM, ATAF1/2, and CUC2) domain transcription factors, which are plant-specific and known to act during biotic and abiotic stress. One NAC gene, *OsNAC10*, was overexpressed specifically in roots. It increased root diameter over nontransgenic plants and significantly increased grain yield under drought conditions in the field (Jeong et al. 2010). Overexpression of another NAC gene that has high root expression levels, *OsNAC45*, increased recovery of seedlings from desiccation stress (Zheng et al. 2009). Globally, the expression levels of several NAC domain transcription factors were shown to be differentially regulated upon drought stress (Wang et al. 2011a).

Maize Root Genomics under Low Water Stress and Simulated Drought

Other major crop plants lag behind rice in studies of root traits under water deficit conditions. Changes to maize (*Zea mays*) morphological traits during drought are of great interest, but genomic analyses of roots under this stress are few. In general, to adapt to drought conditions, maize roots will allocate less resources to growth and water extraction in the upper layers of soil. Development of new crown and lateral roots is decreased, but elongation of established major roots may be promoted (reviewed in Ruta et al. 2010). This allocation of resources in the maize root system under drought is similar to that seen in rice root studies.

Some maize root trait QTL studies have been carried out in the field, but none of these were performed under low water stress. One analysis found a correlation between the number of nodal roots and the concentration of ABA in the xylem (Lebreton et al. 1995). This finding could have significance because of the role of ABA in drought signaling. Besides nodal roots, this study also located QTL for seminal root number and root pulling force. Later, another population was used to locate QTL for nodal root number and thickness on specific internodes (Guingo et al. 1998). Using syntenic relationships between maize and rice chromosomes, a region of maize chromosome 1 that harbors several QTL for root traits, including some from Lebreton et al. (1995), corresponds to a region of rice chromosome 3 that many rice root QTL localize to (Tuberosa et al. 2002a). This is the region of rice grain yield MetaQTL 3.1 (Swamy et al. 2011), which colocalizes with rice root QTL from several studies as detailed in the previous section. So, conserved genes in this chromosomal segment appear to be important for root system morphology and may play roles in yield for multiple crops as well.

Additional root trait QTL studies have been carried out in hydroponics or pouch-growth systems as a way to quickly survey a large number of root traits. In these types of studies, polyethylene glycol (PEG) can be added to the growth medium to induce low water potential. The drawbacks of this method are that it is an artificial environment, and the seedlings are screened at a very young age. However, there is demonstrated colocalization between root trait QTL measured on mature, field-grown maize and QTL from hydroponically grown maize seedling roots (Tuberosa et al. 2002b). From field root, hydroponic root, and grain yield under water stress studies, Tuberosa et al. (2002b) placed a priority on a region of maize chromosome 1 for use in the improvement of drought tolerance. An advantage of hydroponic and pouch-growth systems is that they are non-destructive, and root growth can be measured over time. Ruta et al. (2010) used this method to measure axile and lateral root growth in control and PEG—induced water stress conditions. They found that the elongation rate of axile roots responded significantly to water stress and emphasized that, especially at early stages of growth, it is important to take dynamic measurements into account instead of only static measurements like length. From this study, an area of chromosome 6 associated with decreased growth of lateral roots and an increased amount of axile roots during low water potential may be an important locus for early root adaptation to drought. It corresponds to a region found for drought tolerance of mature plants in the field by Ribaut et al. (1996), but it remains to be seen if root traits of mature plants have a similar trend at this locus.

RNA level changes at low water potential have been looked at in defined sections of the maize root tip (Spollen et al. 2008). The root tip sections were decided based on observations in Sharp et al. (2004) that 3 mm from the root tip, the expansion rate of root cells under water stress begins to decelerate until it completely ceases around 5 mm from the tip. These dynamics of longitudinal expansion are different from those seen in control root cells and show a reduced zone of growth in maize root tips under water stress. By comparing transcript changes in the different root tip regions under control and stress conditions, Spollen et al. (2008) identified several categories of transcripts thought to be involved in primary response to low water stress. The largest group was involved in ROS metabolism, followed by carbon metabolism, signaling, membrane transport, transcription factors, and cell wall structure. In the ROS group, transcripts involved in both consumption and production of these molecules were upregulated, showing that a fine balance is likely required to prevent damage from accumulation yet still allow localized growth. Coinciding with work done in rice (Rabello et al. 2008; Yang et al. 2009; Wang et al. 2011a), several metallothionein-like transcripts were upregulated in maize. A number of additional DEGs identified in this study may be involved in ABA response, including transcripts for a CBL-interacting protein kinase 3 (CIPK3)-like protein, PP2C proteins, bZIP family transcription factors, ABA-response element-binding protein 3 (homolog of OsDREB1a), and dehydrins. In other categories, expansins and inositol phosphate transcripts were increased.

A cell wall proteomics analysis of root tips under water stress was carried out using the same defined sections as in Spollen et al. (2008; Zhu et al. 2007). In Zhu et al. (2007), low water-responsive proteins were categorized into ROS metabolism, defense/detoxification, hydrolases, carbohydrate metabolism, and other/unknown. In corroboration with the Spollen et al. (2008) study, certain oxalate oxidase and peroxidase levels were increased. Again, a balance of ROS levels is hypothesized to be critical for stress response. There were also changes in several xyloglucan endotransglycosylase/hydrolase protein levels (XTH/XET). These are proteins involved in cell wall expansion that have been analyzed in other water-stress studies, but here, as with ROS metabolism proteins, localized increases and decreases in their levels may be important in stress response. Root tip invertase and other carbon metabolism protein levels were altered under stress as were transcript levels in this category in Spollen et al. (2008). Carbon metabolic enzymes can direct the flow of resources for growth to the root system as they elongate in search of water. It should be noted that some cell wall proteins, such as expansins, are too tightly bound to have been present in extracts of this study. So, although expansin transcript levels have been shown to be upregulated under water stress in several studies, different solubilization methods will be needed to look at corresponding protein levels other than those used in Zhu et al. (2007).

Genomics of Cultivated and Wild Wheat Root Traits and Water Deficit

Various types of wheat are grown in a wide range of environments around the world, and drought is a concern in many of these regions. Most mapping studies for bread (*Triticum aestivum*) and durum (*Triticum turgidum* ssp. *durum*) wheat under drought conditions have focused on yield (Quarrie et al. 2005; Maccaferri et al. 2008; Maccaferri et al. 2011). One QTL study has looked for genomic regions contributing to seedling response to ABA, a hormone often implicated in plant response to abiotic stress (Kobayashi et al. 2010). The authors found 5 QTL for relative root growth of ABA-treated seedlings, and they observed that expression levels of several ABA-responsive transcripts could be increased additively by combining certain QTL regions. These included cold-related and LEA transcripts. Only very recently has there been a QTL study in wheat for root traits (Sharma et al. 2011b). The plants were not stressed in this study, but the authors did measure traits like root length and mass in both shallow and deep soil layers, and these characteristics may prove to be important for drought adaptation of wheat. The mapping focused on the short arm of chromosome 1B by using a population with a series of translocations with the short arm of rye chromosome 1. The 1BS region of both hexaploid and durum wheat is predicted to contribute to yield under drought (Quarrie et al. 2005; Maccaferri et al. 2011). In Sharma et al. (2011b), 15 QTL that mapped to four regions of 1BS/1RS were found for the root traits. However, it is clear in this study that the rye portions of the translocation most often contribute to increased root proliferation. Still, it shows regions of wheat that could be targeted for improved root growth whether the favorable genomic segment would be introgressed from a particular wheat or rye variety.

In attempts to breed wheat better able to adapt to drought, synthetic/resynthesized hexaploid lines from crosses between durum wheat and wild *Triticum tauschii* wheat have been crossed to elite cultivars. Under moisture stress, some of these synthetically derived lines showed a shift in root biomass to deeper regions of the soil (Reynolds et al. 2007). This could prove to be a very important adaptation to drought, allowing the plant to focus growth in depths where water may be available and limiting growth in shallow soil layers.

Wild wheat relatives have also been used to look at root transcript differences under drought. Ergen and Budak (2009) surveyed a panel of wild emmer wheat lines (*Triticum turgidum* spp. *dicoccoides* (Korn.) Thell.) during water limitation to find the two lines with the most contrasting responses. These drought-tolerant and sensitive lines of emmer wheat were profiled on Affymetrix arrays under desiccation stress (Ergen et al. 2009). In root tissue, 926 probes had differential expression between the two genotypes during the stress. Among the 30 DEGs that varied the most between the genotypes were transcription factors, heat shock proteins (HSPs), cold-regulated proteins, dehydrins, lipases, and a lysine decarboxylase-like protein. For the transcription factors, several NAC domain proteins as well as other major families like WRKY (named for their characteristic heptapeptide WRKYGQK motif) and MYB (named for the *v-MYB* gene of avian myeloblastosis virus) had altered expressions levels.

Soybean Roots and Drought Tolerance Improvement

Soybean (*Glycine max*) is quickly becoming a more important foodstuff as the world's growing population looks to add protein to their diets. It is grown in many regions prone to periodic drought, but very few studies have been carried out to observe and modify how the roots of this plant cope with low water availability. An estimated 36% lower yield is expected from soy plants that rely only on rain as opposed to regular irrigation (Specht et al. 1999). So, in areas with water shortages, breeding for soybeans with improved yield under unpredictable rain patterns should buffer yield differences between moist and drought years.

There have not been many mapping studies for soybean root traits, and there are none for soybean roots under controlled drought conditions. However, a recent study by Abdel-Haleem et al. (2011) used a RIL population to map QTL for fibrous root growth under rain-fed conditions in the field. More fibrous roots allow for a larger root system surface area and therefore more potential for water uptake (Pantalone et al. 1996a, 1996b). One of the QTL that mapped to chromosome 4 overlapped with a QTL for water use efficiency from other mapping studies (Mian et al. 1996; Mian et al. 1998), and the positive allele for each trait came from the same parent, PI 416937. Introgression of this region into elite soy lines and checks under low water stress may result in higher-yielding, drought-tolerant soybeans. PI 416937 is also known to be tolerant to aluminum stress (Campbell and Carter 1990), and mapped loci for this trait could also be incorporated to make a multi-stress-resistant cultivar (Bianchi-Hall et al. 2000).

For transcript analyses, there have been several recent studies looking at how the expression levels of specific gene families are affected by drought stress. This includes genes involved in the synthesis of isoflavonoids (Gutierrez-Gonzalez et al. 2010), protein phosphorylation kinases (Le et al. 2011a), and NAC domain transcription factors (Le et al. 2011b; Hao et al. 2011). Gutierrez-Gonzalez et al. (2010) showed that progressive water deficit during seed-filling stages caused a decrease in seed isoflavone accumulation. The only isoflavone biosynthetic transcript significantly changed by drought in this study was from the first gene in the pathway, *PAL1*, which is upregulated. The authors hypothesize that during severe water deficit, phenylpropanoid metabolism is shifted toward compounds other than isoflavones. In particular, it may be directed toward creating a precursor for several antioxidant compounds.

Signal transduction pathways are also commonly altered by phosphorylation during abiotic stresses, including drought (for review, see Yang et al. 2010). Le et al. (2011a) studied transcript levels of genes involved in histidine-to-aspartic acid phosphorelays during dehydration stress. Many of the 83 genes in this study had high levels of root expression or were even root specific. Among them, 51 of the transcripts were found to be dehydration-responsive, and 25 of those had changes in the root tissue. This study again emphasizes tissue specificity of transcript changes and that, in some cases, opposite effects are seen in root vs. shoot tissue.

As with rice, there has been a focus on NAC domain transcription factors in soybean. Le et al. (2011b) found 152 NAC transcription factors in soybean and predicted 58 of them to be stress-induced based on homology with *A. thaliana* NACs. Further investigation with a subset of the NACs found that 29 and 6 of them were induced and repressed by dehydration stress, respectively. About 22 of the transcript changes varied by tissue type and 2 of them were root specific. Hao et al. (2011) focused on two of the soybean NACs, *GmNAC20* and *GmNAC11*. They did not test for drought stress, but they found that overexpression of either gene in *A. thaliana* improved root growth and survival under salt stress. *GmNAC20* overexpression also led to enhanced cold tolerance, and even with no stress applied, upregulation of this NAC gave increased lateral root formation, probably through auxin signaling. A similar lateral root promotion was seen in *A. thaliana* with *AtNAC2* overexpression (He et al. 2005). The GmNAC20 and AtNAC2 protein sequences share 70% identity. It would be of interest to see if overexpression of *GmNAC20* also improves low water stress tolerance and how root-specific overexpression affects soybeans under drought.

There has also been a proteomic analysis of soybean roots under low water stress (Yamaguchi et al. 2010). In this study, protein levels in three defined regions of the root tip were observed under water stress conditions in comparison with developmental and temporal controls. In total, they detected 35 proteins with abundance changes in response to water stress, and the majority of the changes were conserved between the first two root tip segments. Three major categories of proteins were discovered in this analysis, including isoflavonoid biosynthesis proteins, proteins involved in lignin accumulation, and proteins regulating ferritin/iron distribution. Isoflavones and the sequestering of free iron in cells could be working to lower ROS accumulation during low water stress. Lignification may be triggered by ROS buildup, but it has possible roles in aiding roots under water stress conditions by targeting water transport to growing tissues while preventing water loss. Isoflavones may have an additional role in determining the distribution of auxin in the root tip, thereby modulating root growth patterns. However, the precise functions for auxin in roots during low water stress need further evaluation.

Insights on Root Drought Response from Crop Plants

Taken together, these studies give some idea of how root morphology and physiology change under low water stress. It is clear that not one type of root system will perform best for drought conditions in every type of plant and in every environment. Soil types in some areas make it necessary for roots to be able to penetrate into hardpan matrices, but thick, mechanically tough roots would be a waste of resources in other areas experiencing drought. The best composition of the root system components will certainly depend on local conditions. In the past, a long maximum root system length was the main focus of root response to drought. Now, the structure and adaptability of the rest of the root system is also under consideration. A root system that can grow efficiently in a well-watered environment and quickly adapt a suitable new growth program under water-deficit conditions is the ultimate goal. However, ideal branching and placement of finer roots in the rhizosphere for the best water uptake will need more evaluation. It may benefit researchers to take fractal analysis of root geometry into account in studies of roots under water deficit. Root branching approaches a fractal shape since small sections of the root system approximate the shape of the root system as a whole. Fractal studies of root architecture have been undertaken to try and understand how branching pattern helps the root plunder heterogeneous soils for water and nutrients (reviewed in Walk et al. 2004). More recently, a fractal analysis of upland and lowland rice under different water regimes has been undertaken (Wang et al. 2009). Continued studies of this nature may increase our knowledge of the relationship between root system structure and water uptake under drought.

Maize, rice, wheat, and soybeans account for approximately 75% of the calories that humans take in either directly or via animal sources (Cassman 1999). Yet, as seen from the research described here, not many root genomics studies under drought have been carried out for these crops. For QTL analyses, there need to be more mapping studies for roots under low water stress conditions. The control plants for such studies will simultaneously add much needed data for root morphologies under average conditions in specific environments. Even though the work is very labor-intensive, more mapping of root traits under drought field conditions is necessary. Rain-out shelters at experimental centers can help to achieve this goal. The field environment will also allow more analysis of plants beyond seedling age in various vegetative and reproductive life stages. In rice, there is enough information to generate introgression lines for various drought root trait and yield QTLs for testing in the field. The results of these tests and continued narrowing of candidate QTL intervals should lead to important gene information that will help in the improvement of all crops to low water stress.

Transcriptome, proteome, and metabolome studies are also contributing to our knowledge about the genomics behind response to water deficit. Recently, it has been proposed that genomic studies of DNA methylation states are important for understanding plant response to abiotic stresses as well also. In rice, it has been seen that drought stress causes differential, tissue-specific methylation between a tolerant and intolerant variety (Wang et al. 2011b). More functional genomics studies of all these types are needed for the major crop plants that include further tissue specificity and more life stages. However, results from current research of this nature in roots and the rest of the plant are giving many leads on what it takes for proper drought response.

The action of plant hormones and the effects they have on gene expression, metabolite fluxes, and cell structural changes are critical to overall plant morphological changes in heterogeneous environments. Functional genomics studies point to the involvement of several plant hormones in drought stress, and their specific roles in roots are under evaluation. For example, cytokinin and ethylene are known to modulate root response to water deficit (Spollen et al. 2000; Werner et al. 2010). However, in particular, ABA levels and distribution are considered highly critical for low water potential responses. The accumulation of ABA in primary root tips of maize is needed for the root's continued elongation during water deficit (Saab et al. 1992), which is of course the main phenotype considered for root adaptation to this stress. This localization of ABA has been shown to be necessary for a variety of responses that occur under water deficit, such as ion transport, osmotic adjustment, changes in cell wall extensibility, and control of ROS (reviewed in Yamaguchi and Sharp 2010).

A metabolite that continually arises in drought response studies and that may be regulated by ABA is proline (Ober and Sharp 1994). Accumulation of proline and other solutes in root tips is part of a process known as osmotic adjustment and occurs in response to water deficit. Even as water potential lowers, these solutes can keep cell turgor pressure at a manageable level. In maize root tips, proline was shown to be a major contributor to this adjustment, and it is most likely transported there from other regions of the plant, including more basal portions of the root and leaves (Voetberg and Sharp 1991; Verslues and Sharp 1999). A recent study using *A. thaliana* has now shown that it is not just accumulation of proline that is important, but a dynamic relationship between synthesis and catabolism that allow for root growth under low water potentials (Sharma et al. 2011a). For this compound, turnover in the shoot and localized transport to the root maintain a balance that is necessary for adaptation to low water potential. This should be kept in mind for future studies of other metabolites so that multiple tissues and time points are observed to gain a proper understanding of compound flux. It should also help to amend discrepancies seen between increased proline synthesis transcripts and perceived lack of proline accumulation seen in studies such as Ergen et al. (2009). Quite possibly in this case, dehydration stress led to increased proline synthesis as evidenced by transcript levels, but they did not see accumulation of proline because only leaves were checked. Portions of the root systems in this experiment may have shown increased proline levels due to transport from the leaves.

Another major category arising in the -omics studies, and also influenced by ABA in some cases, was cell wall modification. For extension of the cell wall, and thereby the ability for the cell to elongate, several studies of roots under low water conditions showed increases in transcripts or proteins for expansins and XTH/XET molecules. XTH activity, but not that of expansins, was shown to be dependent on ABA in roots during water deficit (Wu et al. 1994, 2001). The transcript of soybean expansin *GmEXPB2* was shown to be induced in the root by low water potential and exogenous auxin (Guo et al. 2011). So, interaction between auxin and expansins during water deficit should be explored further. In contrast to changes that are made to promote cell elongation, decreases in cell wall extensibility are also a part of the root water-stress response. In more basal portions of the maize root, the cross-linking and lignification of the cell wall mediated by ROS leads to wall tightening during water deficit (Spollen et al. 2008; reviewed in Yamaguchi and Sharp 2010).

So, in adjacent regions of the root, there are opposite requirements for cell wall extensibility upon application of low water stress.

Clearly, ROS accumulation and control play large roles in response to drought stress. As just mentioned, ROS modify cell wall composition by cross-linking. Conversely, accumulation of ROS, particularly peroxide, is hypothesized to be critical for increased cell expansion and root elongation (reviewed in Yamaguchi and Sharp 2010). Many of the transcriptome, proteome, and metabolome studies for water-stressed roots found changes in superoxide dismutase, oxalate oxidase, and peroxidase levels that could raise H_2O_2 levels and therefore modulate cell elongation. In addition to these roles for ROS, functional genomics studies continually point to genes involved in ROS sequestration to prevent oxidative damage to tissues due to altered metabolism during stress. Because of all of these possible ROS functions, proper root adaptation to water deficit conditions requires that their levels must be controlled in a very localized and nuanced fashion.

Other metabolites that accumulate under low water stress and can act as ROS scavengers and can also affect cell wall cross-linking are flavonoids (including isoflavonoids) and polyamines. Additionally, both categories of metabolites can be modulated by ABA levels (Ithal and Reddy 2004; Alcázar et al. 2006). The -omics evaluations of water-stressed roots pointed to differential regulation of transcripts and proteins involved in the biosynthesis and transport of these compounds (Zhu et al. 2007; Rabello et al. 2008; Ergen et al. 2009; Yamaguchi et al. 2010; Wang et al. 2011a). During water deficit, overexpression of the polyamine biosynthetic gene *arginine decarboxylase* increases overall plant tolerance to the stress and promotes primary root growth (reviewed in Alcázar et al. 2010). As with proline, it is already clear that in future studies of polyamines and abiotic stress, temporal evaluation of various tissues will be important. For example, both heat stress and water deficit create a source–sink relationship for certain polyamines between aerial and underground tissues of plants (Shevyakova et al. 2001; Sziderics et al. 2010). Transport and localization will most likely prove to be specific to the type of stress and the particular polyamine. Observations of flavonoid compounds are also showing that a dynamic control of their abundance and placement is needed to maintain proper cell wall composition during low water stress (Yamaguchi and Sharp 2010; Yamaguchi et al. 2010).

Excess Water—Soil Waterlogging, Flooding, and Submergence

The Effects of Too Much Water on Crop Root Structure and Function

On the other extreme of moisture, heavy rainfall or over-irrigation also lead to crop damage and therefore losses in grain yield. Too much water can come in the form of soil waterlogging, flooding, or complete submergence. Waterlogging occurs when water inundates the soil matrix and completely surrounds the root system, but not other portions of the plant. In flooding, the water plane rises above soil level and envelopes some aerial portions of the plant. Submergence is when water levels are high enough to completely engulf the entire plant in water. These different levels of severity likely elicit both common and distinct stress responses in plant tissues.

As with drought, yield losses by excess water will depend on the severity and duration of the stress. In addition, under conditions of excess water, soil type is crucial in determining how large an impact the stress will have. Some soil matrices drain much faster than others, and soils with high clay content tend to hold water longer. Also, certain nutrients and metals present in the soil may become more soluble as soil redox potential lowers during flooding. Depending on soil pH, this can lead to build up of elements like Mn^{2+} and Fe^{2+} to toxic levels and injure plant tissues by impairing proper cell metabolism (reviewed in Shabala 2011).

There are other components to flooding stress besides phytotoxicity that threaten normal plant physiology. One is the reduction of the plant's ability to exchange gases (reviewed in Bailey-Serres and Voesenek 2008). Gases diffuse much more slowly in water than in the atmosphere, resulting in a lack of available O_2 to plant tissues and problems with aerobic respiration. Ethylene tends to build up in water surrounding plant tissues and interferes with multiple signal transduction pathways. CO_2 levels are tissue- and condition-specific during flooding stress and can also affect plant growth. Another component of flooding and submergence stresses is a decrease in light intensity due to the properties of water surrounding the aerial portions of the plant (reviewed in Bailey-Serres and Voesenek 2008). This lowers the plant's capacity for photosynthesis, depletes carbon stores, and contributes to the energy crisis already brought on by lack of O_2. Because of the importance of O_2 in plant response to excess water, experiments subjecting plant tissues to low O_2 (hypoxia) or completely depriving them of O_2 (anoxia) are relevant to studies of flooding stress. Because of gas diffusion under normal conditions, inner plant tissues and organs with restricted O_2 entry experience hypoxia even without applied stress (Geigenberger 2003). Therefore, hypoxia and anoxia studies have additional value in helping us understand how these tissues maintain cellular performance on an everyday basis.

The complex conditions generated by excesses of water and low O_2 cause a variety of growth responses in aerial portions of the plant. During submergence, a few contrasting strategies can be adopted for shoot elongation. These are quiescence and the low oxygen escape syndrome (LOES) (reviewed in Bailey-Serres and Voesenek 2008; reviewed in Nagai et al. 2010). LOES is an avoidance response that induces rapid elongation of aerial tissues until they emerge above the water's surface. The elongating organs develop with thinner cell walls, thinner cuticles, and with chloroplasts toward the tissue surfaces to further mediate gas exchange and photosynthesis. Flood tolerant crops, such as deepwater rice varieties, employ LOES. The quick elongation response in deepwater rice results from the action of ethylene-responsive factor (ERF) domain-containing transcription factors SNORKEL1 and SNORKEL2 (Hattori et al. 2009). Ethylene accumulation activates these transcription factors, which then employ gibberellic acid (GA) to trigger rapid lengthening.

However, if a plant attempts LOES without proper supply and mobilization of energy reserves, death will result before the plant is able to break the surface. For this reason, some plants have better rates of submergence survival if they adopt a quiescence strategy. This is the method of some lowland rice varieties that conserve energy and resources upon flooding until the water recedes. The quiescence tolerance program was found to be regulated by SUB1A in rice (Fukao et al. 2006; Xu et al. 2006). Like the *SNORKEL* genes, the *SUB1A* locus encodes an ERF domain-containing transcription factor (Xu et al. 2006). In contrast to the *SNORKEL*s, upregulation of certain alleles of *SUB1A* during ethylene accumulation triggers the expression of genes that suppress GA and therefore elongation (Fukao et al. 2006; Fukao and Bailey-Serres 2008).

The root system must also respond to excess water/low O_2 if a plant is able to survive the stress. As with drought, some species including many major crops are able to form cortical aerenchyma during waterlogging or flooding. Root aerenchyma may form through schizogeny (cell wall reorganization and separation), lysigeny (PCD and cell wall autolysis), or a combination of both as proposed for wheat roots (Jiang et al. 2010). A mutant study in *A. thaliana* has found some of the genes involved in PCD for the generation of aerenchyma and the interactions of their development with ethylene, auxin, GA, and ROS (Muhlenbock et al. 2007). Increased root porosity from aerenchyma formation is associated with increased root length and penetration depth (Justin and Armstrong 1987). The aerenchyma increase the amount of O_2 flow throughout the root system to help increase aerobic respiration and energy generation, and they also aid in nutrient absorption (Jackson and Armstrong 1999). In addition, the space provided for gas diffusion allows for dispersal of CO_2 and ethylene. Hypoxic conditions separate from a flooding state can also induce aerenchyma formation (Gunawardena et al. 2001). For additional gas flow, hypertrophy of secondary aerenchyma can form large cracks called lenticels on stem and root surfaces (Shimamura et al. 2010).

Another characteristic of waterlogged plants is the formation of shoot-born roots, known as adventitious roots. Their development is known to help plants survive and recover from flooding stress. If adventitious roots were removed from tomato (*Solanum lycopersicum* cv. "Marglobe") or sunflower (*Helianthus annuus*) shoots, the plants were unable to recover after the cessation of waterlogging stress (Jackson 1955). These roots elongate and can give rise to lateral roots over the progression of waterlogging. Their formation near the surface of the water helps to aerate the rest of the root system. The adventitious roots also contain large pockets of internal gas exchange space due to the formation of aerenchyma (Bacanamwo and Purcell 1999). Besides helping to oxygenate the root system, flood-induced adventitious roots have been shown to be capable of photosynthesis (Rich et al. 2011).

Natural Variation and QTL Mapping Studies of Roots under Excess Water Stress

In crop plants and the model plant *A. thaliana*, there is known to be natural variation in responses and tolerance to stress from excess water. In crop plants, these types of studies are often done to find new breeding germplasm for flood-prone environments. So, yield is often the measurement used to denote tolerance (for example, in soybean, VanToai et al. 1994, 2001). However, many studies of this nature are now looking at multiple plant traits under flooding-type stresses, including characteristics of the root system. In rice, Champoux et al. (1995) tested a group of lines with a range of phenotypic expressions for various root traits like root/shoot ratio, root dry mass, deep root dry mass, and maximum rooting depth under both aerobic and anaerobic conditions. For most of the traits, the ranks of the cultivars with respect to phenotype were maintained between environments. The exception was rooting depth, and lines with a more extensive shallow root system under drained conditions were able to grow the deepest root systems when flooded. This may be because these lines maintained a significant amount of roots near the water's surface during flooding and used the resources gained from them to promote growth of the rest of the root system. For the wild Triticeae member *Hordeum marinum*, it was observed that various accessions from two subspecies had varying adventitious root porosities and rates of radial O_2 loss (Malik et al. 2009). Some of these accessions were used to make hybrids with wheat and resulted in improved waterlogging tolerance (Malik et al. 2011). *A. thaliana* root traits have also been evaluated after submergence stress in multiple accessions (Pigliucci and Kolodynska 2002; Vashisht et al. 2011).

Because of natural variation for response to excess water stresses, there have been several QTL studies looking for tolerance loci in crops. However, most of the studies look at yield traits or rate plants on a tolerance index based on chlorosis and necrosis of aerial tissues. Excess water affects root systems first, and stress signaling cascades that will determine the growth program for the entire plant begin underground. The role of root system components in adapting to these types of stresses is becoming more appreciated, and there have been a few QTL studies that look into root traits. As with drought, rice studies lead the way in the evaluation of roots under anaerobic regimes. Some studies look at only flooded conditions while others evaluated this and roots under upland drought conditions in addition. Some common QTL found for root traits under flooding include chromosome 1 near marker RM315 (Zheng et al. 2006, 2008), chromosome 2 near RZ318 (Zheng et al. 2003,

2008; Zheng 2006), chromosome 3 between RM338 and RZ284 (Zheng et al. 2003, 2008; Zheng 2006), chromosome 5 near RZ70 (Zheng 2006; Zheng et al. 2008), and the end of chromosome 9 (Kamoshita et al. 2002; Zheng et al. 2003). Several of the QTL in these studies colocalize to root trait QTL found in well-drained and drought conditions. This again highlights that certain genomic regions seem to be important for root development regardless of environment. Notably, though, the QTL that localize to chromosome 3 between RM338 and RZ284 do not seem to be associated with root QTL from other studies and so may be specific to flooding adaptation. The QTL that localized there were for flooded adventitious root number (Zheng et al. 2003), flooded root weight (Zheng 2006), and flooded adventitious root length (Zheng et al. 2008). This is an important locus for further studies of adventitious root initiation and proliferation under flooding.

The few QTL studies that observe root traits under flooding often find genetic loci connected with root growth under normal growth conditions as well. In maize, QTL for adventitious root formation under flooding have been located and compared with adventitious root QTL from its wild relative, teosinte (*Zea mays* ssp. huehuetenangensis) (Mano et al. 2005a, 2005b). Mano et al. (2005a, 2005b) found three QTL for adventitious root formation in maize, and one of these QTL on chromosome 8 colocalizes with a QTL found in the teosinte population. Some of the QTL found in these studies also colocalize with maize primary root length QTL found in Tuberosa et al. (2002b). So, it appears these QTL are general root trait QTL that govern root formation and length in multiple environments. A QTL study for roots under waterlogging has also been recently carried out for a pasture grass, perennial ryegrass (*Lolium perenne*) (Pearson et al. 2011). The authors found QTL for root length and mass under both control and waterlogged conditions. For root length, the same region of linkage group 4 was found to contribute under both conditions. A QTL for root mass under waterlogging also localized to this region, so the genes underlying this QTL are most likely important for root growth regardless of environment.

Transcriptome, Proteome, and Metabolome Analyses of Roots during Low Oxygen Stress and Waterlogging

There have been a number of transcript-level studies of plant roots under low O_2 conditions. Exogenously applied sucrose greatly enhances the anoxia tolerance of *A. thaliana*, so Loreti et al. (2005) tested root transcript response to low O_2 with and without the sugar. In particular, they found that *alcohol dehydrogenase* (*ADH*), *pyruvate decarboxylase 1* and *2* (*PDC1* and *2*), and *Arabidopsis nonsymbiotic hemoglobin 1* (*AHB1*) are induced within 40 min of anoxia, and there is a prolonged ability to accumulate *ADH* and *PDC1* when exogenous sucrose is available. *Sucrose synthase* (*SUS*) is usually induced under low O_2 conditions, but here it was not induced under anoxia when exogenous sucrose was supplied. Some general trends under anoxia noted by the authors were that lipid metabolism decreased, glycolysis enzymes increased, and several HSP transcripts increased. Under low O_2, plants shift metabolism to generate energy via glycolysis and fermentation of pyruvate (reviewed in Bailey-Serres and Voesenek 2008), and the transcript changes here are consistent with those cellular programs.

In a combined transcriptome and metabolome study of *A. thaliana* roots under hypoxia, van Dongen et al. (2009) found similar trends to Loreti et al. (2005). Transcripts of *ADH*, *PDC*, *SUS*, and *AHB1* were induced. Generally, there is an increase in transcripts that serve to generate ATP under low O_2 and a decrease in transcripts of proteins associated with ATP-consuming processes. In metabolite analyses, they noted a strong increase in carbon flux to glycolysis brought on by the switch to fermentative metabolism. Some specific metabolites that were induced were proline and GABA, similar to what was seen in root drought studies.

To bring the functional genomic data on plants under low O_2 together, Narsai et al. (2011) performed a synthesis study of transcripts and metabolites under low O_2. Common changes in carbon metabolism were noted. This was also seen in another synthesis study by Mustroph et al. (2010), which looked at transcript responses to low O_2 over four kingdoms of life. Narsai et al. (2011) saw a small number of genes orthologous between rice and *A. thaliana* were regulated in opposite manners, including zinc-finger domain transcription factor and kinases. Also, nonsymbiotic hemoglobin transcript levels increased in *A. thaliana* while they decreased in rice and poplar (*Populus trichocarpa*). This may have to do with nitric oxide (NO) scavenging and signaling. Mustroph et al. (2010) noted that NO reductases are induced in plants and bacteria during hypoxia. So, NO signaling is an area that deserves further consideration in future low O_2 studies.

There are some additional insights on transcriptional regulation under low O_2 from Mustroph et al. (2010). The authors saw that most organisms induce HSP and ROS network genes under low O_2 stress, including some heat shock transcription factors. Specifically in plants, several other transcription factor families had members that were DEGs, including ERF, MYB, CCCH-type zinc finger, WRKY, bZIP, NAC, and MADS. The DEGs in plants also included ethylene biosynthesis and detection transcripts. This is not surprising considering the roles this hormone is known to play in signaling for promotion or repression of underwater elongation as well as the development of aerenchyma. Because part of plant response to low O_2 can be elongation, cell wall expansins and XETs were also DEGs, as they were for roots under low water stress. Another interesting aspect of the DEGs analyzed by Mustroph et al. (2010) is the number of transcripts for proteins of unknown function. About 450 of these hypoxia-responsive unknown proteins (HUPs) are conserved across species, and 89 out of the 200 most highly induced genes are higher plant-specific HUPs. These HUPs make interesting targets for future functional studies of low O_2-responsive genes. Indeed, overexpression of a subset of the HUPs in submerged *A. thaliana* increased survival, although the authors note more precise regulation might give better results (Lee et al. 2011).

Certain transcription factors and signaling components important for species-specific response to low O_2 have been verified in functional studies. As already discussed, expression of ERF transcription factors *SUB1A* or *SNORKEL*s have roles in determining rice response to submergence (Xu et al. 2006; Hattori et al. 2009). Another ERF transcription factor, *hypoxia responsive element 1* (*HRE1*), confers tolerance to low oxygen in *A. thaliana* when it is overexpressed (Licausi et al. 2010). A heat shock transcription factor transcript (*HSFA2*) is also induced by anoxia in *A. thaliana*, and subsequently its protein levels rise, conferring protection of plant tissues under the stress (Banti et al. 2010). This is significant because there is often disconnection between transcript induction and protein levels under anoxia. This may be because translational efficiency during low O_2 can be significantly impaired, possibly because of low ATP and changes in posttranscriptional regulation (Branco-Price et al. 2005, 2008).

There are also a few transcriptome studies of roots under waterlogging. Waterlogged roots of maize had many DEGs involved in signaling, including kinases, ethylene biosynthesis, hemoglobin synthesis, and the transcript of oxygen sensing *prolyl 4-hydrolase* (*P4H*) (Zou et al. 2010). The roles of *P4H* genes in low O_2 sensing in *A. thaliana* have been more extensively studied, and they may be required for regulation of certain transcription factors as they are in mammals (Vlad et al. 2007). Zou et al. (2010) also saw increases in transcripts that lead to accumulation of compounds like alanine and GABA. They aligned their transcriptomic results with QTL maps for traits of submerged maize seedlings, including root length and dry mass (Qiu et al. 2007), but their candidate genes require further evaluation. (VanToai, Valliyodan, and Nguyen, personal communication and/or unpublished data) looked specifically at expression of certain transcription factors in soybean roots under waterlogging. They compared a variety that was rated as tolerant to flooding in field trials to an intolerant line. The tolerant variety showed an enhancement of ethylene biosynthesis genes over the intolerant one. Also differentially expressed between the two at all time points tested were a MYB domain transcription factor, a leucine zipper transcription factor, and a hemoglobin gene.

Proteome analyses of roots under waterlogging stress display similar cellular trends to transcriptome studies of roots under low oxygen or excess water stress. In a study of waterlogged tomato (*Solanum lycopersicum*) roots, Ahsan et al. (2007) found increases in enzymes such as ADH and enolase, pointing to the well-characterized shift to fermentative metabolism and increased glycolytic flux upon low oxygen stress. There were also increases in proteins involved in the synthesis of GA and brassinosteroids, both plant hormones that affect development. Other increases included cell wall degradation proteins and phenylpropanoid/flavonoid synthesis proteins. There was a decrease in *S*-adenosyl-*L*-methionine-synthase (SAM), which is likely to reduce ethylene biosynthesis but may also lead to increased accumulation of polyamines via the linkage of these processes in the AdoMet cycle (reviewed in Roje 2006).

In a series of proteomics experiments on waterlogged soybean roots, changes in proteins involved in fermentation, glycolysis, isoflavonoid production, and increased cell wall extensibility have also been observed (Komatsu et al. 2009, 2010; Alam et al. 2010). Included in these proteins is an ADH gene, *ADH2*, which responds specifically to flooding stress in the root of soybean (Komatsu et al. 2011). In Alam et al. (2010), they noted an increase in coproporphyrinogen oxidase, which again points to the importance of hemes and oxygen signaling during waterlogging stress. They also saw increases in a number of proteins linked to PCD and therefore possibly to aerenchyma formation. There was a decrease in SAM in this study, as seen in tomato (Ahsan et al. 2007). In a cell wall-specific proteome study of flooded wheat roots, there was a notable decrease in proteins involved in cell wall elongation and an increase in chitinases, which could be involved in abiotic stress signaling (Kong et al. 2010). The variable results in these proteomic studies with respect to cell wall composition and extensibility point to the need for further, region-specific protein analysis to find the broad picture of cell wall activity during low oxygen stress.

In the metabolomics portion of the Narsai et al. (2011) analysis, they found that plant species commonly shift to fermentative pathways under low O_2, leading to lactate and ethanol production. Other metabolites that commonly increased were alanine, GABA, succinate, lysine, and tyrosine. Conversely, decreases in aspartate were common. Some proline accumulation was noticeable, especially in later time points of the studies. This study included the metabolite data from van Dongen et al. (2009), and therefore the observations in both are similar. In addition, van Dongen et al. (2009) noted that changes in many metabolite levels upon low oxygen stress can be transient. This may be due in part to an initial increase in levels of a given compound followed by a decrease when the stress induces the breakdown of sucrose and lowers glycolysis rates.

Insights from Root Responses to Excess Water and Low Oxygen

Morphological changes in the aerial portions of plants during submergence are well characterized. Above-ground plant tissues can adopt a quiescent strategy or attempt to renew atmosphere contact through rapid elongation. However, the overall shape the root systems of various plant species take when faced with excess water is less clear. Adventitious roots form and aerenchyma open air chambers between the roots and the shoot, but the molecular program the root tips of the established root system adopt when faced with this stress is not known. More studies are also required to gain a better understanding of elongation rates and branching in soil-bound waterlogged root systems. The response of established root systems following the removal of the stress needs to be determined as well. Further experiments may ascertain if these roots resume growth, if new root branches initiate from these established roots, and if adventitious roots continue to contribute to plant health. Functional genomics data thus far suggest a general quiescence-type strategy in the established root system upon low O_2 stress, which results in a cessation of new growth. Energy is then focused on production of aerenchyma and growth of adventitious roots to mitigate the gas imbalances that occur. However, more studies in individual crop plants are needed to determine how important the composition of the established root system at the time of stress is to flooding tolerance and recovery. Prolific rooting prior to the stress may be helpful to the tolerance of the plant, or it could be tied to an ability to quickly produce a substantial adventitious root system upon low O_2 stress.

QTL studies of root traits during low O_2 stress and flooding are lacking, especially in dicots. The *SUB1* locus of rice represents a success story in flooding QTL mapping for aerial traits (Xu et al. 2006), but QTL for root-related traits are yet to be mapped in many crops, let alone fine mapped. How the anatomy of the root system before, during, and after flooding stress contributes to tolerance is unclear. As with drought, there are some interesting root trait loci in rice to be pursued for introgression and fine mapping. Other major crops require further mapping analysis for interesting flood root trait loci to emerge.

There is more data for root response to low O_2 at the molecular level. From the study of aerial tissues, it is known that the hormones ethylene, ABA, and GA are major regulators of plant response to low O_2 (reviewed in Bailey-Serres and Voesenek 2010). Upon flooding, ethylene accumulates and can activate ERF transcription factors like SUB1A-1 or the SNORKELs, depending on the alleles of the genes that are present. Ethylene represses production of ABA in submerged aerial tissues, which allows for the production of GA (reviewed in Bailey-Serres and Voesenek 2008). In deepwater rice, *SNORKEL*s are activated and GA levels rise, causing rapid growth of internodes. In rice varieties that adopt a quiescent strategy upon submergence, *SUB1A-1* is activated, which in turn drives the expression of transcription factors that inhibit GA and therefore elongation (Fukao and Bailey-Serres 2008). Transcriptome and proteome studies in roots also showed changes in ethylene-responsive transcription factors, so further functional analysis will show how they affect flooded root morphology. Cell wall-modifying genes like GA-regulated expansins were also affected. Especially for these genes, distinctions need to be made between the underground root system, which ceases to elongate, and the adventitious root system, which must develop and grow during the stress. Localized expression studies are essential to understanding the molecular response of root systems to flooding.

As with drought, the control of ROS levels is important in tissues experiencing low O_2 stress. For H_2O_2, G protein signaling is involved in many root responses to low O_2. In *A. thaliana*, RopGAP4 interacts with a small G protein to control Ca^{2+} levels, actin dynamics, and H_2O_2 levels. The H_2O_2 produced acts as a secondary messenger to activate *ADH* expression (Baxter-Burrell et al. 2002). G protein signaling is also a component of ethylene-induced PCD (reviewed in Steffens and Sauter 2010). There is interplay between various G proteins and cell death in parenchyma for aerenchyma formation as well as in epidermal cells for the initiation of adventitious roots. ROS scavenging by certain metabolites also mediates response to low O_2 stress. Localized accumulation of polyamines might be important for this in flooded roots as they are thought to be for roots under low water stress. Waterlogging tolerance of the Welsh onion (*Allium fistulosum*) was enhanced if plant roots were given exogenous spermidine or spermine before the stress (Yiu et al. 2009). The GABA shunt pathway is hypothesized to be responsible for the accumulation of alanine as well as the prevention of ROS accumulation in hypoxic *A. thaliana* roots (Miyashita and Good 2008). The proteomics studies under waterlogging also consistently showed changes in (iso)flavonoid biosynthesis proteins, which can act in scavenging ROS (Ahsan et al. 2007; Komatsu et al. 2009, 2010; Alam et al. 2010). As with drought, these studies confirm that the levels of ROS like H_2O_2 must be finely regulated to control proper cell response to stresses without causing excess tissue damage.

Molecular studies have also shown that the N-end rule pathway of protein degradation is a regulator of hypoxia-responsive proteins (Gibbs et al. 2011; Licausi et al. 2011). Some hypoxia-responsive proteins have an N-terminal tag that leads them to be modified and degraded in the presence of oxygen. However, under low O_2 conditions, such proteins accumulate and activate transcriptional cascades that allow the plant to respond to the stress. Two group VII ERF transcription factors, HRE2 (Gibbs et al. 2011) and RAP2.12 (Licausi et al. 2011), were shown to be regulated in such a manner and their overexpression enhanced survival under low O_2. Modifications to this pathway and proteins subjected to it may be an approach to improve flooding tolerance in crops.

Common Plant Root Responses to Abiotic Stressors

The focus of this chapter has been on plant root response to low and excess water, but many of the changes that occur during exposure to those conditions are common to other abiotic stressors. Excess salt, heavy metals, improper nutrient balance, temperature extremes, and more are also responsible for a great deal of crop losses and prevent certain areas from being used for agriculture. Many crops in the field will be exposed to more than one type of abiotic stress in a given season, so it will be especially useful to focus research on traits and pathways that could afford tolerance to multiple kinds of stress.

Various morphological responses are conserved in reaction to various abiotic stresses. Plant roots will form aerenchyma during waterlogging, drought, and nutrient deficiency (Jackson and Armstrong 1999; Zhu et al. 2010; Postma and Lynch 2011). It has been proposed that modifying PCD and cytokinin levels to prevent early senescence may be helpful in increasing plant abiotic stress tolerance (Xu et al. 2004; Huynh et al. 2005). However, this is yet another manipulation that will most likely serve plants best if it is highly localized, since aerenchyma formation in the roots requires PCD. There is probably a fine balance required for PCD with respect to aerenchyma formation already. The death of too much living tissue to form aerenchyma could feasibly hinder the recovery of the plant upon removal of the stress. However, the regulated formation of aerenchymatous tissue is a possible approach to creating crops resistant to several abiotic stressors. Another root trait, adventitious root formation, is a common response of plants to excess water, but it is also induced during other stresses like phosphorus starvation (Miller et al. 2003) and low levels of other nutrients. Therefore, loci that promote formation of adventitious roots may also lead to multistress-resistant crops.

Another component of response to stress is changes in levels and localization of plant hormones, and the cross talk among them. ABA, ethylene, and GA are central in plant response to water deficits and excesses. ABA also incites gene regulation changes as plants adapt to other forms of osmotic stress such as high salt and low temperature (Zhu 2002). During flooding stress, *SUB1A* is one of the loci that mediate interactions between these hormones (reviewed in Bailey-Serres and Voesenek 2010). Recently, it was shown that *SUB1A* also provides enhanced dehydration tolerance (Fukao et al. 2011). Fukao et al. (2011) showed that the presence of *SUB1A* in rice lines increased their responsiveness to ABA in aerial and seed tissues. This is in contrast to the decreased responsiveness of *SUB1A* lines to GA (Fukao and Bailey-Serres 2008). The SUB1A protein is not subject to N-end rule degradation, and this stability may contribute to its ability to modulate multiple types of stress (Gibbs et al. 2011). Loci acting as hormone-responsiveness hubs in root tissues during stress conditions may also prove to be important for the breeding of abiotic stress tolerance.

Many key biological processes are emerging as important components in plant root stress response. Protein phosphorylation is thought to play a pivotal role in the response of eukaryotes to environmental stress, including plant response to abiotic stress. Phosphorylation is networked with calcium signaling, energy metabolism, and ROS scavenging during stress, but the exact connections still need to be worked out (reviewed in Mittler and Blumwald 2010; reviewed in Kosová et al. 2011). In plants, members of the SNF1-related protein kinase (SnRK) family have been found to be responsive to ABA and affect their tolerances to drought, hyperosmotic, cold, and salt stresses (reviewed in Mao et al. 2010). In particular, SnRK1 appears to be a master regulator during stress by affecting the transcriptional levels of hundreds of genes (Baena-González et al. 2007). With respect to root traits, one wheat SnRK family member, *TaSnRK2.4*, was overexpressed in *A. thaliana* and caused longer primary root growth (Mao et al. 2010). The overexpression plants had increased tolerance to cold, drought, and salt stress. The SnRK family along with many other types of protein kinases such as mitogen-activated protein kinases, calcium-dependent protein kinases, CIPKs, and histidine kinases have emerging roles in plant stress response.

Multiple stresses also involve changes in the levels of metabolite classes such as proline and polyamines. In addition to changes in accumulation already described for drought and flooding, the polyamines spermine and GABA have been shown to have a role in the response of plant roots to salt stress (Yamaguchi et al. 2006; Xing et al. 2007; Renault et al. 2010). In the crucifer *Pringlea antiscorbutica*, the polyamines agmatine, spermidine, and spermine promote primary root growth at low temperatures (Hummel et al. 2002). In wheat, the levels of several polyamines changed with cold and osmotic stresses (Kovács et al. 2010). In this study, a panel of wheat substitution lines with differential freezing tolerance showed freezing tolerance to be associated with accumulation of putrescine and a decrease in spermine and cadaverine. GABA is also linked to the accumulation of alanine, which was a common occurrence in plant roots under low oxygen and flooding (Miyashita and Good 2008) and has also been shown to occur during salt stress (Renault et al. 2010). Proline accumulation in roots is triggered by low O_2, water deficit, salt, heavy metals, nutrient deficiency, UV irradiation, and biotic stress (reviewed in Verbruggen and Hermans 2008). Suggested roles for this metabolite include osmotic adjustment, ROS scavenging, C and N storage, pH balance, and molecular chaperone. Specified accumulation and directed localization of metabolites such as proline and polyamines are important for proper adaptation to stress, and this should be considered in future metabolic studies.

It has recently been suggested that cellular dedifferentiation is a stage in response to both abiotic and biotic stresses (Grafi et al. 2011). Grafi et al. (2011) pointed out that during stress, overall protein synthesis is reduced, chromatin remodeling occurs, and there is an altered expression of a certain set of transcription factors. All of these are trademarks of dedifferentiated cells, such as stem cells. The authors propose that when cells sense a stress, they dedifferentiate into a stem cell-like state until a new fate is programmed. Small RNAs, which are already known to be involved in plant abiotic stress response (reviewed in Sunkar et al. 2007, 2010), may also have a role in stress dedifferentiation and maintenance as they are known to have in animal systems. Micro RNAs targeting the *Argonaute 1 (AGO1)* and *Dicer-like 1 (DCL1)* genes in *Medicago truncatula* had reduced expression during water deficit (Capitão et al. 2011). The authors also showed that several of the *AGO* and *DCL* genes themselves in this species had differential expression during low water stress, especially in root tissues, as has also been seen with certain rice *AGO* genes (Kapoor et al. 2008). *AGO* and *Dicer* genes are both involved in gene silencing, DNA methylation, and chromatin remodeling (Lippman and Martienssen 2004). In both plant and animal systems, the action of these genes is essential for maintenance of a stem cell state (reviewed in Carmell et al. 2002; Rouhana et al. 2010). With these leads, hopefully future studies will help clarify the links between stress, dedifferentiation, and small RNA action in plants.

Plant root interactions with other organisms such as bacteria and fungi are also important for tolerance to a variety of stresses. These interactions supply plants with inorganic nutrients and build resistance against pathogenic attack while the fungal or bacterial symbionts receive carbon from the plant. Class 2 fungal endophytes have been shown to mediate the tolerance to multiple abiotic stresses in both monocots (*Oryza sativa*) and eudicots (*Solanum lycopersicum*) (Rodriguez et al. 2008; Redman et al. 2011). Although more research into the mechanisms of tolerance is needed, the fungi may be affecting osmotic adjustment, water use efficiency, and/or ROS scavenging in the plants. Rhizobacteria and cyanobacteria have also been shown to afford multistress tolerance to plants (Dimkpa et al. 2009; Singh et al. 2011). The bacteria are proposed to act by modulating hormone levels, promoting osmotic adjustment and the accumulation of metabolites such as phenylpropanoids, and by stimulating the development of lateral roots. Concurrent functional genomics studies of plants and their endosymbionts will be helpful to elucidate the whole nature of their interactions during stress. Endosymbiotic interactions may be a valuable resource for the engineering abiotic stress tolerance, although care must be taken since these organisms cycle over time between lifestyles of mutualism and parasitism (reviewed in Rodriguez and Redman 2008).

Continuing Challenges in Breeding for Plant Root Systems Tolerant to Abiotic Stress

Genomic and functional genomic studies of crop abiotic stress resistance have a huge potential to produce life-changing discoveries. It was only recently that the *SUB1A* locus was cloned and incorporated into lowland flood-intolerant rice varieties. Now, these varieties are planted on 12 million hectares in flood-prone regions of India and give over 100,000 farmers up to 40% higher yield than non-*SUB1A* rice if submergence occurs (Ferrer 2010). Other analyses, possibly some focusing on root traits, will hopefully lead to similar yield preservation in the face of various abiotic stressors. Besides the use of only native or, more broadly, plant-derived loci in crop improvement, other kingdoms of life may be able to provide useful input. Introduction of genes from other species may be an important tool for engineering root systems tolerant to abiotic stress. Cold shock proteins from bacteria have been used to increase the yield of maize during water deficit in the field (Castiglioni et al. 2008). In general, cold shock proteins, like HSPs, are known to act as RNA-binding chaperones in times of stress in prokaryotes (Graumann and Marahiel 1998). Cold shock domain-containing proteins native to plants and expressed at the proper time and in the right tissues may also prove to buffer yield against stress in the field, but their distinct roles in stress response require further evaluation. Other cross-kingdom gene transfers may be important in increasing our knowledge of stress tolerance and aid in the characterization of similar plant genes.

Only some abiotic stress studies look at both a plant's action to avoid or tolerate a stress in conjunction with how well it is able to recover from the stress in terms of seed production. As noted, this is particularly difficult to achieve in root trait studies where plants are often sacrificed before maturity. In the end, it must be remembered that survival is not enough when it comes to stress tolerance. The ability of crops to maintain basic health during a stress and then recover and produce adequate yields afterward is of the most importance to farmers (Sinclair 2011). It is also increasingly recognized that the composition of nutrients in seeds produced by stressed plants can be very different from the nutrient content of unstressed plants (reviewed in Wang and Frei 2011). These factors of yield and product quality are important to address in future studies of root contribution to abiotic stress tolerance. Even if initial studies of root traits do not include characterization of yield and seed quality after the stress, they eventually should be looked at. Hopefully this will be made easier by continuing advances in phenotyping technology, which will help create a more holistic picture of the root's role in stress response and recovery.

More experiments designed to look at root biology under combinations of abiotic stresses are needed to make results more applicable to conditions experienced in the field. For drought, it is especially important to look at the combination of heat along with declining soil moisture (Mittler 2006). For drought conditions under rising CO_2 levels, current models predict that changing trends in temperatures will have a larger impact on major crop yields than forecast changes in precipitation, particularly for maize and wheat (Lobell et al. 2011). With waterlogging and flooding, experiments also incorporating excess salt or phytotoxic chemical build-up will be useful for studying the effects on roots in various local environments. Flooding often occurs in coastal environments, where excess salts are a problem. Also, soils around the world contain varying degrees of compounds that can build to toxic levels due to changes brought about by flooding. So, experimenters should be aware of local climates and field soil profiles and modify greenhouse stress conditions to closer mimic target environments.

Appreciation of the root's contribution to whole-plant function and the ever important trait of yield is growing. In a world of constant and shifting stresses, their study is critical to arrive at crops that thrive in a range of environments. The responses of plant roots to abiotic stressors are complex, but with continuing contributions to mapping and functional genomics, the key areas to focus breeding and genetic engineering efforts will emerge.

Acknowledgments

This review was made possible by grants from the United Soybean Board to HTN. The authors would like to thank Theresa Musket, Silvas Prince, and Babu Valliyodan for critical reading and comments on the manuscript.

References

Abdel-Haleem, H. et al. (2011) Identification of QTL for increased fibrous roots in soybean. *Theoretical and Applied Genetics*, 122, 935–946.

Ahsan, N. et al. (2007) A proteomic screen and identification of waterlogging-regulated proteins in tomato roots. *Plant and Soil*, 295, 37–51.

Alam, I. et al. (2010) Proteome analysis of soybean roots under waterlogging stress at an early vegetative stage. *Journal of Biosciences*, 35, 49–62.

Alcázar, R. et al. (2010) Polyamines: molecules with regulatory functions in plant abiotic stress tolerance. *Planta*, 231, 1237–1249.

Alcázar, R. et al. (2006) Abscisic acid modulates polyamine metabolism under water stress in Arabidopsis thaliana. *Physiologia Plantarum*, 128, 448–455.

Azhiri-Sigari, T. et al. (2000) Genotypic variation in response of rainfed lowland rice to drought and rewatering: II. Root growth. *Plant Production Science*, 3, 180–188.

Bacanamwo, M. and Purcell, L.C. (1999) Soybean root morphological and anatomical traits associated with acclimation to flooding. *Crop Science*, 39, 143–149.

Baena-González, E. et al. (2007) A central integrator of transcription networks in plant stress and energy signaling. *Nature*, 448, 938–942.

Bailey-Serres, J. and Voesenek, L.A. (2008) Flooding stress: acclimations and genetic diversity. *Annual Review of Plant Biology*, 59, 313–339.

Bailey-Serres, J. and Voesenek, L.A. (2010) Life in the balance: a signaling network controlling survival of flooding. *Current Opinion in Plant Biology*, 13, 489–494.

Banti, V. et al. (2010) The heat-inducible transcription factor HsfA2 enhances anoxia tolerance in Arabidopsis. *Plant Physiology*, 152, 1471–1483.

Baxter-Burrell, A. et al. (2002) RopGAP4-dependent Rop GTPase rheostat control of Arabidopsis oxygen deprivation tolerance. *Science*, 296, 2026–2028.

Bianchi-Hall, C.M. et al. (2000) Aluminum tolerance associated with quantitative trait loci derived from soybean PI 416937 in hydroponics. *Crop Science*, 40, 538–545.

Branco-Price, C. et al. (2008) Selective mRNA translation coordinates energetic and metabolic adjustments to cellular oxygen deprivation and reoxygenation in *Arabidopsis thaliana*. *Plant Journal*, 56, 743–755.

Branco-Price, C. et al. (2005) Genome-wide analysis of transcript abundance and translation in Arabidopsis seedlings subjected to oxygen deprivation. *Annals of Botany*, 96, 647–660.

Campbell, K.A.G. and Carter, T.E. (1990) Aluminum tolerance in soybean: I. genotypic correlation and repeatability of solution culture and greenhouse screening methods. *Crop Science*, 30, 1049–1054.

Capitao, C. et al. (2011) In *Medicago truncatula*, water deficit modulates the transcript accumulation of components of small RNA pathways. *BMC Plant Biology*, 11, 79.

Carmell, M.A. et al. (2002) The Argonaute family: tentacles that reach into RNAi, developmental control, stem cell maintenance, and tumorigenesis. *Genes and Development*, 16, 2733–2742.

Cassman, K.G. (1999) Ecological intensification of cereal production systems: yield potential, soil quality, and precision agriculture. *Proceedings of the National Academy of Sciences of the United States of America*, 96, 5952–5959.

Castiglioni, P. et al. (2008) Bacterial RNA chaperones confer abiotic stress tolerance in plants and improved grain yield in maize under water-limited conditions. *Plant Physiology*, 147, 446–455.

Cattivelli, L. et al. (2008) Drought tolerance improvement in crop plants: an integrated view from breeding to genomics. *Field Crops Research*, 105, 1–14.

Champoux, M.C. et al. (1995) Locating genes associated with root morphology and drought avoidance in rice via linkage to molecular markers. *Theoretical and Applied Genetics*, 90, 969–981.

Clark, L.J. et al. (2000) Screening the ability of rice roots to overcome the mechanical impedance of wax layers: importance of test conditions and measurement criteria. *Plant and Soil*, 219, 187–196.

Courtois, B. et al. (2009) Rice root genetic architecture: meta-analysis from a drought QTL database. *Rice*, 2, 115–128.

Dimkpa, C. et al. (2009) Plant-rhizobacteria interactions alleviate abiotic stress conditions. *Plant, Cell & Environment*, 32, 1682–1694.

Ergen, N.Z. and Budak, H. (2009) Sequencing over 13,000 expressed sequence tags from six subtractive cDNA libraries of wild and modern wheats following slow drought stress. *Plant, Cell & Environment*, 32, 220–236.

Ergen, N.Z. et al. (2009) Transcriptome pathways unique to dehydration tolerant relatives of modern wheat. *Functional & Integrative Genomics*, 9, 377–396.

Ferrer, P.B. (2010) *Squeezing the Most Out of Flood-proof Rice*. In: *IRRI Anual Report*.

Foley, J.A. et al. (2011) Solutions for a cultivated planet. *Nature*, 478, 337–342.

Fukao, T. and Bailey-Serres, J. (2008) Submergence tolerance conferred by Sub1A is mediated by SLR1 and SLRL1 restriction of gibberellin responses in rice. *Proceedings of the National Academy of Sciences of the United States of America*, 105, 16814–16819.

Fukao, T. et al. (2006) A variable cluster of ethylene response factor-like genes regulates metabolic and developmental acclimation responses to submergence in rice. *Plant Cell*, 18, 2021–2034.

Fukao, T. et al. (2011) The submergence tolerance regulator SUB1A mediates crosstalk between submergence and drought tolerance in rice. *Plant Cell*, 23, 412–427.

Geigenberger, P. (2003) Response of plant metabolism to too little oxygen. *Current Opinion in Plant Biology*, 6, 247–256.

Gewin, V. (2010) Food: an underground revolution. *Nature*, 466, 552–553.

Gibbs, D.J. et al. (2011) Homeostatic response to hypoxia is regulated by the N-end rule pathway in plants. *Nature*, 479, 415–418.

Gowda, V.R.P. et al. (2011) Root biology and genetic improvement for drought avoidance in rice. *Field Crops Research*, 122, 1–13.

Grafi, G. et al. (2011) Plant response to stress meets dedifferentiation. *Planta*, 233, 433–438.

Graumann, P.L. and Marahiel, M.A. (1998) A superfamily of proteins that contain the cold-shock domain. *Trends in Biochemical Sciences*, 23, 286–290.

Guingo, E. et al. (1998) Genetic analysis of root traits in maize. *Agronomie*, 18, 225–235.

Gunawardena, A.H.L.A.N. et al. (2001) Characterisation of programmed cell death during aerenchyma formation induced by ethylene or hypoxia in roots of maize(*Zea mays* L.). *Planta*, 212, 205–214.

Guo, W. et al. (2011) A soybean beta-expansin gene GmEXPB2 intrinsically involved in root system architecture responses to abiotic stresses. *Plant Journal*, 66, 541–552.

Gutierrez-Gonzalez, J.J. et al. (2010) Differential expression of isoflavone biosynthetic genes in soybean during water deficits. *Plant and Cell Physiology*, 51, 936–948.

Hao, Y.J. et al. (2011) Soybean NAC transcription factors promote abiotic stress tolerance and lateral root formation in transgenic plants. *Plant Journal*, 68, 302–313.

Hattori, Y. et al. (2009) The ethylene response factors SNORKEL1 and SNORKEL2 allow rice to adapt to deep water. *Nature*, 460, 1026–1030.

He, X.J. et al. (2005) AtNAC2, a transcription factor downstream of ethylene and auxin signaling pathways, is involved in salt stress response and lateral root development. *Plant Journal*, 44, 903–916.

Hemamalini, G.S. et al. (2000) Molecular marker assisted tagging of morphological and physiological traits under two contrasting moisture regimes at peak vegetative stage in rice (*Oryza sativa* L.). *Euphytica*, 112, 69–78.

Hummel, I. et al. (2002) Involvement of polyamines in root development at low temperature in the subantarctic cruciferous species *Pringlea antiscorbutica*. *Journal of Experimental Botany*, 53, 1463–1473.

Huynh, L.N. et al. (2005) Regulation of flooding tolerance of SAG12:ipt Arabidopsis plants by cytokinin. *Journal of Experimental Botany*, 56, 1397–1407.

Ithal, N. and Reddy, A.R. (2004) Rice flavonoid pathway genes, OsDfr and OsAns, are induced by dehydration, high salt and ABA, and contain stress responsive promoter elements that interact with the transcription activator, OsC1-MYB. *Plant Science*, 166, 1505–1513.

Jackson, M.B. and Armstrong, W. (1999) Formation of aerenchyma and the processes of plant ventilation in relation to soil flooding and submergence. *Plant Biology*, 1, 274–287.

Jackson, W.T. (1955) The role of adventitious roots in recovery of shoots following flooding of the original root systems. *American Journal of Botany*, 42, 816–819.

Jeong, J.S. et al. (2010) Root-specific expression of OsNAC10 improves drought tolerance and grain yield in rice under field drought conditions. *Plant Physiology*, 153, 185–197.

Jiang, Z. et al. (2010) Aerenchyma formation: programmed cell death in adventitious roots of winter wheat (*Triticum aestivum*) under waterlogging. *Functional Plant Biology*, 37, 748–755.

Justin, S.H.F.W. and Armstrong, W. (1987) The anatomical characteristics of roots and plant response to soil flooding. *New Phytologist*, 106, 465–495.

Kamoshita, A. et al. (2002) Mapping QTLs for root morphology of a rice population adapted to rainfed lowland conditions. *Theoretical and Applied Genetics*, 104, 880–893.

Kamoshita, A. et al. (2000) Genotypic variation in response of rainfed lowland rice to drought and rewatering: III. Water extraction during the drought period. *Plant Production Science*, 3, 189–196.

Kapoor, M. et al. (2008) Genome-wide identification, organization and phylogenetic analysis of Dicer-like, Argonaute and RNA-dependent RNA polymerase gene families and their expression analysis during reproductive development and stress in rice. *BMC Genomics*, 9, 451.

Khowaja, F.S. et al. (2009) Improved resolution in the position of drought-related QTLs in a single mapping population of rice by meta-analysis. *BMC Genomics*, 10, 276.

Kobayashi, F. et al. (2010) Identification of quantitative trait loci for ABA responsiveness at the seedling stage associated with ABA-regulated gene expression in common wheat. *Theoretical and Applied Genetics*, 121, 629–641.

Komatsu, S. et al. (2010) Identification of flooding stress responsible cascades in root and hypocotyl of soybean using proteome analysis. *Amino Acids*, 38, 729–738.

Komatsu, S. et al. (2011) Characterization of a novel flooding stress-responsive alcohol dehydrogenase expressed in soybean roots. *Plant Molecular Biology*, 77, 309–322.

Komatsu, S. et al. (2009) A comprehensive analysis of the soybean genes and proteins expressed under flooding stress using transcriptome and proteome techniques. *Journal of Proteome Research*, 8, 4766–4778.

Kong, F.-J. et al. (2010) Cell wall proteome of wheat roots under flooding stress using gel-based and LC MS/MS-based proteomics approaches. *Biochimica et Biophysica Acta (BBA)—Proteins & Proteomics*, 1804, 124–136.

Kosová, K. et al. (2011) Plant proteome changes under abiotic stress—contribution of proteomics studies to understanding plant stress response. *Journal of Proteomics*, 74, 1301–1322.

Kovács, Z. et al. (2010) Differential effects of cold, osmotic stress and abscisic acid on polyamine accumulation in wheat. *Amino Acids*, 38, 623–631.

Le, D.T. et al. (2011a) Genome-wide expression profiling of soybean two-component system genes in soybean root and shoot tissues under dehydration stress. *DNA Research*, 18, 17–29.

Le, D.T. et al. (2011b) Genome-wide survey and expression analysis of the plant-specific NAC transcription factor family in soybean during development and dehydration stress. *DNA Research*, 18, 263–276.

Lebreton, C. et al. (1995) Identification of QTL for drought responses in maize and their use in testing causal relationships between traits. *Journal of Experimental Botany*, 46, 853–865.

Lee, S.C. et al. (2011) Molecular characterization of the submergence response of the *Arabidopsis thaliana* ecotype Columbia. *New Phytologist*, 190, 457–471.

Licausi, F. et al. (2011) Oxygen sensing in plants is mediated by an N-end rule pathway for protein destabilization. *Nature*, 479, 419–422.

Licausi, F. et al. (2010) HRE1 and HRE2, two hypoxia-inducible ethylene response factors, affect anaerobic responses in *Arabidopsis thaliana*. *Plant Journal*, 62, 302–315.

Lippman, Z. and Martienssen, R. (2004) The role of RNA interference in heterochromatic silencing. *Nature*, 431, 364–370.

Lobell, D.B. et al. (2011) Climate trends and global crop production since 1980. *Science*, 333, 616–620.

Loreti, E. et al. (2005) A genome-wide analysis of the effects of sucrose on gene expression in arabidopsis seedlings under anoxia. *Plant Physiology*, 137, 1130–1138.

Ludlow, M.M. (1989) Strategies of response to water stress. In: K.H. Kreeb, H. Richter, and T.M. Hinckley (eds) *Structural and Functional Responses to Environmental Stresses*. SPB Academic, The Hague, pp. 269–281

Lynch, J. (2008) The second green revolution takes root. *Australian Journal of Botany*, 55, 493–512.

Maccaferri, M. et al. (2008) Quantitative trait loci for grain yield and adaptation of durum wheat (*Triticum durum* Desf.) across a wide range of water availability. *Genetics*, 178, 489–511.

Maccaferri, M. et al. (2011) Association mapping in durum wheat grown across a broad range of water regimes. *Journal of Experimental Botany*, 62, 409–438.

Macmillan, K. et al. (2006) Assessing the importance of genotype × environment interaction for root traits in rice using a mapping population II: conventional QTL analysis. *Theoretical and Applied Genetics*, 113, 953–964.

Malik, A.I. et al. (2009) Tolerance of *Hordeum marinum* accessions to O_2 deficiency, salinity and these stresses combined. *Annals of Botany*, 103, 237–248.

Malik, A.I. et al. (2011) Transfer of the barrier to radial oxygen loss in roots of *Hordeum marinum* to wheat (*Triticum aestivum*): evaluation of four *H. marinum*-wheat amphiploids. *New Phytologist*, 190(2), 499–508.

Manavalan, L.P. et al. (2009) Physiological and molecular approaches to improve drought resistance in soybean. *Plant and Cell Physiology*, 50, 1260–1276.

Mano, Y. et al. (2005a) Identification of QTL controlling adventitious root formation during flooding conditions in teosinte (*Zea mays* ssp. huehuetenangensis) seedlings. *Euphytica*, 142, 33–42.

Mano, Y. et al. (2005b) QTL mapping of adventitious root formation under flooding conditions in tropical maize (*Zea mays* L.) seedlings. *Breeding Science*, 55, 343–347.

Mao, X. et al. (2010) TaSnRK2.4, an SNF1-type serine/threonine protein kinase of wheat (*Triticum aestivum* L.), confers enhanced multistress tolerance in Arabidopsis. *Journal of Experimental Botany*, 61, 683–696.

Mian, M.A.R. et al. (1998) An additional QTL for water use efficiency in soybean. *Crop Science*, 38, 390–393.

Mian, M.A.R. et al. (1996) Molecular markers associated with water use efficiency and leaf ash in soybean. *Crop Science*, 36, 1252–1257.

Miller, C.R. et al. (2003) Genetic variation for adventitious rooting in response to low phosphorus availability: potential utility for phosphorus acquisition from stratified soils. *Functional Plant Biology*, 30, 973–985.

Mittler, R. (2006) Abiotic stress, the field environment and stress combination. *Trends in Plant Science*, 11, 15–19.

Mittler, R. and Blumwald, E. (2010) Genetic engineering for modern agriculture: challenges and perspectives. *Annual Review of Plant Biology*, 61, 443–462.

Miyashita, Y. and Good, A.G. (2008) Contribution of the GABA shunt to hypoxia-induced alanine accumulation in roots of *Arabidopsis thaliana*. *Plant and Cell Physiology*, 49, 92–102.

Muhlenbock, P. et al. (2007) Lysigenous aerenchyma formation in Arabidopsis is controlled by LESION SIMULATING DISEASE1. *Plant Cell*, 19, 3819–3830.

Mustroph, A. et al. (2010) Cross-kingdom comparison of transcriptomic adjustments to low-oxygen stress highlights conserved and plant-specific responses. *Plant Physiology*, 152, 1484–1500.

Nagai, K. et al. (2010) Stunt or elongate? Two opposite strategies by which rice adapts to floods. *Journal of Plant Research*, 123, 303–309.

Narsai, R. et al. (2011) Comparative analysis between plant species of transcriptional and metabolic responses to hypoxia. *New Phytologist*, 190, 472–487.

Nguyen, T.T. et al. (2004) Saturation mapping of QTL regions and identification of putative candidate genes for drought tolerance in rice. *Molecular Genetics and Genomics*, 272, 35–46.

Ober, E.S. and Sharp, R.E. (1994) Proline accumulation in maize (*Zea mays* L.) primary roots at low water potentials. I. Requirement for increased levels of abscisic acid. *Plant Physiology*, 105, 981–987.

Pantalone, V.R. et al. (1996a) Soybean fibrous root heritability and genotypic correlations with agronomic and seed quality traits. *Crop Science*, 36, 1120–1125.

Pantalone, V.R. et al. (1996b) Phenotypic evaluation of root traits in soybean and applicability to plant breeding. *Crop Science*, 36, 456–459.

Pearson, A. et al. (2011) Identification of QTLs for morphological traits influencing waterlogging tolerance in perennial ryegrass (*Lolium perenne* L.). *Theoretical and Applied Genetics*, 122, 609–622.

Pigliucci, M. and Kolodynska, A. (2002) Phenotypic plasticity and integration in response to flooded conditions in natural accessions of *Arabidopsis thaliana* (L.) Heynh (Brassicaceae). *Annals of Botany*, 90, 199–207.

Postma, J.A. and Lynch, J.P. (2011) Root cortical aerenchyma enhances the growth of maize on soils with suboptimal availability of nitrogen, phosphorus, and potassium. *Plant Physiology*, 156, 1190–1201.

Price, A.H. et al. (2002) Upland rice grown in soil-filled chambers and exposed to contrasting water-deficit regimes. II. Mapping quantitative trait loci for root morphology and distribution. *Field Crops Research*, 76, 25–43.

Qiu, F. et al. (2007) Mapping of QTL associated with waterlogging tolerance during the seedling stage in maize. *Annals in Botany*, 99, 1067–1081.

Quarrie, S.A. et al. (2005) A high-density genetic map of hexaploid wheat (*Triticum aestivum* L.) from the cross Chinese Spring × SQ1 and its use to compare QTLs for grain yield across a range of environments. *Theoretical and Applied Genetics*, 110, 865–880.

Rabello, A.R. et al. (2008) Identification of drought-responsive genes in roots of upland rice (*Oryza sativa* L). *BMC Genomics*, 9, 485.

Redman, R.S. et al. (2011) Increased fitness of rice plants to abiotic stress via habitat adapted symbiosis: a strategy for mitigating impacts of climate change. *PLoS One*, 6(7).

Renault, H. et al. (2010) The Arabidopsis *pop2–1* mutant reveals the involvement of GABA transaminase in salt stress tolerance. *BMC Plant Biology*, 10, 20.

Reynolds, M. et al. (2007) Drought-adaptive traits derived from wheat wild relatives and landraces. *Journal of Experimental Botany*, 58, 177–186.

Reynolds, M. and Tuberosa, R. (2008) Translational research impacting on crop productivity in drought-prone environments. *Current Opinion in Plant Biology*, 11, 171–179.

Ribaut, J.M. et al. (1996) Identification of quantitative trait loci under drought conditions in tropical maize. 1. Flowering parameters and the anthesis-silking interval. *Theoretical and Applied Genetics*, 92, 905–914.

Rich, S.M. et al. (2011) Aquatic adventitious roots of the wetland plant *Meionectes brownii* can photosynthesize: implications for root function during flooding. *New Phytologist*, 190, 311–319.

Rodriguez, R. and Redman, R. (2008) More than 400 million years of evolution and some plants still can't make it on their own: plant stress tolerance via fungal symbiosis. *Journal of Experimental Botany*, 59, 1109–1114.

Rodriguez, R.J. et al. (2008) Stress tolerance in plants via habitat-adapted symbiosis. *ISME Journal*, 2, 404–416.

Roje, S. (2006) *S*-Adenosyl-*L*-methionine: beyond the universal methyl group donor. *Phytochemistry*, 67, 1686–1698.

Rouhana, L. et al. (2010) Different requirements for conserved post-transcriptional regulators in planarian regeneration and stem cell maintenance. *Developmental Biology*, 341, 429–443.

Ruta, N. et al. (2010) QTLs for the elongation of axile and lateral roots of maize in response to low water potential. *Theoretical and Applied Genetics*, 120, 621–631.

Saab, I.N. et al. (1992) Effect of inhibition of abscisic acid accumulation on the spatial distribution of elongation in the primary root and mesocotyl of maize at low water potentials. *Plant Physiology*, 99, 26–33.

Schnall, J.A., and Quatrano, R.S. (1992) Abscisic acid elicits the water-stress response in root hairs of *Arabidopsis thaliana*. *Plant Physiology*, 100, 216–218.

Shabala, S. (2011) Physiological and cellular aspects of phytotoxicity tolerance in plants: the role of membrane transporters and implications for crop breeding for waterlogging tolerance. *New Phytologist*, 190, 289–298.

Sharma, S. et al. (2011a) Essential role of tissue-specific proline synthesis and catabolism in growth and redox balance at low water potential. *Plant Physiology*, 157, 292–304.

Sharma, S. et al. (2011b) Dissection of QTL effects for root traits using a chromosome arm-specific mapping population in bread wheat. *Theoretical and Applied Genetics*, 122, 759–769.

Sharp, R.E. and Davies, W.J. (1989) Regulation of growth and development of plants growing with a restricted supply of water. In: H.G. Jones, T.L. Flowers, and M.B. Jones (eds) *Plants Under Stress*. Cambridge University Press, Cambridge, pp. 71–93.

Sharp, R.E. et al. (2004) Root growth maintenance during water deficits: physiology to functional genomics. *Journal of Experimental Botany*, 55, 2343–2351.

Sharp, R.E. et al. (1988) Growth of the maize primary root at low water potentials. I. Spatial distribution of expansive growth. *Plant Physiology*, 87, 50–57.

Shevyakova, N.I et al. (2001) Heat shock-induced cadaverine accumulation and translocation throughout the plant. *Plant Science*, 161, 1125–1133.

Shimamura, S. et al. (2010) Stem hypertrophic lenticels and secondary aerenchyma enable oxygen transport to roots of soybean in flooded soil. *Annals of Botany*, 106, 277–284.

Siebert, S. and Döll, P. (2010) Quantifying blue and green virtual water contents in global crop production as well as potential production losses without irrigation. *Journal of Hydrology*, 384, 198–217.

Sinclair, T.R. (2011) Challenges in breeding for yield increase for drought. *Trends in Plant Science*, 16, 289–293.

Singh, D.P. et al. (2011) Cyanobacteria-mediated phenylpropanoids and phytohormones in rice (*Oryza sativa*) enhance plant growth and stress tolerance. *Antonie Van Leeuwenhoek*, 100, 557–568.

Specht, J.E. et al. (1999) Soybean yield potential—a genetic and physiological perspective. *Crop Science*, 39, 1560–1570.

Spollen, W.G. et al. (2000) Abscisic acid accumulation maintains maize primary root elongation at low water potentials by restricting ethylene production. *Plant Physiology*, 122, 967–976.

Spollen, W.G. et al. (2008) Spatial distribution of transcript changes in the maize primary root elongation zone at low water potential. *BMC Plant Biology*, 8, 32.

Steele, K.A. et al. (2006) Marker-assisted selection to introgress rice QTLs controlling root traits into an Indian upland rice variety. *Theoretical and Applied Genetics*, 112, 208–221.

Steffens, B. and Sauter, M. (2010) G proteins as regulators in ethylene-mediated hypoxia signaling. *Plant Signalling and Behavior*, 5, 375–378.

Sunkar, R. (2010) MicroRNAs with macro-effects on plant stress responses. *Seminars in Cell and Developmental Biology*, 21, 805–811.

Sunkar, R. et al. (2007) Small RNAs as big players in plant abiotic stress responses and nutrient deprivation. *Trends in Plant Science*, 12, 301–309.

Swamy, B.P. et al. (2011) Meta-analysis of grain yield QTL identified during agricultural drought in grasses showed consensus. *BMC Genomics*, 12, 319.

Sziderics, A.H. et al. (2010) Organ-specific defence strategies of pepper (*Capsicum annuum* L.) during early phase of water deficit. *Plant Cell Reports*, 29, 295–305.

Tuberosa, R. et al. (2002a) Mapping QTLS regulating morpho-physiological traits and yield: case studies, shortcomings and perspectives in drought-stressed maize. *Annals of Botany*, 89, 941–963.

Tuberosa, R. et al. (2002b) Identification of QTLs for root characteristics in maize grown in hydroponics and analysis of their overlap with QTLs for grain yield in the field at two water regimes. *Plant Molecular Biology*, 48, 697–712.

Valliyodan, B. and Nguyen, H.T. (2006) Understanding regulatory networks and engineering for enhanced drought tolerance in plants. *Current Opinion in Plant Biology*, 9, 189–195.

Van Dongen, J.T. et al. (2009) Transcript and metabolite profiling of the adaptive response to mild decreases in oxygen concentration in the roots of arabidopsis plants. *Annals of Botany*, 103, 269–280.

Vantoai, T.T. et al. (1994) Genetic variability for flooding tolerance in soybeans. *Crop Science*, 34, 1112–1115.

Vantoai, T.T. et al. (2001) Identification of a QTL associated with tolerance of soybean to soil waterlogging. *Crop Science*, 41, 1247–1252.

Vartanian, N. et al. (1994) Drought rhizogenesis in *Arabidopsis thaliana* (differential responses of hormonal mutants). *Plant Physiology*, 104, 761–767.

Vashisht, D. et al. (2011) Natural variation of submergence tolerance among *Arabidopsis thaliana* accessions. *New Phytologist*, 190, 299–310.

Venuprasad, R. et al. (2002) Tagging quantitative trait loci associated with grain yield and root morphological traits in rice (*Oryza sativa* L.) under contrasting moisture regimes. *Euphytica*, 128, 293–300.

Verbruggen, N. and Hermans, C. (2008) Proline accumulation in plants: a review. *Amino Acids*, 35, 753–759.

Verslues, P.E. and Sharp, R.E. (1999) Proline accumulation in maize (*Zea mays* L.) primary roots at low water potentials. II. Metabolic source of increased proline deposition in the elongation zone. *Plant Physiology*, 119, 1349–1360.

Vinod, M.S. et al. (2006) Candidate genes for drought tolerance and improved productivity in rice (*Oryza sativa* L.). *Journal of Biosciences*, 31, 69–74.

Vlad, F. et al. (2007) Arabidopsis prolyl 4-hydroxylases are differentially expressed in response to hypoxia, anoxia and mechanical wounding. *Physiologia Plantarum*, 130, 471–483.

Voetberg, G.S. and Sharp, R.E. (1991) Growth of the maize primary root at low water potentials: III. Role of increased proline deposition in osmotic adjustment. *Plant Physiology*, 96, 1125–1130.

Walk, T.C. et al. (2004) Modelling applicability of fractal analysis to efficiency of soil exploration by roots. *Annals of Botany*, 94, 119–128.

Wang, D. et al. (2011a) Genome-wide temporal-spatial gene expression profiling of drought responsiveness in rice. *BMC Genomics*, 12, 149.

Wang, H. et al. (2009) Fractal analysis on root systems of rice plants in response to drought stress. *Environmental and Experimental Botany*, 65, 338–344.

Wang, W.S. et al. (2011b) Drought-induced site-specific DNA methylation and its association with drought tolerance in rice (*Oryza sativa* L.). *Journal of Experimental Botany*, 62, 1951–1960.

Wang, Y. and Frei, M. (2011) Stressed food—the impact of abiotic environmental stresses on crop quality. *Agriculture, Ecosystems & Environment*, 141, 271–286.

Werner, T. et al. (2010) Root-specific reduction of cytokinin causes enhanced root growth, drought tolerance, and leaf mineral enrichment in Arabidopsis and tobacco. *Plant Cell*, 22, 3905–3920.

Wu, Y. et al. (1994) Root growth maintenance at low water potentials. Increased activity of xyloglucan endotransglycosylase and its possible regulation by abscisic acid. *Plant Physiology*, 106, 607–615.

Wu, Y. et al. (2001) Modification of expansin transcript levels in the maize primary root at low water potentials. *Plant Physiology*, 126, 1471–1479.

Xing, S.G. et al. (2007) Higher accumulation of γ-aminobutyric acid induced by salt stress through stimulating the activity of diamine oxidases in *Glycine max* (L.) Merr. roots. *Plant Physiology and Biochemistry*, 45, 560–566.

Xiong, L. et al. (2006) Identification of drought tolerance determinants by genetic analysis of root response to drought stress and abscisic acid. *Plant Physiology*, 142, 1065–1074.

Xu, K. et al. (2006) Sub1A is an ethylene-response-factor-like gene that confers submergence tolerance to rice. *Nature*, 442, 705–708.

Xu, P. et al. (2004) Expression of antiapoptotic genes bcl-xL and ced-9 in tomato enhances tolerance to viral-induced necrosis and abiotic stress. *Proceedings of the National Academy of Sciences of the United States of America*, 101, 15805–15810.

Yadav, R. et al. (1997) Mapping genes controlling root morphology and root distribution in a doubled-haploid population of rice. *Theoretical and Applied Genetics*, 94, 619–632.

Yamaguchi, K. et al. (2006) The polyamine spermine protects against high salt stress in *Arabidopsis thaliana*. *FEBS Letters*, 580, 6783–6788.

Yamaguchi, M. and Sharp, R.E. (2010) Complexity and coordination of root growth at low water potentials: recent advances from transcriptomic and proteomic analyses. *Plant, Cell & Environment*, 33, 590–603.

Yamaguchi, M. et al. (2010) Regulation of growth response to water stress in the soybean primary root. I. Proteomic analysis reveals region-specific regulation of phenylpropanoid metabolism and control of free iron in the elongation zone. *Plant, Cell & Environment*, 33, 223–243.

Yang, L. et al. (2004) Analysis of transcripts that are differentially expressed in three sectors of the rice root system under water deficit. *Molecular Genetics and Genomics*, 272, 433–442.

Yang, S. et al. (2010) Narrowing down the targets: towards successful genetic engineering of drought-tolerant crops. *Molecular Plant*, 3, 469–490.

Yang, Z. et al. (2009) OsMT1a, a type 1 metallothionein, plays the pivotal role in zinc homeostasis and drought tolerance in rice. *Plant Molecular Biology*, 70, 219–229.

Yiu, J.C. et al. (2009) Waterlogging tolerance of Welsh onion (*Allium fistulosum* L.) enhanced by exogenous spermidine and spermine. *Plant Physiology and Biochemistry*, 47, 710–716.

Yu, L.-X. et al. (1995) Use of wax-petrolatum layers for screening rice root penetration. *Crop Science*, 35, 684–687.

Zhang, J. et al. (2001) Locating genomic regions associated with components of drought resistance in rice: comparative mapping within and across species. *Theoretical and Applied Genetics*, 103, 19–29.

Zheng, B. (2006) QTLs and candidate genes for rice root growth under flooding and upland conditions. *Acta Genetica Sinica*, 33, 141–151.

Zheng, B. et al. (2008) Comparison of QTLs for rice seedling morphology under different water supply conditions. *Journal of Genetics and Genomics*, 35, 473–484.

Zheng, B.S. et al. (2003) Mapping QTLs and candidate genes for rice root traits under different water-supply conditions and comparative analysis across three populations. *Theoretical and Applied Genetics*, 107, 1505–1515.

Zheng, H.G. et al. (2000) Quantitative trait loci for root-penetration ability and root thickness in rice: comparison of genetic backgrounds. *Genome*, 43, 53–61.

Zheng, X. et al. (2009) Overexpression of a NAC transcription factor enhances rice drought and salt tolerance. *Biochemistry and Biophysics Research Communication*, 379, 985–989.

Zhu, J. et al. (2007) Cell wall proteome in the maize primary root elongation zone. II. Region-specific changes in water soluble and lightly ionically bound proteins under water deficit. *Plant Physiology*, 145, 1533–1548.

Zhu, J. et al. (2010) Root cortical aerenchyma improves the drought tolerance of maize (*Zea mays* L.). *Plant Cell Environment*, 33, 740–749.

Zhu, J. et al. (2011) From lab to field, new approaches to phenotyping root system architecture. *Current Opinion in Plant Biology*, 14, 310–317.

Zhu, J.K. (2002) Salt and drought stress signal transduction in plants. *Annual Review of Plant Biology*, 53, 247–273.

Zou, X. et al. (2010) Identification of transcriptome induced in roots of maize seedlings at the late stage of waterlogging. *BMC Plant Biology*, 10, 189.

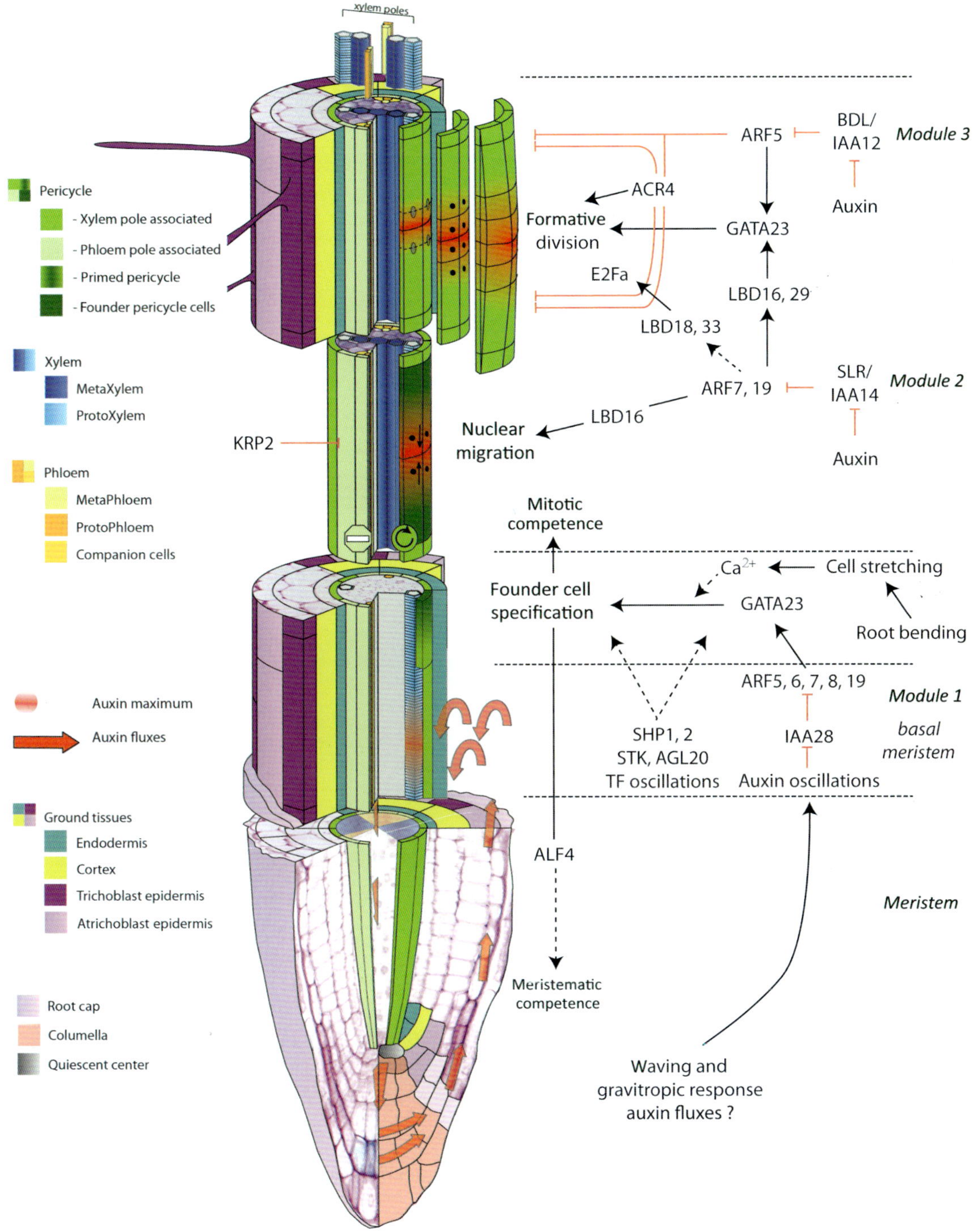

Figure 1.1 Structure of the primary root and different steps of lateral root initiation. See the text for detailed description.

Root Genomics and Soil Interactions, First Edition. Edited by Martin Crespi.

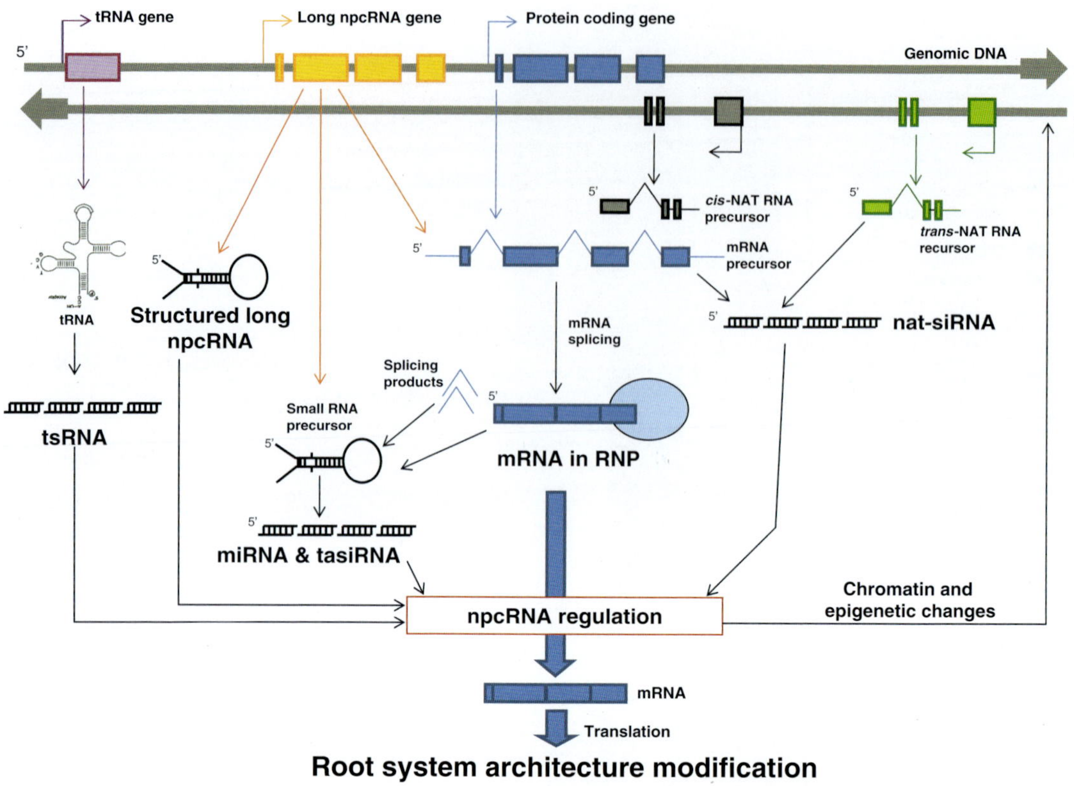

Figure 2.1 The nonprotein-coding RNA network in plants. Long npcRNAs can be precursors of small RNAs or *cis/trans*-antisense RNAs (NATs) of other transcripts. Once incorporated into the RISC effector complex, small RNAs from different pathways of RNA silencing (miRNA, siRNA, or nat-siRNA) can act directly on the target mRNA or other npcRNAs. These regulatory pathways can lead to epigenetic changes and DNA methylation (TGS). Hence, long and short npcRNAs form a network of ribonucleoproteins within the cell that may affect the expression patterns of coding mRNAs, affecting root development.

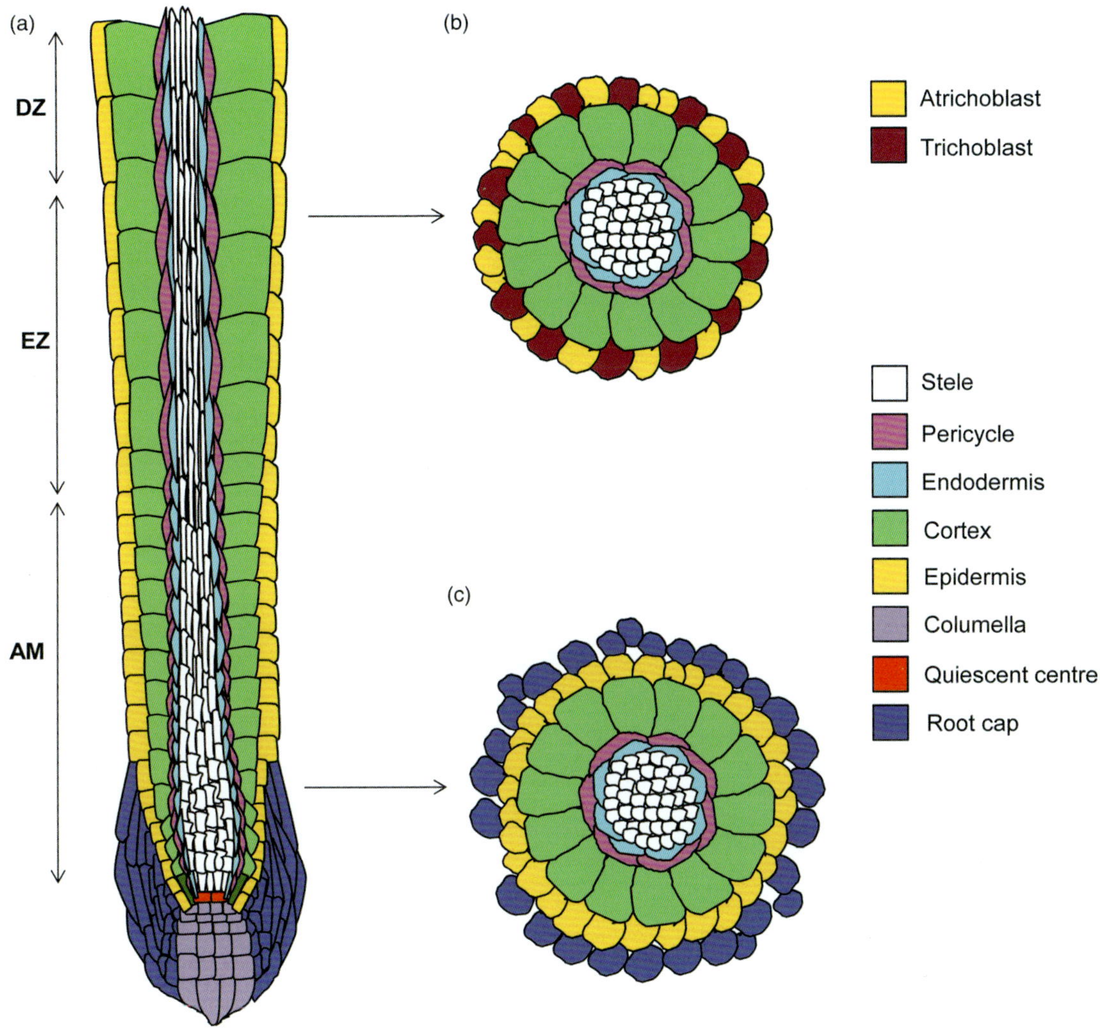

Figure 3.1 Cellular organization of the Arabidopsis root. (a) The root tissue develops from cell division and differentiation of the four different types of daughter cells, situated adjacently to the quiescent center. The resulting cell divisions give rise to the proximal distal organization of the root, with the root cap/epidermal initials forming the root cap and the epidermal cells, the columella initials establish the columella cells, central to the root cap, cortex/endodermis initials give rise to the formation of ground tissue, and the vascular initials produce the vascular and pericycle cells. The apical meristem (AP), elongation zone (EZ), and differentiation zone (DZ) of the root are indicated on the left of the figure. (b) The cross section of the elongation zone highlighting the patterning of the trichoblasts and the atrichoblasts responsible for hair cell fate. (c) Radial organization of the immature root is shown in a cross section.

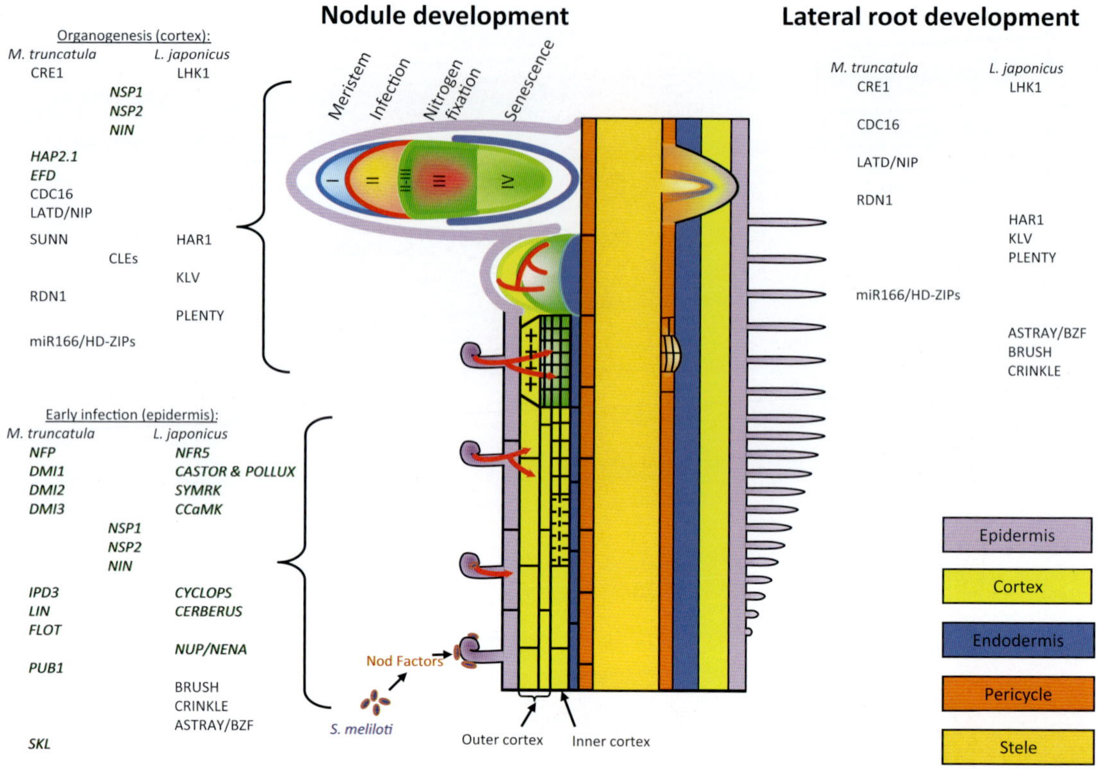

Figure 8.1 Mutants affected in nodule and/or root development in the model legumes Medicago truncatula and Lotus japonicus. Schematic representation of the different steps of symbiotic nodule (left) and lateral root (right) development in *M. truncatula*. Briefly, for nodulation, bacterial infection (*Sinorhizobium meliloti*) is initiated by root hair deformations, followed by the formation of an infection thread progressing toward the epidermis and then inner root cell layers. In parallel, nodule organogenesis is initiated in the cortex, where initial cell divisions lead to a primordium formation, which will differentiate into an indeterminate growing nodule consisting of a persistent apical meristem (I), a rhizobial infection and differentiation zone (II), a nitrogen-fixing active zone (III), and a senescence zone (IV). Concerning lateral root formation, divisions are initiated in the pericycle and will lead with a contribution of endodermis and cortex layers to a primordium, which will differentiate into a lateral root, following the parental root pattern. bZF, basic zipper and zinc finger transcription factor; CCaMK, calcium–calmodulin-dependent protein kinase; CDC16, cell division cycle 16; CLE, CLAVATA3/endosperm-surrounding region; CRE1, cytokinin response 1; DMI, does not make infections; EFD, ERF (ethylene-responsive element) required for nodule differentiation; FLOT, flotillin; HAR1, hypernodulation aberrant root 1; HD-ZIPs, homeodomain-leucine zipper type III transcription factors; IPD, interacting protein of DMI; KLV, klavier; LATD, root lateral organs deficient; LHK1, lotus histidine kinase 1; LIN, lumpy infections; NFP, nod factor perception; NFR, nod factor receptor kinase; NIN, nodule inception; NIP, numerous infections and polyphenolics; NUP, nucleoporin; NSP, nodulation signaling pathway; NUP, nucleoporin; PUB1, plant U-box protein 1; RDN1, root determined nodulation 1; SKL, sickle; SUNN, super numeric nodules; SYMRK, SYMbiosis receptor-like kinase. In regular font, genes common to nodule and lateral root development are indicated; in italic font, genes specific for nodulation.

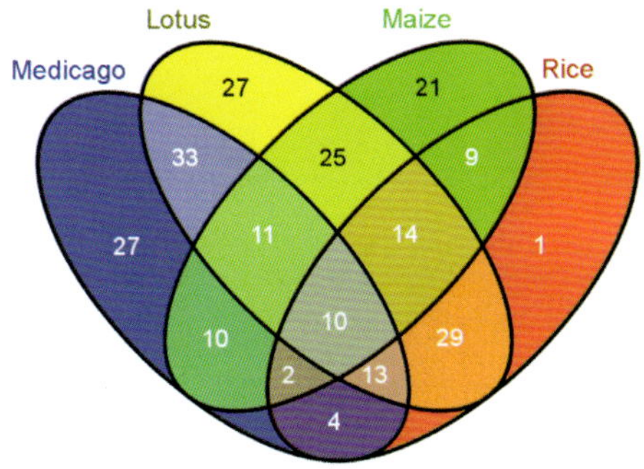

Figure 9.1 Four-way Venn diagram illustrating orthologous gene clusters (OGC) generated using OrthoMCL analysis (for method, see Li et al. 2003) on cDNA consensus sequences of regulated genes during AM symbiosis in *Medicago truncatula* (blue circle; Hohnjec et al. 2005), *Lotus japonicus* (yellow circle; Guether et al. 2009), *Zea mays* (green circle; Jourda et al. unpublished), and *Oryza sativa* (red circle; Güimil et al. 2005). The Venn diagram was generated using the Venny tool (Oliveros 2007).

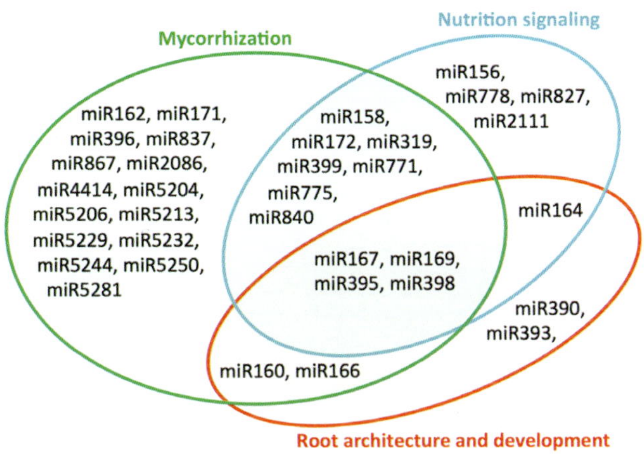

Figure 9.2 Venn diagram of the distribution and overlapping of microRNA families. The blue circle contains the microRNAs involved in nutrition signaling, the red circle contains the microRNAs involved in root architecture and development, and green circle contains the microRNAs involved in mycorrhization (see the text for references).

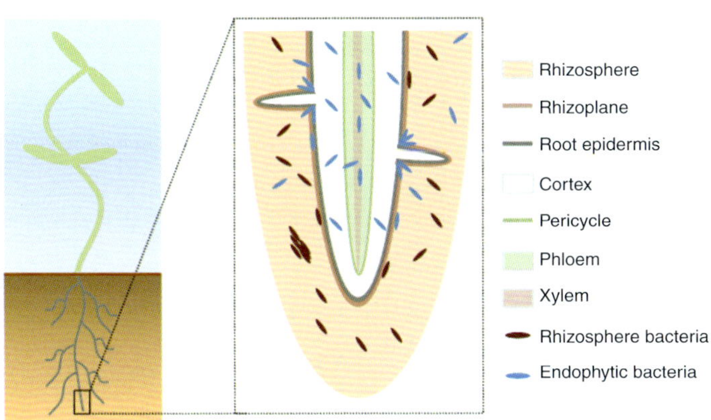

Figure 12.1 Root niches for PGPR colonization. Rhizospheric PGPRs (garnet cells) colonize rhizosphere soil area and roots surface (rhizoplane), but they cannot invade internal plant tissues. Endophytic PGPRs (blue cells) colonize any region within the epidermis of the plant root, and they can reside in apoplastic intercellular spaces and xylem vessel apoplast. In general, the endophytes invade the internal plant tissues through sites of injury in the epidermis, root tips, and root cracks formed at the sites of lateral root emergence. Some endophytic PGPRs can spread through distinct plant organs (stem, leaves, seeds, fruits).

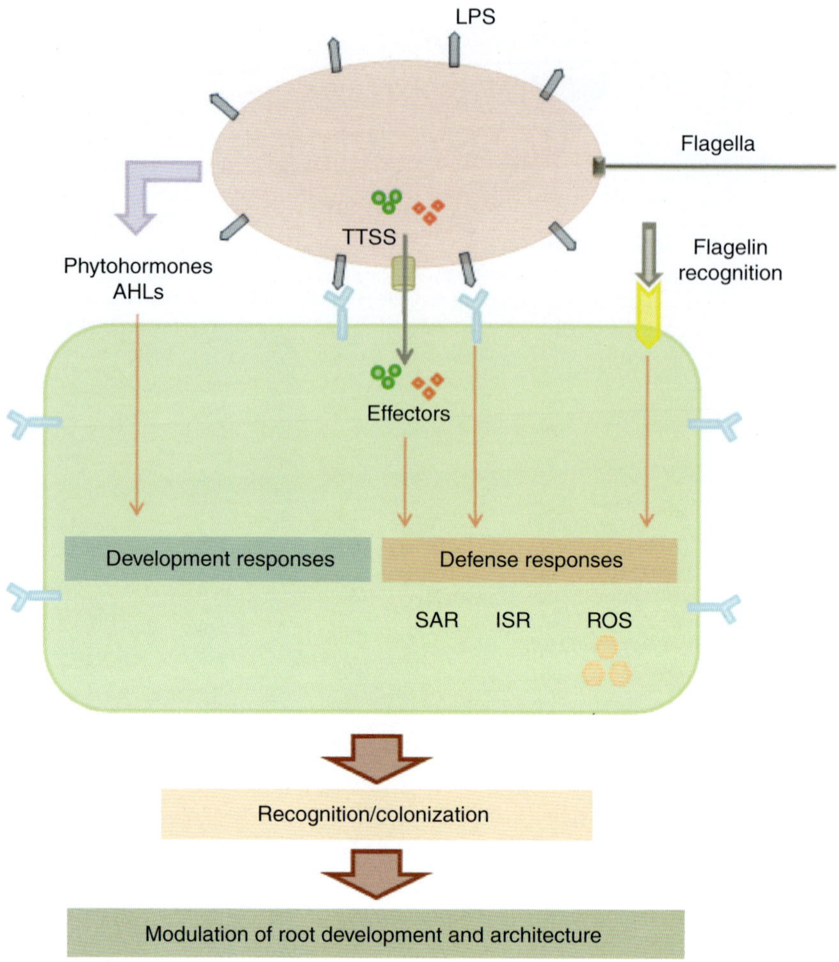

Figure 12.2 Overview of mechanisms involved in PGPRs recognition by plant cells. The figure illustrates a range of factors described to be produced by PGPRs, which can be recognized by plant cells, triggering specific responses. Lipopolysaccharides (LPS), constituents of bacterial envelope, are recognized by plant hosts. Flagellins, subunits of the polar flagella, are produced by motile PGPRs and are recognized by FLS2, a membrane-associated kinase with an extracellular LRR domain. PGPRs type III protein secretion system (TTSS) is used to deliver effector proteins directly into cytoplasm of host cells. The recognition process and the complex signal transduction pathways lead to activation of defense responses such as ISR, SAR, and ROS production. Host plant colonization depends on plant–bacteria recognition. This colonization together with phytohormones production by PGPR will result in root growth promotion and modulation of root architecture.

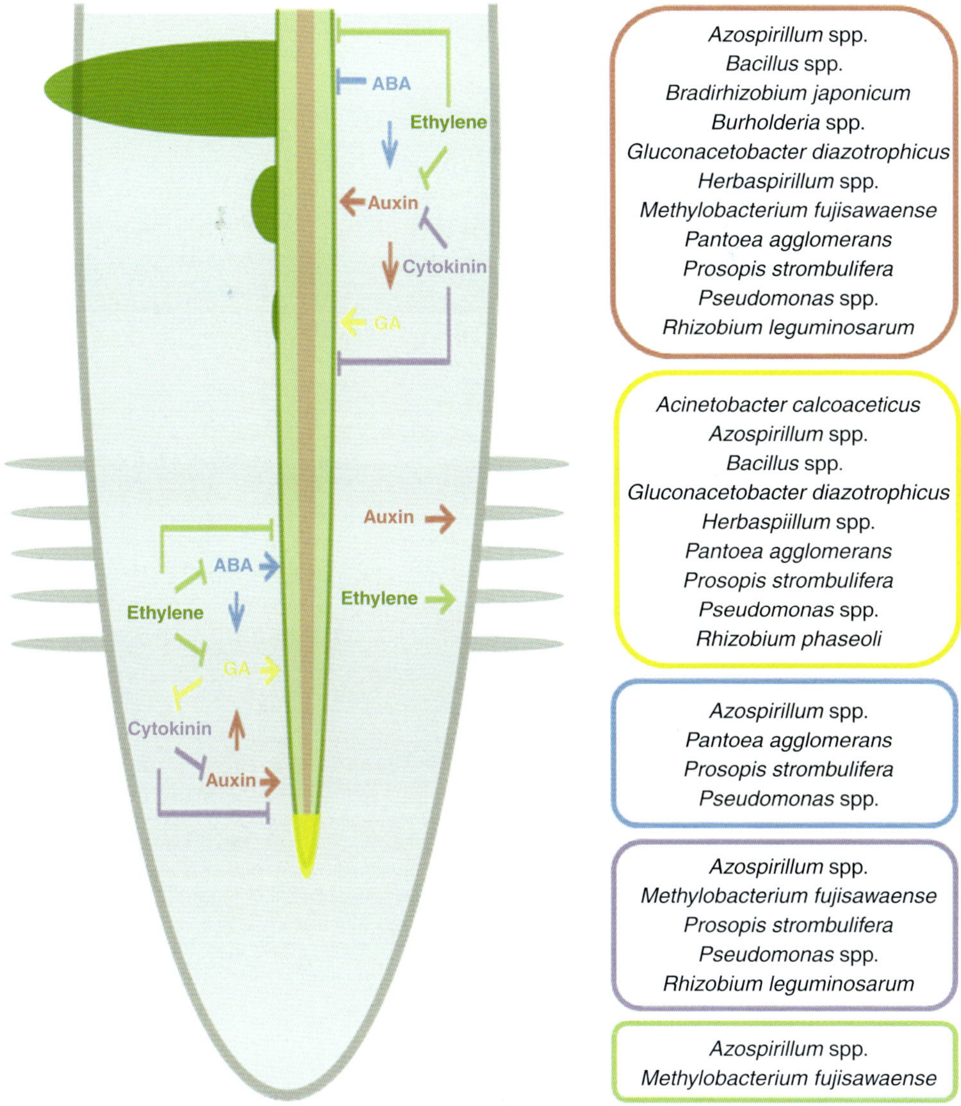

Figure 12.3 Proposed model for different mechanisms regulating root architecture by PGPRs. PGPR inoculations regulate different aspects of root architecture, such as primary root growth, lateral root formation, and root hair development. These processes are controlled by intricate regulatory networks signaled by phytohormones. Several PGPRs produce plant growth regulators such as auxin (red), ethylene (green), gibberellins (yellow), cytokinin (purple), and abscisic acid (blue). Each balloon contains the PGPRs already described as phytohormone producers, and the colors of the balloons represent the different phytohormones.

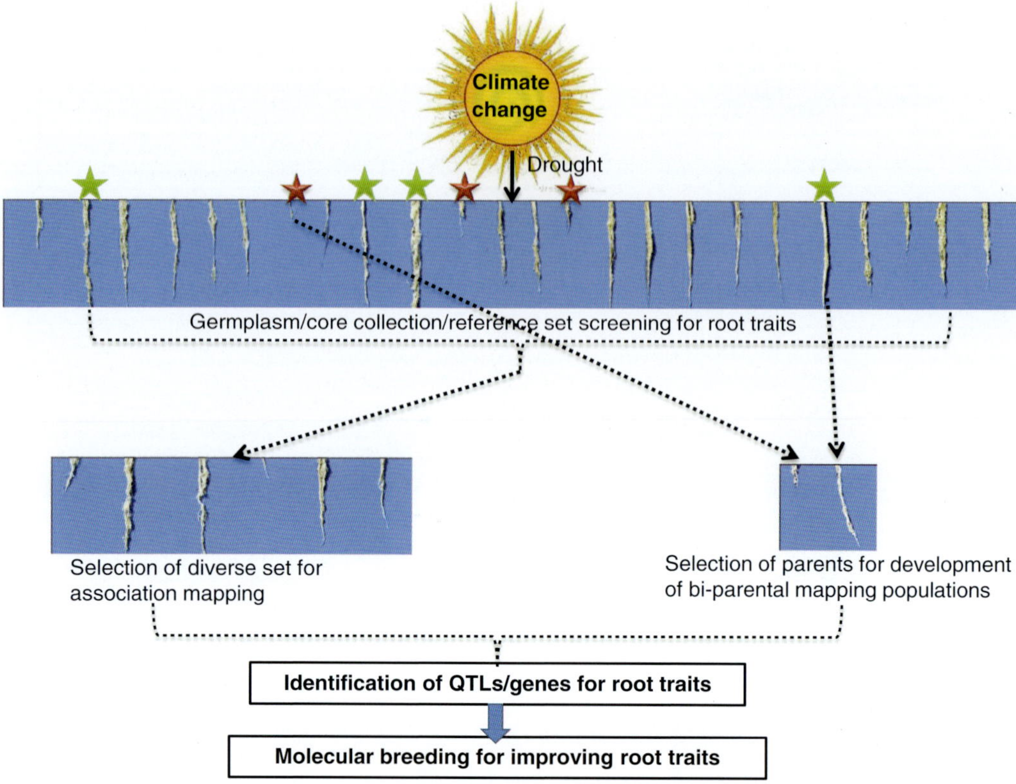

Figure 13.1 A schematic figure showing discovery and translation of root trait QTLs for crop improvement. The figure shows screening of germplasm under natural conditions for root traits. While a diverse set of germplasm showing a variation of root traits is selected and used for association mapping, contrasting lines comprising of higher root, and lower root biomass are selected for developing biparental mapping populations and linkage mapping-based QTL analysis. Once major QTLs are identified, molecular breeding for these QTLs can lead to development of superior lines for improved root traits. Red stars indicated lines with short roots (mostly drought susceptible), while green lines indicated lines with long roots (mostly drought tolerant).

8 Legume Roots and Nitrogen-Fixing Symbiotic Interactions

Philippe Laporte, Andreas Niebel, and Florian Frugier

Genetic Dissection of the Legume Root System

Plants adapt their root system architecture depending on biotic and abiotic conditions they encounter in the soil. In legumes (*Fabaceae* Family), two types of root lateral organs can be formed: lateral roots and nitrogen-fixing nodules. Nodules are established following a mutualistic interaction with soil bacteria from the *Rhizobiaceae* Family (collectively referred to as rhizobia). These new root lateral organs offer an adapted environment for the bacteria to fix atmospheric nitrogen, allowing legumes to grow in poor nitrogen soils as pioneer plants. This specificity has been used early on in agriculture to replenish nitrogen–depleted soils, and to select widespread forage and seed food crops producing proteins with a low energy cost. Hence, mechanisms underlying the developmental regulation of lateral roots and nodules are often intrinsically linked. A thorough understanding of legume root architecture thus requires parallel studies of the development of both root lateral organs.

In this chapter, we present recent advances in the dissection of legume root development and nodule organogenesis. We focus on initial steps of the nitrogen fixing symbiosis and on common pathways recruited during root and nodule organogeneses. Recent progress in forward genetic and reverse functional genomic analyses concerning legume nodules and roots will be both considered.

Forward and Reverse Genetic Dissection of Initial Steps of Rhizobium-Induced Nitrogen-Fixing Symbiosis in Legumes

Legume roots can develop two types of secondary root organs: lateral roots and nitrogen-fixing nodules. Lateral root formation is common to all higher plants; however, nodules are present on legume roots as a result of the symbiotic interaction with nitrogen-fixing rhizobia. Interestingly, mechanisms underlying the developmental regulation of lateral roots and nodules are often interrelated, and nodules may have evolved from preexisting lateral root organs that got infected by rhizobia (Hirsch and LaRue 1997). Thus, a thorough understanding of legume root architecture requires parallel studies of the development of both root lateral organs.

Nodule organogenesis and rhizobial invasion inside roots arise from a molecular dialog, implying flavonoid compounds exuded by the host plant that activate the synthesis and secretion by rhizobia of signaling molecules, called Nod factors (NFs; Gough and Cullimore 2011). NFs are lipochitooligosaccharides decorated with strain-specific lipid moieties and side chain decorations that bring host specificity into the interaction (Denarie et al. 1996). Perception of these NF by the

Root Genomics and Soil Interactions, First Edition. Edited by Martin Crespi.

host legume initiates a series of rapid morphological and physiological changes in emerging root hairs, beginning with a rapid influx of calcium followed by a KCl efflux, leading to a depolarization of the plasma membrane (Felle et al. 1995); and minutes later by an oscillatory calcium signal (Calcium spiking; Ehrhardt et al. 1996; Cardenas et al. 1999); root hair deformation and curling; remodeling of actin filaments near the root hair tips; and specific gene expression (Oldroyd and Downie 2004). Subsequently, bacterial colonization is initiated via a plant-derived tubular structure, the infection thread (IT). Concomitantly to rhizobial infection, NFs stimulate pericycle cells and inner cortical cells in front of a protoxylem pole, to reenter the cell cycle, leading to the formation of a nodule primordium (Timmers et al. 1999). Then, intracellular bacterial growth in the IT involving a host-driven cellular mechanism (Fournier et al. 2008) allows bacteria to progress through the root tissues toward the growing nodule primordium. Symbionts differentiate into nitrogen-fixing bacteroids, released from IT into the cytosol of the nodule primordium cells and surrounded by a peribacteroid membrane to form the symbiosome compartment (Gaude et al. 2004). In parallel to bacteroid differentiation, the nodule primordium develops into a mature nodule. Two major developmental types of legume nodules are noteworthy. In determinate nodules (in legumes like *Lotus japonicus*, *Phaseolus vulgaris,* and *Glycine max*), cell divisions are initiated in the outer cortex of the root but rapidly cease, resulting in the formation of spherical nodules with a transient nodule meristem. In contrast, in indeterminate nodules (in legumes like *Medicago truncatula* or *Pisum sativum*), a persistent apical meristem is initiated by cell divisions in the inner cortex where a primordium will arise and will allow continuous growth of the nodule. These nodules are elongated and contain a differentiation gradient of plant and bacterial cells from the distal (apical) meristem to the proximal region attached to the main root. Different zones have been defined: the meristematic region (zone I); the infection zone (II), where bacteria are released from ITs into plant cells; a region characterized by an accumulation of amyloplasts, the interzone (II–III); the fixation zone (III), where symbiosomes are differentiated, atmospheric dinitrogen is fixed by the bacteroids into ammonium, which is assimilated by plant cells; and the senescence zone (IV), present in older nodules, where both bacteroids and plant cells degenerate (Vasse et al., 1990). Peripheral cell layers are found, including a parenchyma, an outer and an inner cortex, an endodermis, as well as vascular bundles connecting the nodule to the parental root.

The NF signaling pathway has been dissected through the identification of mutants showing early defects in nodulation (*nod*$^-$). Most of these genetic studies have been carried out in the model legumes *L. japonicus* and *M. truncatula* (Barker et al. 1990; Handberg and Stougaard 1992; Desbrosses and Stougaard 2011; Oldroyd et al. 2011).

Several receptor-like kinases (RLK) have been identified as putative NF receptors: the LysM-type serine/threonine receptor kinases RLK NFR5/NFR1 in *L. japonicus* (Madsen et al. 2003; Radutoiu et al. 2003), NFP/LYK3/LYK4 in *M. truncatula* (Ben Amor et al. 2003; Limpens et al. 2003; Arrighi et al. 2006; Smit et al. 2007), SYM10/SYM2 in *P. sativum* (Madsen et al. 2003; Zhukov et al. 2008), and NFR5$\alpha\beta$/NFR1$\alpha\beta$ in soybean (Indrasumunar and Gresshoff 2010; Indrasumunar et al. 2011). The *Lotus nfr1* and *nfr5* (*nod factor response*), pea *sym2* (*symbiotic locus*), and *Medicago nfp* (*Nod factor perception*) mutants are affected in all the earliest NF responses, such as root hair deformation, curling, and calcium spiking (Ben Amor et al. 2003; Madsen et al. 2003; Radutoiu et al. 2003). The corresponding genes encode transmembrane LysM-RLK.

The LysM domain corresponds to a peptidoglycan-binding motif shown to bind chitin, a well-known pathogenic elicitor (Bateman and Bycroft 2000; Kaku et al. 2006). Therefore, LysM RLKs might be good candidates for NF receptors that contain *N*-acetylglucosamine backbones. In *M. truncatula*, two of these genes, *MtLYK3* and *MtLYK4*, are closely related to *LjNFR1*. RNA interference (RNAi) constructs for each of these genes reduce ITs formation (Limpens et al. 2003).

Lately, study of a weak *Medicago hcl* (*hair curling*) mutant allele indicated that it was affected in a component of the NF entry receptor. Indeed, the cloned *HCL* gene corresponded to MtLYK3 (Smit et al. 2007). Cloning of *MtNFP* revealed that it is probably an ortholog of *LjNFR5* (Arrighi et al. 2006).

NFR1 and its homologs contain an active kinase domain whose conserved subdomains VII and VIII are modified in NFP/NFR5, especially in the activation loop leading to a nonactive kinase domain (Madsen et al. 2003). These data suggest that LysM kinases might act as heteromeric NF receptors. The interaction between NFR1 and NFR5 has been shown using agroinfection in *Nicotiana benthamiana* leaves (Madsen et al. 2010). In this study, NFR1 and NFR5 were simultaneously expressed, which induced a cell death response mediated by the presence of both proteins and an active NFR1 kinase domain. Two recent *in vitro* studies also showed that NFR1/LYK3 autophosphorylate and are able to phoshorylate NFR5/NFP (Klaus-Heisen et al. 2011; Madsen et al. 2011). Complementation experiments revealed that phosphorylation of a single residue in the activation loop is required for NF receptor signaling (Klaus-Heisen et al. 2011; Madsen et al. 2011). The LysM II domain of *Lotus* NFR5 has been identified as being likely implicated in the determination of NF host-range specificity (Radutoiu et al. 2007). Although three different NF-binding sites have been biochemically identified and characterized (Hogg et al. 2006), a direct interaction between NFs and their putative receptors remains to be demonstrated. Moreover, the three different LysM domains of NFP were revealed to be critical for the recognition of NF substitutions at different steps of symbiosis (Bensmihen et al. 2011).

Various LYK complexes may be required at different steps of the interaction (Radutoiu et al. 2003; Arrighi et al. 2006), as suggested by the contrasted expression patterns of LYK/NFR genes: *LjNFR1* seems root specific, and both *LjNFR1* and *LjNFR5* expression is not affected by *Mesorhizobium loti* inoculation. In *M. truncatula*, *MtLYK3* is expressed in root tissues, but not in nodules, whereas *MtLYK4* is not detected using Northern analysis (Limpens et al. 2003). More detailed studies revealed that *MtNFP* is expressed only in growing root hairs susceptible to *Rhizobium* and that, upon rhizobial inoculation, its expression follows nodule primordia formation and is found later in the infection zone of mature nodules (Madsen et al. 2003). These results indicate that several heterodimeric NF receptors might be involved throughout the infection process, for IT development and bacterial release. In addition, several interactors of the LysM kinases have been identified. For example, "plant U-box protein 1" that interacts with and is phosphorylated by the LYK3 intracellular region, containing its kinase domain, may play a role as a negative regulator of NF signaling (Mbengue et al. 2010). Another U-box protein, LIN (for "Lumpy Infections"), acts in rhizobial colonization (Kiss et al. 2009).

Other components of the NF signaling pathway, acting downstream of NF perception, have been identified in Medicago by characterizing mutants that are unable to form normal ITs, that is, the *dmi1*, *dmi2*, and *dmi3* mutants (*doesn't make infection* mutants). These mutants are affected in nodulation as well as in symbiotic interactions with mycorrhizal fungi (Weidmann et al. 2004; Kistner et al. 2005; Olah et al. 2005). *dmi1* and *dmi2* mutants show wild-type root hair deformations, but no calcium spiking, in response to rhizobia. The *MtDMI1* and *MtDMI2* genes code for a putative cation channel and a leucine-rich-repeats receptor-like kinase, respectively, and might act downstream of the putative NF receptor (Ane et al. 2004; Levy et al. 2004). MtDMI1 was shown to transport monovalent cations but not Ca^{2+}, suggesting that this channel may be indirectly linked to calcium oscillation induced in response to NF and Myc factors (Peiter et al. 2007). *PsSYM8* in pea, and *CASTOR* and *POLLUX* in *Lotus* are orthologs of *MtDMI1* (Imaizumi-Anraku et al. 2005; Edwards et al. 2007). ProMtDMI1:MtDMI1:GFP fusions localize in the nuclear envelope of Medicago (Riely et al. 2007), accordingly with the *Lotus* homologs (Charpentier et al. 2008).

Recently, Capoen et al. (2011) showed that DMI1 is preferentially located to the inner nuclear membrane. This subcellular localization of DMI1 is consistent with the involvement of components of the nuclear pore in NF response. Indeed, the *Lotus* NUCLEOPORIN mutants (*nup133*, *nup85*, and *nena*) show thermosensitive mycorrhiza and nodulation phenotypes, affecting, in the latter case, calcium spiking and root hair curling in *nup133* (Kanamori et al. 2006; Saito et al. 2007; Groth et al. 2010). These nucleoporin-encoding genes are expressed in all plant organs, and NUP133 and NENA are localized in the nuclear envelope of root cells. NUP133 and NUP85 are predicted components of the NUP107-NUP160 subcomplex located on both sides of the nuclear pore (Meier and Brkljacic 2009). This nuclear pore subcomplex implicated in symbiotic interactions might be involved in transport or localization required for the calcium-spiking response. The fact that no nucleoporin mutant has been identified so far in *M. truncatula* could point to gene redundancy in this model plant, which could prevent successful forward genetic approaches. The reloading of calcium into the nucleus may require an ATPase. MCA8, a sarco/endoplasmic reticulum (ER) calcium ATPase, localized in the inner and outer nuclear membrane as well as in the ER, is also required for symbiotic calcium oscillations (based on an RNAi phenotype; Capoen et al. 2011). Thus, MCA8 could play a role in reloading of calcium released into the nucleus during symbiotic signaling.

DMI2 orthologs have been identified in *Lotus* (*LjSYMRK* for SYMbiosis Receptor-like Kinase; Schauser et al. 1998; Stracke et al. 2002; Kistner et al. 2005), in *S. rostrata* (*SrSYMRK*; Capoen et al. 2005), in alfalfa (*MsNORK* for NOdulation Receptor Kinase; Endre et al. 2002), and in pea (*PsSYM19* for *SYMbiosis locus 19*; Stracke et al. 2002). All these SYMRK genes are able to complement the mycorrhizal defect. Together with the observation that the SYMRK gene structure is conserved among Eurosids but not in other phylogenic groups, this suggests that SYMRK has been recruited from a preexisting signaling pathway linked to mycorrhiza (Markmann et al. 2008). However, the *Medicago* MtDMI2 homolog has also been associated to touch sensitivity of root hairs and pathogenic responses, suggesting that this receptor may have broader functions in root perception of nonsymbiotic environmental conditions (Esseling et al. 2004). Like many of its orthologs, the *MtDMI2* gene is expressed at high levels in roots and also in the IT membrane. MtDMI2 is in addition found in the infection zone of nodules, also suggesting a possible function during later stages of nodule development in restricting IT growth and internalizing rhizobia (Bersoult et al. 2005; Limpens et al. 2005). Kevei et al. (2007) have identified a plasma membrane protein interacting with the DMI2 kinase domain and corresponding to a key enzyme for mevalonate synthesis (3-hydroxy-3-methylglutaryl CoA reductase 1; MtHMGR1). Although this pathway seems to play an essential role for nodule development, it remains to be precisely shown which isoprenoid compound(s) is (are) really involved. Another protein interacting with LjSYMRK has been identified by Zhu et al. (2008) as being an ARID-type DNA-binding protein (SIP1). This nuclear protein interacts with the kinase domain of SYMRK and itself via its C-terminal domain. Its ARID domain (helix-turn-helix) located in the N-terminal region can bind the promoter of the early *Nodulin* gene *NIN* (see later). More recently, the *Lotus symRK-14* mutant affected in the extracellular domain of the RLK is affected in early symbiotic responses in the epidermis but not in nodule development, thereby uncoupling the epidermal and cortical symbiotic programs and illustrating the essential role of SymRK at different steps of the symbiotic interaction (Kosuta et al. 2011). Finally, an RNAi targeting the *Phaseolus* SymRK homolog revealed an additional role for this pathway in root and nodule vascular development (Sanchez-Lopez et al. 2011).

A symbiotic remorin protein (SYMREM1), which is required for bacterial infection in *M. truncatula*, interacts with LYK3, NFP, and DMI2 (Lefebvre et al. 2010). MtSYMREM1 is localized in the IT membrane and may play a role in recruiting or retaining RLKs in membrane rafts at the growing IT apex, allowing calcium signaling and IT elongation through the root cortex. These protein relocalizations allowing rearrangement of the plasma membrane in subdomains may involve flotillin proteins (FLOT; Haney and Long 2010). One of these proteins is indeed relocalized to the apex of growing root hairs upon inoculation with *S. meliloti*. In addition, silencing of different *FLOT* genes leads to IT and nodule defects affecting both its development and function. In addition, a LYK3:GFP fusion was relocalized in the same subcellular compartment as a FLOT4:mCherry fusion, and this localization was altered in *lyk3* mutants. These data suggest that FLOT4 and LYK3 are part of a same complex recruited during *S. meliloti* infection.

Further downstream in the NFs signaling pathway is the *Medicago dmi3* mutant, which still shows calcium spiking in response to NFs but is defective during later stages of the symbiotic response and also in the mycorrhizal interaction (Catoira et al. 2000). Similarly, the *sym9* mutant in pea is altered at the same nodulation stage, suggesting that *PsSYM9* is a putative ortholog of *MtDMI3* (Duc et al. 1989; Walker et al. 2000). MtDMI3 encodes a calcium-calmodulin-dependent protein kinase (CCaMK; Levy et al. 2004; Mitra et al. 2004), located in the nucleus (Kalo et al. 2005), and is a putative decoder of calcium signaling. This gene is strongly expressed in roots and nodules (Levy et al. 2004). A modified MtDMI3 protein lacking the predicted auto-inhibition domain and a gain-of-function mutant in orthologous gene in *Lotus* (*snf1*) have been shown to induce the formation of nodules in the absence of rhizobia (spontaneous nodules; Gleason et al. 2006; Tirichine et al. 2006). These results reveal a key regulatory position for DMI3 in the NF signaling pathway, upstream of all components necessary for the reinitiation of the cell cycle, leading to symbiotic nodule development.

A nuclear protein called IPD3 (interacting protein of DMI3), exhibiting a conserved coiled-coil motif, has been shown to interact with DMI3 (Messinese et al. 2007). Later, Yano et al. (2008) have identified *CYCLOPS*, the *Lotus* ortholog of *IPD3*. The *cyclops* mutants are infected neither by AM fungi nor by rhizobia, but calcium spiking and nodule primordium development can still be activated, positioning *CYCLOPS* downstream of *CCaMK*. Accordingly, these authors also showed that CYCLOPS interacts *in planta* and can be phosphorylated *in vitro* by CCaMK. *Medicago ipd3* mutants revealed the role of IPD3 in IT growth, nodule invasion, and mycorrhizal colonization (Horvath et al. 2011). In parallel, Ovchinnikova et al. (2011) identified that the Medicago *Mtsym1* and pea *Pssym33* mutants, impaired in their ability to form symbiosomes, were both affected in IPD3. These data revealed an additional late role for IPD3 in symbiosome formation.

Recently, another CCaMK-interacting protein was identified in *Lotus*: CIP73, encoding an ubiquitin protein containing a Scythe_N ubiquitin-like domain (Kang et al. 2011a). Knockdown of CIP73 using RNAi led to a decrease in nodule formation (Kang et al. 2011a).

The *nodulation signaling pathway 1* and 2 mutants *nsp1* and *nsp2* were initially identified as specifically affected in nodulation, but a more careful examination of *nsp2* recently revealed a mycorrhizal phenotype (Maillet et al. 2011). The double mutant *nsp1/nsp2* is in addition affected in the synthesis of strigolactones, a signal important for the mycorrhizal symbiosis (Liu et al. 2011). Furthermore, strigolactones have been recently described as promoting nodulation in pea (Foo and Davies 2011). *MtNSP1* and *MtNSP2* genes encode putative GRAS-family transcription factors (Kalo et al. 2005; Smit et al. 2005) and through the synthesis of strigolactones may affect both symbiotic interactions. MtNSP1, preferentially expressed in roots, is located in the nucleus, whereas MtNSP2, expressed in roots, shoots, and leaves, is predominantly localized in the nuclear envelope and in the ER. The pea *SYM7* gene might be an ortholog of *MtNSP2* based on its genetic map location (Kalo et al. 2005). Both have similar nodulation phenotypes, showing wild-type calcium spiking, but reduced root hair deformations, blocked IT formation in the epidermis, drastically reduced expression of early *Nodulin* genes, and no initiation of nodule primordia, thus suggesting that these genes operate downstream of the *DMI* genes, but upstream of NF-induced early *Nodulin* genes (such as *MtRIP1* (*Rhizobium-induced peroxydase*) and *MtEnod11*) (Catoira et al. 2000; Wais et al. 2000; Oldroyd and Long 2003). Indeed, MtNSP1 can bind the *MtEnod11, MtERN1,* and *MtNIN* promoters, forming a DNA-binding heteropolymer complex with NSP2 (Hirsch et al. 2009).

Another gene that probably acts as a transcriptional regulator during early symbiotic steps is *LjNIN* (for *nodule inception*). The *Lotus nin* mutant showed excessive root hair curling, but aborted ITs, no cortical cell division nor nodule primordia formation (Schauser et al. 1999). The fact that NFs can still induce calcium spiking in *nin* as well as early *Nodulin* expression patterns in this mutant place *nin* downstream of the early signal exchange between symbionts. *LjNIN* expression is strongly upregulated during nodulation, and transcripts are localized in dividing cells of nodule primordia as well as in mature nodule parenchyma, nodule vascular bundles, and nitrogen-fixing zones (Schauser et al. 1999). An ortholog is encoded by the *sym35* locus in pea (Borisov et al. 2003) and by *MtNIN* in Medicago (Marsh et al. 2007). Interestingly, it was shown that *MtNIN* negatively regulates the spatial expression domain of *MtEnod11* in Rhizobium-infected roots and thus acts in the MtNFP-dependent NF pathway, leading to early *Nodulin* expression (Marsh et al. 2007). LjNIN contains membrane-spanning helices and a nuclear localization signal, similar to Notch and SREBEP transcription factors in animals (Schauser et al. 1999).

A mutant affected in a transcription factor of the AP2/ERF family (Apetala 2/ethylene response element), referred to as ERN (ERF required for nodulation), is blocked in IT formation and NF-induced gene expression (Middleton et al. 2007). In parallel, a small family of ERN proteins (MtERN1 to 3; ERN1 corresponding to the previously mentioned ERN) has been identified in a yeast one hybrid approach as able to bind the NF-responsive regulatory unit (the NF-box) present in the *MtENOD11* promoter, but outside the NSP1-binding region. This NF box is necessary and sufficient to confer a NF-specific response in root hairs (Andriankaja et al. 2007). In this tissue, MtENOD11 expression is upregulated after NF treatment via an MtERN1- and 2-dependent pathway, and MtERN 3 is hypothesized to function as a repressor of MtERN1 and 2 activities. This suggests that antagonistic ERN transcription factor activities may be involved in fine-tuning of gene expression in response to NFs.

From the data presented earlier, it is clear that most of the genes involved in NF-signaling, and even sometimes initially described as specific for this interaction, also act in the mycorrhizal symbiosis. In addition, besides the early roles in NF signaling highlighted by mutant phenotypes (Figure 8.1), most of these genes can also act at later nodulation stages. These recent findings thereby challenge the initial linear genetic pathways focused on NF perception to early *Nodulin* gene expression, rhizobial infection, and nodule organogenesis. Alternative branched models are now proposed (e.g. Madsen et al. 2010; Murray 2011).

Genetic Dissection of Shared Pathways between Root and Nodule Organogeneses

In general, legume roots are composed by a single layer of epidermal cells at the surface, with no specific root hair patterning, followed by three to five irregularly shaped cortical cell layers, usually defined as outer, middle, and inner cortex (Gage 2004). A single layer of endodermal cells is then found, followed by a pericycle layer, which delimits the stele region enclosing the vascular bundles.

Lateral root development shows different tissue layers requirement depending on species (Berthon 1943). In contrast to *Arabidopsis,* where lateral root primordia arise only from pericycle cells, legumes activate pericycle, endodermal, and cortical cells. These lateral (or adventitious) roots can be used by Rhizobia, depending on species (e.g., tropical legumes) and environmental conditions (e.g., hydroponics conditions), to directly colonize the root cortex via a crack in the epidermis

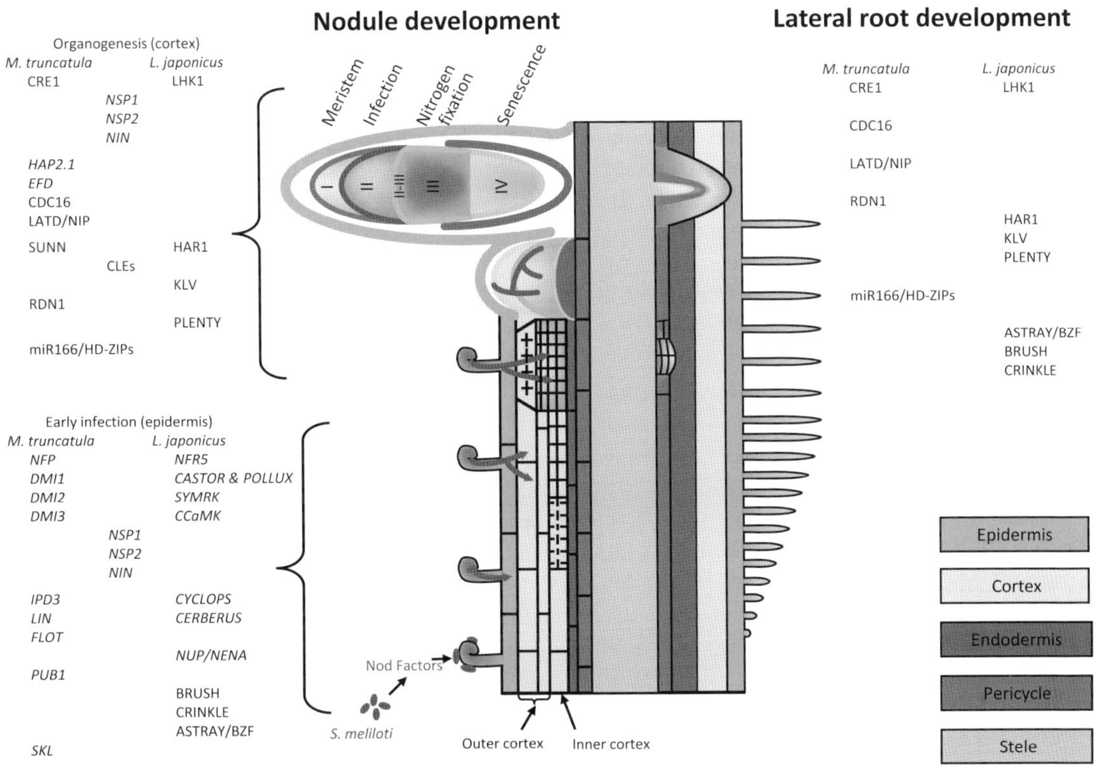

Figure 8.1 Mutants affected in nodule and/or root development in the model legumes *Medicago truncatula* and *Lotus japonicus*. Schematic representation of the different steps of symbiotic nodule (left) and lateral root (right) development in *M. truncatula*. Briefly, for nodulation, bacterial infection (*Sinorhizobium meliloti*) is initiated by root hair deformations, followed by the formation of an infection thread progressing toward the epidermis and then inner root cell layers. In parallel, nodule organogenesis is initiated in the cortex, where initial cell divisions lead to a primordium formation, which will differentiate into an indeterminate growing nodule consisting of a persistent apical meristem (I), a rhizobial infection and differentiation zone (II), a nitrogen-fixing active zone (III), and a senescence zone (IV). Concerning lateral root formation, divisions are initiated in the pericycle and will lead with a contribution of endodermis and cortex layers to a primordium, which will differentiate into a lateral root, following the parental root pattern. bZF, basic zipper and zinc finger transcription factor; CCaMK, calcium–calmodulin-dependent protein kinase; CDC16, cell division cycle 16; CLE, CLAVATA3/endosperm-surrounding region; CRE1, cytokinin response 1; DMI, does not make infections; EFD, ERF (ethylene-responsive element) required for nodule differentiation; FLOT, flotillin; HAR1, hypernodulation aberrant root 1; HD-ZIPs, homeodomain-leucine zipper type III transcription factors; IPD, interacting protein of DMI; KLV, klavier; LATD, root lateral organs deficient; LHK1, lotus histidine kinase 1; LIN, lumpy infections; NFP, nod factor perception; NFR, nod factor receptor kinase; NIN, nodule inception; NIP, numerous infections and polyphenolics; NUP, nucleoporin; NSP, nodulation signaling pathway; NUP, nucleoporin; PUB1, plant U-box protein 1; RDN1, root determined nodulation 1; SKL, sickle; SUNN, super numeric nodules; SYMRK, SYMbiosis receptor-like kinase. In regular font, genes common to nodule and lateral root development are indicated; in italic font, genes specific for nodulation. (For a color version of this figure, see the color plate section.)

layer (crack entry) during emergence (and so-called lateral root-based or lateral root-associated nodulation; James et al. 1992; Boogerd and Van Rossum 1997; Mathesius et al. 2000; Goormachtig et al. 2004).

Several hypotheses have been proposed for the evolutionary origin of nodules, among which one postulates that they could derive from preexisting organs such as lateral roots (Nutman 1948.

Accordingly, several nodulin genes are also expressed in lateral roots (Hirsch et al. 1989; Yang et al. 1993; Crespi et al. 1994; Journet et al. 1994; Papadopoulou et al. 1996; De Carvalho Niebel et al. 1998; de Billy et al. 2001). In addition, nodules and lateral roots share some aspects of their development, such as their initiation in front of protoxylem poles (Hirsch and LaRue 1997; Mathesius et al. 2000), even though different tissues are predominantly involved (respectively, cortex and pericycle; Brewin 1991; Hirsch 1992). Intermediate root lateral organs have been identified, so-called root-nodule hybrids, supporting the theory that nodule formation evolved from lateral root developmental pathways (Vandenbosch et al. 1985; Dudley and Long 1989; McIver et al. 1997; Ferraioli et al. 2004). Usually, these plant or bacterial mutants initiate a wild-type nodulation (such as root hair infection), but later the primordia development is altered and ectopic roots emerge from the distal meristematic region.

Several plant mutants, initially identified based on their nodulation phenotype, were then found to be affected in root development. Two classes can be defined: (1) mutants similarly affecting nodule and root development, either positively or negatively, and (2) mutants affecting antagonistically root and nodule development. This collectively indicates that root and nodule organogenesis share common regulatory pathways.

Mutants Inhibiting or Stimulating Both Root and Nodule Development

A mutant defective in nodule formation and also in root development is the *M. truncatula latd* mutant, standing for "*lateral root organ-defective*" (Bright et al. 2005). Even though the main root of *latd* grows normally after germination, it is then blocked and lateral root formation is impaired. Lateral roots of *latd* are disorganized, lacking notably their root caps. Nodulation in *latd* is also arrested at a stage where nodules remain globular, white, and undifferentiated. Therefore, the *MtLATD* gene is required for the function of root and nodule meristems. Interestingly, the shoot meristem is in contrast not affected in these mutants. Recently, the *LATD/NIP* gene was cloned, and it encodes a member of the NRT1/PTR transporter family expressed throughout the plant, notably in root and nodule meristems (Yendrek et al. 2010). Expression of LATD/NIP is regulated by abscisic acid (ABA), which was previously shown to rescue the primary and lateral root phenotype of *latd* mutants. As *Arabidopsis* NRT1 genes are nitrate transporters (Gojon et al. 2011), a *latd* nitrate (but not ammonium) sensitivity defect was identified, and LATD/NIP was recently shown to encode a high affinity nitrate transporter (Bagchi et al. 2012).

Root and nodule deficiency can be also induced by the overexpression of a microRNA166 precursor shown to downregulate class III HD-ZIP transcription factors (Boualem et al. 2008). An abnormal patterning of root vascular bundles and of the xylem pole number is observed, which is correlated with defective lateral root and nodule formation.

Another type of nodulation mutants are affected in systemic or local negative feedback regulations that controls the number of nodules formed on the root system and referred to as autoregulation of nodulation (AON). Some of these hypernodulating mutants are also affected in the regulation of root meristems and, consequently, in the architecture of their root system. One of those is a shoot-dependent hypernodulating mutant that has been identified in several legumes: *har1* in *Lotus* (*hypernodulation aberrant root formation 1*; Nishimura et al. 2002a), *nodule autoregulation receptor kinase* (*nark*) in soybean (Searle et al. 2003), *sym29* in pea (Krusell et al. 2002), and *sunn* in *M. truncatula (super numeric nodules*; Penmetsa et al. 2003; Schnabel et al. 2005). Although these mutants display a similar hypernodulation phenotype and nitrate insensitivity, the *Lotus har1* mutant also displays a drastically altered root development, similarly dependent on the shoot (Wopereis et al. 2000; Krusell et al. 2002). Compared to wild-type, the mutant root system consists of a shorter main root associated to extra pericycle cell divisions and to an increased number of lateral roots.

In contrast, the *sunn Medicago* mutant displays a shortened primary root growth in presence or absence of rhizobia, but no significant lateral root phenotype (Penmetsa et al. 2003; Schnabel et al. 2005). These mutants are, however, all affected in a leucine-rich repeat (LRR) type receptor kinase similar to the *Arabidopsis CLAVATA1* (*CLV1*) gene, suggesting a diversification of this gene family even among the legumes. Interestingly, no shoot phenotype was reported, in contrast to *Arabidopsis*, where CLV1 is required for the restriction of shoot meristem proliferation through a short-distance negative feedback loop (Clark et al. 1997).

Changes in root morphology have been also associated to other hypernodulating mutants. The recently reported *L. japonicus* root-dependent hypernodulating mutant *plenty* also has a larger number of lateral roots even in the absence of rhizobium (Yoshida et al. 2010). However, in contrast to *har1*, primary and lateral root length is reduced compared to the wild-type, leading to a "compact root architecture" phenotype. The pea *nod3* mutant displays a root-dependent hypernodulation phenotype (Gremaud and Harper 1989) and might be affected in a signaling pathway transducing a signal produced by roots or nodule primordia that would move to the shoot to activate systemic negative feedback mechanisms. Accordingly, this mutant also has an altered root architecture phenotype, with a reduced main root growth and an increased lateral root formation (Jacobsen and Feenstra 1984). In *M. truncatula*, the *root determined nodulation 1* (*rdn1*) mutant was recently shown to be an ortholog of *nod3* and to be affected in a plant-specific protein with an unknown function (Schnabel et al. 2011). Finally, the soybean *nts382* mutant shows a shoot-dependent hypernodulation phenotype and has an increased lateral root formation, independently of rhizobia and/or nitrate application (Day et al. 1986).

Mutants Antagonistically Affecting Root and Nodule Development

In *M. truncatula*, specific RNAi of one of the cytokinin receptors, MtCRE1, leads to cytokinin-insensitive roots, increased lateral root formation, and inhibition of nodulation (Gonzalez-Rizzo et al. 2006). More recently, stable *cre1* mutants were obtained displaying similar phenotypes (Plet et al. 2011, and F. Frugier, unpublished data). The symbiotic interaction is blocked at an early stage, with delayed growth of ITs in the outer root cell layers that are frequently abortive, and lack of inner cortical cell divisions. A few nodules, which are irregularly shaped and contain multiple lobes, can however form, suggesting a maintained role for cytokinins during indeterminate nodulation (Plet et al. 2011). In *L. japonicus*, the *hit1* (*hyperinfected 1*) mutant, having a mutation in the MtCRE1 orthologous gene (*Lotus* Histidine Kinase 1, LHK1), similarly shows a drastically reduced nodulation, associated with a proliferation of ITs in the epidermis (Murray et al. 2007). Potential consequences on root development have however not been described in *Lotus*. Genetic interactions together with transcriptional regulation of nodulation signaling genes by cytokinins indicate that CRE1/LHK1 acts downstream of *NFP/LYK/NFR*s and *DMI* genes, and upstream of *NSP*s, *NIN*, and *ERN* genes (Gonzalez-Rizzo et al. 2006; Murray et al. 2007; Tirichine et al. 2007; Plet et al. 2011). A gain of function mutation in the same *Lotus* LHK1 cytokinin receptor yields plants developing spontaneous nodules in the absence of rhizobia (*snf2*; Tirichine et al. 2007), indicating that cytokinins are necessary and sufficient to induce cortical cell divisions and nodule organogenesis. Analysis of the *snf2 ccamk* double mutant additionally reveals that activation of cytokinin signaling is insufficient to allow rhizobial infection (Madsen et al. 2010). *snf2* root organization is also perturbed, with extra layers of vascular cell files and pericycle cell divisions, but it is not known if this consequently affects lateral root formation. Collectively, these data suggest a role for local accumulation of cytokinins and/or regulation of cytokinin sensitivity in relation with root lateral organs development, with a negative role in lateral root formation, and a positive role in symbiotic nodule organogenesis (Frugier et al. 2008).

ABA acts as an inhibitor of lateral root formation in some species (such as *Arabidopsis*; De Smet et al. 2003), but it increases lateral root development in several nodulated legume species, suggesting that a negative role of ABA on root branching is correlated with the ability to nodulate (Liang and Harris 2005). In contrast, ABA inhibits nodulation in various legumes and genetic inhibition of ABA signaling through the use of a dominant-negative allele of ABSCISIC ACID INSENSITIVE1 leads to a hypernodulation phenotype in *M. truncatula* (Ding et al. 2008). These results suggest a negative role for ABA in nodulation and a positive role in lateral root formation, as opposed to the cytokinin function. Accordingly, the *Medicago latd* mutants that are less sensitive to ABA can only be rescued by this phytohormone for their short root phenotype but not for their defective nodulation (Liang et al. 2007). These results suggest that ABA together with cytokinins might allow maintaining the balance between lateral roots and nodules in legume roots.

The *Medicago skl* (*sickle*) mutant has a root-dependent hyperinfection by Rhizobium that is ethylene-insensitive and also has an increased root length (Penmetsa and Cook 1997; Penmetsa et al. 2003; Prayitno et al. 2006). This mutant is affected in the *EIN2* (*Ethylene Insensitive 2*) gene involved in ethylene signaling (Penmetsa et al. 2008). Surprisingly, the corresponding *Lotus* mutant *enigma/ein2* is not hypernodulating (Gresshoff et al. 2009), suggesting that ethylene may affect root system development differently depending on the legume considered.

Considering other hormones, the situation is less clear. In *M. truncatula*, MtCDC16 RNAi roots, reducing the expression of a core component of the anaphase-promoting complex, have a reduced sensitivity to auxin and lead to an inhibition of lateral root formation and to an increased nodulation (Kuppusamy et al. 2009). Pea mutants affected in gibberellin (GA) synthesis are dwarf and consequently exhibit a reduction in lateral root and nodule formation (Reid et al. 2004; Ferguson et al. 2005; Weston et al. 2008). Nevertheless, constitutive GA signaling mutants also form significantly fewer nodules than wild-type plants (Ferguson et al. 2011). Similarly in *L. japonicus*, overexpression of the gain-of-function mutation SLEEPY1, a positive regulator of GA signaling, results in a reduced nodule number, without affecting other aspects of root development (Maekawa et al. 2009). These data suggest a negative role for GAs in nodulation, but its effect on lateral roots remains to be examined independently of the dwarf (shoot) phenotype. Reduction of endogenous SA levels by expressing a salicylate hydroxylase (*NahG*) gene either in *Lotus* or in *Medicago* leads to an increase in nodulation (Stacey et al. 2006). Interestingly, *Lotus* transgenic lines also showed an increased root growth, but not *Medicago* lines. Finally, brassinosteroids (BRs) might participate in shoot-dependent AON (Ferguson et al. 2005). Indeed, mutants deficient in BRs showed a reduction in nodulation depending on shoots. Interestingly, a correlation between nodules and lateral root numbers has been observed in the lines affected in BRs metabolism.

Some nodulation mutants are also developmentally pleiotropic. The *Lotus klavier* (*klv*) shoot-dependent hypernodulating mutant is dwarf and exhibits short roots with a decreased number of lateral roots (Oka-Kira et al. 2005). KLAVIER encodes another LRR receptor-like kinase similar to CLV1 (Miyazawa et al. 2010). A double-mutant analysis indicated that KLV and HAR1 function in the same genetic pathway to control nodule numbers, whereas KLV also has multiple roles in shoot development, including maintenance of shoot apical meristem, vascular continuity, shoot growth, and promotion of flowering. Interaction between HAR1 and KLV in lateral root development remains to be analyzed in detail, as well as the consequences of the pleiotropic shoot phenotypes on root development. The *Lotus astray* mutant (*Ljsym77*) shows enhanced nodulation and has pleiotropic effects on root, shoot, and hypocotyl elongation under different light conditions (Nishimura et al. 2002b). The mutant name *astray* is derived from the agravitropic lateral roots phenotype (going "astray" against gravity). The gene responsible for these phenotypes, *LjBZF*, encodes a homolog of the *Arabidopsis* bZIP transcription factor HY5 (long HYpocotyl 5) involved in photomorphogenesis

(Hardtke et al. 2000; Nishimura et al. 2002c). Unlike the *hy5* mutant, *astray* did not show an enhancement of lateral root formation. The *L. japonicus crinkle* mutant (*Ljsym79*) is perturbed in root growth and root hairs, and also in trichome, seed, and pollen development (Tansengco et al. 2003; Tansengco et al. 2004). After inoculation with rhizobia, shoot growth and lateral root formation are significantly decreased when compared to the wild type. The temperature-sensitive *L. japonicus brush* mutant also exhibits defective nodule, root, and shoot development (Maekawa-Yoshikawa et al. 2009). Roots have a brush-like apical meristem with abnormal cellular organization and reduced cell expansion, and reciprocal grafting indicated that the shoot phenotype was a secondary effect.

Besides root mutants initially identified based on their nodulation phenotype, there are very few studies aiming to identify legume mutants primarily affected in root development. Recently, the *M. truncatula compact root architecture 1* (*cra1*) mutant was reported as having short and thick roots due to defects in cell elongation. Density of lateral roots or symbiotic nodules was however similar to the wild type (Laffont et al. 2010). Grafting experiments showed that pleiotropic shoot phenotypes were independent of the short root system. Analysis of the *cra1* root transcriptome revealed mainly cell wall metabolism defects, notably affecting lignin biosynthesis, and accordingly whole lignin content was decreased in mutant roots. This was associated to a differential accumulation of specific flavonoids and to a decreased polar auxin transport, suggesting that the *CRA1* gene may therefore control legume root growth through the regulation of lignin and flavonoid profiles, consequently modifying polar auxin transport.

Collectively, these data suggest that a single gene can regulate several aspects of legume root architecture (rhizobial infection, nodule development, root growth, and lateral root formation) as well as other plant developmental programs (in hypocotyls, leaves, flowers, or seeds). Certain genes can have either a positive role on the formation of both types of root lateral organs (such as *LATD/NRT1*), opposite effects depending on the organ considered (such as *CRE1* or *CDC16*), or inhibitory effects on the two organogenesis (such as *HAR1*).

Functional Genomic Analyses of Legume Nodules and Roots

Genomic Analyses in Legume Roots

Diverse transcriptomic strategies have been used to identify differentially expressed genes in symbiotic nodules or infected root hairs. Before genome data were available, differential display, differential and cold-plaque screenings, subtractive hybridizations (SSHs) as well as systematic sequencing of symbiotic cDNA libraries were used: expressed sequence tag (EST) collections are available at the "Medicago EST Navigation System," http://medicago.toulouse.inra.fr/Mt/EST/, and the *M. truncatula* Gene Index, http://compbio.dfci.harvard.edu/tgi/cgi-bin/tgi/gimain.pl?gudb=medicago).

Among the different genomic studies performed in legumes, roots are frequently used as controls for mycorrhizal or rhizobial infections, as well as for biotic or abiotic stresses. Integrated transcriptomic databases, including root samples, are available for at least three legumes: *M. truncatula*, through the Gene Expression Atlas (MtGEA; http://mtgea.noble.org/v2/; Benedito et al. 2008) or the "Truncatulix" database (http://lily.cebitec.uni-bielefeld.de/truncatulix/; Henckel et al. 2009); the *L. japonicus* database (http://cgi-www.cs.au.dk/cgi-compbio/Niels/index.cgi; Hogslund et al. 2009); and soybean, through the transcriptome atlas (http://digbio.missouri.edu/soybean_atlas; Libault et al. 2010b), the "Soybase" (http://www.soybase.org/soyseq/; Severin et al. 2010), or the SoyXpress database (http://soyxpress.agrenv.mcgill.ca; Cheng and Stromvik 2008). Interestingly, Holmes et al. (2008) did a transcriptional profiling of *M. truncatula* meristematic root cells, by comparing root tips (3 mm meristems) to the region above (1 cm adjacent from root tip), corresponding to the differentiation zone where the rhizobial interaction is initiated. In *L. japonicus*, transcriptome of whole roots inoculated or not by *Rhizobium* was analyzed, initially using a Serial Analysis of Gene Expression (SAGE) approach (Asamizu et al. 2005). More recently, among other organs and mutants analyzed, the transcriptome of *Lotus* whole roots was described using Gene Chips, in comparison to the one of roots tips and of the region susceptible to *Rhizobium* (Hogslund et al. 2009; http://www.ebi.ac.uk/microarray-as/ae/). In soybean, the transcriptome of root tips as well as root hair cells has been carried out, as a control to study rhizobial infection (Libault et al. 2010a, 2010b).

Root proteomic data are available in *M. truncatula* (Mathesius et al. 2001; Watson et al. 2003) and soybean. In this later case, the root apex proteome (1 mm apex) has been recently compared to the differentiated root zone responding to *Rhizobia* (Mathesius et al. 2011). In addition, the root hair proteome has also been analyzed in soybean (Wan et al. 2005; Brechenmacher et al. 2009). Recent studies focused on in depth characterizations of the *M. truncatula* proteome have been reviewed by Colditz and Braun (Colditz and Braun 2010).

Small RNA populations of *M. truncatula*, soybean, and *P. vulgaris* (common bean) have been also characterized (Subramanian et al. 2008; Arenas-Huertero et al. 2009; Lelandais-Briere et al. 2009; Valdes-Lopez et al. 2010; Devers et al. 2011; Zhai et al. 2011). These data are available at the miRBase database (http://www.mirbase.org/) or through dedicated websites for *M. truncatula* (MIRMED; http://medicago.toulouse.inra.fr/MIRMED or the Medicago next gene sequence database, http://mpss.udel.edu/mt_sbs/).

Legume Root Genomics under the Symbiotic Interaction

Few putative regulatory genes identified as differentially regulated during nodulation and involved in early nodule developmental stages have been functionally studied (Figure 8.1). One of the first studies concerned the *Medicago Enod40s*, which are early *Nodulin* genes rapidly induced by rhizobia (Crespi et al. 1994; Fang and Hirsch 1998; Compaan et al. 2003; Wan et al. 2007). The *Enod40* genes encode RNAs with highly conserved secondary structures (Girard et al. 2003) and containing short open reading frame (sORF; Crespi et al. 1994). Both sORFs encoded peptides and the RNA itself seem to be involved in gene function (Sousa et al. 2001; Rohrig et al. 2002; Rohrig et al. 2004). *Enod40* genes are expressed in the root pericycle as well as in nodule primordia, and in the infection zone and vascular bundles of indeterminate nodules (Crespi et al. 1994; Gronlund et al. 2005; Wan et al. 2007; Laporte et al. 2010). Overexpression of one of the *M. truncatula Enod40* genes resulted in acceleration of nodulation (Charon et al. 1999). Among the transgenic plants obtained, two lines in which this gene was cosuppressed developed few nodules with arrested meristems, suggesting that *MtEnod40* function is required for nodule morphogenesis. More recently, reduced expression of the two *MtEnod40* genes led to plants with few and modified nodule-like structures devoid of bacteroids (Wan et al. 2007), confirming a role in nodule initiation. In *Lotus*, *Enod40* RNAi revealed an inhibition of nodule primordia development but normal IT formation (Kumagai et al. 2006). *Enod40* sORFs encode two small peptides able to bind a cytosolic sucrose synthase enzyme that may be required for nodulation (Rohrig et al. 2002; Baier et al. 2007). The *MtEnod40* RNA can also interact with a RNA-binding protein, MtRBP1, which is relocalized by the *ENOD40* RNA from nuclear speckles to the cytoplasm (Campalans et al. 2004; Laporte et al. 2007). The *MtEnod40* RNA can also interact with two small peptides named "small nodulin acidic RNA-binding proteins"

(SNARPs; Laporte et al. 2010). These complex interactions may participate in regulating various ENOD40 functions in nodulation.

An early nodulation-induced transcription factor of the CCAAT-box binding factor family (CBF; also called nuclear factor Y (NF-Y) or HAP complex), MtHAP2-1, was associated with nodule organogenesis possibly by positively controlling nodule meristem function. In addition, a role for *MtHAP2-1* transcript stability regulation by microRNA169 in controlling nodule meristematic activity versus differentiation has been identified (Combier et al. 2006). Moreover, alternative splicing of an intron in the *MtHAP2-1* 5′ leader sequence (LS) becomes predominant during the development of nodules, leading to the production of a small peptide, uORF1p (Combier et al. 2008a). The binding of uORF1p to *MtHAP2-1* 5′LS mRNA leads to reduced accumulation of the *MtHAP2-1* transcript and may contribute to spatial restriction of its expression within the nodule. Thus, miR169 and uORF1p play sequential and nonredundant roles in regulating *MtHAP2-1* expression.

During a large-scale microarray analysis to identify NF responsive genes, Combier et al. (Combier et al. 2008b) identified two genes encoding small peptides, MtRALFL1 and MtDVL1, playing a negative role in rhizobial infection and nodule development, as shown based on an overexpression analysis.

An early nodulation-induced ERF (ethylene response factor) transcription factor, called EFD (ERF required for nodule differentiation), was linked to the differentiation of functional nodules (Vernie et al. 2008). Interestingly, EFD regulates the expression of the primary cytokinin response regulator gene *MtRR4*. EFD also seems to participate in an ethylene-independent feedback inhibition of nodulation.

Two CLAVATA3/endosperm-surrounding region (CLE) peptides (*LjCLE-RS1* and *LjCLE-RS2*), over the 39 *CLE* genes identified in *L. japonicus*, are transcriptionally highly accumulated in roots in response to rhizobia inoculation (Okamoto et al. 2009). *LjCLE-RS2* is also upregulated in response to nitrate. Overexpression of both peptide precursors in *Lotus* inhibits nodulation in a *har1*-dependent way, consistently with the AON model. A similar regulation of nodulation involving CLE-peptides and SUNN has been subsequently identified in *M. truncatula* and soybean (Mortier et al. 2010; Reid et al. 2011). This suggests that the HAR1/SUNN CLV-like RLK is linked to CLE peptide signaling. In *Arabidopsis thaliana*, CLE peptides are implicated in meristem persistence, thus legume CLE peptides may play a similar role in nodule meristems. Their expression pattern is indeed consistent with such function (Mortier et al. 2010).

In the inverted repeat-lacking clade of legumes (IRLC; Wojciechowski et al. 2004; e.g., *Medicago*, *Pisum*, or *Trifolium spp.*), bacteria undergo a peculiar irreversible differentiation. Bacteria enlarge spectacularly, become polyploid, change their membrane permeability, and lose their reproductive capacity (Mergaert et al. 2006). Mergaert et al. (2003) identified a large family of nodule-specific genes (more than 400 NCRs, for nodule cysteine-rich proteins), encoding proteins similar to defensin-type antimicrobial peptides (AMPs). Peculiarly, this family is specifically present in IRLC legumes, so NCRs could play a key role in bacteroid differentiation. More recently, it was demonstrated that some of these NCR have *in vitro* AMP activity, and at least NCR035 and NCR001 were strictly targeted to the bacteroids (Van de Velde et al. 2010). Besides, the ectopic expression of some NCRs in the non-IRLC legume *L. japonicus* is able to induce a partial terminal differentiation of bacteroids. These results have been strengthened by the study of the *dnf-1* mutant, defective for differentiation of bacteria into bacteroids (Wang et al. 2010). In fact, the DNF1 gene encodes a subunit of a signal peptidase complex of the secretory pathway, highly expressed in nodules. This signal peptidase complex, located in the ER, removes the signal peptide of secretory proteins during their translocation into the ER lumen. The observed phenotype may thus be due to a wrong targeting of NCRs.

How the Environment Affects Legume Root Genomic Responses

Several environmental factors have been reported to influence the development of root-derived organs in legumes. These factors include availability of nutrients such as nitrate or phosphate, and abiotic stress conditions, such as salt, water deficit/, drought, or low pH combined with heavy metals. Besides lateral root development, the ability of legume roots to interact with symbiotic microorganisms, such as rhizobia or mycorrhizal fungi, constitutes an adaptation to specific nutrient starvation conditions. These interactions are themselves influenced by environmental conditions of the soil (such as abiotic stresses).

Nitrate (NO_3^-) is particularly relevant for legume root architecture, because its availability modulates plant growth and exerts complex effects on root growth, lateral root formation, and also on nodulation. Indeed, nitrate locally stimulates lateral root formation but systemically inhibits nodulation (Carroll et al. 1985; Streeter 1985a; 1985b; Zhang and Forde 1998). In addition, hypernodulating mutants, such as *har1* in *Lotus*, *sunn* in *Medicago*, or several *nitrate tolerant symbiosis* (*nts*) mutants in soybean, are also affected in their nitrate regulation, indicating that nitrate signaling pathways are tightly interconnected to AON (Carroll et al. 1985; Wopereis et al. 2000; Penmetsa et al. 2003). More recently, a *M. truncatula* transcriptomic study based on a split-root system revealed that depending on the nitrogen source used (NO_3^-, NH_4^+, or N_2 through symbiosis), specific transcriptome responses are induced (Ruffel et al. 2008). Among the genes systemically regulated by the plant N status, differential gene expression patterns were observed depending on the N source. In *Lotus*, the effect of low and high nitrate (10 μM NH_4NO_3 versus 10 mM NH_4NO_3) on nodulation efficiency was analyzed using 90K custom arrays (Omrane et al. 2009).

Phosphate (PO_4^{3-}) availability is, together with nitrogen, a major environmental factor affecting root development and nodulation (Pereira and Bliss 1989). Except for mycorrhization, that is essential for phosphate nutrition, very few molecular or genomic data are available in legume roots. One of the few examples is the cloning from a Pi starvation-induced cDNA library of a soybean β-expansin gene, Gm*EXPB2* (Guo et al. 2011). This gene, primarily expressed in roots and highly induced by Pi starvation, is involved in root elongation, as shown both by overexpression and RNAi studies. Subsequently, this enhances P uptake, especially at low P levels. Interestingly, Gm*EXPB2* was also induced by iron and mild water deficiencies in addition to Pi starvation. Another critical phosphate response element, initially identified in *Arabidopsis*, is the miR399 targeting genes involved in the acquisition of this nutrient. Similar functions have been shown in soybean and bean as in *Arabidopsis* (Valdés-López et al. 2008; Liu et al. 2010). More generally, a study was performed in order to identify common bean (*P. vulgaris*) miRNAs regulated in roots or symbiotic nodules by different nutrient deficiencies (phosphorus, nitrogen, or iron) or stress (acidic pH or manganese toxicity; Valdes-Lopez et al. 2010). Five miRNAs were only expressed under stress conditions, whereas a group of miRNAs responded to nearly all stresses; others showed organ- and stress-specific responses.

Concerning abiotic stresses, the first molecular and functional studies were initially performed in legumes using salt stress. Increasing salt concentrations affect root system architecture and symbiotic nodulation (Singleton and Bohlool 1984; McKay and Djordjevic 1993; Arrese-Igor et al. 1999; Duzan et al. 2004). The first salt-regulated transcription factor functionally linked to root salt stress response, the C_2H_2 zing-finger protein Alfin1, was identified in alfalfa (Winicov 1993).

Overexpression of this zinc-finger protein predominantly expressed in roots enhances growth under control and saline conditions (Winicov 2000). The lack of root formation of alfalfa transgenic lines expressing *MsAlfin1* in an antisense orientation also supports a crucial role for this gene in root development. Another C_2H_2 zinc-finger transcription factor (*ZPT2-1*) has been identified in alfalfa and subsequently characterized in *M. truncatula* (Frugier et al. 1998; Frugier et al. 2000). This gene, expressed in vascular tissues of roots and nodules and induced in roots by salt and cold stresses, is essential for symbiotic nodule development, and *MtZPT2-1* antisense lines are also less able to recover from a salt stress than the wild-type plant (Merchan et al. 2003; Merchan et al. 2007). This suggests that this transcription factor might be involved both in nodule and root-adaptive responses to osmotic and salt stresses. In addition, *MtZPT2-1* and the related *MtZPT2-2* transcription factor showed a differential expression in response to salt between two contrasting *M. truncatula* genotypes: Jemalong A17 and 108-R (de Lorenzo et al. 2007). Overexpression of either *MtZPT2-1* or *MtZPT2-2* in the more salt-sensitive 108-R genotype allowed significant increase in root growth under salt stress, suggesting a role for this regulatory pathway in the adaptive response to salt stress. This also allowed identifying potential MtZPT2-1 target genes acting in roots, corresponding to a cold-regulated A (CORA) homolog, a flower-promoting factor homolog, and an auxin-induced proline-rich protein gene (Merchan et al. 2007). However, their putative direct regulation by a ZPT2 TRANSCRIPTION FACTOR remains to be shown.

More recently, transcriptomic studies have been carried out in several legume roots treated by salt. In *L. japonicus*, an acclimatization procedure was used to analyze the effect of a nonlethal long-term salinity stress (using an initial single step of 25–75 mM NaCl for 28 days or gradual steps of 50–150 mM NaCl for 4 days each; Sanchez et al. 2008). Analyses were performed at the ionomic, transcriptomic, and metabolomic levels. Salt stress induced reductions in potassium, phosphorus, sulfur, zinc, and molybdenum contents, and metabolic changes were characterized by a general increase in the steady-state levels of many amino acids, sugars, and polyols, associated to a decrease in organic acids. These combined observations suggest a successive and increasingly global requirement for the reprogramming of gene expression and metabolic pathways to maintain ionic and osmotic homeostasis. These results suggested that beyond the simple dose-dependence effect, dynamics of concentration changes are essential. In a following study, the same authors analyzed the relevance of transcriptomic studies in model species (*L. japonicus*) for translational genomics by searching for robust responses (Sanchez et al. 2010). For this, the previous long-term, nonlethal salt stress experiments were carried out over two consecutive years in the "best possible controlled greenhouse conditions." A surprisingly large fraction of the transcriptional and metabolic responses to salt stress were not reproducible between experiments, even though a core set of robust changes shared between experiments could be defined. Many of these robust responses were qualitatively and quantitatively conserved between different accessions of the same species, indicating that these genes may be a relevant starting point for translational genomics. In *M. truncatula*, transcriptional changes in responses to a short-term NaCl stress (100 mM for 1 h) were analyzed initially using a SSH approach associated to a dedicated macroarray (Merchan et al. 2007) and later on based on two complementary transcriptomic approaches (16 K+ microarray and real-time RT-PCR on a collection of transcription factors encoding genes (Gruber et al. 2009; microarray data available through the "Truncatulix" database: http://lily.cebitec.uni-bielefeld.de/truncatulix/). Several salt-induced transcription factor genes, belonging to various families, also responded to other abiotic stresses such as osmotic stress, cold, and heat. In addition, spatial differences in the regulation of transcription factor encoding genes by salt were observed depending on the root region considered (root tips versus whole roots). This suggests that specific pathways are activated by salt in root apical meristems. In parallel, another *M. truncatula* study focused on a salt stress time-course experiment (NaCl 180–200 mM for 1–24 h) using Affymetrix Medicago Gene Chips (Li et al. 2009; MtED database, available at http://bioinformatics.cau.edu.cn/MtED/ and also through the MtGEA: http://mtgea.noble.org/v2/). In soybean, the proteome of salt-stressed roots (100 mM NaCl) was analyzed, and among the seven proteins differentially regulated was a basic helix–loop–helix transcription factor (Aghaei et al. 2009). In chickpea, a first study reports drought- and salinity-responsive ESTs from roots (Varshney et al. 2009). In addition, the short-term moderate (25 mM NaCl for 2 h) salt-responsive transcriptome of chickpea roots and nodules was recently studied using a deep SuperSAGE technology (Molina et al. 2011). Finally, the effect of an alkaline stress ($NaHCO_3$ 50 mmol/L for 0.5–24 h) was analyzed in soybean root apices (3 cm) using Affymetrix chips (Ge et al. 2010).

On the basis of these studies, selected genes have been functionally analyzed to determine their potential impact on root system architecture in response to salinity. Although several studies have used heterologous systems (e.g. *Arabidopsis*, *Nicotiana*, or yeast) for overexpression of abiotic stress-regulated genes, we will focus here on functional analyses performed in roots of the homologous legume. In *M. truncatula*, the Merchan et al. (2007) transcriptomic analyses identified a leucine-rich repeat receptor-like protein kinase (RLK) rapidly induced by salt stress in the root epidermis, which was analyzed further (de Lorenzo et al. 2009). An RNAi strategy specifically targeting this *Srlk* (for *Salt-regulated RLK*) gene yielded transgenic roots whose growth was less inhibited by the presence of salt. Two *srlk* mutant alleles also failed to limit root growth in response to salt stress and accumulated fewer sodium ions. In addition, early salt-regulated genes were downregulated in *srlk* mutant roots submitted to a salt stress. More recently, the Gruber et al. (2009) transcriptomic studies revealed that a *M. truncatula* HD-Zip I transcription factor, referred to MtHB1, was expressed in primary and lateral root meristems and induced by salt stress (Ariel et al. 2010). Constitutive expression of MtHB1 altered root system architecture, whereas *hb1* mutants showed an increased lateral root emergence. A target of this transcription factor was subsequently identified, corresponding to a lateral organ boundaries-like gene, Mt*LBD1*, which is repressed by MtHB1 during lateral root emergence. A model was therefore proposed where the MtHB1 transcription factor would allow, in response to a salt stress (or ABA), to minimize the root surface exposed to the abiotic constraint. Recently, a NAC transcription factor (MtNAC969) induced by salt stress in roots was shown to be recruited during nodule senescence (De Zelicourt et al. 2012). Depending on the organ considered (root versus nodule), salt and nitrate regulations of *MtNAC969* were different. Finally, based on Li et al. (Li et al. 2009) transcriptomic data, one of the highly induced transcription factor genes, MtCBF4 (C-repeat binding factor 4), was functionally studied (Li et al. 2011). This AP2-EREBP transcription factor was induced by salt, drought, or cold stress, as well as ABA. Overexpression of MtCBF4 in *M. truncatula* enhanced salt tolerance and induced expression of genes containing drought-responsive elements in their promoters. In soybean, two NAC transcription factors previously identified as ESTs (Tian et al. 2004), GmNAC11 and GmNAC20, are regulated by salt and drought stresses (Hao et al. 2011). Although GmNAC20 overexpression induced soybean transgenic root growth, GmNAC11 overexpression decreased it. Both genes however led to an increased root growth under salt stress. Heterologous expression of GmNAC20 in *A. thaliana* also promoted lateral root formation, but potential effects on lateral roots or nodules in soybean were not reported.

Concerning drought stress, in alfalfa, a recent study analyzed the effect of long-term drought and re-watering in two drought-contrasting varieties (*M. sativa* ssp. *falcata* var. *Wisfal* and *M. sativa* ssp. *sativa* var. *Chilean*; Kang et al. 2011b). At a qualitative level, molecular, biochemical, and physiological responses to drought stress were similar in the two varieties, suggesting that the same strategies are involved to survive to a drought stress. However, quantitative differences that may contribute to greater drought tolerance in Wisfal were revealed. Among those, the greater Wisfal root growth following a drought episode was noticed, even though functional pathways associated were not identified based on either the differential transcriptomic data (obtained based on the Affymetrix Medicago chip) or the metabolite profiling. A similar metabolic study was performed in different *Lotus* species in response to salt and drought, revealing a weak overlap between drought-responsive metabolites (Sanchez et al. 2012). In chickpea, a superSAGE approach associated to a dedicated microarray was used to analyze the root transcriptome 6 h after drought stress (Molina et al. 2008). In *P. vulgaris*, miRNAs regulated by various abiotic stresses were analyzed in whole seedlings (Arenas-Huertero et al. 2009). Finally, in *Vigna radiata*, a study analyzed the effect of progressive drought stress (3–4 days) and recovery (6 days of re-watering) on the root proteome (Sengupta et al. 2011). All these studies therefore led to the discovery of interesting markers of legume root adaptation to drought, even though no functional analysis was initiated.

Heavy metals stresses have also been recently the focus of genomic studies in legumes. Soil acidity affects crop yield in about 40 % of the world arable land, mainly because of aluminum (Al), proton, and manganese (Mn) toxicity (Marschner 1995). Al toxicity causes inhibition of root growth by inhibiting root apices, notably in legumes (Rangel et al. 2007). In *M. truncatula*, 16 K^+ microarrays were used to analyze changes in transcript accumulation in root tips after an Al treatment (Chandran et al. 2008a; 2008b). A screen for Al resistance in 54 *M. truncatula* accessions identified eight Al-resistant and eight Al-sensitive lines. Among those, transcriptomic response of an Al-resistant (T32) and an Al-sensitive (S70) genotype were analyzed after a 2.5 μM Al treatment for 12 or 48 h (Chandran et al. 2008a). Fewer genes were differentially regulated in response to Al in T32 compared to S70. Among the novel genes linked to Al stress response were cell wall loosening enzymes and genes related to ethylene production (Chandran et al. 2008b). In addition, analysis of candidate gene expression in root apices versus the mature region of the root revealed that Al-induced gene expression was not restricted to the tip, suggesting that differentiated regions may also play a role in the tolerance to this heavy metal. Finally, RNAi of two Al-induced genes, coding for a pectin acetyl esterase and for an annexin, slightly increased root sensitivity to Al, suggesting that these genes may play a role in Al resistance. A transcriptomic study was also conducted in common bean to analyze the effect of short-term Al treatments (20 μm $AlCl_3$ for 4 h) on root tips (1 cm long), using a SSH approach and Al-resistant and -sensitive genotypes (Etch et al. 2010). A recent study analyzed the population of miRNAs regulated in roots by an aluminum stress (10 μM for 4 and 24 h; Chen et al. 2011), and a similar study was done with a mercury stress (10 μM HgC12 for 6–48 h; Zhou et al. 2012). As for drought, this collection of marker genes opens the way for identifying pathways allowing an increased tolerance of legume roots to heavy metals.

Concluding Remarks

The availability of the recently sequenced genomes for both model legumes *Medicago* and *Lotus* associated to genetic resources and global genomic studies open new perspectives for the understanding of nodule and root development as well as organ physiology in response to their environment.

Laser-capture micro dissection methods or fluorescence-activated cell sorting combined to next-generation sequencing will be a logical next step to take to further understand root and nodule development and to decipher the coordinated expression of genes at the tissue or cell type level. The use of this dissection process should lead to the discovery of highly specific subsets of gene networks.

Another big challenge for future will be the transfer, with the help of phylogenetic approaches, of all this knowledge from models to the most important legume crops.

Acknowledgments

F.F. acknowledges funding from the LEGUROOT and ROOT ANR projects. A.N. and P.L. acknowledge funding from the HAPIHUB ANR project.

References

Aghaei, K. et al. (2009) Proteome analysis of soybean hypocotyl and root under salt stress. *Amino Acids*, 36, 91–8.

Andriankaja, A. et al. (2007) AP2-ERF transcription factors mediate Nod factor dependent Mt ENOD11 activation in root hairs via a novel *cis*-regulatory motif. *Plant Cell*, 19, 2866–2885.

Ane, J.M. et al. (2004) Medicago truncatula DMI1 required for bacterial and fungal symbioses in legumes. *Science*, 303, 1364–1367.

Arenas-Huertero, C. et al. (2009) Conserved and novel miRNAs in the legume *Phaseolus vulgaris* in response to stress. *Plant Molecular Biology*, 70, 385–401.

Ariel, F. et al. (2010) Environmental regulation of lateral root emergence in Medicago truncatula requires the HD-Zip I transcription factor HB1. *Plant Cell*, 22, 2171–2183.

Arrese-Igor, C. et al. (1999) Sucrose synthase and nodule nitrogen fixation under drought and other environmental stresses. *Symbiosis*, 27, 189–212.

Arrighi, J.F. et al. (2006) The *Medicago truncatula* lysine motif-receptor-like kinase gene family includes NFP and new nodule-expressed genes. *Plant Physiology*, 142, 265–279.

Asamizu, E. et al. (2005) Comparison of the transcript profiles from the root and the nodulating root of the model legume *Lotus japonicus* by serial analysis of gene expression. *Molecular Plant–Microbe Interactions*, 18, 487–498.

Bagchi, R., Salehin, M., Adeyemo, O.S., Salazar, C., Shulaev, V., Sherrier, D.J., Dickstein, R. (2012) Functional assessment of the Medicago NIP/LATD protein demonstrates that it is a high affinity nitrate transporter. *Plant Physiol*, Aug 2. (in press).

Baier, M.C. et al. (2007) Antisense repression of the *Medicago truncatula* nodule-enhanced sucrose synthase leads to a handicapped nitrogen fixation mirrored by specific alterations in the symbiotic transcriptome and metabolome. *Plant Physiology*, 145, 1600–1618.

Barker, D.G. et al. (1990) *Medicago truncatula*, a model plant for studying the molecular genetics of the rhizobium-legume symbiosis. *Plant Molecular Biology Reporter*, 8, 40–49.

Bateman, A. and Bycroft, M. (2000) The structure of a LysM domain from *E. coli* membrane-bound lytic murein transglycosylase D (MltD). *Journal of Molecular Biology*, 299, 1113–1119.

Ben Amor, B. et al. (2003) The NFP locus of *Medicago truncatula* controls an early step of Nod factor signal transduction upstream of a rapid calcium flux and root hair deformation. *Plant Journal*, 34, 495–506.

Benedito, V.A. et al. (2008) A gene expression atlas of the model legume *Medicago truncatula*. *Plant J*, 55, 504–513.

Bensmihen, S. et al. (2011) Contribution of NFP LysM domains to the recognition of Nod factors during the *Medicago truncatula/Sinorhizobium meliloti* symbiosis. *PLoS One*, 6, e26114.

Bersoult, A. et al. (2005) Expression of the *Medicago truncatula* DM12 gene suggests roles of the symbiotic nodulation receptor kinase in nodules and during early nodule development. *Molecular Plant–Microbe Interaction*, 18, 869–876.

Berthon, R. (1943) *Sur l'origine des radicelles chez les angiospermes*. Academie des Sciences, Paris.

Boogerd, F.C. and Van Rossum, D. (1997) Nodulation of groundnut by Bradyrhizobium: a simple infection process by crack entry. *FEMS Microbiology Review*, 21, 5–27.

Borisov, A.Y. et al. (2003) The Sym35 gene for root nodule development in *Pisum sativum* is an orthologue of Nin from *Lotus japonicus*. *Plant Physiology*, 131, 1009–1017.

Boualem, A. et al. (2008) MicroRNA166 controls root and nodule development in *Medicago truncatula*. *Plant Journal*, 54, 876–887.

Brechenmacher, L. et al. (2009) Establishment of a protein reference map for soybean root hair cells. *Plant Physiology*, 149, 670–682.

Brewin, N.J. (1991) Development of the legume root nodule. *Annual Review of Cell Biology*, 7, 191–226.

Bright, L.J. et al. (2005) The LATD gene of *Medicago truncatula* is required for both nodule and root development. *Molecular Plant–Microbe Interactions*, 18, 521–532.

Campalans, A. et al. (2004) Enod40, a short open reading frame-containing mRNA, induces cytoplasmic localization of a nuclear RNA binding protein in *Medicago truncatula*. *Plant Cell*, 16, 1047–1059.

Capoen, W. et al. (2005) SrSymRK, a plant receptor essential for symbiosome formation. *Proceedings of the National Academy of Science of the United States of America*, 102, 10369–10374.

Capoen, W. et al. (2011) Nuclear membranes control symbiotic calcium signaling of legumes. *Proceedings of the National Academy of Science of the United States of America*, 108, 14348–14353.

Cardenas, L. et al. (1999) Rhizobium nod factors induce increases in intracellular free calcium and extracellular calcium influxes in bean root hairs. *Plant Journal*, 19, 347–352.

Carroll, B.J. et al. (1985) A supernodulation and nitrate-tolerant symbiotic (nts) soybean mutant. *Plant Physiology*, 78, 34–40.

Catoira, R. et al. (2000) Four genes of *Medicago truncatula* controlling components of a nod factor transduction pathway. *Plant Cell*, 12, 1647–1666.

Chandran, D. et al. (2008b) Transcriptome profiling identified novel genes associated with aluminum toxicity, resistance and tolerance in *Medicago truncatula*. *Planta*, 228, 151–166.

Chandran, D. et al. (2008a) Physiological and molecular characterization of aluminum resistance in *Medicago truncatula*. *BMC Plant Biology*, 8, 89.

Charon, C. et al. (1999) Alteration of enod40 expression modifies *Medicago truncatula* root nodule development induced by *Sinorhizobium meliloti*. *Plant Cell*, 11, 1953–1966.

Charpentier, M. et al. (2008) *Lotus japonicus* CASTOR and POLLUX are ion channels essential for perinuclear calcium spiking in legume root endosymbiosis. *Plant Cell*, 20, 3467–3479.

Chen, L. et al. (2011) Identification of aluminum-responsive microRNAs in *Medicago truncatula* by genome-wide high-throughput sequencing. *Planta*, 235, 375–386.

Cheng, K.C. and Stromvik, M.V. (2008) SoyXpress: a database for exploring the soybean transcriptome. *BMC Genomics*, 9, 368.

Clark, S.E. et al. (1997) The CLAVATA1 gene encodes a putative receptor kinase that controls shoot and floral meristem size in arabidopsis. *Cell*, 89, 575–585.

Colditz, F. and Braun, H.P. (2010) *Medicago truncatula* proteomics. *Journal of Proteomics*, 73, 1974–1985.

Combier, J.P. et al. (2008a) Trans-regulation of the expression of the transcription factor MtHAP2–1 by a uORF controls root nodule development. *Genes Development*, 22, 1549–1559.

Combier, J.P. et al. (2006) MtHAP2–1 is a key transcriptional regulator of symbiotic nodule development regulated by microRNA169 in *Medicago truncatula*. *Genes Development*, 20, 3084–3088.

Combier, J.P. et al. (2008b) Evidence for the involvement in nodulation of the two small putative regulatory peptide-encoding genes MtRALFL1 and MtDVL1. *Molecular Plant–Microbe Interactions*, 21, 1118–1127.

Compaan, B. et al. (2003) Identification and characterization of a *Zea mays* line carrying a transposon-tagged ENOD40. *Biochimica et Biophysica Acta*, 1629, 84–91.

Crespi, M.D. et al. (1994) Enod40, a gene expressed during nodule organogenesis, codes for a non-translatable RNA involved in plant growth. *Embo Journal*, 13, 5099–5112.

Day, D.A. et al. (1986) Growth comparisons of a supernodulating soybean (*Glycine max*) mutant and its wild-type parent. *Physiologia Plantarum*, 68, 375–382.

De Billy, F. et al. (2001) Expression studies on AUX1-like genes in *Medicago truncatula* suggest that auxin is required at two steps in early nodule development. *Molecular Plant–Microbe Interactions*, 14, 267–277.

De Carvalho Niebel, F. et al. (1998) The *Medicago truncatula MtAnn1* gene encoding an annexin is induced by Nod factors and during the symbiotic interaction with *Rhizobium meliloti*. *Molecular Plant–Microbe Interactions*, 11, 504–513.

De Lorenzo, L. et al. (2007) Differential expression of the TFIIIA regulatory pathway in response to salt stress between *Medicago truncatula* genotypes. *Plant Physiology*, 145, 1521–1532.

De Lorenzo, L. et al. (2009) A novel plant leucine-rich repeat receptor kinase regulates the response of *Medicago truncatula* roots to salt stress. *Plant Cell*, 21, 668–680.

De Smet, I. et al. (2003) An abscisic acid-sensitive checkpoint in lateral root development of Arabidopsis. *Plant Journal*, 33, 543–555.

De Zelicourt, A. et al. (2012) Dual implication of a *Medicago truncatula* NAC transcription factor in root abiotic stress response and symbiotic nodule senescence. *Plant Journal*, 70, 220–230.

Denarie, J. et al. (1996) Rhizobium lipo-chitooligosaccharide nodulation factors: signaling molecules mediating recognition and morphogenesis. *Annual Review of Biochemistry*, 65, 503–535.

Desbrosses, G.J. and Stougaard, J. (2011) Root nodulation: a paradigm for how plant–microbe symbiosis influences host developmental pathways. *Cell Host Microbe*, 10, 348–358.

Devers, E.A. et al. (2011) Stars and symbiosis: microRNA- and microRNA∗-mediated transcript cleavage involved in arbuscular mycorrhizal symbiosis. *Plant Physiology*, 156, 1990–2010.

Ding, Y. et al. (2008) Abscisic acid coordinates nod factor and cytokinin signaling during the regulation of nodulation in *Medicago truncatula*. *Plant Cell*, 20, 2681–2695.

Duc, G. et al. (1989) First report of non-mycorrhizal plant mutants (Myc-) obtained in pea (*Pisum sativum* L.) and fababean (*Vicia faba* L.). *Plant Science*, 60, 215–222.

Dudley, M.E. and Long, S.R. (1989) A non-nodulating alfalfa mutant displays neither root hair curling nor early cell division in response to *Rhizobium meliloti*. *Plant Cell*, 1, 65–72.

Duzan, H.M. et al. (2004) Perception of Bradyrhizobium japonicum Nod factor by soybean [*Glycine max* (L.) *Merr.*] root hairs under abiotic stress conditions. *Journal of Experimental Botany*, 55, 2641–2646.

Edwards, A. et al. (2007) Structural implications of mutations in the pea SYM8 symbiosis gene, the DMI1 ortholog, encoding a predicted ion channel. *Molecular Plant–Microbe Interactions*, 20, 1183–1191.

Ehrhardt, D.W. et al. (1996) Calcium spiking in plant root hairs responding to Rhizobium nodulation signals. *Cell*, 85, 673–681.

Endre, G. et al. (2002) A receptor kinase gene regulating symbiotic nodule development. *Nature*, 417, 962–966.

Esseling, J.J. et al. (2004) A nonsymbiotic root hair tip growth phenotype in NORK-mutated legumes: implications for nodulation factor-induced signaling and formation of a multifaceted root hair pocket for bacteria. *Plant Cell*, 16, 933–944.

Eticha, D. et al. (2010) Transcriptomic analysis reveals differential gene expression in response to aluminium in common bean (*Phaseolus vulgaris*) genotypes. *Annals of Botany*, 105, 1119–1128.

Fang, Y.W. and Hirsch, A.M. (1998) Studying early nodulin gene *ENOD40* expression and induction by nodulation factor and cytokinin in transgenic alfalfa. *Plant Physiology*, 116, 53–68.

Felle, H.H. et al. (1995) Nod signal-induced plasma membrane potential changes in alfalfa root hairs are differentially sensitive to structural modifications of the lipochitooligosaccharide. *Plant Journal*, 7, 939–947.

Ferguson, B.J. et al. (2011) Relationship between gibberellin, ethylene and nodulation in *Pisum sativum*. *New Phytologist*, 189, 829–842.

Ferguson, B.J. et al. (2005) Nodulation phenotypes of gibberellin and brassinosteroid mutants of pea. *Plant Physiology*, 138, 2396–2405.

Ferraioli, S. et al. (2004) Development of ectopic roots from abortive nodule primordia. *Molecular Plant–Microbe Interactions*, 17, 1043–1050.

Foo, E. and Davies, N.W. (2011) Strigolactones promote nodulation in pea. *Planta*, 234, 1073–1081.

Fournier, J. et al. (2008) Mechanism of infection thread elongation in root hairs of *Medicago truncatula* and dynamic interplay with associated rhizobial colonization. *Plant Physiology*, 148, 1985–1995.

Frugier, F. et al. (1998) Identification of novel putative regulatory genes induced during alfalfa nodule development with a cold-plaque screening procedure. *Molecular Plant–Microbe Interactions*, 11, 358–366.

Frugier, F. et al. (2008) Cytokinin: secret agent of symbiosis. *Trends in Plant Science*, 13, 115–120.

Frugier, F. et al. (2000) A Kruppel-like zinc finger protein is involved in nitrogen-fixing root nodule organogenesis. *Genes Development*, 14, 475–482.

Gage, D.J. (2004) Infection and invasion of roots by symbiotic, nitrogen-fixing rhizobia during nodulation of temperate legumes. *Microbiology and Molecular Biology Reviews*, 68, 280–300.

Gaude, N. et al. (2004) The galactolipid digalactosyldiacylglycerol accumulates in the peribacteroid membrane of nitrogen-fixing nodules of soybean and Lotus. *Journal of Biological Chemistry*, 279, 34624–34630.

GE, Y. et al. (2010) Global transcriptome profiling of wild soybean (*Glycine soja*) roots under $NaHCO_3$ treatment. *BMC Plant Biology*, 10, 153.

Girard, G. et al. (2003) Structural motifs in the RNA encoded by the early nodulation gene enod40 of soybean. *Nucleic Acids Research*, 31, 5003–5015.

Gleason, C. et al. (2006) Nodulation independent of rhizobia induced by a calcium-activated kinase lacking autoinhibition. *Nature*, 441, 1149–1152.

Gojon, A. et al. (2011) Nitrate transceptor(s) in plants. *Journal of Experimental Botany*, 62, 2299–2308.

Gonzalez-Rizzo, S. et al. (2006) The *Medicago truncatula* CRE1 cytokinin receptor regulates lateral root development and early symbiotic interaction with *Sinorhizobium meliloti*. *Plant Cell*, 18, 2680–2693.

Goormachtig, S. et al. (2004) Rhizobium infection: lessions from the versatile nodulation behaviour of water-tolerant legumes. *Trends in Plant Science*, 9, 518–522.

Gough, C. and Cullimore, J. (2011) Lipo-chitooligosaccharide signaling in endosymbiotic plant–microbe interactions. *Molecular Plant–Microbe Interactions*, 24, 867–878.

Gremaud, M.F. and Harper, J.E. (1989) Selection and initial characterization of partially nitrate-tolerant mutants of soybean. *Plant Physiology*, 89, 169–173.

Gresshoff, P.M. et al. (2009) Genetic analysis of ethylene regulation of legume nodulation. *Plant Signal and Behavior*, 4, 818–823.

Gronlund, M. et al. (2005) Analysis of promoter activity of the early nodulin Enod40 in *Lotus japonicus*. *Molecular Plant–Microbe Interactions*, 18, 414–427.

Groth, M. et al. (2010) NENA, a *Lotus japonicus* homolog of Sec13, is required for rhizodermal infection by arbuscular mycorrhiza fungi and rhizobia but dispensable for cortical endosymbiotic development. *Plant Cell*, 22, 2509–2526.

Gruber, V. et al. (2009) Identification of transcription factors involved in root apex responses to salt stress in *Medicago truncatula*. *Molecular Genetics and Genomics*, 281, 55–66.

Guo, W. et al. (2011) A soybean beta-expansin gene GmEXPB2 intrinsically involved in root system architecture responses to abiotic stresses. *Plant Journal*, 66, 541–552.

Handberg, K. and Stougaard, J. (1992) *Lotus japonicus*, an autogamous, diploid legume species for classical and molecular genetics. *Plant Journal*, 2, 487–496.

Haney, C.H. and Long, S.R. (2010) Plant flotillins are required for infection by nitrogen-fixing bacteria. *Proceedings of the National Academy of Science of the United States of America*, 107, 478–483.

Hao, Y.J. et al. (2011) Soybean NAC transcription factors promote abiotic stress tolerance and lateral root formation in transgenic plants. *Plant Journal*, 68, 302–313.

Hardtke, C.S. et al. (2000) HY5 stability and activity in arabidopsis is regulated by phosphorylation in its COP1 binding domain. *EMBO Journal*, 19, 4997–5006.

Henckel, K. et al. (2009) TRUNCATULIX—a data warehouse for the legume community. *BMC Plant Biology*, 9, 19.

Hirsch, A.M. (1992) Developmental biology of legume nodulation. *New Phytologist*, 122, 211–237.

Hirsch, A.M. et al. (1989) Early nodulin genes are induced in alfalfa root outgrowths elicited by auxin transport inhibitors. *Proceedinsg of the National Academy of Science of the United States of America*, 86, 1244–1248.

Hirsch, A.M. and Larue, T.A. (1997) Is the legume nodule a modified root or stem or an organ sui generis? *Critical Reviews in Plant Science*, 16, 361–392.

Hirsch, S. et al. (2009) GRAS proteins form a DNA binding complex to induce gene expression during nodulation signaling in *Medicago truncatula*. *Plant Cell*, 21, 545–557.

Hogg, B.V. et al. (2006) The DMI1 and DMI2 early symbiotic genes of *Medicago truncatula* are required for a high-affinity nodulation factor-binding site associated to a particulate fraction of roots. *Plant Physiology*, 140, 365–373.

Hogslund, N. et al. (2009) Dissection of symbiosis and organ development by integrated transcriptome analysis of *Lotus japonicus* mutant and wild-type plants. *PLoS One*, 4, e6556.

Holmes, P. et al. (2008) Transcriptional profiling of *Medicago truncatula* meristematic root cells. *BMC Plant Biology*, 8, 21.

Horvath, B. et al. (2011) *Medicago truncatula* IPD3 is a member of the common symbiotic signaling pathway required for rhizobial and mycorrhizal symbioses. *Molecular Plant–Microbe Interactions*, 24, 1345–1358.

Imaizumi-Anraku, H. et al. (2005) Plastid proteins crucial for symbiotic fungal and bacterial entry into plant roots. *Nature*, 433, 527–531.

Indrasumunar, A. and Gresshoff, P.M. (2010) Duplicated nod-factor receptor 5 (NFR5) genes are mutated in soybean (*Glycine max* L. *Merr*.). *Plant Signaling and Behavior*, 5, 535–536.

Indrasumunar, A. et al. (2011) Nodulation factor receptor kinase 1alpha controls nodule organ number in soybean (*Glycine max* L. *Merr*). *Plant Journal*, 65, 39–50.

Jacobsen, E. and Feenstra, W.J. (1984) A new pea mutant with efficient nodulation in the presence of nitrate. *Plant Science Letters*, 33, 337–344.

James, E.K. et al. (1992) The structure of nitrogen fixing root nodules on the aquatic mimosoid legume *Neptunia plena*. *Annals of Botany*, 69, 173–180.

Journet, E.P. et al. (1994) *Rhizobium meliloti* Nod factors elicit cell-specific transcription of the ENOD12 gene in transgenic alfalfa. *Plant Journal*, 6, 241–249.

Kaku, H. et al. (2006) Plant cells recognize chitin fragments for defense signaling through a plasma membrane receptor. *Proceedings of the National Academy of Science of the United States of America*, 103, 11086–11091.

Kalo, P. et al. (2005) Nodulation signaling in legumes requires NSP2, a member of the GRAS family of transcriptional regulators. *Science*, 308, 1786–1789.

Kanamori, N. et al. (2006) A nucleoporin is required for induction of Ca^{2+} spiking in legume nodule development and essential for rhizobial and fungal symbiosis. *Proceedings of the National Academy of Science of the United States of America*, 103, 359–364.

Kang, H. et al. (2011a) A novel interaction between CCaMK and a protein containing the Scythe_N ubiquitin-like domain in *Lotus japonicus*. *Plant Physiology*, 155, 1312–1324.

Kang, Y. et al. (2011b) System responses to long-term drought and re-watering of two contrasting alfalfa varieties. *Plant Journal*, 68, 871–889.

Kevei, Z. et al. (2007) 3-Hydroxy-3-methylglutaryl coenzyme a reductase 1 interacts with NORK and is crucial for nodulation in *Medicago truncatula*. *Plant Cell*, 19, 3974–3989.

Kiss, E. et al. (2009) LIN, a novel type of U-box/WD40 protein, controls early infection by rhizobia in legumes. *Plant Physiology*, 151, 1239–1249.

Kistner, C. et al. (2005) Seven *Lotus japonicus* genes required for transcriptional reprogramming of the root during fungal and bacterial symbiosis. *Plant Cell*, 17, 2217–2229.

Klaus-Heisen, D. et al. (2011) Structure–function similarities between a plant receptor-like kinase and the human interleukin-1 receptor-associated kinase-4. *Journal of Biological Chemistry*, 286, 11202–11210.

Kosuta, S. et al. (2011) *Lotus japonicus* symRK-14 uncouples the cortical and epidermal symbiotic program. *Plant Journal*, 67, 929–940.

Krusell, L. et al. (2002) Shoot control of root development and nodulation is mediated by a receptor-like kinase. *Nature*, 420, 422–426.

Kumagai, H. et al. (2006) RNAi knock-down of ENOD40s leads to significant suppression of nodule formation in *Lotus japonicus*. *Plant and Cell Physiology*, 47, 1102–1111.

Kuppusamy, K.T. et al. (2009) Knockdown of CELL DIVISION CYCLE16 reveals an inverse relationship between lateral root and nodule numbers and a link to auxin in *Medicago truncatula*. *Plant Physiology*, 151, 1155–1166.

Laffont, C. et al. (2010) The compact root architecture1 gene regulates lignification, flavonoid production, and polar auxin transport in *Medicago truncatula*. *Plant Physiology*, 153, 1597–1607.

Laporte, P. et al. (2007) Riboregulators in plant development. *Biochemistry Society Transactions*, 35, 1638–1642.

Laporte, P. et al. (2010) A novel RNA-binding peptide regulates the establishment of the *Medicago truncatula–Sinorhizobium meliloti* nitrogen-fixing symbiosis. *Plant Journal*, 62, 24–38.

Lefebvre, B. et al. (2010) A remorin protein interacts with symbiotic receptors and regulates bacterial infection. *Proceedings of the National Academy of Science of the United States of America*, 107, 2343–2348.

Lelandais-Briere, C. et al. (2009) Genome-wide *Medicago truncatula* small RNA analysis revealed novel microRNAs and isoforms differentially regulated in roots and nodules. *Plant Cell*, 21, 2780–2796.

Levy, J. et al. (2004) A putative Ca^{2+} and calmodulin-dependent protein kinase required for bacterial and fungal symbioses. *Science*, 303, 1361–1364.

Li, D. et al. (2009) An expression database for roots of the model legume *Medicago truncatula* under salt stress. *BMC Genomics*, 10, 517.

Li, D. et al. (2011) Transcriptional profiling of *Medicago truncatula* under salt stress identified a novel CBF transcription factor MtCBF4 that plays an important role in abiotic stress responses. *BMC Plant Biology*, 11, 109.

Liu, J.Q., Allan, D.L., Vance, C.P. (2010) Systemic signaling and local sensing of phosphate in common bean: cross-talk between photosynthate and microRNA399. *Mol Plant*, 3(2), 428–437.

Liang, Y. and Harris, J. (2005) Response of root branching to abscisic acid is correlated with nodule formation both in legumes and nonlegumes. *American Journal of Botany*, 92, 1675–1683.

Liang, Y. et al. (2007) Abscisic acid rescues the root meristem defects of the *Medicago truncatula latd* mutant. *Developmental Biology*, 304, 297–307.

Libault, M. et al. (2010b) Complete transcriptome of the soybean root hair cell, a single-cell model, and its alteration in response to *Bradyrhizobium japonicum* infection. *Plant Physiology*, 152, 541–552.

LIBAULT, M. et al. (2010a) An integrated transcriptome atlas of the crop model *Glycine max*, and its use in comparative analyses in plants. *Plant Journal*, 63, 86–99.

Limpens, E. et al. (2003) LysM domain receptor kinases regulating rhizobial Nod factor-induced infection. *Science*, 302, 630–633.

Limpens, E. et al. (2005) Formation of organelle-like N_2-fixing symbiosomes in legume root nodules is controlled by DMI2. *Proceedings of the National Academy of Science of the United States of America*, 102, 10375–10380.

Liu, W. et al. (2011) Strigolactone biosynthesis in *Medicago truncatula* and rice requires the symbiotic GRAS-type transcription factors NSP1 and NSP2. *Plant Cell*, 23, 3853–3865.

Madsen, E.B. et al. (2011) Autophosphorylation is essential for the in vivo function of the *Lotus japonicus* Nod factor receptor 1 and receptor-mediated signalling in cooperation with Nod factor receptor 5. *Plant Journal*, 65, 404–417.

Madsen, E.B. et al. (2003) A receptor kinase gene of the LysM type is involved in legume perception of rhizobial signals. *Nature*, 425, 637–640.

Madsen, L.H. et al. (2010) The molecular network governing nodule organogenesis and infection in the model legume *Lotus japonicus*. *Nature Communications*, 1, 10.

Maekawa-Yoshikawa, M. et al. (2009) The temperature-sensitive brush mutant of the legume *Lotus japonicus* reveals a link between root development and nodule infection by rhizobia. *Plant Physiology*, 149, 1785–1796.

Maekawa, T. et al. (2009) Gibberellin controls the nodulation signaling pathway in *Lotus japonicus*. *Plant Journal*, 58, 183–194.

Maillet, F. et al. (2011) Fungal lipochitooligosaccharide symbiotic signals in arbuscular mycorrhiza. *Nature*, 469, 58–63.

Markmann, K. et al. (2008) Functional adaptation of a plant receptor-kinase paved the way for the evolution of intracellular root symbioses with bacteria. *PLoS Biology*, 6, e68.

Marschner, H. (1995) *Mineral Nutrition of Higher Plants*. Academic Press, London.

Marsh, J.F. et al. (2007) *Medicago truncatula* NIN is essential for rhizobial-independent nodule organogenesis induced by autoactive calcium/calmodulin-dependent protein kinase. *Plant Physiology*, 144, 324–335.

Mathesius, U. et al. (2011) Comparative proteomic profiles of the soybean (*Glycine max*) root apex and differentiated root zone. *Proteomics*, 11, 1707–1719.

Mathesius, U. et al. (2001) Establishment of a root proteome reference map for the model legume *Medicago truncatula* using the expressed sequence tag database for peptide mass fingerprinting. *Proteomics*, 1, 1424–1440.

Mathesius, U. et al. (2000) Rhizobia can induce nodules in white clover by "hijacking" mature cortical cells activated during lateral root development. *Molecular Plant–Microbe Interactions*, 13, 170–182.

Mbengue, M. et al. (2010) The *Medicago truncatula* E3 ubiquitin ligase PUB1 interacts with the LYK3 symbiotic receptor and negatively regulates infection and nodulation. *Plant Cell*, 22, 3474–3488.

Mciver, J. et al. (1997) Influence of *Rhizobium leguminosarum* biovar *trifolii* host specific nodulation genes on the ontogeny of clover nodulation. *Protoplasma*, 172, 166–179.

McKay, I.A. and Djordjevic, M.A. (1993) Production and excretion of nod metabolites by *Rhizobium leguminosarum* bv. *trifolii* are disrupted by the same environmental factors that reduce nodulation in the field. *Applied and Environmental Microbiology*, 59, 3385–3392.

Meier, I. and Brkljacic, J. (2009) The nuclear pore and plant development. *Current Opinions in Plant Biology*, 12, 87–95.

Merchan, F. et al. (2003) A Krüppel-like transcription factor gene is involved in salt stress responses in Medicago spp. *Plant and Soil*, 257, 1–9.

Merchan, F. et al. (2007) Identification of regulatory pathways involved in the reacquisition of root growth after salt stress in *Medicago truncatula*. *Plant Journal*, 51, 1–17.

Mergaert, P. et al. (2003) A novel family in Medicago truncatula consisting of more than 300 nodule-specific genes coding for small, secreted polypeptides with conserved cysteine motifs. *Plant Physiology*, 132, 161–173.

Mergaert, P. et al. (2006). Eukaryotic control on bacterial cell cycle and differentiation in the *Rhizobium*-legume symbiosis. *Proceeding of the National Academy of Sciences of USA*, 103, 5230–5235.

Messinese, E. et al. (2007) A novel nuclear protein interacts with the symbiotic DMI3 calcium- and calmodulin-dependent protein kinase of *Medicago truncatula*. *Molecular Plant–Microbe Interactions*, 20, 912–921.

Middleton, P.H. et al. (2007) An ERF transcription factor in *Medicago truncatula* that is essential for Nod factor signal transduction. *Plant Cell*, 19, 1221–1234.

Mitra, R.M. et al. (2004) A Ca^{2+}/calmodulin-dependent protein kinase required for symbiotic nodule development: gene identification by transcript-based cloning. *Proceedings of the National Academy of Sciences of the United States of America*, 101, 4701–4705.

Miyazawa, H. et al. (2010) The receptor-like kinase KLAVIER mediates systemic regulation of nodulation and non-symbiotic shoot development in *Lotus japonicus*. *Development*, 137, 4317–4325.

Molina, C. et al. (2008) SuperSAGE: the drought stress-responsive transcriptome of chickpea roots. *BMC Genomics*, 9, 553.

Molina, C. et al. (2011) The salt-responsive transcriptome of chickpea roots and nodules via deepSuperSAGE. *BMC Plant Biology*, 11, 31.

Mortier, V. et al. (2010) CLE peptides control *Medicago truncatula* nodulation locally and systemically. *Plant Physiology*, 153, 222–237.

Murray, J.D. et al. (2007) A cytokinin perception mutant colonized by Rhizobium in the absence of nodule organogenesis. *Science*, 315, 101–104.

Murray, J.D. (2011) Invasion by invitation: rhizobial infection in legumes. *Molecular Plant-Microbe Interactions*, 24 (6), 631–639.

Nishimura, R. et al. (2002a) HAR1 mediates systemic regulation of symbiotic organ development. *Nature*, 420, 426–429.

Nishimura, R. et al. (2002c) A Lotus basic leucine zipper protein with a RING-finger motif negatively regulates the developmental program of nodulation. *Proceedings of the National Academy of Sciences of the United States of America*, 99, 15206–15210.

Nishimura, R. et al. (2002b) The novel symbiotic phenotype of enhanced-nodulating mutant of *Lotus japonicus*: astray mutant is an early nodulating mutant with wider nodulation zone. *Plant and Cell Physiology*, 43, 853–859.

Nutman, P.S. (1948) Physiological studies on nodule formation. The relation between nodulaiotn and lateral root formation in red clover. *Annuals of Botany (London)*, 7, 81–96.

Oka-Kira, E. et al. (2005) Klavier (klv), a novel hypernodulation mutant of *Lotus japonicus* affected in vascular tissue organization and floral induction. *Plant Journal*, 44, 505–515.

Okamoto, S. et al. (2009) Nod factor/nitrate-induced CLE genes that drive HAR1-mediated systemic regulation of nodulation. *Plant and Cell Physiology*, 50, 67–77.

Olah, B. et al. (2005) Nod factors and a diffusible factor from arbuscular mycorrhizal fungi stimulate lateral root formation in *Medicago truncatula* via the DMI1/DMI2 signalling pathway. *Plant Journal*, 44, 195–207.

Oldroyd, G.E. and Downie, J.A. (2004) Calcium, kinases and nodulation signalling in legumes. *Nature Reviews Molecular Cell Biology*, 5, 566–576.

Oldroyd, G.E. and Long, S.R. (2003) Identification and characterization of nodulation-signaling pathway 2, a gene of *Medicago truncatula* involved in Nod actor signaling. *Plant Physiology*, 131, 1027–1032.

Oldroyd, G.E. et al. (2011) The rules of engagement in the legume-rhizobial symbiosis. *Annual Review of Genetics*, 45, 119–144.

Omrane, S. et al. (2009) Symbiotic competence in *Lotus japonicus* is affected by plant nitrogen status: transcriptomic identification of genes affected by a new signalling pathway. *New Phytologist*, 183, 380–394.

Ovchinnikova, E. et al. (2011) IPD3 controls the formation of nitrogen-fixing symbiosomes in pea and Medicago Spp. *Molecular Plan–Microbe Interactions*, 24, 1333–1344.

Papadopoulou, K. et al. (1996) Phaseolus ENOD40 is involved in symbiotic and non-symbiotic organogenetic processes: expression during nodule and lateral root development. *Plant Molecular Biology*, 30, 403–417.

Peiter, E. et al. (2007) The Medicago truncatula DMI1 protein modulates cytosolic calcium signaling. *Plant Physiology*, 145, 192–203.

Penmetsa, R.V. and Cook, D.R. (1997) A legume ethylene-insensitive mutant hyperinfected by its rhizobial symbiont. *Science*, 275, 527–530.

Penmetsa, R.V. et al. (2003) Dual genetic pathways controlling nodule number in *Medicago truncatula. Plant Physiology*, 131, 998–1008.

Penmetsa, R.V. et al. (2008) The *Medicago truncatula* ortholog of Arabidopsis EIN2, sickle, is a negative regulator of symbiotic and pathogenic microbial associations. *Plant Journal*, 55, 580–595.

Pereira, P.A. and Bliss, F.A. (1989) Selection of common bean (*Phaseolus vulgaris* L) for nitrogen fixation at differents levels of avaiable phosphorus under field and environmentally controlled conditions. *Plant Soil*, 115, 75–82.

Plet, J. et al. (2011) MtCRE1-dependent cytokinin signaling integrates bacterial and plant cues to coordinate symbiotic nodule organogenesis in *Medicago truncatula. Plant Journal*, 65, 622–633.

Prayitno, J. et al. (2006) The ethylene-insensitive sickle mutant of *Medicago truncatula* shows altered auxin transport regulation during nodulation. *Plant Physiology*, 142, 168–180.

Radutoiu, S. et al. (2003) Plant recognition of symbiotic bacteria requires two LysM receptor-like kinases. *Nature*, 425, 585–592.

Radutoiu, S. et al. (2007) LysM domains mediate lipochitin–oligosaccharide recognition and Nfr genes extend the symbiotic host range. *EMBO Journal*, 26, 3923–3935.

Rangel, A.F. et al. (2007) Spatial aluminium sensitivity of root apices of two common bean (*Phaseolus vulgaris* L.) genotypes with contrasting aluminium resistance. *Journal of Experimental Botany*, 58, 3895–3904.

Reid, D.E. et al. (2011) Inoculation- and nitrate-induced CLE peptides of soybean control NARK-dependent nodule formation. *Molecular Plant–Microbe Interactions*, 24, 606–618.

Reid, J.B. et al. (2004) *Regulation of Gibberellin and Brassinosteroid Biosynthesis By Genetic, Environmental and Hormonal Factors*. Kluwer Academic Publishers, Dordrecht, The Netherlands.

Riely, B.K. et al. (2007) The symbiotic ion channel homolog DMI1 is localized in the nuclear membrane of *Medicago truncatula* roots. *Plant Journal*, 49, 208–216.

Rohrig, H. et al. (2004) Modification of soybean sucrose synthase by *S*-thiolation with ENOD40 peptide A. *Biochemical and Biophysical Research Communications*, 325, 864–870.

Rohrig, H. et al. (2002) Soybean ENOD40 encodes two peptides that bind to sucrose synthase. *Proceedings of the National Academy of Sciences of the United States of America*, 99, 1915–1920.

Ruffel, S. et al. (2008) Systemic signaling of the plant nitrogen status triggers specific transcriptome responses depending on the nitrogen source in *Medicago truncatula. Plant Physiology*, 146, 2020–2035.

Saito, K. et al. (2007) NUCLEOPORIN85 is required for calcium spiking, fungal and bacterial symbioses, and seed production in *Lotus japonicus. Plant Cell*, 19, 610–624.

Sanchez-Lopez, R. et al. (2011) Down-regulation of SymRK correlates with a deficiency in vascular bundle development in *Phaseolus vulgaris* nodules. *Plant, Cell & Environment*, 34, 2109–2121.

Sanchez, D.H. et al. (2008) Integrative functional genomics of salt acclimatization in the model legume *Lotus japonicus. Plant Journal*, 53, 973–987.

Sanchez, D.H. et al. (2012) Comparative metabolomics of drought acclimation in model and forage legumes. *Plant, Cell & Environment*, 35, 136–149.

Sanchez, D.H. et al. (2010) Mining for robust transcriptional and metabolic responses to long-term salt stress: a case study on the model legume *Lotus japonicus. Plant, Cell & Environment*, 33, 468–480.

Schauser, L. et al. (1998) Symbiotic mutants deficient in nodule establishment identified after T-DNA transformation of *Lotus japonicus. Molecular and General Genetics*, 259, 414–423.

Schauser, L. et al. (1999) A plant regulator controlling development of symbiotic root nodules. *Nature*, 402, 191–195.

Schnabel, E. et al. (2005) The *Medicago truncatula* SUNN gene encodes a CLV1-like leucine-rich repeat receptor kinase that regulates nodule number and root length. *Plant Molecular Biology*, 58, 809–822.

Schnabel, E.L. et al. (2011) The *ROOT DETERMINED NODULATION1* gene regulates nodule number in roots of *Medicago truncatula* and defines a highly conserved, uncharacterized plant gene family. *Plant Physiology*, 157, 328–340.

Searle, I.R. et al. (2003) Long-distance signaling in nodulation directed by a CLAVATA1-like receptor kinase. *Science*, 299, 109–112.

Sengupta, D. et al. (2011) A root proteomics-based insight reveals dynamic regulation of root proteins under progressive drought stress and recovery in *Vigna radiata* (L.) *Wilczek*. *Planta*, 233, 1111–1127.

Severin, A.J. et al. (2010) RNA-Seq atlas of *Glycine max*: a guide to the soybean transcriptome. *BMC Plant Biology*, 10, 160.

Singleton, P.W. and Bohlool, B. (1984) Effect of salinity on nodule formation by soybean. *Plant Physiology*, 74, 72–76.

Smit, P. et al. (2007) Medicago LYK3, an entry receptor in rhizobial nodulation factor signaling. *Plant Physiology*, 145, 183–191.

Smit, P. et al. (2005) NSP1 of the GRAS protein family is essential for rhizobial Nod factor-induced transcription. *Science*, 308, 1789–1791.

Sousa, C. et al. (2001) Translational and structural requirements of the early nodulin gene enod40, a short-open reading frame-containing RNA, for elicitation of a cell-specific growth response in the alfalfa root cortex. *Molecular and Cell Biology*, 21, 354–366.

Stacey, G. et al. (2006) Effects of endogenous salicylic acid on nodulation in the model legumes *Lotus japonicus* and *Medicago truncatula*. *Plant Physiology*, 141, 1473–1481.

Stracke, S. et al. (2002) A plant receptor-like kinase required for both bacterial and fungal symbiosis. *Nature*, 417, 959–962.

Streeter, J.G. (1985a) Nitrate inhibition of legume nodule growth and activity: I. Long term studies with a continuous supply of nitrate. *Plant Physiology*, 77, 321–324.

Streeter, J.G. (1985b) Nitrate inhibition of legume nodule growth and activity: II. Short term studies with high nitrate supply. *Plant Physiology*, 77, 325–328.

Subramanian, S. et al. (2008) Novel and nodulation-regulated microRNAs in soybean roots. *BMC Genomics*, 9, 160.

Tansengco, M.I. et al. (2003) Crinkle, a novel symbiotic mutant that affects the infection thread growth and alters the root hair, trichome, and seed development in *Lotus japonicus*. *Plant Physiology*, 131, 1054–1063.

Tansengco, M.L. et al. (2004) Pollen development and tube growth are affected in the symbiotic mutant of *Lotus japonicus*, crinkle. *Plant and Cell Physiology*, 45, 511–520.

Tian, A.G. et al. (2004) Characterization of soybean genomic features by analysis of its expressed sequence tags. *Theoretical and Applied Genetics*, 108, 903–913.

Timmers, A.C. et al. (1999) Refined analysis of early symbiotic steps of the Rhizobium–Medicago interaction in relationship with microtubular cytoskeleton rearrangements. *Development*, 126, 3617–3628.

Tirichine, L. et al. (2006) Spontaneous root-nodule formation in the model legume *Lotus japonicus*: a novel class of mutants nodulates in the absence of rhizobia. *Molecular Plant–Microbe Interactions*, 19, 373–382.

Tirichine, L. et al. (2007) A gain-of-function mutation in a cytokinin receptor triggers spontaneous root nodule organogenesis. *Science*, 315, 104–107.

Valdés-López, O., Arenas-Huertero, C., Ramírez, M., Girard, L., Sánchez, F., Vance, C.P., Luis Reyes, J., Hernández, G. (2008) Essential role of MYB transcription factor: PvPHR1 and microRNA: PvmiR399 in phosphorus-deficiency signalling in common bean roots. *Plant Cell Environ*, 31(12), 1834–1843.

Valdes-Lopez, O. et al. (2010) MicroRNA expression profile in common bean (*Phaseolus vulgaris*) under nutrient deficiency stresses and manganese toxicity. *New Phytologist*, 187, 805–818.

Vandenbosch, K.A. et al. (1985) Nodule initiation elicited by noninfective mutants of *Rhizobium phaseoli*. *Journal of Bacteriology*, 162, 950–959.

Van de velde, W. et al. (2010) Plant peptides govern terminal differentiation of bacteria in symbiosis. *Science*, 327, 1122–1126.

Vasse, J., de Billy, F., Camut, S., Truchet, G. (1990) Correlation between ultrastructural differentiation of bacteroids and nitrogen fixation in alfalfa nodules. *J Bacteriol*, 172(8), 4295–4306.

Varshney, R.K. et al. (2009) A comprehensive resource of drought- and salinity- responsive ESTs for gene discovery and marker development in chickpea (*Cicer arietinum* L.). *BMC Genomics*, 10, 523.

Vernie, T. et al. (2008) EFD is an ERF transcription factor involved in the control of nodule number and differentiation in *Medicago truncatula*. *Plant Cell*, 20, 2696–2713.

Wais, R.J. et al. (2000) Genetic analysis of calcium spiking responses in nodulation mutants of *Medicago truncatula*. *Proceedings of the National Academy of the Sciences of the United States of America*, 97, 13407–13412.

Walker, S.A. et al. (2000) Dissection of nodulation signaling using pea mutants defective for calcium spiking induced by nod factors and chitin oligomers. *Proceedings of the National Academy of the Sciences of the United States of America*, 97, 13413–13418.

Wan, J. et al. (2005) Proteomic analysis of soybean root hairs after infection by *Bradyrhizobium japonicum*. *Molecular Plant–Microbe Interactions*, 18, 458–467.

Wan, X. et al. (2007) *Medicago truncatula* ENOD40–1 and ENOD40–2 are both involved in nodule initiation and bacteroid development. *Journal of Experimental Botany*, 58, 2033–2041.

Wang, D. et al. (2010) A nodule-specific protein secretory pathway required for nitrogen-fixing symbiosis. *Science*, 327, 1126–1129.

Watson, B.S. et al. (2003) Mapping the proteome of barrel medic (*Medicago truncatula*). *Plant Physiology*, 131, 1104–1123.

Weidmann, S. et al. (2004) Fungal elicitation of signal transduction-related plant genes precedes mycorrhiza establishment and requires the dmi3 gene in *Medicago truncatula*. *Molecular Plant–Microbe Interactions*, 17, 1385–1393.

Weston, D.E. et al. (2008) The Pea DELLA proteins LA and CRY are important regulators of gibberellin synthesis and root growth. *Plant Physiology*, 147, 199–205.

Winicov, I. (1993) cDNA encoding putative zinc finger motifs from salt-tolerant alfalfa (*Medicago sativa* L.) cells. *Plant Physiology*, 102, 681–682.

Winicov, I. (2000) Alfin transcriptor factor overexpression enhances plants root growth under normal and saline conditions and improves salt tolerance in alfalfa. *Planta*, 210, 416–422.

Wojciechowski, M.F., Lavin, M., and Sanderson, M.J. (2004) A phylogeny of legumes (Leguminosae) based on analysis of the plastid matK gene resolves many well-supported subclades within the family. *The American Journal of Botany*, 91, 1846–1862.

Wopereis, J. et al. (2000) Short root mutant of *Lotus japonicus* with a dramatically altered symbiotic phenotype. *Plant Journal*, 23, 97–114.

Yang, W.C. et al. (1993) Characterization of GmENOD40, a gene showing novel patterns of cell-specific expression during soybean nodule development. *Plant Journal*, 3, 573–585.

Yano, K. et al. (2008) CYCLOPS, a mediator of symbiotic intracellular accommodation. *Proceedings of the National Academy of Sciences of United States of America*, 105, 20540–20545.

Yendrek, C.R. et al. (2010) A putative transporter is essential for integrating nutrient and hormone signaling with lateral root growth and nodule development in *Medicago truncatula*. *Plant Journal*, 62, 100–112.

Yoshida, C. et al. (2010) Plenty, a novel hypernodulation mutant in *Lotus japonicus*. *Plant Cell Physiology*, 51, 1425–1435.

Zhai, J. et al. (2011) MicroRNAs as master regulators of the plant NB-LRR defense gene family via the production of phased, trans-acting siRNAs. *Genes & Development*, 25, 2540–2553.

Zhang, H.M. and Forde, B.G. (1998) An *Arabidopsis* MADS box gene that controls nutrient-induced changes in root architecture. *Science*, 279, 407–409.

Zhou, Z.S. et al. (2012) Genome-wide identification of *Medicago truncatula* microRNAs and their targets reveals their differential regulation by heavy metal. *Plant, Cell & Environment*, 35, 86–99.

Zhu, H. et al. (2008) A novel ARID DNA-binding protein interacts with SymRK and is expressed during early nodule development in *Lotus japonicus*. *Plant Physiology*, 148, 337–347.

Zhukov, V., Radutoiu, S., Madsen, L.H. et al. (2008) The pea Sym37 receptor kinase gene controls infection-thread initiation and nodule development. *Molecular Plant–Microbe Interactions*, 21, 1600–1608.

9 What the Genomics of Arbuscular Mycorrhizal Symbiosis Teaches Us about Root Development

Damien Formey, Cyril Jourda, Christophe Roux, and Pierre-Marc Delaux

The soil is a complex biotic environment where plant roots are exposed to a myriad of microbes. It has been estimated that 1 g of soil contains up to 10^{10}–10^{11} bacteria (Horner-Devine et al. 2003) and up to 200 mg of fungal mycelium (Leake et al. 2004). This biotic environment has a great incidence on plant diversity and productivity (van der Heijden et al. 2008). Most soil microbes are saprobic decomposing organic litter; others are pathogenic or symbiotic and directly interact with plants. Arbuscular mycorrhizal (AM) fungi are prevalent symbiotic soil microorganisms that belong to a particular monophyletic fungal group called Glomeromycota (Schüβler et al. 2001; Schwarzott et al. 2001). This fungal group has been described as a basal sister clade of the Dikarya (Helgason et al. 2003; James et al. 2006; Liu et al. 2006, 2009; Redecker and Raab 2006). AM symbiosis is considered as the most widespread interaction on earth between plants and microbes. AM fungi are present in most ecosystems where they are of great incidence in the organization of plant communities (van der Heijden et al. 1998; Klironomos et al. 2011; Schnitzer and Klironomos 2011). They interact with a majority of plant species and *ca* 80% of land plants (Smith and Reads 2008). The plant species that do not interact with AM fungi can be considered as exceptions. They belong to a small number of plant families such as Caryophyllaceae, Chenopodiaceae, or Brassicaceae. In this last family, *Arabidopsis thaliana*, the most studied model plant, is nonhost, and no symbiotic structures were observed after inoculation, although some species of Brassicaceae were described as possibly infected (see DeMars and Boerner 1995). Except for these nonmycotrophic plant families, the ancestral character of AM fungi and their extremely large plant host spectrum led several authors to hypothesize that these fungi have played a critical role in plant evolution, participating, for instance, in the appearance of root tissues (Simon et al. 1993; Brundrett 2002). Even if these assumptions are highly speculative, the coevolution of the AM symbiosis with the first land plants leading to the current flora is well supported by fossil records showing that these fungi were already living in close association with the first land plants (Remy et al. 1994; Redecker et al. 2000).

The establishment of AM symbiosis starts with the germination of AM fungal spores or, more frequently in undisturbed land, with the extraradical mycelium extending out of mycorrhizal roots (Finlay 2008). In the vicinity of a host root, both partners exchange symbiotic signals. The plant releases a mix of molecules that stimulate hyphal branching and growth. A class of molecules participating in this stimulation are strigolactones (Akiyama et al. 2005; Besserer et al. 2006). On the fungal side, diffusible compounds that activate the symbiotic program of the host plant are produced (Kosuta et al. 2003; Weidmann et al. 2004; Olah et al. 2005; Gutjahr et al. 2009;

Root Genomics and Soil Interactions, First Edition. Edited by Martin Crespi.

Kuhn et al. 2010). A family of lipo-chito-oligosaccharides (Myc-LCOs) has recently been identified (Maillet et al. 2011). These molecules stimulate the establishment of symbiosis and the formation of lateral roots (LRs). Following this presymbiotic phase, contact between the two partners occurs: the AM fungus forms a hyphopodium on the root epidermis. In response, the host cell develops a prepenetration apparatus through which the fungus colonizes the root (Genre et al. 2005). Intraradical hyphae then grow inter- and intracellularly where the fungus forms highly branched structures called arbuscules in the root cortical cells. As an obligate biotroph, the AM fungus is dependent on its host to obtain its carbon (Helber et al. 2011); in return, it provides mineral nutrients, such as phosphate, and water to the host plant (Smith and Read 2008). These exchanges are intense and plant dependence on AM fungi can be important as in some cases all plant phosphate can be provided by the fungal partner (Pearson and Jacobson 1993). This symbiotic phosphate uptake requires specific phosphate transporters localized in the peri-arbuscular membrane of the host cell (Rausch et al. 2001; Harrison et al. 2002). Highlighting the significance of these exchanges, it has been proposed that AM symbiosis have been widely conserved during the evolution of the green lineage as its establishment is tightly tuned by both partners (Kiers et al. 2011). The fact that AM symbiosis can enhance plant growth and productivity makes AM fungi extremely interesting for the development of sustainable agriculture. Different works have been developed to enhance the infectivity of commercial mycorrhizal inoculants (Smith and Read 2008; Corkidi 2004). However, the biology of AM fungi is still full of enigmas. The tricky cultivation of these obligate symbionts, insufficient knowledge of their genetics and sexuality, the impossibility to transform them in a stable way make the study of these multinucleate organisms extremely difficult. Recent works showing that transient genetic transformation (Helber and Requena 2008) and host-induced RNAi (Helber et al. 2011) were possible with AM fungi are encouraging. They open the way for functional analyses of AM fungal genes. Although different species of Glomeromycetes are currently used in laboratories (*Rhizophagus irregularis*—formerly *Glomus intraradices-*, *Glomus mosseae, Glomus versiforme, Gigaspora rosea, Gigaspora gigantea, Gigaspora margarita*), one isolate of *Rhizophagus irregularis* (DAOM197198) was chosen as a model due to its aggressiveness and easy multiplication in root organ cultures (Bécard and Fortin 1988; Chabot et al. 1992). Moreover, an international program was launched to sequence its genome (Martin et al. 2008).

Besides the trophic incidence of AM symbiosis, or associated to it, the observation was made long ago that AM fungi modify plant root systems (Berta et al. 2002; Guether et al. 2009). These studies argue for cross talks between the mechanisms involved in AM symbiosis and root development. Here, we explore genetics and genomics data on AM fungus–plant interactions that could provide new insights into the cross talk.

Forward and Reverse Genetics for Identifying Myc Mutants

The difficulty to identify the discrete structures involved in AM symbiosis (intraradical arbuscules), which require careful root staining and microscopic observation, has strongly limited high-throughput mycorrhizal phenotyping and development of forward genetics approaches. In the last decade, an overlap has been observed in legume plants between the symbiotic pathways leading to nodulation and mycorrhization. This has been shown by the characterization of plant mutants unable to interact either with rhizobia (Nod$^-$ phenotype) or with AM fungi (Myc$^-$ phenotype) (Marsh and Schultze 2001; Parniske 2008). This overlap was of great interest to identify the first Myc$^-$ mutants and important symbiotic genes in legumes as well as their orthologs in other plant families. Thereafter, reverse genetics approaches were developed.

Common Symbiotic Pathways among Legumes

Because the signaling pathways leading to nodulation and mycorrhization shared common genes, they were called the "common symbiosis pathway" (CSP, Parniske 2008). To date, eight members of the CSP have been identified in *Medicago truncatula* and/or *Lotus japonicas*: a receptor-like kinase (MtDMI2/LjSymRK; Endre et al. 2002; Stracke et al. 2002); a cation channel (MtDMI1/LjCASTOR-LjPOLLUX; Ané et al. 2004; Imaizumi-Anraku et al. 2005); three nucleoporins (LjNUP85, LjNUP133, and LjNENA; Kanamori et al. 2006; Saito et al. 2007; Groth et al. 2010) involved together with the cation channel in the generation of specific nuclear calcium spiking (Oldroyd and Downie 2006); a calcium and calmodulin-dependent protein kinase (CCAMK), which has been proposed to decode this specific calcium signature (MtDMI3/LjSYM15, Levy et al. 2004; Tirichine et al. 2006); MtIPD3/LjCYCLOPS, which interact with CCAMK (Yano et al. 1996; Messinese et al. 2007; Parniske 2008; Bonfante and Genre 2010; Maillet et al. 2011); and NSP2, a transcription factor of the GRAS family (Kaló et al. 2005; Heckmann et al. 2006; Maillet et al. 2011). Mutants of the CSP show a strong AM symbiosis defective phenotype with alteration in root penetration, intraradical colonization, and arbuscule formation.

Screening of Mutant Collections to Phenotype AM Symbiosis Defective Mutants

Systematic phenotyping of a large collection of mutants to identify AM symbiosis-defective phenotype (Myc-) is a difficult approach as symbiotic markers are discrete microscopic intraradical structures. However, the accumulation of a yellow pigment during AM symbiosis, identified as the apocarotenoid-derived mycorradicin, can be used for rapid mycorrhization phenotyping in some monocot species like maize (Walter et al. 2000) and *Brachypodium distachyon* (C. Roux, pers. commun.). The use of monocots is of great interest because of easy transposon mutagenesis. Using this approach, two maize mutants—*nop1* and *taci1*—were identified (Paszkowski et al. 2006). With *nop1*, the fungus was unable to differentiate hyphopodia, highlighting the role played by the host plant in the control of its fungal partner. With *taci1*, root colonization was affected in a later stage: the fungus penetrated via hyphopodia formation but intraradical hyphal growth was restricted near to the infection point. Mutagenesis strategies were also developed in dicots. In *M. truncatula* screening of an ethyl methanesulfonate collection enabled the identification of two half-ABC transporters STR1 and STR2 required for AM symbiosis (Zhang et al. 2010). In tomato, two mutants (*pmi* and *pmi2*) were characterized (David-Schwartz et al. 2001, 2003). These mutants showed a defect in presymbiotic steps, resulting in lower level of colonization. Finally, a transposon-mutagenized population of *Petunia hybrida* was also screened for Myc phenotype (Reddy et al. 2007). The *pam1* mutant issued from this screening was strongly affected in root penetration, colonization, in arbuscule formation, and in the expression of the specific phosphate transporter PhPT3 (Feddermann et al. 2010). Very interestingly, the *pam1* insertion was map-cloned in a gene encoding an ortholog of the VAPYRIN protein of *M. truncatula* (see later).

Candidate Genes Identified through Specific Transcriptomic and Metabolic Analyses

SbtM1 and SbtM3 are two subtilisin-like proteases upregulated in the mycorrhizal roots of *L. japonicus* and RNAi of these genes reduce the number of arbuscules and intraradical hyphae (Takeda et al. 2009). The CDPK1, a calcium-dependent protein kinase (Ivashuta et al. 2005), and VAPYRIN, a protein containing two protein-binding domains (Pumplin et al. 2010), were found to be highly upregulated in *M. truncatula* during AM symbiosis (Gomez et al. 2009). Mycorrhization of the corresponding RNAi lines was strongly altered (Ivashuta et al. 2005; Pumplin et al. 2010). Other genes previously identified in transcriptomic analyses of mycorrhized *M. truncatula* and required in the early steps of mycorrhizal interaction have also been identified by RNAi. Thus, *MSBP1*, which encodes a steroid-binding protein, was upregulated in the vicinity of highly branched fungal hyphae, and the corresponding RNAi lines were strongly affected in arbuscule formation (Kuhn et al. 2010).

Another strategy was to phenotype mutants already characterized for a specific metabolic function. Phosphate transport from the fungus to the host plant is for instance a crucial metabolic trait of AM symbiosis. Since the first description of a mycorrhiza-specific plant phosphate transporter in potato, StPT3 (Rausch et al. 2001), similar transporters have been characterized in other dicots like *Populus trichocarpa* (Loth-Pereda et al. 2011) and tomato (Nagy et al. 2005) and also in monocots like rice (Paszkowski et al. 2002) and maize (Glassop et al. 2005). In *M. truncatula,* the phosphate transporter MtPT4 is specifically expressed in arbuscules of mycorrhizal roots (Harrison et al. 2002). The corresponding mutant *Mtpt4* was impaired both in root colonization and arbuscule formation (Javot et al. 2007). Interestingly, the arbuscules formed were aborted and extraradical fungal growth was strongly limited. These works provided the first evidence of the undeniable requirement of "symbiotic" phosphate uptake for the maintenance of the AM symbiosis. Investigations on another, completely different, metabolism have recently been carried out, which illustrates the intimate link between plant developmental processes and AM formation. Carotenoid cleavage dioxygenase (CCD) 7 and CCD8 have been previously identified in *Pisum sativum*, *Oryza sativa*, *A. thaliana,* and *Petunia hybrida* for their highly branched shoot phenotype (Morris et al. 2001; Sorefan et al. 2003; Booker et al. 2004; Snowden et al. 2005; Arite et al. 2007; Drummond et al. 2009). Independently, Matusova et al. (2005) proposed that the biosynthesis of strigolactones, compounds released by host plants that induce seed germination of parasitic weeds like Striga species, involved the cleavage of some carotenoid substrates by CCD enzymes. Strigolactones were later described as inducers of fungal growth and branching (Akiyama et al. 2005; Besserer et al. 2006). On the basis of these observations, Gomez-Roldan et al. (2008) and Umehara et al. (2008) assessed the ability of the *CCD7* and *CCD8* mutants of pea (*ramosus* mutants of *Pisum sativum Psrms1*, *Psrms5*, respectively—Foo et al. 2005) and rice (rice dwarf mutants *D10, D17*, respectively) to produce strigolactones. They demonstrated that strigolactone synthesis is dependent on these two enzymes. In addition, Gomez-Roldan et al. (2008) showed that mycorrhization of both *ccd* pea mutants was severely reduced. These works pointed out the role played by strigolactones as a plant hormone in the establishment of the AM symbiosis.

Mutant Analysis Reveals a Biological Integration between AM Symbiosis and Root Development

As previously mentioned, the establishment of AM symbiosis modifies root architecture. This modification of root architecture is not limited to AM interaction. Ectomycorrhizal fungi are also able to induce modification of LR density by producing plant hormones like auxin and ethylene (Felten et al. 2009; Splivallo et al. 2009). By contrast, the modifications of LR density induced by AM fungi seem to be under the control of AM fungal-specific signals. Treatment of A17 WT lines of *M. truncatula* with spore exudates (Kosuta et al. 2003; Olah et al. 2005; Mukerjhe and Ané 2011) or synthetic Myc-LCOs (Maillet et al. 2011) stimulates the development of new LR. The same treatments performed on different mutants of the CSP had no effect, arguing that the LR response is dependent on the perception and transduction pathway of LCOs. These observations suggest the occurrence of a symbiotic program targeting the mechanisms directing root development.

The biological overlap of AM symbiosis and LR formation is illustrated by the maize mutant *lrt1* (Paszkowski et al. 2002). This mutant displays growth defects, including lack of LRs. Interestingly, this growth defect can be fully reversed after AM fungal inoculation. Stimulation of LR density was fairly well described in several other plant species like rice, leek, or cherry plum (Berta et al. 1990, 1995; Gutjahr et al. 2009). A complementary illustration of this overlap is that several of the AM-defective mutants showed altered root architecture. For instance, the root length of the *cdpk1* mutant of *M. truncatula* was described as significantly lower than WT plants, and root hairs were also shorter (Ivashuta et al. 2005). This growth defect is correlated to shorter cortical cells and altered cell wall (CW) organization and composition. As root colonization by AM fungi requires strong CW remodeling (Genre et al. 2005), an altered CW could interfere with the mycorrhization process.

As previously mentioned, the discovery of strigolactone involvement is a paradigm of the integration of AM symbiosis and plant development mechanisms. Strigolactones were first described for their role in shoot branching, while subsequent data argued for a role in root development too. The root growth of the nonmycotrophic *A. thaliana* was analyzed using the *ccd7* and *ccd8* mutants, defective in strigolactone biosynthesis. Both mutants displayed a higher LR density (Kapulnik et al. 2011) and shorter primary root (Ruyter-Spira et al. 2011). In addition, exogenous treatment with the synthetic strigolactone analog GR24, on the two mutants, restored a WT phenotype. The LR phenotype was also observed on corresponding pea mutants (Delaux and Combier, unpublished data). It must be pointed out that LRs are the main entry point for AM fungal colonization. The increase of the global LR density would result in more abundant AM symbiotic interactions by increasing the probability of contact between the two partners. The *ccd7* and *ccd8* pea mutants displayed lower root colonization, suggesting that the Myc- phenotype of these mutants, that must produce more LRs, was more related to the defect of strigolactone excretion and a lack of fungal stimulation. Recently, Liu et al. (2011) demonstrated that strigolactone biosynthesis is mediated via the symbiotic NSP1/NSP2 transcription factors, involved in LCO perception. Strigolactone root-exudation decreases in the plant during symbiosis establishment (López-Ráez et al. 2011) and increases in phosphorus deficiency (Yoneyama et al. 2007). All these results argued for a molecular dialog between plants and AM fungi for root development involving LCOs, phosphate, and strigolactone, and the exact role of the endogenous and exogenous pools of each of these molecules on root development is still not defined.

Comparative Transcriptomics of AM Symbiosis: Toward Identification of Genes Involved in Root Development

The first comparative transcriptomes to investigate root symbioses were performed on *M. truncatula* in nodule and mycorrhizal roots (Journet et al. 2002). Because of the overlap of signaling mechanisms leading to nodulation and mycorrhization in legumes, a specific evolution of this pathway might have occurred in this plant family compared to others. For this reason, comparative transcriptomics on legumes and nonlegumes would provide information about the species-related and more general mechanisms involved in AM symbiosis. The transcriptomic profiles of AM symbiosis were produced in whole root systems using different strategies—EST analysis, suppressive-subtractive cDNA libraries, and cDNA array hybridizations—during interaction with different AM fungi such as *Rhizophagus irregularis (Glomus intraradices)*, *Glomus mosseae*, and *Gigaspora margarita*

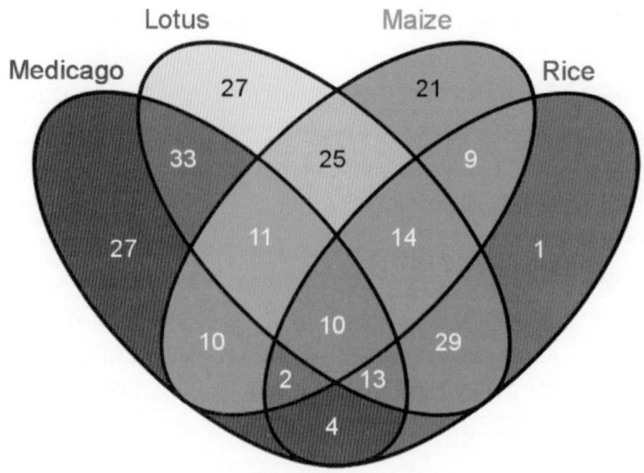

Figure 9.1 Four-way Venn diagram illustrating orthologous gene clusters (OGC) generated using OrthoMCL analysis (for method, see Li et al. 2003) on cDNA consensus sequences of regulated genes during AM symbiosis in *Medicago truncatula* (blue circle; Hohnjec et al. 2005), *Lotus japonicus* (yellow circle; Guether et al. 2009), *Zea mays* (green circle; Jourda et al. unpublished), and *Oryza sativa* (red circle; Güimil et al. 2005). The Venn diagram was generated using the Venny tool (Oliveros 2007). (For a color version of this figure, see the color plate section.)

(Journet et al. 2002; Küster et al. 2004; Siciliano et al. 2007; López-Ráez et al. 2010 as examples). More recently, specific gene expression profiles during AM symbiosis have been established in arbuscule-containing cortical cells of mycorrhizal roots using laser capture microdissection combined with microarray hybridization (Gaude et al. 2012). Although the data obtained are informative, a comparative analysis of transcriptomes during AM symbiosis is a complementary approach that could allow the identification of specific and shared mechanisms among plant species. We here present an original comparison of transcriptomes of mycorrhizal roots from different plant species by selecting data obtained in similar experimental settings: transcriptomes in response to *Rhizophagus irregularis* DAOM197198 from whole root system of *M. truncatula* (Hohnjec et al. 2005), *L. japonicus* (Guether et al. 2009), *Oryza sativa* (Güimil et al. 2005), and *Zea mays* (Jourda et al., unpublished). Comparing gene expression of different species needs the identification of orthologous genes. The comparative analysis was performed using OrthoMCL approach, leading to the identification of 236 orthologous gene clusters (OGC; see Figure 9.1 for details). It must be underlined that these data are partial as the microarrays used for the different plant species were not always representative of the complete genomes. Results illustrated in Figure 9.1 showed that 10 clusters are common to the 4 species, 40 clusters are shared by 3 species, 110 clusters are shared by 2 species, and 76 were found only in 1 species. In the 2 legume species, 205 clusters were identified, 67 were common to Medicago and Lotus, and 87 were legume- or dicot-specific (not found in Poaceae). In maize and rice, 139 clusters were found among which 35 were common to the 2 species and 31 were Poaceae specific (not found in Medicago and Lotus).

We selected OGC putatively involved in root development, considering that root development involves genes belonging to functional classes as diverse as transport activity, transcription factors, CW metabolism, and hormone metabolism (see Table 9.1). This subjective choice led to the selection of 38 OGC comprising 105 ESTs among which only one was common to the four plant species (Table 9.1). This OGC, formed by four orthologous ESTs, correspond to a phosphate transporter specifically induced during AM symbiosis in plants, including MtPT4 (MT009707), OsPT11, Ljwgs_014433.2_at, and Zm.1921.1.S1_at. As previously noticed, the Medicago MtPT4 phosphate transporter is a mycorrhiza-specific gene required for arbuscule development and function

Table 9.1 List of a subset of 38 clusters of orthologous gene (OGC) generated using OrthoMCL analysis (from method, see Li et al. 2003) on cDNA consensus sequences of regulated genes putatively involved in root development during AM symbiosis in *Medicago truncatula* (Hohnjec et al. 2005), *Lotus japonicus* (Guether et al. 2009), *Oryza sativa* (Güimil et al. 2005), and *Zea mays* (Jourda et al., unpublished). Cluster annotation (metabolism and function), plant species with gene identity, annotation, and regulation during AM symbiosis are indicated. Signs + and − indicate gene is upregulated or downregulated, respectively. Gene ID corresponds to Affymetrix Gene Chip identification available on www.affymetrix.com.

Metabolism	Annotation (OGC)	Plant	Gene ID	Regulation
Transport	Aquaporin	Maize	Zm.606.1.A1_at	−
		Maize	Zm.607.1.A1_at	−
	Aquaporin	Lotus	Ljwgs_149324.1_at	+
		Maize	Zm.612.1.A1_at	
	Aquaporin	Lotus	TM0748.11_at	+
		Maize	Zm.602.1.A1_a_at	−
	High affinity nitrate transporter/membrane transporter	Medicago	MT002501	+
		Medicago	MT009589	+
	Oligopeptide transporter-like protein/putative nitrate transporter	Lotus	chr6.CM0118.43_at	+
		Medicago	MT006556	+
		Oryza	OsAM60	+
	Peptide transporter	Lotus	chr1.CM0295.1_s_at	+
		Lotus	chr1.CM0295.2.1_at	+
	Peptide transporter like	Lotus	chr2.CM0903.44_at	−
		Maize	Zm.17744.1.A1_at	+
	Phosphate transporter	Lotus	Ljwgs_014433.2_at	+
		Maize	Zm.1921.1.S1_at	+
		Medicago	MT009707	+
		Oryza	OsPT11	+
	Putative ammonium transporter	Lotus	Ljwgs_016680.1_at	+
		Oryza	OsAM76	+
	Putative sucrose transport protein	Lotus	chr5.CM0344.52_at	−
		Maize	Zm.199.1.S1_s_at	+
	Sugar transporter-like protein	Lotus	Ljwgs_024490.1_s_at	−
		Medicago	MT000349	−
	Sulfate transporter	Lotus	chr6.CM0314.34_at	+
		Lotus	Ljwgs_011755.1_at	+
		Maize	Zm.11651.1.A1_at	−
Transcription factor	GRAS family	Lotus	Ljwgs_016263.1_at	+
		Lotus	Ljwgs_023888.1_at	+
		Lotus	Ljwgs_027761.2_at	+
		Maize	Zm.6402.2.A1_at	+
	MYB family	Lotus	chr1.BM1732.4_at	+
		Medicago	MT001930	+
	MYB family	Lotus	Ljwgs_014616.1_at	−
		Medicago	MT007392	+
	MYB family	Maize	Zm.13885.1.S1_s_at	−
		Medicago	MT002218	−
	NAC family	Medicago	MT000799	−
		Medicago	MT009487	−
Cell wall metabolism	4-Coumarate–CoA ligase-like protein	Lotus	Ljwgs_075474.1_at	+
		Maize	Zm.15085.1.A1_at	+
		Medicago	MT000912	−
	Caffeic acid *O*-methyltransferase	Medicago	MT005666	−
		Medicago	MT007074	−

(*Continued*)

Table 9.1 (*Continued*)

Metabolism	Annotation (OGC)	Plant	Gene ID	Regulation
	Cinnamoyl CoA reductase-like protein	Lotus	Ljwgs_100777.1_s_at	+
		Medicago	MT000507	−
		Oryza	OsAM163	+
		Oryza	OsAM49	+
	Endo-1,3(4)-β-glucanase	Lotus	chr4.CM0119.5_at	+
		Medicago	MT003194	+
	Expansin protein	Lotus	Ljwgs_041638.1_s_at	+
		Maize	Zm.17372.1.A1_at	−
	Expansin-related protein	Lotus	Ljwgs_012501.1_at	−
		Medicago	MT000967	−
	Extensin-like protein	Medicago	MT007031	+
		Medicago	MT007032	+
	Osmotin-like protein/putative thaumatin	Medicago	MT001796	−
		Medicago	MT007463	+
	Pectin esterase-like protein	Lotus	Ljwgs_031539.1_at	+
		Medicago	MT001341	+
		Medicago	MT007664	+
	Pectin esterase-like protein	Lotus	chr1.TM1573.5_at	−
		Maize	Zm.7152.1.A1_at	+
		Medicago	MT006917	+
	Putative endo-1,3;1,4-β-D-glucanase	Medicago	MT000950	+
		Oryza	OsAM222	−
	Putative xyloglucan endotransglycosylase	Lotus	Ljwgs_022629.1_at	−
		Maize	Zm.704.1.S1_at	+
		Oryza	OsAM220	−
	Zeamatin/thaumatin-like protein	Maize	Zm.281.1.S1_s_at	+
		Medicago	MT000966	−
Hormone metabolism	Cytokinin-*O*-glucosyltransferase 2	Maize	Zm.9570.1.A1_a_at	+
		Maize	Zm.9570.2.A1_x_at	+
	Gibberellin 20 oxidase 2	Lotus	chr5.CM0953.24_at	+
		Lotus	Ljwgs_018141.1_at	+
		Maize	Zm.13480.1.S1_at	+
		Medicago	MT003693	+
	Gibberellin regulatory protein like	Lotus	chr5.CM0239.51_at	+
		Lotus	chr6.CM0539.6_at	+
	Zeatin *O*-xylosyltransferase	Lotus	Ljwgs_037161.1_at	+
		Lotus	Ljwgs_037161.1_x_at	+
MISC	MtN19-like protein	Lotus	chr6.CM0437.7_at	+
		Medicago	MT015318	−
		Medicago	MT015656	−
		Oryza	OsAM197	+
		Lotus	chr4.CM0337.32.7_at	+
		Medicago	MT001044	+
		Medicago	MT001597	+
	Multifunctional nodulin 26-like aquaporin	Lotus	chr4.CM0046.51_at	+
		Medicago	MT007526	+
		Oryza	OsAM143	+
	Peroxidase	Lotus	Ljwgs_018430.1_s_at	−
		Maize	Zm.11214.1.S1_at	+
		Maize	Zm.10660.1.A1_at	+

Table 9.1 (*Continued*)

Metabolism	Annotation (OGC)	Plant	Gene ID	Regulation
		Maize	Zm.5170.1.S1_at	–
		Medicago	MT001052	–
		Medicago	MT010277	–
		Medicago	MT002659	+
		Medicago	MT007740	+
		Medicago	MT007030	+
		Medicago	MT008600	+
	Putative nodulin	Lotus	Ljwgs_119620.1_at	+
		Medicago	MT013567	+
		Oryza	OsAM129	+

(Harrison et al. 2002; Javot et al. 2007). High-affinity phosphate transporters specifically induced during AM symbiosis were also characterized in rice (OsPt11; Paszkowski et al. 2002), Lotus, (Ljwgs_014433.2; Maeda et al. 2006), and maize (Glassop et al. 2005). Improvement of phosphate nutrition by phosphate transporters specific to AM symbiosis could be involved in LR formation. In maize for instance, a mutant lacking LRs can be complemented by high phosphate nutrition and by AM symbiosis (Paszkowski and Boller 2002). How AM symbiosis modulates LR development is still unresolved, but the incidence of AM symbiosis on soil phosphate uptake is undoubtedly one of the mechanisms involved in. It has indeed been demonstrated in *A. thaliana* that phosphate availability affects root hair length, primary root growth, and LR density in an auxin-dependent way (Nacry et al. 2005; Perez-Torres et al. 2008).

In addition to phosphate transporters, nutrient transporter coding genes are an important subset of regulated genes during root growth and AM symbiosis with 12 OGCs identified. In Lotus, a putative ammonium transporter-encoding gene (Ljwgs_016680.1_at) is the strongest upregulated gene during AM symbiosis, but this observation is not common to the four plant species. Other OGC-encoding hexose transporters, oligopeptide transporters, water transporters, sulfate transporters, and nitrate transporters have been found. Overall, OGC involved in transport could be involved in improving root nutrient exchanges during AM symbiosis, and this could partly explain the stimulation of root branching during AM symbiosis (Berta et al. 1995).

Among the 38 OGC, five code for transcription factor families (GRAS, NAC, and MYB families) potentially involved in the control of root development, comprising 12 genes. In Lotus, for example, three copies of OGC encoding for GRAS members are scarecrow protein coding genes. Scarecrow members are considered as genetic determinants of root identity (Dolan 2007). Different GRAS family members were induced during AM symbiosis and could be involved in modulation of root development during AM symbiosis. The OGC-encoding NAC transcription factors comprise only two repressed Medicago paralogs. Members of NAC transcription factor family could have conserved functions in secondary CW biosynthesis and in root development (Christiansen et al. 2011) and so could play a role in root development by secondary CW regulation during AM symbiosis. Three OGC encoding MYB transcription factors have been detected with different expression profiles during AM symbiosis. In *Arabidopsis*, MYB transcription factors are involved in developmental processes as diverse as secondary CW biosynthesis (Zhong et al. 2008), phosphate starvation (Bustos et al. 2010), and LR formation (Shin et al. 2007).

Root development is related to cell elongation and thus to CW modifications. A set of 13 OGC that could be involved in degradation, modification, biosynthesis, and structure of CW have been identified. The OGC related to CW metabolism encode for proline-rich CW protein (extensin-like), 4-coumarate:coenzyme A ligase, β-glucanase, cinnamoyl-CoA reductase-like protein, expansin-related protein, thaumatin-like protein, caffeic acid *O*-methyltransferase, putative xyloglucan endotransglycosylase, and pectinesterase-like protein with different expression profiles according to the host plant. These differences could arise from the diversity of CW structures observed among plant species (Popper et al. 2011), especially between dicotyledon and monocotyledon species. As an illustration, OGC encoding for proline-rich CW protein (extensin-like) are only formed by two Medicago paralogous genes induced during AM symbiosis. Yokoyama and Nishitani (2004) have described the presence of extensin in type I CW but its absence in type II CW, respectively, represented by dicot and monocot plant CW. Structural CW proteins (such as expansins and extensin-like proteins) are involved in the cleavage and the reassembly of CW polymers necessary for cell elongation (Cosgrove 2005). Regulation of these potential agents of CW loosening (Cosgrove 2001) could drive root elongation during AM symbiosis. An important subset of plant peroxidase coding genes clustered in an OGC comprising 10 elements, but their orthologs were absent in rice. However, in mycorrhizal roots of rice, the strongest induced gene corresponded to a type III peroxidase and its expression was specific of the AM condition (Güimil et al. 2005). Peroxidase enzymes are involved in hydrogen peroxide production and are reported to play a key role in CW organization and root elongation (Liszkay et al. 2004). These proteins could then be involved in CW reorganization and LR formation during AM symbiosis. Lastly, nodulin genes have been identified in these transcriptomic approaches. They were identified as molecular markers of root nodule organogenesis (Crespi and Galvez 2000; Moreau et al. 2011) and could be common elements of the signaling pathway between mycorrhization and nodulation of plant roots (van Rhijn et al. 1997). However, the function of several nodulins is yet to be described. Recently, MtN21-like proteins were described as potential transporters required for secondary CW formation (Ranocha et al. 2010). Except in rice, at least one MtN21-like protein encoding gene was found to be regulated during AM symbiosis. Three other OGC putatively encoding for nodulin proteins were identified. For example, NOD26-like membrane intrinsic protein (NIP) coding genes have been clustered. NIPs belong to the aquaporin superfamily and are plant-specific with different functions and expression profiles (Liu et al. 2009).

Root development is obviously under hormone metabolism control (see Osmont et al. 2007). Four OGC involved in hormone metabolism have been found and encode for gibberellin-20-oxidase 2, zeatin *O*-xylosyltransferase, cytokinin-*O*-glucosyltransferase 2, and gibberellins regulatory protein-like families. In this context, it is interesting to note that no common OGC-encoding auxin-specific genes have been found. However, it could indicate the existence of different auxin response patterns according to plant species. The transcript Zm.5919.1.S1_at encodes a putative auxin response factor and is induced during AM symbiosis in maize. Similar observations have been reported for OsAM173 and MT000634 induced during AM symbiosis in rice and Medicago, respectively, but no putative auxin response factor has been found to be regulated during AM symbiosis in Lotus.

This comparative analysis of transcriptomes obtained with four plant species gives a first overview of potentially conserved mechanisms involved in root development during AM symbiosis. We must keep in mind that the above transcriptomes were not exhaustive and were performed using well-established mycorrhizal root systems. Although new infection sites are continuously formed on developing roots, the genes involved in the early steps of mycorrhizal establishment might have been diluted in the root systems used. The recent and exponential development of next-generation sequencing technology gives new perspectives to perform global comparative transcriptome analysis of plant root system using RNA-seq applications (Ozsolak and Milos 2011).

Small RNAs in AM Symbiosis

It can be suggested that master regulator genes, and also fine-tuning regulators, can be involved in the regulation of nutrient uptake and root development during AM symbiosis (Krajinski and Frenzel 2007). The microRNAs are small (about 21nt) negative regulators of gene expression present in plants. These small RNAs are involved in developmental processes, hormonal signaling, nutrient balance, abiotic, and biotic stresses (Bartel 2004; Jones-Rhoades et al. 2006; Mallory and Vaucheret 2006; Ruiz-Ferrer and Voinnet 2009). MicroRNAs also play a role in plant–microorganism interactions (Navarro et al. 2006; Subramanian et al. 2008). During nodule development in *M. truncatula*, miR169 is a regulator acting by targeting the transcriptional factor *MtHAP2–1* (Combier et al. 2006). It is also known that miR166 plays a crucial role in root and nodule development and its overexpression in *M. truncatula* leads to a reduced number of LR and symbiotic nodules (Boualem et al. 2008). As mentioned earlier, nodulation and mycorrhizal symbioses share a common gene pathway called CSP. Several analogies are found in the presymbiotic and symbiotic phases of these two symbioses. Hence, it can be proposed that the regulation pathway could be either shared or discriminant between the symbioses according to the regulators. Recently, Maillet et al. (2011) demonstrated that *MtNsp2*, a GRAS transcription factor essential for signaling during root nodule development (Oldroyd and Long 2003), is also involved in AM fungal colonization. As *MtNsp2* transcripts were proposed to be cleaved by the miR171h (Devers et al. 2011), the role of this regulatory loop in early and late steps of AM symbiosis was assessed on *M. truncatula* (Lauressergues et al., submitted). These results support the hypothesis that, like nodulation, a regulatory net of microRNAs could finely regulate AM symbiosis. The microRNA-mediated phosphate (Pi) regulation can illustrate this hypothesis. Pi regulation has been widely studied by the scientific community working on AM fungus because of the role of this symbiosis in Pi acquisition. As illustrated in pea or petunia, AM symbiosis is suppressed in high phosphate concentration assays (Balzergue et al. 2011; Breuillin et al. 2010). Some Pi transporters are specifically expressed in Pi-starvation condition, while some are specifically expressed during the AM symbiosis (Chiou et al. 2001; Rausch et al. 2001; Harrison et al. 2002). The Pi and AM signaling pathways share upstream regulatory factors and downstream structural genes as revealed previously by transcriptomic analyses. One of these regulators is the MYB transcription factor PHR1 that is a key element in signaling of primary Pi responses (Rubio et al. 2001). PHR1 binds to a *cis* element and causes the transcription of many Pi starvation-inducible genes, including members of the miR399 family (Bari et al. 2006). Some of these microRNAs target *PHO2* transcript—coding for an ubiquitin-conjugating enzyme—and negatively regulate its expression (Allen et al. 2005). *PHO2* is responsible for the downregulation of a subset of genes induced in response to Pi starvation, like phosphate transporters (Aung et al. 2006; Bari et al. 2006; Chiou et al. 2006). As a consequence, miR399 members can be considered as Pi-starvation signals involved in Pi homeostasis, de-repressing the Pi starvation responses and the Pi uptake mechanisms. Branscheid and coauthors (2010) showed the correlation between miR399 expression, Pi homeostasis, and AM development in plants. Interestingly, they found that *A. thaliana*, a nonmycotrophic species, possesses less than half the miR399 family members of species capable of enhancing Pi uptake through AMF symbiosis, like *M. truncatula*, rice, or poplar. In parallel, these authors observed an accumulation of mature miR399 in mycorrhizal roots compared with nonmycorrhizal roots in *M. truncatula* and tobacco. This suggests that, in mycorrhizal roots, PHO2 is involved in the suppression of AM symbiosis in Pi-depleted plants because miR399 prevents its accumulation and activity in response to fungal Pi uptake. Thus, miR399 could be a regulator in maintaining AM colonization. Gu et al. (2010) completed these analyses using a microarray-based approach in tomato. They identified a total of 14 miRNAs differentially expressed in mycorrhizal compared to nonmycorrhizal conditions. Seven of these microRNAs were upregulated in both Pi sufficient level or AM symbiosis (miR158, miR169g*, miR172, miR172b*, miR319, miR771, and miR775), while two were downregulated (miR319 and miR394). Four were specifically upregulated by AM symbiosis and their expression pattern was similar in Pi sufficient or deficient conditions (miR395, miR779.1, miR840, and miR867). Some of these miRNAs have been identified as regulators of sulphate and water transport, like miR395 (Kawashima et al. 2009) and miR840 (Gu et al. 2010), respectively. Interestingly, Gu et al. (2010) observed that miR837–3p was completely suppressed by AMF inoculation in leaves. In the MTGI10 (*M. truncatula* gene index 10) database regrouping nonredundant ESTs and consensus genes of *M. truncatula*, the putative miR837–3p targeted transcript (TC143374) is similar to a pathogenesis-related transcriptional factor and ERF1. These data suggest a role of systemic defence of the miR837–3p that permits the accumulation of the TC143374 in tomato during AMF colonization.

The development of powerful genomic approaches in the world of small-RNAs allowed defining global miRNA expression patterns in plants (Lu et al. 2005, 2007). Recently (Denvers et al. 2011), an analysis of the miRNAome in AM symbiotic condition has been performed in parallel to the analysis by deep sequencing of the overall RNA degradation (degradome). This approach has allowed comparison of miRNA populations and the degradation of the corresponding potential targets between AM and mock condition in the *M. truncatula* model. As expected, many miRNAs are involved in AM symbiosis, targeting hormonal, architectural, and signaling root processes. These authors propose that mycorrhizal symbiosis "leads to a reprogramming of the miRNA target network in roots, including miRNA strand preference." They also showed that many miRNA-targeted genes were relevant to this symbiosis, like proteins involved in cellular phosphate homeostasis. The nutrient uptake network is not the only regulatory system that overlaps with the mechanisms implemented by AM symbiosis. As mentioned before in this chapter, AM fungi alter the root architecture of the host plant at both presymbiotic and symbiotic stages. The *in planta* mechanisms required for these alterations lead us to wonder whether some miRNAs involved in AM symbiosis can also regulate root architecture. The incidence of microRNA-mediated regulation on symbiosis and root development has already been demonstrated. MicroRNA166 for instance has a role on nodule development and also in LR formation (Boualem et al. 2008). Devers et al. (2011) found that miR166 is differentially regulated during AM symbiosis, suggesting that miR166 plays a role in this symbiosis, probably by regulating the LR formation. In parallel, Devers et al. (2011) also found microRNAs involved in root architecture and development that are differentially expressed in the mycorrhizal condition. A set of six microRNAs is common to the two processes (Figure 9.2). These microRNAs target transcription factors (miR169 for HAP2 and miR166 for HD-ZipIII), auxin response factor (miR160 for ARF 6, 8, miR167 for 10, 16, 17), and nutrient metabolism genes (miR 395 for SULTR2; APS1,4; miR398 for CSD1,2). Some of these microRNAs also overlap with mineral nutrition signaling, like miR169 (P), miR167 (N), miR395 (P, S), and miR398 (Zn, P). In previous studies, microRNAs that regulate root development and architecture were identified and reviewed for *Arabidopsis* (dicotyledonous) and rice (monocotyledonous) (Meng et al. 2010), and also in a more global report (Khan et al. 2011).

As illustrated in Figure 9.2, the microRNA expression pattern defined during AM symbiosis largely overlapped with the specific miRNAome of root developmental processes, as does nutrient homeostasis or root development and architecture. Other metabolic pathways that are perturbed by AMF colonization, like strigolactone regulations, remain to be explored at the miRNA regulation level. To investigate the relationship between mycorrhization and root development, the fine description of presymbiotic AM symbiosis steps will bring the opportunity to test the involvement of microRNA regulation in these steps. Similarly, specific investigations in arbuscule-hosting cells and noncolonized cortical cells, using laser dissection, cell sorting, or reporter gene strategy, will define

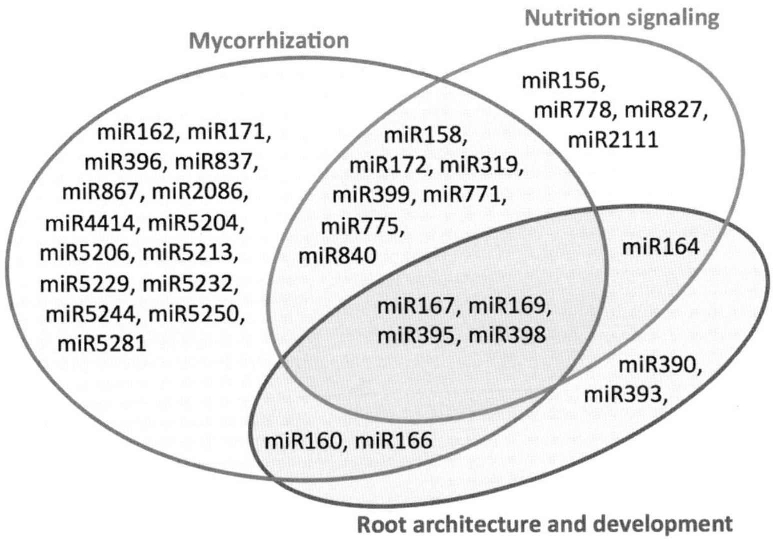

Figure 9.2 Venn diagram of the distribution and overlapping of microRNA families. The blue circle contains the microRNAs involved in nutrition signaling, the red circle contains the microRNAs involved in root architecture and development, and green circle contains the microRNAs involved in mycorrhization (see the text for references). (For a color version of this figure, see the color plate section.)

and identify locally expressed and repressed microRNAs. Finally, as informative as the global approach of microRNA identification is, a long road remains before the functional role of microRNAs in the fine-tuning of AM symbiosis is described.

As a conclusion, because of the ancestral nature of AM symbiosis interaction, it was suggested that root development of early land plants was accompanied by the acquisition of this symbiotic interaction (see Brundrett 2002). The evolutionary origin of these traits is often discussed by authors and it will be difficult to resolve whether AM fungi promoted root tissue differentiation as hypothesized by mycorrhizologists or hijacked mechanisms previously formed for root development. Whatever the evolutionary origin, in the same way as physiopathology teaches us about cell physiology by highlighting perturbed cell mechanisms, the study of AM symbiosis, dealing with universal mechanisms of plant development as revealed by the strigolactone discovery, will bring us new insights into root development.

Acknowledgments

D.F. is granted by Agronutrition S.A. (France) and C.J. by a CIRAD program. The authors thank Gaëtan Droc (Montpellier, CIRAD) for bioinformatics support in comparative transcriptomics and G. Bécard for their helpful comments on the manuscript.

References

Akiyama, K. et al. (2005) Plant sesquiterpenes induce hyphal branching in arbuscular mycorrhizal fungi. *Nature*, 435, 824–827.
Allen, E. et al. (2005) MicroRNA-directed phasing during trans-acting siRNA biogenesis in plants. *Cell*, 121, 207–221.
Ané, J-M. et al. (2004) *Medicago truncatula* DMI1 required for bacterial and fungal symbioses in legumes. *Science*, 303, 1364–1367.

Arite, T. et al. (2007) DWARF10, an RMS1/MAX4/DAD1 ortholog, controls lateral bud outgrowth in rice. *Plant Journal*, 51, 1019–1029.

Aung, K. et al. (2006) Pho2, a phosphate over-accumulator, is caused by a nonsense mutation in a MicroRNA399 target gene. *Plant Physiology*, 141, 1000–1011.

Balzergue, C. et al. (2011) The regulation of arbuscular mycorrhizal symbiosis by phosphate in pea involves early and systemic signalling events. *Journal of Experimental Botany*, 62, 1049–1060.

Bari, R. et al. (2006) PHO2, microRNA399, and PHR1 define a phosphate-signaling pathway in plants. *Plant Physiology*, 141, 988–999.

Bartel, D.P. (2004) MicroRNAs: genomics, biogenesis, mechanism, and function. *Cell*, 116, 281–297.

Bécard, G. and Fortin, J.A. (1988) Early events of vesicular–arbuscular mycorrhiza formation on Ri T-DNA transformed roots. *New Phytologist*, 108, 211–218.

Berta, G. et al. (1990) Morphogenetic modifications induced by the mycorrhizal fungus Glomus strain E3 in the root system of *Allium porrum* L. *New Phytologist*, 114, 207–215.

Berta, G. et al. (1995) Arbuscular mycorrhizal induced changes to plant growth and root system morphology in *Prunus cerasifera*. *Tree Physiology*, 15, 281–293.

Berta, G. et al. (2002) Arbuscular mycorrhizal modifications to plant root systems: scale, mechanisms and consequences. In: S. Gianinazzi, H. Schuepp, J.M. Barea and K. Haselwandter (eds) *Mycorrhizal Technology in Agriculture: from Genes to Bioproducts*. Birkhäuser Verlag, Basel-Boston-Berlin, pp. 71–85.

Besserer, A. et al. (2006) Strigolactones stimulate arbuscular mycorrhizal fungi by activating mitochondria. *PLoS Biology*, 4, e226.

Bonfante, P. and Genre, A. (2010) Mechanisms underlying beneficial plant-fungus interactions in mycorrhizal symbiosis. *Nature communications*, 1, 48.

Booker, J. et al. (2004) MAX3/CCD7 is a carotenoid cleavage dioxygenase required for the synthesis of a novel plant signaling molecule. *Current Biology*, 14, 1232–1238.

Boualem, A. et al. (2008) MicroRNA166 controls root and nodule development in *Medicago truncatula*. *Plant Journal*, 54, 876–887.

Branscheid, A. et al. (2010) Expression pattern suggests a role of mir399 in the regulation of the cellular response to local pi increase during arbuscular mycorrhizal symbiosis. *Molecular Plant–Microbe Interactions*, 23, 915–926.

Breuillin, F. et al. (2010) Phosphate systemically inhibits development of arbuscular mycorrhiza in *Petunia hybrida* and represses genes involved in mycorrhizal functioning. *Plant Journal*, 64, 1002–1017.

Brundrett, M.C. (2002) Coevolution of roots and mycorrhizas of land plants. *New Phytologist*, 154, 275–304.

Bustos, R. et al. (2010) A central regulatory system largely controls transcriptional activation and repression responses to phosphate starvation in Arabidopsis. *PLoS Genetics*, 6, e1001102.

Chabot, S. et al. (1992) Life cycle of *Glomus intraradix* in root organ culture. *Mycologia*, 84, 315–321.

Chiou, T.J. et al. (2006) Regulation of phosphate homeostasis by microRNA in Arabidopsis. *Plant Cell*, 18, 412–421.

Chiou, T.J. et al. (2001) The spatial expression patterns of a phosphate transporter (MtPT1) from *Medicago truncatula* indicate a role in phosphate transport at the root/soil interface. *Plant Journal*, 25, 281–293.

Christiansen, M.W. et al. (2011) Characterization of barley (*Hordeum vulgare*, L.) NAC transcription factors suggests conserved functions compared to both monocots and dicots. *BMC Research Notes*, 4, 302–314.

Combier, J.-P. et al. (2006) MtHAP2-1 is a key transcriptional regulator of symbiotic nodule development regulated by microRNA169 in *Medicago truncatula*. *Genes & Development*, 20, 3084–3088.

Corkidi, L. et al. (2004) Assessing the infectivity of commercial mycorrhizal inoculants in plant nursery conditions. *Journal of Environmental Horticulture*, 22, 149–154.

Cosgrove, D.J. (2001) Wall structure and wall loosening. A look backwards and forwards. *Plant Physiology*, 125, 131–134.

Cosgrove, D.J. (2005) Growth of the plant cell wall. *Nature Reviews Molecular Cell Biology*, 6, 850–861

Crespi, M. and Galvez, S. (2000) Molecular mechanisms in root nodule development. *Journal of Plant Growth Regulation*, 19, 155–166.

David-Schwartz, R. et al. (2001) Identification of a novel genetically controlled step in mycorrhizal colonization: plant resistance to infection by fungal spores but not extra-radical hyphae. *The Plant Journal*, 27, 561–569.

David-Schwartz, R. et al. (2003) Isolation of a premycorrhizal infection (*pmi2*) mutant of tomato, resistant to arbuscular mycorrhizal fungal colonization. *Molecular Plant-Microbe Interactions*, 16, 382–388.

DeMars, B.G. and Boerner, R.E.J. (1995) Arbuscular mycorrhizal development in three crucifers. *Mycorrhiza*, 5, 405–408.

Devers, E.A. et al. (2011) Stars and symbiosis: microRNA- and microRNA*-mediated transcript cleavage involved in arbuscular mycorrhizal symbiosis. *Plant Physiology*, 156, 1990–2010.

Dolan, L. (2007) Plant science. SCARECROWs at the border. *Science*, 316, 377–378.

Drummond, R.S.M. et al. (2009) *Petunia hybrida* CAROTENOID CLEAVAGE DIOXYGENASE7 is involved in the production of negative and positive branching signals in petunia. *Plant Physiology*, 151, 1867–1877.

Endre, G. et al. (2002) A receptor kinase gene regulating symbiotic nodule development. *Nature*, 417, 962–966.

Feddermann, N. et al. (2010) The PAM1 gene of petunia, required for intracellular accommodation and morphogenesis of arbuscular mycorrhizal fungi, encodes a homologue of VAPYRIN. *The Plant Journal*, 64, 470–481.

Felten, J. et al. (2009) The ectomycorrhizal fungus *Laccaria bicolor* stimulates lateral root formation in poplar and Arabidopsis through auxin transport and signaling. *Plant Physiology*, 151, 1991–2005.

Finlay, R.D. (2008) Ecological aspects of mycorrhizal symbiosis: with special emphasis on the functional diversity of interactions involving the extraradical mycelium. *Journal of Experimental Botany*, 59, 1115–1126.

Foo, E. et al. (2005) The branching gene RAMOSUS1 mediates interactions among two novel signals and auxin in pea. *The Plant Cell*, 17, 464–474.

Gaude, N. et al. (2012) Arbuscule-containing and non-colonized cortical cells of mycorrhizal roots undergo a massive and specific reprogramming during arbuscular mycorrhizal development. *The Plant Journal*, 69, 510–528.

Genre, A. et al. (2005) Arbuscular mycorrhizal fungi elicit a novel intracellular apparatus in *Medicago truncatula* root epidermal cells before infection. *The Plant Cell*, 17, 3489–3499.

Glassop, D. et al. (2005) Cereal phosphate transporters associated with the mycorrhizal pathway of phosphate uptake into roots. *Planta*, 222, 688–698.

Gomez, S.K. et al. (2009) *Medicago truncatula* and *Glomus intraradices* gene expression in cortical cells harboring arbuscules in the arbuscular mycorrhizal symbiosis. *BMC Plant Biology*, 9, 10.

Gomez-Roldan, V. et al. (2008) Strigolactone inhibition of shoot branching. *Nature*, 455, 189–194.

Groth, M. et al. (2010) NENA, a *Lotus japonicus* homolog of Sec13, is required for rhizodermal infection by arbuscular mycorrhiza fungi and rhizobia but dispensable for cortical endosymbiotic development. *The Plant Cell*, 22, 2509–2526.

Gu, M. et al. (2010) Expression analysis suggests potential roles of microRNAs for phosphate and arbuscular mycorrhizal signaling in *Solanum lycopersicum*. *Physiologia Plantarum*, 138, 226–237.

Guether, M. et al. (2009) Genome-wide reprogramming of regulatory networks, transport, cell wall and membrane biogenesis during arbuscular mycorrhizal symbiosis in *Lotus japonicas*. *The New Phytologist*, 182, 200–212.

Güimil, S. et al. (2005) Comparative transcriptomics of rice reveals an ancient pattern of response to microbial colonization. *Proceedings of the National Academy of Sciences USA*, 102, 8066–8070.

Gutjahr, C. et al. (2009) *Glomus intraradices* induces changes in root system architecture of rice independently of common symbiosis signaling. *The New Phytologist*, 182, 829–837.

Harrison, M.J. et al. (2002) A phosphate transporter from *Medicago truncatula* involved in the acquisiton of phosphate released by arbuscular mycorrhizal fungi. *The Plant Cell*, 14, 2413–2429.

Heckmann, A.B. et al. (2006) *Lotus japonicus* nodulation requires two GRAS domain regulators, one of which is functionally conserved in a non-legume. *Plant Physiologyi*, 142, 1739–1750.

Helber, N. and Requena, N. (2008) Expression of the fluorescence markers DsRed and GFP fused to a nuclear localization signal in the arbuscular mycorrhizal fungus *Glomus intraradices*. *The New Phytologist*, 177, 537–548.

Helber, N.K. et al. (2011) A versatile monosaccharide transporter that operates in the arbuscular mycorrhizal fungus *Glomus* sp is crucial for the symbiotic relationship with plants. *The Plant Cell*, 23, 3812–3823.

Helgason, T. et al. (2003) Phylogeny of the Glomerales and Diversisporales (fungi: Glomeromycota) from actin and elongation factor 1-alpha sequences. *FEMS Microbiol Letters*, 229, 127–132.

Hohnjec, N. et al. (2005) Overlaps in the transcriptional profiles of *Medicago truncatula* roots inoculated with two different glomus fungi provide insights into the genetic program activated during arbuscular mycorrhiza. *Plant Physiology*, 137, 1283–1301

Horner-Devine, M.C. et al. (2003) Bacterial diversity patterns along a gradient of primary productivity. *Ecology Letters*, 6, 613–622.

Imaizumi-Anraku, H. et al. (2005) Plastid proteins crucial for symbiotic fungal and bacterial entry into plant roots. *Nature*, 433, 527–531.

Ivashuta, S. et al. (2005) RNA interference identifies a calcium-dependent protein kinase involved in *Medicago truncatula* root development. *Plant Cell*, 17, 2911–2921.

James, T.Y. et al. (2006) Reconstructing the early evolution of fungi using a six-gene phylogeny. *Nature*, 443, 818–822.

Javot, H. et al. (2007) A *Medicago truncatula* phosphate transporter indispensable for the arbuscular mycorrhizal symbiosis. *Proceedings of the National Academy of Sciences of the United States of America*, 104, 1720–1725.

Jones-Rhoades, M.W. et al. (2006) MicroRNAs and their regulatory roles in plants. *Annual Review of Plant Biology*, 57, 19–53.

Journet, E.-P. et al. (2002) Exploring root symbiotic programs in the model legume *Medicago truncatula* using EST analysis. *Nucleic Acids Research*, 30, 5579–5592.

Kaló, P. et al. (2005) Nodulation signaling in legumes requires NSP2, a member of the GRAS family of transcriptional regulators. *Science*, 308, 1786–1789.

Kanamori, N. et al. (2006) A nucleoporin is required for induction of Ca^{2+} spiking in legume nodule development and essential for rhizobial and fungal symbiosis. *Proceedings of the National Academy of Sciences of the United States of America*, 103, 359–364.

Kapulnik, Y. et al. (2011) Strigolactones affect lateral root formation and root-hair elongation in Arabidopsis. *Planta*, 233, 209–216.
Kawashima, C.G. et al. (2009) Sulphur starvation induces the expression of microRNA-395 and one of its target genes but in different cell types. *Plant Journal*, 57, 313–321.
Khan, G.A. et al. (2011) MicroRNAs as regulators of root development and architecture. *Plant Molecular Biology*, 77, 47–58.
Kiers, E.T. et al. (2011) Reciprocal rewards stabilize cooperation in the mycorrhizal symbiosis. *Science*, 333, 880–882
Klironomos, J. et al. (2011) Forces that structure plant communities: quantifying the importance of the mycorrhizal symbiosis. *The New Phytologist*, 189, 366–370
Kosuta, S. et al. (2003) A diffusible factor from arbuscular mycorrhizal fungi induces symbiosis-specific MtENOD11 expression in roots of *Medicago truncatula*. *Plant Physiology*, 131, 952–962.
Krajinski, F. and Frenzel, A. (2007) Towards the elucidation of AM-specific transcription in *Medicago truncatula*. *Phytochemistry*, 68, 75–81.
Kuhn, H. et al. (2010) Membrane steroid-binding protein 1 induced by a diffusible fungal signal is critical for mycorrhization in *Medicago truncatula*. *The New Phytologist*, 185, 716–733.
Küster, H. et al. (2004) Construction and validation of cDNA-based Mt6k-RIT macro- and microarrays to explore root endosymbioses in the model legume *Medicago truncatula*. *Journal of Biotechnology*, 108, 95–113.
Leake, J.R. et al. (2004) Networks of power and influence: the role of mycorrhizal mycelium in controlling plant communities and agroecosystem functioning. *Canadian Journal of Botany*, 82, 1016–1045.
Lévy, J. et al. (2004) A putative Ca^{2+} and calmodulin-dependent protein kinase required for bacterial and fungal symbioses. *Science*, 303, 1361–1364.
Li, L. et al. (2003) OrthoMCL: identification of ortholog groups for eukaryotic genomes. *Genome Research*, 13, 2178–2189.
Liszkay, A. et al. (2004) Production of reactive oxygen intermediates (O(2)(·−), H(2)O(2), and (·)OH) by maize roots and their role in wall loosening and elongation growth. *Plant Physiology*, 136, 3114–3123.
Liu, Q. et al. (2009) Divergence in function and expression of the NOD26-like intrinsic proteins in plants. *BMC Genomics*, 10, 313–325.
Liu, Y. et al. (2009) Phylogenomic analyses support the monophyly of Taphrinomycotina, including Schizosaccharomyces fission yeasts. *Molecular Biology and Evolution*, 26, 27–34.
Liu, Y.J. et al. (2006) Loss of the flagellum happened only once in the fungal lineage: phylogenetic structure of kingdom Fungi inferred from RNA polymerase II subunit genes. *BMC Evolutionnary Biology*, 6, 74.
Liu, W. et al. (2011) Strigolactone biosynthesis in *Medicago truncatula* and rice requires the symbiotic GRAS-type transcription factors NSP1 and NSP2. *Plant Cell*, 23, 3853–3865.
López-Ráez, J.A. et al. (2010) Hormonal and transcriptional profiles highlight common and differential host responses to arbuscular mycorrhizal fungi and the regulation of the oxylipin pathway. *Journal of Experimental Botany*, 61, 2589–2601.
López-Ráez, J.A. et al. (2011) Arbuscular mycorrhizal symbiosis decreases strigolactone production in tomato. *Journal of Plant Physiology*, 168, 294–297.
Loth-Pereda, V. et al. (2011) Structure and expression profile of the phosphate Pht1 transporter gene family in mycorrhizal *Populus trichocarpa*. *Plant Physiology*, 156, 2141–2154.
Lu, C. et al. (2005) Elucidation of the small RNA component of the transcriptome. *Science*, 309, 1567–1569.
Lu, C. et al. (2006) MicroRNAs and other small RNAs enriched in the Arabidopsis RNA-dependent RNA polymerase-2 mutant. *Genome Research*, 16, 1276–1288.
Lu, C. et al. (2007) Construction of small RNA cDNA libraries for deep sequencing. *Methods*, 43, 110–117.
Maeda, D. et al. (2006) Knockdown of an arbuscular mycorrhiza-inducible phosphate transporter gene of *Lotus japonicus* suppresses mutualistic symbiosis. *Plant Cell Physiology*, 47, 807–817.
Maillet, F. et al. (2011) Fungal lipochitooligosaccharide symbiotic signals in arbuscular mycorrhiza. *Nature*, 469, 58–63.
Mallory, A.C. and Vaucheret, H. (2006) Functions of microRNAs and related small RNAs in plants. *Nature Genetics*, 38, S31–S36.
Marsh, J. F. and Schultze, M. (2001) Analysis of arbuscular mycorrhizas using symbiosis-defective plant mutants. *New Phytologist*, 150, 525–532.
Martin, F. et al. (2008) The long hard road to a completed *Glomus intraradices* genome. *New Phytologist*, 180, 747–750.
Matusova, R. et al. (2005) The strigolactone germination stimulants of the plant-parasitic Striga and Orobanche spp. are derived from the carotenoid pathway. *Plant Physiology*, 139, 920–934.
Meng, Y. et al. (2010) MicroRNA-mediated signaling involved in plant root development. *Biochemical and Biophysical Research Communications*, 393, 345–349.
Messinese, E. et al. (2007) A novel nuclear protein interacts with the symbiotic DMI3 calcium- and calmodulin-dependent protein kinase of *Medicago truncatula*. *Molecular Plant–Microbe Interactions*, 20, 912–921.
Moreau, S. et al. (2011) Transcription reprogramming during root nodule development in *Medicago truncatula*. *PLoS One*, 6, e16463.

Morris, S.E. et al. (2001) Mutational analysis of branching in pea. Evidence that Rms1 and Rms5 regulate the same novel signal. *Plant Physiology*, 126, 1205–1213.

Mukherjee, A. and Ané, J-M. (2011) Germinating spore exudates from arbuscular mycorrhizal fungi: molecular and developmental responses in plants and their regulation by ethylene. *Molecular Plant–Microbe Interactions*, 24, 260–270.

Nacry, P. et al. (2005) A role for auxin redistribution in the responses of the root system architecture to phosphate starvation in *Arabidopsis*. *Plant Physiology*, 138, 2061–2074.

Nagy, R. et al. (2005) The characterization of novel mycorrhiza-specific phosphate transporters from *Lycopersicon esculentum* and *Solanum tuberosum* uncovers functional redundancy in symbiotic phosphate transport in solanaceous species. *The Plant Journal*, 42, 236–250.

Navarro, L. et al. (2006) A plant miRNA contributes to antibacterial resistance by repressing auxin signaling. *Science*, 312, 436–439.

Olah, B. et al. (2005) Nod factors and a diffusible factor from arbuscular mycorrhizal fungi stimulate lateral root formation in *Medicago truncatula* via the DMI1/DMI2 signalling pathway. *Plant Journal*, 44, 195–207.

Oldroyd, G. and Downie, A. (2006) Nuclear calcium changes at the core of symbiosis signaling. *Current Opinion in Plant Biology*, 9, 351–357.

Oldroyd, G.E.D. and Long, S.R. (2003) Identification and characterization of nodulation-signaling pathway 2, a gene of *Medicago truncatula* involved in Nod factor signaling. *Plant Physiology*, 131, 1027–1032.

Oliveros, J.C. (2007) VENNY. An interactive tool for comparing lists with Venn diagrams. http://bioinfogp.cnb.csic.es/tools/venny/index.html.

Osmont, K.S. et al. (2007) Hidden branches: developments in root system architecture. *Annual Review in Plant Biology*, 58, 93–113.

Ozsolak, F. and Milos, P.M. (2011) RNA sequencing: advances, challenges and opportunities. *Nature Review Genetics*, 12, 87–98.

Parniske, M. (2008) Arbuscular mycorrhiza: the mother of plant root endosymbioses. *Nature Review Microbiology*, 26, 763–775.

Paszkowski, U. and Boller, T. (2002) The growth defect of lrt1, a maize mutant lacking lateral roots, can be complemented by symbiotic fungi or high phosphate nutrition. *Planta*, 214, 584–590.

Paszkowski, U. et al. (2002) Rice phosphate transporters include an evolutionarily divergent gene specifically activated in arbuscular mycorrhizal symbiosis. *Proceedings of the National Academy of Sciences of the United States of America*, 99, 13324–13329.

Pearson, J.N. and Jakobsen, I. (1993) The relative contribution of hyphae and roots to phosphorus uptake by arbuscular mycorrhizal plants measured by dual labelling with 32P and 33P. *The New Phytologist*, 124, 489–494.

Paszkowski, U. et al. (2006) Maize mutants affected at distinct stages of the arbuscular mycorrhizal symbiosis. *Plant Journal*, 47, 165–173.

Perez-Torres, C.-A. et al. (2008) Phosphate availability alters lateral root development in Arabidopsis by modulating auxin sensitivity via a mechanism involving the TIR1 auxin receptor. *Plant Cell*, 20, 3258–3272.

Popper, Z.A. et al. (2011) Evolution and diversity of plant cell walls: from algae to flowering plants. *Annual Review of Plant Biology*, 62, 567–590.

Pumplin, N. et al. (2010) *Medicago truncatula* Vapyrin is a novel protein required for arbuscular mycorrhizal symbiosis. *The Plant Journal*, 61, 482–494.

Ranocha, P. et al. (2010) Walls are thin 1 (WAT1), an Arabidopsis homolog of *Medicago truncatula* NODULIN21, is a tonoplast-localized protein required for secondary wall formation in fibers. *Plant Journal*, 63, 469–483.

Rausch, C. et al. (2001) A phosphate transporter expressed in arbuscule-containing cells in potato. *Nature*, 414, 462–466.

Redecker, D. and Raab, P. (2006) Phylogeny of the glomeromycota (arbuscular mycorrhizal fungi): recent developments and new gene markers. *Mycologia*, 98, 885–895.

Reddy, D.M.R.S. et al. (2007) A petunia mutant affected in intracellular accommodation and morphogenesis of arbuscular mycorrhizal fungi. *The Plant Journal*, 51, 739–750.

Redecker, D. et al. (2000) Glomalean fungi from the Ordovician. *Science*, 289, 1920–1921.

Remy, W. et al. (1994) Four hundred-million-year-old vesicular arbuscular mycorrhizae. *Proceedings of the National Academy of Sciences of the United States of America*, 91, 11841–11843.

Rubio, V. et al. (2001) A conserved MYB transcription factor involved in phosphate starvation signaling both in vascular plants and in unicellular algae. *Genes & Development*, 15, 2122–2133.

Ruiz-Ferrer, V. and Voinnet, O. (2009) Roles of plant small RNAs in biotic stress responses. *Annual Review of Plant Biology*, 60, 485–510.

Ruyter-Spira, C. et al. (2011) Physiological effects of the synthetic strigolactone analog GR24 on root system architecture in Arabidopsis: another belowground role for strigolactones? *Plant Physiology*, 155, 721–734.

Saito, K. et al. (2007) NUCLEOPORIN85 is required for calcium spiking, fungal and bacterial symbioses, and seed production in *Lotus japonicus*. *The Plant Cell*, 19, 610–624.

Schnitzer, S.A. and Klironomos, J. (2011) Soil microbes regulate ecosystem productivity and maintain species diversity. *Plant Signaling and Behaviour*, 6, 1240–1243.

Schüßler, A. et al. (2001) A new fungal phylum, the Glomeromycota: phylogeny and evolution. *Mycological Research*, 105, 1413–1421.

Schwarzott, D. et al. (2001) *Glomus*, the largest genus of the arbuscular mycorrhizal fungi (Glomales), is nonmonophyletic. *Molecular Phylogenetics and Evolution*, 21, 190–197.

Shin, R. et al. (2007) The Arabidopsis transcription factor MYB77 modulates auxin signal transduction. *Plant Cell*, 19, 2440–2453

Siciliano, V. et al. (2007) Transcriptome analysis of arbuscular mycorrhizal roots during development of the prepenetration apparatus. *Plant Physiology*, 144, 1455–1466.

Simon, L.J. et al. (1993) Origin and diversification of endomycorrhizal fungi and coincidence with vascular land plants. *Nature*, 362, 67–69.

Smith, S.E. and Read, D.J. (2008) *Mycorrhizal Symbiosis*, Third edition, Vol. 2. Academic Press, London, pp. 525–611.

Snowden, K.C. et al. (2005) The secreased apical dominance1/*Petunia hybrida* CAROTENOID CLEAVAGE DIOXYGENASE8 gene affects branch production and plays a role in leaf senescence, root growth, and flower development. *The Plant Cell*, 17, 746–759.

Sorefan, K. et al. (2003) MAX4 and RMS1 are orthologous dioxygenase-like genes that regulate shoot branching in Arabidopsis and pea. *Genes & Development*, 17, 1469–1474.

Splivallo, R. et al. (2009) Truffles regulate plant root morphogenesis via the production of auxin and ethylene. *Plant Physiology*, 150, 2018–2029.

Stracke, S. et al. (2002) A plant receptor-like kinase required for both bacterial and fungal symbiosis. *Nature*, 417, 959–962.

Subramanian, S. et al. (2008) Novel and nodulation-regulated microRNAs in soybean roots. *BMC Genomics*, 10, 160.

Takeda, N. et al. (2009) Apoplastic plant subtilases support arbuscular mycorrhiza development in *Lotus japonicus*. *Plant Journal*, 58, 766–777.

Tirichine, L. et al. (2006) Deregulation of a Ca^{2+}/calmodulin-dependent kinase leads to spontaneous nodule development. *Nature*, 441, 1153–1156.

Umehara, M. A. et al. (2008) Inhibition of shoot branching by new terpenoid plant hormones. *Nature*, 455, 195–200.

van der Heijden, M.G.A. et al. (1998) Mycorrhizal fungal diversity determines plant biodiversity, ecosystem variability and productivity. *Nature*, 396, 69–72

van der Heijden, M.G.A. et al. (2008) The unseen majority: soil microbes as drivers of plant diversity and productivity in terrestrial ecosystems. *Ecology Letters*, 11, 296–310.

van Rhijn, P. et al. (1997) Expression of early nodulin genes in alfalfa mycorrhizae indicates that signal transduction pathways used in forming arbuscular mycorrhizae and Rhizobium-induced nodules may be conserved. *Proceedings of the National Academy of Sciences of the United States of America*, 94, 5467–5472.

Walter, M.H. et al. (2000) Arbuscular mycorrhizal fungi induce the non-mevalonate methylerythritol phosphate pathway of isoprenoid biosynthesis correlated with accumulation of the "yellow pigment" and other apocarotenoids. *The Plant Journal*, 21, 571–578.

Weidmann, S. et al. (2004) Fungal elicitation of signal transduction-related plant genes precedes mycorrhizal establishment and requires the dmi3 gene in *Medicago truncatula*. *Molecular Plant-Microbe Interactions*, 17, 1385–1393.

Yano, K. et al. (1996) Localized alteration in lateral root development in roots colonized by an arbuscular mycorrhizal fungus. *Mycorrhiza*, 6, 409–415.

Yano, K. et al. (2008) CYCLOPS, a mediator of symbiotic intracellular accommodation. *Proceedings of the National Academy of Sciences*, 105, 20540–20545.

Yokoyama, R. and Nishitani, K. (2004) Genomic basis for cell-wall diversity in plants. A comparative approach to gene families in rice and Arabidopsis. *Plant Cell Physiology*, 45, 1111–1121.

Yoneyama, K. et al. (2007) Phosphorus deficiency in red clover promotes exudation of orobanchol, the signal for mycorrhizal symbionts and germination stimulant for root parasites. *Planta*, 225, 1031–1038.

Zhang, Q. et al. (2010) Two *Medicago truncatula* half-ABC transporters are essential for arbuscule development in arbuscular mycorrhizal symbiosis. *Plant Cell*, 22, 1483–1497.

Zhong, R. et al. (2008) A battery of transcription factors involved in the regulation of secondary cell wall biosynthesis in Arabidopsis. *Plant Cell*, 20, 2763–2782.

10 How Pathogens Affect Root Structure

Michaël Quentin, Tarek Hewezi, Isabelle Damiani, Pierre Abad, Thomas Baum, and Bruno Favery

Introduction

Several groups of soil microorganisms cause diseases to the roots. These "soilborne" plant pathogens are particularly challenging since they often survive in soil for many years and each crop may be susceptible to several species. They belong to various systematic groups. Major groups are oomycetes, fungi, bacteria, protists, and nematodes. Few soilborne viruses affect vegetable crops. The impact of these pathogens on the growth of plants can be massive. They may lead to the complete destruction of the roots and subsequently of the plants. Soilborne pathogens cause three major types of disease: (i) damping off caused by oomycetes such as *Phytophthora*, *Pythium*, and by fungi such as *Rhizoctonia* species, when seeds or seedlings are killed before or after germination; (ii) vascular wilt when the pathogens (e.g., *Fusarium* or soilborne *Ralstonia* bacteria) penetrate the roots and infect the xylem vessels, which become plugged; and (iii) root and crown rots caused by species of *Phytophthora*, *Rhizoctonia*, *Fusarium*, *Gaeumannomyces*, and *Sclerotinia* (Katan 1996). In addition, plant-parasitic nematodes and biotrophic protists (*Plasmodiophora*) are obligate parasites able to induce dramatic morphological and physiological changes in host roots. Infected roots undergo a developmental switch that results in the formation of aberrant root structures (clubs or root-knots) (Figure 10.1). The most intriguing pathogens in this group are sedentary endoparasitic nematodes. Two of the most economically damaging groups are the root-knot nematodes (RKN, *Meloidogyne* spp.) and the cyst nematodes (CN). CN are mainly represented by the two genera *Globodera* and *Heterodera*. These microscopic roundworms invade the roots and engage in sophisticated interactions with their host plants inducing the formation of permanent feeding sites. These nematodes only feed after they have reached their destination in the host roots and have become sedentary by selecting a feeding site. These feeding sites consist of cells within the vascular cylinder, which become completely reorganized, hypertrophied, and metabolically highly active and serve as food sources throughout the nematode life cycle. It remains unclear how these pathogens cause such root alterations, but increasing evidence showed that nematode parasitism proteins, secreted from the esophageal gland cells through a hollow protrusive stylet, have direct effects on plant metabolic and developmental pathways to reprogram cells for feeding cell ontogenesis (Davis et al. 2008; Bird et al. 2009; Gheysen and Mitchum 2011).

This chapter reviews biology and molecular insights into mechanisms and plant developmental pathways manipulated by RKN, CN, and *Plasmodiophora brassicae*, elucidated in the last years

Root Genomics and Soil Interactions, First Edition. Edited by Martin Crespi.

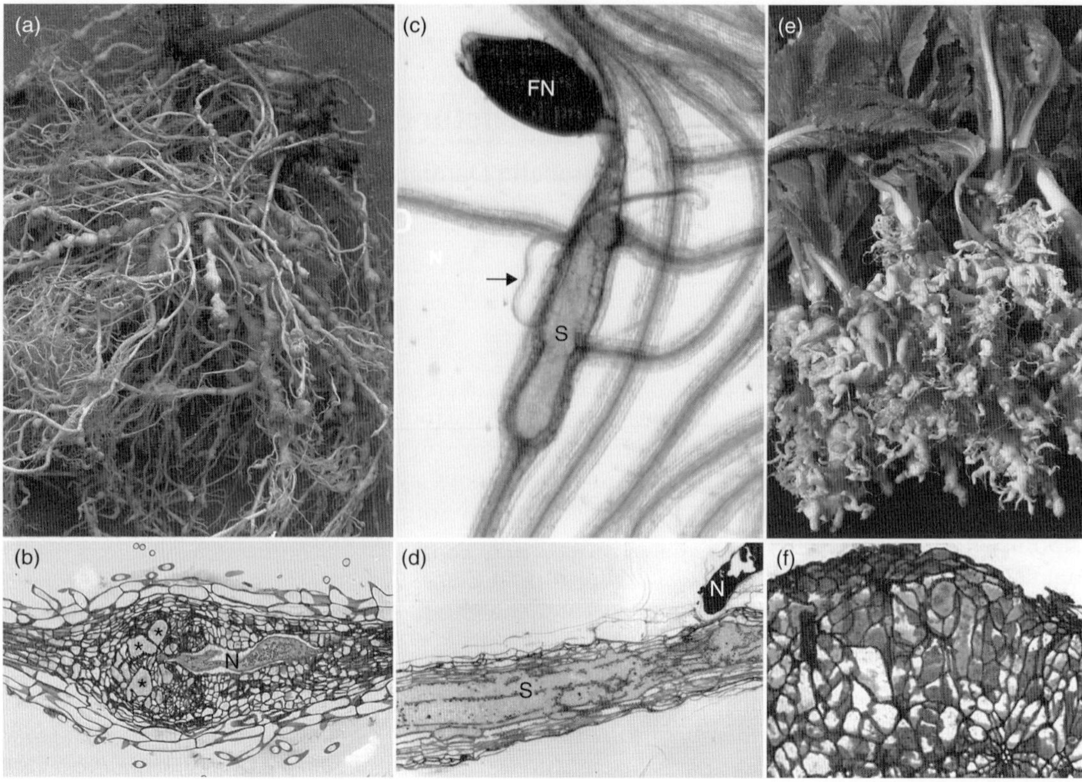

Figure 10.1 Root responses to RKN, CN, and *Plasmodiophora brassicae* infection. (a) Galls on tomato roots infected by *M. incognita*. (b) Multinucleate and hypertrophied giant-cells induced by *M. incognita* in *Arabidopsis* roots. Sections through a gall 15 days post infection. Asterisks, giant cells; N, nematode. (c) Feeding site induced by the sugar beet cyst nematode *H. schachtii* in *Arabidopsis* roots. Shown is a sedentary adult female nematode (FN) feeding from the syncytium (S) and a nonfeeding male (indicated by arrow). (d) Section through the head of the nematode (N) and the syncytium (S) showing partial cell wall degradation and cell-to-cell fusions. (e) Illustration of severe clubroot symptoms on young *Brassica rapa* roots. (f) Section through *Arabidopsis* Col-0 root 21 days postinoculation with *P. brassicae*. Note cell proliferation and enlargement resulting in a disorganization of tissue structure. Bars = 100 μm.

by transcriptome profiling and functional analysis. Recent characterizations of nematode parasitism protein plant targets are also presented.

Root Infection and Feeding Cell Ontogenesis

The RKN are able to form root galls (Figure 10.1a) resulting from cell hypertrophy and hyperplasia in nearly all flowering plants (Abad and Williamson 2010). The wide host range is associated with a worldwide distribution in all temperate and tropical areas, making them among the most damaging crop pests (Trudgill and Blok 2001). These nematodes infect roots as microscopic vermiform second-stage juveniles (J2) hatched in the soil. J2s are attracted to the host root, penetrate the root close to the root apex, and migrate between cells to reach the root apex and then enter into the plant vascular cylinder. To further develop and molt into a pear-shaped female that will release hundreds of eggs on the root surface in a protective gelatinous matrix, J2 need to successfully establish and maintain specialized feeding sites within the root. RKN induce the redifferentiation of five to seven selected pro-vascular root cells into elaborate feeding structures called "giant-cells" (GC; Figure 10.1b) from which they withdraw water and nutrients, allowing their sedentary biotrophic lifestyle (Caillaud et al. 2008b; Abad et al. 2009). These cells undergo successive nuclear division (karyokinesis) without complete cytokinesis, leading to the formation of multinucleate cells that can contain up to 100 nuclei (Jones and Payne 1978; Caillaud et al. 2008c). Meanwhile, GC expands tremendously and can reach 400 times the size of a normal parenchyma root cell. From 3 days postinfection (dpi), cell walls (CWs) thicken extensively. CW ingrowths typical of transfer cells develop to increase cell membrane surface and facilitate solute flow from vascular cells to GC, thus allowing nutrient acquisition by the parasitic nematode (Jones and Dropkin 1976). Once mature, the highly metabolically active multinucleate and hypertrophied GC present typical features such as dense granular cytoplasm filled with numerous organelles and small vacuoles, and high osmotic potential (Jones and Payne 1978; Jones 1981). Concomitant to feeding cell ontogenesis, a proliferation and expansion of cells surrounding the GC is observed, leading to the formation of a gall, the visible disease symptom (Figure 10.1a).

The CN are also very specialized and successful plant-parasitic pests in agriculture. However, their host range is usually very narrow. Root modifications are also a hallmark of infections by CN (Figure 10.1c) (Sobczak and Golinowski 2011). Macroscopically, the effects of CN attack appear less severe than those of the RKN or clubroot disease at the first glance. However, roots frequently are stunted and, more interestingly, show increased branching with lateral roots. The latter is most dramatically illustrated by the root "beard" of sugar beets infected by the sugar beet CN *Heterodera schachtii*. However, the whole extend of root changes becomes evident when observing infected roots microscopically or molecularly. At the heart of the CN infection lays the feeding site. Preparasitic J2s penetrate the root system of their hosts and migrate intracellularly through root cortical cells until they reach the differentiating vascular cylinder where they induce the formation of permanent feeding sites, so-called syncytia. In contrast to GC, and as the name implies, syncytia are formed by the fusion of independent cells. CN initiate contact with the initial feeding cell by inserting its stylet through the CW and invaginate the plasmalemma. Following this nematode–plant cell contact, dramatic cellular changes are set in motion at whose end the formation of the syncytium as a novel plant root organ is accomplished. In the course of these changes, the initial feeding cell loses its central vacuole and replaces it with cytoplasm densely filled with organelles. Early on in the interaction, the cell's nucleus undergoes changes by reentering a shunted version of the cell cycle. While chromosome complements are replicated, the nucleus does not divide but sets out on a course of repeated endoreduplication cycles. Most impressively, these changes are replicated in neighboring cells, which ultimately fuse with each other and the initial feeding cell through CW dissolution to form the continuous cytoplasm of fused and modified root cells that make up the syncytium (Figure 10.1d). Up to several hundred cells are modified and incorporated into this structure, and together with modified cells surrounding the syncytium proper, these modifications literally result in the production of a novel plant organ (Grundler et al. 1998, Sobczak and Golinowski 2011).

Impressive impact on root structure is also caused by *P. brassicae* (Figure 10.1e). Its disease symptom is called clubroot as distortions give the shape of a club or spindle to the root. These gall-like structures represent strong sinks for assimilates from which the pathogens are completely dependent for feeding. Clubroot is a common disease of cruciferous plants, although *P. brassicae* can also infect noncruciferous plants in Gramineae, Rosaceae, Papaveraceae, Polygonaceae, Resedaceae, and Leguminosae (Ludwig-Müller and Schuller 2008). *P. brassicae* is an obligate parasite within the class Phytomyxea of the protist subgroup Rhizaria and is characterized by a complex life cycle (Kageyama and Asano 2009). Motile primary zoospores released from germinated resting spores, the survival stage in the soil, reach the root hairs. They penetrate the CW through a specialized device called "Stachel" and "Rohr" that injects their cytoplasm into the host during the primary phase of the infection, which is restricted to root hairs (Ludwig-Müller 2009). The parasite nuclei divide to produce a multinucleate primary plasmodium, which then cleaves to release the secondary zoospores into the soil. These zoospores enter a secondary infection cycle of the hypocotyls and the root cortex before migrating in direction of the central stele. Again, multinucleate plasmodia are formed. These phenomena are accompanied by plant cell nuclear divisions followed by cell elongation, leading to abnormal tissue proliferation forming the gall symptom (Figure 10.1f). When resting spores develop, the host tissue undergoes senescence and the resting spores are released into the soil.

Genome-Wide Analysis of the Plant Response to Infection

To unravel the plant molecular responses involved in the RKN or CN infection process, different molecular approaches have been undertaken, ranging from cDNA subtraction or differential display, promoter–reporter gene constructs, reverse-transcription polymerase chain reaction, mRNA *in situ* hybridization, to promoter-trap strategies (Gheysen and Fenoll 2002; Caillaud et al. 2008b; Abad et al. 2009).

Comprehension of the host gene expression patterns benefited from the development of plant microarrays allowing acquisition of large-scaled transcriptomic data. These studies were conducted during compatible interactions between *Meloidogyne* spp. and susceptible *Arabidopsis thaliana* (Hammes et al. 2005; Jammes et al. 2005; Fuller et al. 2007; Barcala et al. 2010), tomato (*Solanum lycopersicum*) (Bar-Or et al. 2005; Schaff et al. 2007; Bhattarai et al. 2008; Fosu-Nyarko et al. 2009; Portillo et al. 2009), cowpea (*Vigna unguiculata*) (Das et al. 2010), and soybean (*Glycine max*) (Ibrahim et al. 2011). Additional data were obtained during incompatible interaction between the RKN and resistant tomato (Schaff et al. 2007; Bhattarai et al. 2008) or cowpea (Das et al. 2010). Some of these genome-wide scaled studies were done using whole infected roots (Hammes et al. 2005; Ibrahim et al. 2011). Further manual gall dissection allowed sample enrichment in GC transcripts (Bar-Or et al. 2005; Jammes et al. 2005; Fuller et al. 2007; Das et al. 2010). From these experiments, it remained, however, difficult to identify genes specifically expressed within the feeding cells. Galls are indeed multicellular structures composed of different tissues and cell types. A first attempt on cytoplasmic content microaspiration revealed GC-specific gene regulation (Wang et al. 2003). Ongoing technical advances in laser-assisted microdissection makes it a powerful tool to explore the complexity of plant–microbe interactions (Ramsay et al. 2006), and it has been applied to decipher mechanisms involved in GC ontogenesis more accurately (Ramsay et al. 2004; Fosu-Nyarko et al. 2009; Portillo et al. 2009; Barcala et al. 2010; Ithal and Mitchum 2011). These studies confirmed the molecular distinctiveness of the GC within the gall. A remarkable finding is that nascent GC are characterized by a main gene repression and that the majority of the differentially regulated genes have not, until now, been attributed a function (Barcala et al. 2010). To date, little attention has been given to GC-surrounding cells, and we do not know if they could contribute to the signaling leading to GC ontogenesis or to nutrients flow, but regulation of some genes, such as the *Medicago truncatula ENOD40,* has been reported to be restricted to these surrounding cells. It is obvious that laser-assisted microdissection approaches could also provide information on these particular tissue specificities and functions.

Similar approaches have been applied, allowing a global view on changes in gene expression in CN feeding sites. Experiments were done on whole roots or dissected root segments of *Arabidopsis* plants infected with *Heterodera schachtii* (Puthoff et al. 2003) and soybean infected with *Heterodera glycines* (Alkharouf et al. 2006; Ithal et al. 2007a; Klink et al. 2007b; Puthoff et al. 2007). Laser capture microdissection has also been successfully applied to study the transcriptome of syncytia induced in soybean roots by *H. glycines* (Alkharouf et al. 2004; Ithal et al. 2007b; Klink et al. 2007a). A recent paper described the transcriptome of microaspirated syncytia induced by *H. schachtii* on *Arabidopsis* roots (Szakasits et al. 2009). Furthermore, extensive microarray studies were conducted to analyze the soybean transcriptome during incompatible interactions (Klink et al. 2007b, 2009, 2010; Kandoth et al. 2011). Altogether, available data highlight that the development of the nematode-induced feeding structures is associated with massive changes in the expression of genes linked to fundamental aspects of plant development such as cell cycle regulation, cytoskeleton organization, hormone response, CW remodeling, transport or regulation of gene transcription, and plant defense (Gheysen and Fenoll 2002; Caillaud et al. 2008b; Gheysen and Mitchum 2009).

Genomic studies on plant responses to *P. brassicae* remain scarce. To study pathogen-induced changes in host physiology during club formation, proteomic approaches have been used (Devos et al. 2006; Cao et al. 2008), and, to date, only one microarray analysis, using the *Arabidopsis* full-genome Affymetrix chip, has been reported (Siemens et al. 2006). Data show differential expression of genes related to cell cycle, sugar metabolism, and defense. A major trend emerging from these studies is the transcriptional regulation of genes involved in the plant growth regulator homeostasis, particularly auxin and cytokinin (Ludwig-Müller et al. 2009).

The Plant Cytoskeleton Is Targeted by Root Pathogens

The plant cytoskeleton is a highly dynamic and flexible intracellular scaffold composed mainly of microtubules (MTs) and actin filaments. It plays a central role in intracellular transport, cell division, cell differentiation, and morphogenesis. Its dynamics are controlled by diverse actin-binding proteins (ABPs), microtubule-associated proteins (MAPs), and their regulating kinases and phosphatases. The manipulation of the plant cytoskeleton is a main step in nematode feeding site formation and is necessary for successful completion of the nematode life cycle. Chemical blocking (stabilization) or depolymerization of the actin or MT cytoskeleton influences nematode feeding site formation and consequently nematode development (de Almeida Engler et al. 2004; de Almeida Engler and Favery 2011). During compatible plant–nematode interactions, the actin network suffers structural changes (de Almeida Engler et al. 2004; de Almeida Engler et al. 2010; de Almeida Engler and Favery 2011), and the MT network is also rearranged for nematode feeding site ontogenesis (de Almeida Engler et al. 2004; Caillaud et al. 2008a; de Almeida Engler and Favery 2011), as revealed by histochemical analysis and the use of fluorescent markers. But cytoskeleton remodelings were different between RKN and CN feeding sites: while MTs and actin microfilaments were disrupted in syncytia, in GC a functional mitotic apparatus and disorganized cortical MTs and actin filaments were still present. These cytoskeleton changes may be triggered by ABPs and MAPs, whose transcript abundance appears to vary following nematode infection. Indeed, upregulation of membrane-anchored actin-nucleating formins AtFH1, AtFH6, and AtFH10, that may participate in actin cytoskeleton remodeling, was observed in developing feeding sites in *Arabidopsis* (Favery et al. 2004; Jammes et al. 2005; Barcala et al. 2010). Upregulation of five of the seven *Arabidopsis* actin-depolymerizing factor (ADF) promoters also occurs in expanding GC and/or surrounding cells

(Fuller et al. 2007; Clément et al. 2009), and increased expression of ADF3 was observed in CN-induced syncytia (Fuller et al. 2007). Experimental evidence has validated the direct involvement of the *ADF2* gene in GC formation and RKN development in *Arabidopsis* (Clément et al. 2009). The *Arabidopsis* MAP65–3 is essential for the development of GC induced by *M. incognita* (Caillaud et al. 2008c). Detailed functional analyses of MAP65–3, which plays a key role in the organization of microtubule arrays during both mitosis (spindle morphogenesis) and cytokinesis (phragmoplast expansion) in dividing plant cells, showed that this protein was associated with mini cell plates formed between daughter nuclei during cytokinesis initiation in developing GC. In the absence of MAP65–3, GC started to develop but accumulation of mitosis defects (CW stubs and connected nuclei) during repeated events of nuclear division prevented the development of functional feeding cells. GC did not complete their differentiation, impairing the maturation of the infecting nematodes (Caillaud et al. 2008c).

To our knowledge, cytoskeleton rearrangements in gall cells following infection with *P. brassicae* are not documented. Strong downregulation of both α and β-tubulins was, however, observed following infection, suggesting MT rearrangements are associated with clubroot disease (Devos et al. 2006). And, the upregulation in inoculated *Brassica napus* of a profilin, an ABP having effects on the dynamics of actin microfilaments, suggest that the plant actin cytoskeleton may also be targeted by *P. brassicae* (Cao et al. 2008).

Root Pathogens Hijack Cell Cycle Regulators

As mentioned earlier, GC initiation implies cell cycle reentry of a group of selected provascular cells that will undergo successive mitotic events leading to multinucleated cells. In parallel, GC-surrounding parenchyma vascular cells and xylem elements will divide extensively, causing a root swelling leading to gall formation. The CN-induced feeding site formation also involves reentry of the cell cycle of neighboring cells, which are then incorporated into the syncytium through CW dissolution. The use of the cell cycle inhibitors hydroxyurea and oryzalin arrested nematode feeding site development (de Almeida Engler et al. 1999). Promoter-driven GUS expression and *in situ* hybridization in *Arabidopsis* confirmed the RKN and CN recruitment of cell cycle regulators, two mitotic cyclins (AtCYCA2;1 *and AtCYCB1;1),* and two cyclin-dependent kinases (AtCDKA;1 *and AtCDKB1;1*), at early stages of the interactions (Niebel et al. 1996; de Almeida Engler et al. 1999). The upregulation of mitotic cyclins of the A- and B-type by RKN, which are involved in S-phase progression and control G_2/M and mitotic transitions, respectively (Inzé and De Veylder 2006), and CDKs has been confirmed in different microarray analysis (Jammes et al. 2005, Barcala et al. 2010; Ibrahim et al. 2011). Recent results show that silencing of CDKA;1 reduced *Arabidopsis* susceptibility to both RKN and CN (Van de Cappelle et al. 2008). In plants, mitotic cyclins of the D-type control progression through the G1 and into the S phases (Inzé and De Veylder 2006). Transcriptome analysis revealed upregulation of the *Arabidopsis* cyclin D3;2 in hand-dissected galls (Jammes et al. 2005), and in tomato, the cyclin D3;2 encoding gene was shown to be the only mitotic cyclin induced specifically in microdissected GC (Ramsay et al. 2004). The *Arabidopsis* CKS2, a regulator of CDKs that controls cell division, is also expressed in both GC and syncytia (de Almeida Engler et al. 2011). Induction of several other positive cell cycle regulator genes such as the *Arabidopsis PROLIFERA*(Huang et al. 2003; Barcala et al. 2010), the tomato, and cowpea *PHI-1* (Fosu-Nyarko et al. 2009; Das et al. 2010), or soybean nuclear DBF2-related protein kinases (Ibrahim et al. 2011), have been reported during early stages of the interaction with RKN. Plant infection by *P. brassicae* leads to the neo-formation of a meristem-like tissue. *Arabidopsis* transcriptome analysis confirmed that cell cycle genes, including several cyclins, were upregulated during the earlier time points following *P. brassicae* infection (Siemens et al. 2006). And, the use of the *AtCYCB1;1::GUS* construct illustrated the local re-initiation of cell division in the root cortex from the first days after *P. brassicae* inoculation (Devos et al. 2006).

In both GC and syncytia, nuclei are enlarged and amoeboid, and DNA endoreduplication may occur to sustain the enhanced metabolic demands associated with the interaction (de Almeida Engler et al. 1999; Wildermuth 2010; de Almeida Engler et al. 2011). Strong accumulation of the mitotic inhibitor *CCS52a* transcripts has been reported in RKN-induced GC (Koltai et al. 2001; Favery et al. 2002). Preliminary experiments suggest a role for *CCS52a* in CN growth and development (Gheysen and Mitchum 2009). The *Arabidopsis CPR5* gene, which also regulates endoreduplication and cell division (Kirik et al. 2001), is upregulated in root-knots as well (Jammes et al. 2005). But to date, endoreduplication processes associated with plant–nematode interactions remain poorly understood.

Severe Cell Wall Remodeling Is Associated with Feeding Site Formation

The plant CW is a dynamic, highly organized extra-cytoplasmic matrix made of various polysaccharides (mainly cellulose, hemicelluloses, and pectin), structural proteins, and aromatic substances and is constantly remodeled and restructured during growth and development (Carpita and Gibeaut 1993). Extensive CW modifications are required to build up a GC and syncytia (Sobczak et al. 2011). In addition to CW changes mediated by the plant enzymes during GC or syncytium formation, RKN and CN themselves secrete a number of CW-modifying enzymes. For example, the nematodes encode an array of CW-modifying proteins that function in the host cell and host tissues during nematode penetration, migration, and the early stages of feeding site formation.

In order to sustain the remarkable cell expansion associated with the development of the hypertrophied GC, CW relaxation by CW-loosening enzymes is required. Whole transcriptome analysis confirmed that several classes of genes coding for CW loosening enzymes including expansins, endoglucanases, xyloglucan endotransglycosylases (XETs), pectate lyases, and polygalacturonases are activated early in response to infection (Jammes et al. 2005; Fosu-Nyarko et al. 2009; Barcala et al. 2010; Ibrahim et al. 2011). In *Arabidopsis*, α-expansins and pectate lyases account for the most strongly upregulated genes within the gall. The use of promoter fusions with the GUS reporter gene and *in situ* hybridization confirmed expression in GC of *Arabidopsis* and tobacco endoglucanases (Goellner et al. 2001; Mitchum et al. 2004; Wang et al. 2007), of an *Arabidopsis* pectin acetylesterase (Vercauteren et al. 2002), and a tomato expansin (Gal et al. 2006). Silencing expression of tomato *EXPA5* affected the ability of the nematode to complete its life cycle (Gal et al. 2006). Other characteristic features of GC are an important CW thickening and formation of CW ingrowths, meaning synthesis and deposition of new CW material are also necessary. Several genes encoding cellulose synthases, CW structural proteins, or other CW component synthesizing enzymes are also differentially regulated during plant–RKN interactions (Jammes et al. 2005, Ibrahim et al. 2011; Sobczak et al. 2011). As an example, the upregulation of the cross-linking structural extensin proteins has been well documented (Niebel et al. 1993; Van der Eycken et al. 1996).

Syncytium development is also associated with extensive CW modifications. The CN feeding structure forms by the dissolution of plant CW and membranes between adjacent cells, and CW expansion. In addition, syncytial CW thicken and CW ingrowths form, which implies CW synthesis (Sobczak et al. 2011). Transcriptome analysis has confirmed tremendous regulation in the expression of genes encoding CW-degrading enzymes in *Arabidopsis* and soybean infected with *Heterodera* spp. (Ithal et al. 2007b; Tucker et al. 2007, 2011; Szakasits et al. 2009). The crucial role played in syncytium development by the plant host expansins (Wieczorek et al. 2006; Wieczorek and Grundler 2006; Fudali et al. 2008) and endoglucanases (Karczmarek et al. 2008; Wieczorek et al. 2008) has been particularly well documented.

During *P. brassicae*-induced club formation, cell division is followed by an extension of the newly formed cells. *Arabidopsis* clubroot microarray analysis highlighted upregulation of genes encoding CW-loosening enzymes such as α-expansins (Siemens et al. 2006). The action in gall cells of the CW-loosening XETs on CW expansion has also been documented (Devos et al. 2005).

Phytohormones Regulating Development and Defense May Control Feeding Site Formation

Plant hormones are involved in all aspects of plant growth and developmental processes as well as responses to environmental biotic and abiotic stresses. Auxin is a key player of plant organogenesis through regulation of cell division, polarity and extension, and cell fate determination. Auxin content may be increased in GC and this hormone would play a critical role during feeding site formation. Nematodes themselves could secrete auxin within the host cell (De Meutter et al. 2005). Microarray analyses reveal activation of genes responsible for auxin homeostasis and auxin-responsive genes, while repressors of auxin responses are downregulated in *Arabidopsis*, supporting the role of auxin in the successful establishment of RKN (Barcala et al. 2010). Histochemically, activation of auxin-responsive promoter GH3 and DR5 could be confirmed in nascent feeding cells (Hutangura et al. 1999; Karczmarek et al. 2004). And induction in RKN feeding sites of *AUX1* and *AtAUX4/LAX3* encoding putative auxin transporters has been reported consistently (Mazarei et al. 2003; Hammes et al. 2005; Jammes et al. 2005). In addition, depletion in *Medicago truncatula* roots of flavonoids, which are auxin transport regulators playing a potential role in secondary root organogenesis, affected *M. javanica* feeding site morphology (Wasson et al. 2009). But to date, only one auxin mutant, the tomato *diageotropica,* has been reported to alter RKN parasitism (Richardson and Price 1984). Promoting cell division and regulating cell differentiation, cytokinins have been associated with important roles in shoot and root development. More particularly, cytokinins may be involved in cyclin induction and favor cell cycle reentry. Nematodes appear to secrete biologically active cytokinins (Bird and Loveys 1980; De Meutter et al. 2003), and increase in the general content in cytokinins in GC could be measured (Bird and Loveys 1980), suggesting these hormones could play crucial functions in GC formation. Cytokinin-responsive genes, such as *ARR5,* are activated in the nascent gall (Lohar et al. 2004). However, this gene is repressed in GC sensu stricto and in the mature gall. Thus, cytokinins can play a role either as stimulator or inhibitor of cell division. In addition, hairy roots overexpressing a cytokinin oxidase, responsible for cytokinin degradation, were shown to be more resistant to RKN (Lohar et al. 2004). More recently, a general downregulation of a negative regulator of cytokinin signaling was observed in microdissected GC (Barcala et al. 2010).

Other mitogenic phytohormones, the small signaling peptides phytosulfokines (PSKs), could also trigger GC differentiation. While a downregulation of genes encoding PSKs was detected in *Arabidopsis* (Jammes et al. 2005), a strong induction of a gene coding for a PSK has been observed in RKN-challenged soybean roots (Ibrahim et al. 2011). Functional analysis of PSK and their receptor PSKR1 in *Arabidopsis* showed that the expression of these genes were restricted to the GC surrounding cells. Interestingly *pskr1* mutants presented a defect in GC expansion, leading to increased nematode resistance (Favery, unpublished).

Signaling pathways of the phytohormones salicylic acid (SA), ethylene (ET), and jasmonic acid (JA) that control aspects of plant development and defense responses may also be targeted by RKN for successful induction or maintenance of feeding sites. External application of SA, JA, ET, or analogs, or chemical treatments to affect synthesis of these hormones, influenced host susceptibility to RKN (Cooper et al. 2005; Bhattarai et al. 2008; Fujimoto et al. 2011; Nahar et al. 2011). Transgenic tobacco plants expressing NPR1, a positive regulator of the SA pathway showed a decrease in the number of root galls and egg masses following *M. incognita* inoculation, when compared to wild-type plants (Priya et al. 2011), and accordingly, the overexpression in rice plant of *NahG*, expressing a bacterial salicylate hydroxylase that degrade SA, are slightly more sensitive to *M. graminicola* (Nahar et al. 2011). Silencing the rice *OsEIN2b*, a component of the ET-signaling pathway, also enhanced susceptibility to RKN (Nahar et al. 2011). In rice, the JA biosynthesis mutant *hebiba* is more susceptible toward *M. graminicola* (Nahar et al. 2011). In tomato, results suggest, however, that host susceptibility requires an intact JA signaling pathway for early steps of the feeding site formation (Bhattarai et al. 2008). Generally, the different transcriptomic analyses show a repression of transcription factors of the WRKY family that act downstream of these hormones and of several SA-, ET-, and JA-responsive genes (such as the pathogenesis-related genes, PRs) to suppress plant defenses and allow RKN biotrophy (Jammes et al. 2005, Barcala et al. 2010; Hamamouch et al. 2011; Ibrahim et al. 2011).

Because of the complexity of hormone signaling pathways, the existing synergistic and antagonistic cross talks within these pathways, and the diversity of the genes they may control, it remains difficult to elucidate the exact role played by phytohormones during feeding cell ontogenesis and gall formation in the different plant hosts.

Similarly, genome-wide gene expression profiling in CN-induced syncytia revealed a number of genes associated with the biosynthesis and signaling of auxin and to a lesser extent ET, gibberellins, JA, and cytokinin (Ithal et al. 2007b; Szakasits et al. 2009). An increase of local auxin in the syncytium was first demonstrated by increasing GUS activity in the *AUX1::GUS* transgenic *Arabidopsis* plants upon *H. schachtii* or infection (Mazarei et al. 2003). In support, the artificial auxin-responsive promoter element DR5 was highly activated in the syncytium after *H. schachtii* infection (Karczmarek et al. 2004). The central role that auxin plays in the establishment of the syncytium was revealed by data showing that mutants deficient in auxin signaling and transport pathways negatively impact parasitism success and syncytium development (Grunewald et al. 2009). Similarly, ET was found to be necessary for proper syncytium formation and mediating plant susceptibility. Functional characterizations of *Arabidopsis* mutants that overproduce ET (*eto1–1*, *eto2*, and *eto3*) were found to be hypersusceptible to *H. schachtii*. In contrast, mutants that are ET-insensitive (*etr1–1, ein2–1, ein3–1, eir1–1,* and *axr2*) were less susceptible to *H. schachtii*, indicating that ET signal transduction positively impacts plant susceptibility to CN (Wubben et al. 2001). The role of ET in syncytium development was supported by the histological analysis of infected *eto2* roots showing expansion of the feeding site structure and extensive CW dissolution between fused cells (Goverse et al. 2000). However, direct evidence supporting roles of other phytohormones in syncytium formation and function are lacking.

In *P. brassicae*, it has been shown that plasmodia are able to produce cytokinin, which are released into the host cytoplasm, inducing host cell division (Devos et al. 2006). Availability of hormone-responsive promoters DR5 and ARR5 allowed detection of strong auxin and cytokinin responses, respectively, in the root gall on *Arabidopsis*. An increase in concentration of active forms of both auxin and cytokinin could be quantified in *B. rapa* (Devos et al. 2005). Furthermore, data from transcriptome analysis of *Arabidopsis* clubroots show strong regulation of genes involved in auxin and cytokinin homeostasis (Siemens et al. 2006). In *Arabidopsis*, genes involved in auxin synthesis, transport, and responses are strongly upregulated. Regulation of nitrilase genes involved in auxin synthesis has also been reported in *B. rapa* (Ando et al. 2008). The *alh1* mutant of

Arabidopsis, which may be affected in auxin transport, shows resistance responses to *P. brassicae* (Devos et al. 2006). As observed in root-knot, cytokinin oxidases have been shown to be particularly downregulated in infected tissues. Again, overexpression of *CKXs* in transgenic *Arabidopsis* results in increased resistance to *P. brassicae* (Siemens et al. 2006). It is accepted that early cytokinin accumulation promotes cell division, allowing club ontogenesis. Auxin could in the first phases of the interaction induce cell elongation through upregulation of XETs, and in the later phase, the impressive upregulation of auxin biosynthesis pathways and transport might lead to cell degradation for resting spore release. However, results remain contradictory and hormone actions may vary depending on plant host, protist ecotypes, and stage of disease development. In addition, abscisic acid, ET, and jasmonate may also play roles during root disease. Two recent reviews summarized the current knowledge on hormonal changes and signals during clubroot development (Ludwig-Müller and Schuller 2008; Ludwig-Müller et al. 2009).

Role of miRNAs in Feeding Site Formation and Function

Genome-wide gene expression profiling has revealed that unique changes of plant cell fate associated with nematode-induced feeding site formation involve extensive gene expression changes, including substantial downregulation of genes at the mRNA level in parasitized plant cells and roots (Jammes et al. 2005; Alkharouf et al. 2006; Ithal et al. 2007b; Klink et al. 2009; Barcala et al. 2010). While the genetic networks underlying gene expression regulation in feeding sites are unknown, increasing evidence supports fundamental roles of small RNA molecules as powerful regulators of gene expression in living cells. Endogenous small RNAs are 20–24 nucleotide molecules that regulate gene expression through posttranscriptional gene silencing, translational repression and heterochromatin modification (Vaucheret 2006). On the basis of their precursor structures and biogenesis, small RNAs can be divided into two main classes, microRNAs (miRNAs) and small interfering RNAs (siRNAs). siRNAs were identified in animals, plants, bacteria, and fungi (Grosshans and Filipowicz 2008), an indication of both the significance and the ancient origin of these regulatory molecules. With the advent of high-throughput sequencing technology of small RNA libraries, a new era for small RNA exploration has arrived. Analysis of small RNA components at genomic scales revealed the diversity and the complexity of these molecules (Schwach et al. 2009). Complexity of siRNAs arises from their diverse origins, which include coding genes, pseudogenes, repetitive sequences such as transposons and retrotransposons, tandem and inverted repeats, and intergenic regions (Lu et al. 2005).

miRNAs initially have been shown to be involved in the regulation of a variety of plant developmental processes including phase transition, hormone synthesis and signaling, pattern formation, and morphogenesis (Chen 2009). Recent studies indicate that miRNAs and small endogenous RNAs also are involved in biotic stress responses in plants. For example, *Arabidopsis* miRNA393 contributes to resistance against the virulent *Pseudomonas syringae* pv. *tomato* strain DC3000 (*Pto* DC3000) by repressing auxin signaling (Navarro et al. 2006). In contrast, transgenic plants overexpressing miRNA398b or miRNA773 showed increased susceptibility to both virulent and nonvirulent strains of *P. syringae* (Li et al. 2010). In addition, alteration of miRNA gene expression was observed upon infection by various plant pathogens, including viruses (He et al. 2008), fungi (Lu et al. 2007), bacteria (Fahlgren et al. 2007), and herbivorous insects (Pandey et al. 2008). Pathogen-induced accumulation of a natural antisense transcript-associated siRNA was found to inhibit a negative regulator of the resistance gene *RPS2*-mediated race-specific resistance pathway (Katiyar-Agarwal et al. 2006). Similarly, endogenous long siRNAs (30–40 nt) were induced upon bacterial infection and contributed to host defense responses to the pathogen (Katiyar-Agarwal et al. 2007). Consistent with the role of small RNAs in the regulation of plant immune responses, *Arabidopsis* mutants deficient in siRNA or miRNA biogenesis were affected in their susceptibility to bacteria (Navarro et al. 2008). Collectively, these emerging data indicate that small RNA-mediated gene regulation is a fundamental regulatory mechanism during plant–pathogen interactions.

Recent studies revealed essential roles of small RNA pathways in cell specification. For example, it has been demonstrated that small RNA pathways control male and female gamete formation (Olmedo-Monfil et al. 2010; Ron et al. 2010) and phloem function (Buhtz et al. 2008). Interestingly, small RNAs have been shown to control the formation and development of the nitrogen-fixing root nodule, a de novo organ formed in response to Rhizobium bacteria (Combier et al. 2006; Boualem et al. 2008; Lelandais-Briere et al. 2009). As a first step of investigating the potential implication of small RNA molecules in mediating gene regulation processes that are of importance for the formation of the syncytium, CN susceptibility of *Arabidopsis* mutants of genes involved in the biogenesis and production of small RNAs was examined (Hewezi et al. 2008a). Interestingly, several DICER-like (*dcl*) and RNA-dependent RNA polymerase (*rdr*) mutants showed significant decreases in nematode susceptibility relative to the wild type, suggesting a crucial role of small RNAs in mediating gene regulation processes that are involved in plant–nematode interactions. In addition, generation of small RNA libraries from *Arabidopsis* roots harvested at 4 and 7 dpi post *H. schachtii* infection led to the identification of new endogenous siRNAs and known miRNAs that are differentially expressed in response to *H. schachtii* infection (Hewezi et al. 2008a). Similar to temporal changes of miRNA expression levels in response to biotic and abiotic stress treatments (Bari et al. 2006; Navarro et al. 2006; Sunkar et al. 2006; Navarro et al. 2008), the differentially expressed miRNAs identified in that study were found to be either downregulated or unchanged at an early stage of infection, that is, 4 dpi, but at the 7 dpi time point, these miRNAs showed significant up- or downregulation. However, it is unlikely that the observed downregulation of miRNA at the early stage of CN infection is a general characteristic of miRNA response to CN parasitism because recently, two additional *Arabidopsis* miRNAs were found to be induced in the early developing *H. schachtii*-induced syncytia (Hewezi and Baum 2012; Hewezi et al. 2012).

Consistent with the critical role that the phytohormone auxin plays in syncytium initiation and formation (Goverse et al. 2000), it was found that miR160 and miR167, which target five auxin response factor genes (*ARF6, ARF8, ARF10, ARF16,* and *ARF17*), were downregulated in response to *H. schachtii* at the early stage of infection and that this downregulation was accompanied by significant upregulation of the ARF target genes, enabling the control of early auxin-responsive gene expression in the feeding sites. One of the common alterations in root systems associated with syncytium development is the lateral root formation around the syncytium. The miR164 could be one of the main players in this context. It acts as a negative regulator of auxin-mediated lateral root development by controlling mRNA abundance of NAC1, which functions in transmitting auxin signals for *Arabidopsis* lateral root development (Xie et al. 2000). Downregulation of miR164 coupled with strong upregulation of *NAC1* up to 7 days post *H. schachtii* infection suggests that active auxin signal transduction associated with lateral root induction around the syncytium is connected to localized auxin increase in the feeding sites (Hewezi et al. 2008a). The finding that miRNAs are involved in developmental and biological processes related to those implicated in syncytium formation and function gives rise to the question whether spatial regulation of miRNA expression contributes to syncytial cell differentiation and development. In this context, miR396 was identified as a potentially interesting regulator of target gene expression during nematode parasitism (Hewezi et al. 2008a). miR396 regulates the expression of seven growth regulating transcription factor genes (*GRFs*) (Jones-Rhoades and Bartel 2004), which are known to act in a functionally redundant fashion to positively control cell proliferation and size in leaves (Kim et al. 2003; Kim and Kende 2004; Horiguchi et al. 2005; Kim and Lee 2006). Very recently, the miR396/GRF regulatory module was shown to have vital roles in controlling developmental events during plant–pathogen interactions (Hewezi and Baum 2012; Hewezi et al. 2012). Promoter activity analyses together with qPCR quantification revealed that miR396 expression characteristics in the syncytium clearly delineate the induction/formation phase from the maintenance phase of the syncytium and that the switch of one phase to the other appears to be mediated through miR396. *GRF1* and *GRF3* as the main targets of miR396 in roots were found to be predominantly active in syncytia during all parasitic stages but were posttranscriptionally silenced when miR396 abundance increased during the onset of the maintenance phase. The importance of the coordinated regulation of miR396 and *GRF1* and *GRF3* for normal root development is supported by the data showing that overexpression of miR396 in *Arabidopsis* as well as its target genes *GRF1* and *GRF3* significantly reduces root length. The significance of this synchronized regulation for correct cell fate specification and differentiation in the developing syncytium is further supported by the data showing that miR396 overexpression in *Arabidopsis* as well as modulations of *GRF1* and *GRF3* expression dramatically limit plant susceptibility to the *H. schachtii*. Reduced susceptibility was not associated with low nematode penetration rates but rather was associated with arrested nematode development during an early stage of infection and with reduced syncytium size. In addition, genome-wide expression profiling revealed a large number of GRF-regulated genes that are involved in a wide range of biological and developmental processes. Interestingly, the miR396/GRF regulatory module can change the expression of 44% of the more than 7000 genes that change expression in the *Arabidopsis* syncytium (Szakasits et al. 2009). In conclusion, these data establish miR396 as a powerful regulator targeted by the parasitic nematodes to drive presyncytial cells toward novel developmental pathways required for successful syncytium formation and maintenance.

Nematode Effectors That Alter Root Cell Development during Parasitism

Various molecular proteomic and transcriptomic analysis on nematode esophageal secretions have been used to discover genes encoding effector proteins potentially involved in parasitism (Bellafiore et al. 2008; Davis et al. 2008). And new perspectives arise from the availability of nematode genomes (Bird et al. 2009). Among nematode secreted proteins are many CW-degrading or -modifying enzymes. Others may be nucleus-targeted and could bind to nucleotides and influence gene expression. These secretions could also interact with host proteins or mimic plant proteins or other signaling molecules, allowing substantial alterations of plant physiology (Gheysen and Mitchum 2011).

Plant CW represents the main barrier for successful penetration and migration of nematodes throughout root tissues. CW-degrading or -modifying enzymes may account for the RKN and CN parasitic ability, allowing the impressive CW modifications or degradation required for GC or syncytia ontogenesis. Thus, there is no surprise that these proteins compose a substantial portion of nematodes secretions; and their diversity is unique in the animal kingdom (Davis et al. 2008; Davies et al. 2011). Genes encoding these enzymes could have been acquired through different events of lateral gene transfers from different bacteria (Danchin et al. 2010). One remarkable example of a functionally characterized CN effector protein with function in host CW modification is the cellulose-binding protein (CBP) from *H.* schachtii (Hewezi et al. 2008b). This CBP has a unique structural feature in that it consists only of a signal peptide for secretion and a cellulose-binding domain (CBD) without a catalytic domain. The CBD is usually found in cellulolytic enzymes, where it mediates the attachment of the enzymes to the insoluble cellulose substrate, while a catalytic domain is responsible for hydrolytic activity. Because Heterodera CBPs do not contain catalytic domains, their functions were enigmatic, but a multifaceted functional characterization approach uncovered their mode of action. The *CBP* gene is expressed in the nematode's secretory subventral gland cells, consistent with a role during migration or early syncytium formation. In planta overexpression of the *CBP* gene produced plants with increased root length and increased susceptibility to H. schachtii infection but producing feeding sites of the same size as the wild type. Because of the absence of the catalytic domain, CBP has been hypothesized to act in concert with host proteins to exert its function. Using yeast two-hybrid screens, CBP was found to interact with *Arabidopsis* pectin methylesterase protein 3 (PME3), a CW-localized enzyme. Interestingly, PME activity increased in *CBP*-overexpressing plants, indicating that CBP acts in activating PME3 during the infection process. As hypothesized, overexpression of *PME3* mimicked the phenotypes of *CBP*-overexpressing plants of longer roots and increased nematode susceptibility, whereas a pme3 mutant exhibited the opposite phenotypes. In addition, expression of *CBP* in the *pme3* mutant background confirmed that PME3 is required for the CBP overexpression phenotypes, but it also revealed that PME3 is not the sole host factor responsible. In short, after studying PME3 functions, it seems most likely that a reduction of CW pectin methyl esterification through increasing PME3 activity as a result of its interaction with CBP augments the access of other CW-modifying enzymes to the plant CW of presyncytial cells, thereby facilitating parasitic establishment in the root system.

Another CN effector that acts in activating CW-modifying enzymes through its interaction with host factors is the 19C07 protein from H. schachtii. 19C07 was shown to interact with the auxin influx transporter LAX3 (Lee et al. 2011), known to activate CW-digesting enzymes (Swarup et al. 2008). LAX3 plays a key role in lateral root emergence by directing auxin signaling that activates the expression of CW-modifying enzymes to root cells overlaying lateral root primordial where it is predominantly expressed, permitting lateral roots to emerge. The interaction with LAX3 seems to be one mechanism by which CN alters auxin flow into root cells for syncytium formation. Consistent with the suggestion that 19C07 functions in increasing the activity of LAX3, overexpression of *19C07* in *Arabidopsis* increased the rate of lateral root emergence, an indication of increasing auxin influx. Moreover, nematode infection promotes lateral root formation in the *aux1/lax3* double and *aux1/lax1/lax2/lax3* quadruple mutants at the site of infection (Lee et al. 2011). The novel *H. schachtii* effector gene *10A06* also was shown to impact root development. Constitutive expression of *10A06* in *Arabidopsis* produced plants with up to 28% increase in root length (Hewezi et al. 2010). Interaction of 10A06 with *Arabidopsis* spermidine synthase 2 (SPDS2), a key enzyme involved in polyamine biosynthesis, induce changes in polyamine metabolism and catabolic pathway in the syncytium. These changes resulted in both activation of cellular antioxidant machinery and inhibition of SA-dependent defense signaling concomitant with increased susceptibility of *10A06*-overexpressing plants to *H. schachtii*, *Pseudomonas* (*Pst* DC3000), and cucumber mosaic virus Y (Hewezi et al. 2010). It appears most likely that inhibition of host defense responses in the developing syncytium along with initiation of a protective antioxidant environment to protect the developing syncytia against reactive oxygen species produced by the host as common feature of plant responses to plant-parasitic nematode infection in both compatible and incompatible interactions (Waetzig et al. 1999) are mechanisms by which 10A06 contributes to CN parasitism.

Despite the fact that both *H. schachtii and H. glycines* have very similar effector repertoires, *H. schachtii* infects and efficiently reproduces on *Arabidopsis* roots, while *H. glycines* is rarely able to complete the life cycle on this model plant. The molecular basis for these differences remains unknown. A recent report has provided new evidence that sequence divergence between certain

CN effector protein orthologs could contribute to host specificity (Hewezi and Baum 2010). In this study, it was found that ectopic expression of *H. glycines 10A06* (*Hg-10A06*) in the nonhost *Arabidopsis* plants affected root length to a much lesser degree when compared with *Arabidopsis* plants overexpressing *10A06* form *H. schachtii* (*Hs-10A06*), although both orthologs share high sequence identity of 86%. In addition, Hg-10A06 was found to interact with *Arabidopsis* SPDS2, while its ability to interact with SPDS2 decreased by about fivefold when compared with Hs-10A06. Taken together, these data suggest that effector functions could be involved in determining host range and that the root phenotypes obtained in *10A06* overexpressing plants are associated with host factor recognition and more specifically polyamine signaling pathways. The influence of polyamines on root emergence, formation, and development has been shown in numerous plant species (Kakkar et al. 2000; Bais and Ravishankar 2002).

It has been shown that RKN and CN secrete effectors with sequence similarity to endogenous host plant proteins. These include, for example, chorismate mutases (Lambert et al. 1999; Bekal et al. 2003; Huang et al. 2005; Vanholme et al. 2009), the expansin-like proteins (Qin et al. 2004; Abad et al. 2008; Hewezi et al. 2008b; Danchin et al. 2010), the CN CLAVATA3-like (CLE) peptides (Wang et al. 2005, 2010; Lu et al. 2009) and annexins (Patel et al. 2010). The *CLE* genes were identified first in plant species where they regulate a wide range of biological processes including meristem function, organ size, apical dominance, and root development (Strabala et al. 2006; Fiers et al. 2007; Oelkers et al. 2008). Direct evidence for the implication of *CLE* genes in *Arabidopsis* root growth and differentiation were provided through overexpression and/or direct application of synthetic 14-amino acid peptides corresponding to the CLE motif (Casamitjana-Martinez et al. 2003; Hobe et al. 2003; Fiers et al. 2004, 2005). While it was assumed that *CLE* genes are a plant-specific gene family, surprisingly sequence analyses of cDNA clones generated from microaspirated esophageal gland cell cytoplasm of *H. glycines* led to the discovery of the first *CLE*-like gene (*HgSYV46*) outside of plants (Wang et al. 2001). Subsequent functional analysis of *HgSYV46* through overexpression in *Arabidopsis* wild-type and *clv3* mutant backgrounds provided the first evidence that CN *HgSYV46* has functional similarity to *Arabidopsis CLE* genes (Wang et al. 2005). In addition, recent studies have reported the existence of *CLE*-like genes in G. rostochiensis (*GrCLE*s) (Lu et al. 2009) and *H. schachtii* (*HsCLEs*) (Wang et al. 2011). While the *HsCLE* genes are structurally similar to those identified in *H. glycines*, GrCLEs have a novel structural feature consisting of several CLE motifs organized in tandem repeats (Lu et al. 2009). This gives rise to the possibility that multiple CLE motif-containing proteins could be processed in a way to generate a variety of CLE peptides that stimulate distinct CLE signaling pathways that act antagonistically or cooperatively as in the case of plant CLEs (Whitford et al. 2008). Similar to *HgSYV46*, overexpression of *GrCLEs* produced phenotypes resembling those of plant *CLE* genes in *Arabidopsis*, and Gr-CLE proteins could complement the *Arabidopsis clv3–2* mutant phenotype (Lu et al. 2009). Interestingly, overexpression of *Gr-CLEs* containing single or multiple CLE motifs in *Arabidopsis* and potato hairy roots produced a short root phenotype. The short root phenotype can also be generated *in vitro* by exogenous application of synthetic Gr-CLE peptides to *Arabidopsis* and potato roots (Lu et al. 2009)

Overall, these data raise the question of how CLE peptides contribute to the redifferentiation of mature root cells to form the very specialized and novel syncytial cell type. It has been proposed that the secreted CLE-like peptides function as peptide mimics of particular host CLE factors to drive the presyncytial cells toward specific and unique developmental programs required for the initiation of the feeding cells (Mitchum et al. 2008). This suggestion is further supported by recent studies reporting that signaling of stem cell maintenance in *Arabidopsis* shoot and root meristems is regulated by common regulatory mechanisms (Sarkar et al. 2007), including a CLAVATA

(CLV)-like pathway (Casamitjana-Martinez et al. 2003), and that the receptor kinase family members CORYNE (CRN) and CLV2 implicate CLE signaling in *Arabidopsis* roots (Fiers et al. 2005; Muller et al. 2008). In support, it has been recently shown that CLV2 and CRN are involved in mediating nematode CLE signaling in *Arabidopsis* roots (Replogle et al. 2011). Interestingly, *clv2–1* and *crn-1* mutants were found to be resistant to exogenous application of the synthetic nematode peptides HgCLE, HsCLE1, and HsCLE2. Likewise, overexpression of *HgCLE*, *HsCLE1*, and *HsCLE2* genes in the *clv2–1* and *crn-1* mutant backgrounds abolished the phenotypes observed in the transgenic plants overexpressing these *CLE* genes in wild-type backgrounds (Wang et al. 2005, 2011). Both receptors were found to be expressed in the nematode-induced syncytia and manipulation of their expression impacted nematode susceptibility and syncytium size (Replogle et al. 2011). While these results suggest that perception of nematode CLEs by CLV2 and CRN is required for successful parasitism, biochemical data showing the localization of actual perception are lacking. Despite the fact that the exact trafficking mechanism of nematode CLE peptides to the apoplast where they function is unknown, it has been recently suggested that nematode CLEs are processed in host cells rather than in the nematodes and that the propeptides are delivered into the cytoplasm of syncytial cells, but ultimately function in the apoplast (Wang et al. 2010), consistent with their anticipated roles as ligand mimics of host CLE peptides. Further studies to identify the underlying mechanisms that integrate CLE signaling and downstream components in the developmental reprogramming of selected root cells to form the syncytium merit detailed functional studies.

The role of CLEs in plant RKN interactions is not as clear, but genome analysis reveals the presence of five and seven candidate CLE genes in *M. incognita* and *M. hapla*, respectively (Abad et al. 2008; Opperman et al. 2008). However, *clv1* mutants were not found to impacted RKN susceptibility (Favery, unpublished). Huang and collaborators presented the first evidence for a direct interaction of a *M. incognita*-secreted peptide, 16D10, having sequence similarity to the CLE motif, with a plant SCARECROW-like transcription factor (Huang et al. 2006a). SCARECROW are members of the GRAS family of transcription factors, regulating root cell division and are required for root growth and morphogenesis (Pysh et al. 1999). Expression of the 16D10 effector in *Arabidopsis* stimulated root growth (Huang et al. 2006a). And silencing RKN 16D10 through host-induced gene silencing resulted in reduced nematode infectivity (Huang et al. 2006b). However, this peptide was not able to rescue the *clv3* mutant, and its function as a plant CLE ligand mimic remains to be proven.

Conclusion

Growth and performance of root determines crop yield and is significantly influenced by microorganisms colonizing the rhizosphere. In the past few years, molecular and genetic approaches allowed substantial progress in our understanding of how pathogens manipulate plant hosts and their roots. The analyses at a genome-wide scale using transcriptomic approaches revealed pathogen-responsive plant genes. However, their functional analyses remain a major obstacle. In the future, a new challenge will be to integrate transcriptomic data with proteomic and metabolomic information to better identify aspects of hosts' physiology corrupted by pathogens. One other future aspect of research has been opened up by the increasing number of studies on pathogen secretomes. While microbial effectors of parasitism are being identified, significant progress remains to be made in characterizing the direct host targets of these secreted proteins and studying whether small RNAs function as key players. The genome sequences of pathogens such as *M. incognita* (Abad et al. 2008; http://www.inra.fr/meloidogyne_incognita), *M. hapla* (Opperman et al. 2008;

http://www.hapla.org), *H. glycines,* and *G. pallida* will provide new horizons for studying plant–nematode interactions. Finally, determining how a pathogen modifies root cells to serve as feeding cells will provide unique insights into fundamental cellular mechanisms in plants and will contribute to offer novel approaches to protect plants.

Acknowledgments

Our gratitude goes to Maria Manzanares-Dauleux at UMR118 INRA-Agrocampus Ouest-Université de Rennes 1, France, for the clubroot pictures, and Marie-Cécile Caillaud and Philippe Castagnone for RKN images. P.A. and B.F. team was supported by INRA, CNRS, Nice Sophia Antipolis University, the Agence Nationale de la Recherche (ANR-05-GPLA-020 "AFINDIS" and ANR-08-GENM-014 "SCRIPS") grant. We apologize to all colleagues and researchers whose work has not been cited directly.

References

Abad, P. et al. (2009) Invasion, feeding and development. In: R.N. Perry, M. Moens, and J.L. Starr (eds) *Root-Knot Nematodes.* CAB International, Wallingford, pp. 163–181.

Abad, P. et al. (2008) Genome sequence of the metazoan plant-parasitic nematode *Meloidogyne incognita. Nature Biotechnology*, 26, 909–915.

Abad, P. and Williamson, V.M. (2010) Plant nematode interaction: a sophisticated dialogue. In: J.-C. Kader and M. Delseny (eds) *Advances in Botanical Research*. Academic Press, New York, pp. 147–192.

Alkharouf, N. et al. (2004) Analysis of expressed sequence tags from roots of resistant soybean infected by the soybean cyst nematode. *Genome*, 47, 380–388.

Alkharouf, N.W. et al. (2006) Timecourse microarray analyses reveal global changes in gene expression of susceptible *Glycine max* (soybean) roots during infection by *Heterodera glycines* (soybean cyst nematode). *Planta*, 224, 838–852.

Ando, S. et al. (2008) *Plant Molecular Biology*, Vol. 68. Springer, Netherlands, pp. 557–569.

Bais, H.P. and Ravishankar, G.A. (2002) Role of polyamines in the ontogeny of plants and their biotechnological applications. *Plant Cell Tissue and Organ Culture*, 69, 1–34.

Barcala, M. et al. (2010) Early transcriptomic events in microdissected Arabidopsis nematode-induced giant cells. *The Plant Journal*, 61, 698–712.

Bari, R. et al. (2006) PHO2, microRNA399, and PHR1 define a phosphate-signaling pathway in plants. *Plant Physiology*, 141, 988–999.

Bar-Or, C. et al. (2005) A broad characterization of the transcriptional profile of the compatible tomato response to the plant parasitic root knot nematode *Meloidogyne javanica. European Journal of Plant Pathology*, 111, 181–192.

Bekal, S. et al. (2003) A chorismate mutase from the soybean cyst nematode *Heterodera glycines* shows polymorphisms that correlate with virulence. *Molecular Plant–Microbe Interactions*, 16, 439–446.

Bellafiore, S. et al. (2008) Direct identification of the *Meloidogyne incognita* secretome reveals proteins with host cell reprogramming potential. *PLoS Pathogens*, 4, e1000192.

Bhattarai, K.K. et al. (2008) Tomato susceptibility to root-knot nematodes requires an intact jasmonic acid signaling pathway. *Molecular Plant–Microbe Interactions*, 21, 1205–1214.

Bird, A.F. and Loveys, B.R. (1980) The involvement of cytokinins in a host-parasite relationship between the tomato (*Lycopersicon esculentum*) and a nematode (*Meloidogyne javanica*). *Parasitology*, 80, 497–505.

Bird, D.M. et al. (2009) The genomes of root-knot nematodes. *Annual Review of Phytopathology*, 47, 333–351.

Boualem, A. et al. (2008) MicroRNA166 controls root and nodule development in *Medicago truncatula. The Plant Journal*, 54, 876–887.

Buhtz, A. et al. (2008) Identification and characterization of small RNAs from the phloem of *Brassica napus. The Plant Journal*, 53, 739–749.

Caillaud, M.-C. et al. (2008a) Cytoskeleton reorganization: a key process in root-knot nematode-induced giant cell ontogenesis. *Plant Signaling and Behavior*, 3, 816–818.

Caillaud, M.-C. et al. (2008b) Root-knot nematodes manipulate plant cell functions during a compatible interaction. *Journal of Plant Physiology*, 165, 104–113.

Caillaud, M.-C. et al. (2008c) MAP65–3 Microtubule-associated protein is essential for nematode-induced giant cell ontogenesis in Arabidopsis. *The Plant Cell*, 20, 423–437.

Cao, T. et al. (2008) Proteome-level changes in the roots of *Brassica napus* as a result of *Plasmodiophora brassicae* infection. *Plant Science*, 174, 97–115.

Carpita, N.C. and Gibeaut, D.M. (1993) Structural models of primary cell walls in flowering plants: consistency of molecular structure with the physical properties of the walls during growth. *The Plant Journal*, 3, 1–30.

Casamitjana-Martinez, E. et al. (2003) Root-specific CLE19 overexpression and the sol1/2 suppressors implicate a CLV-like pathway in the control of Arabidopsis root meristem maintenance. *Current Biology*, 13, 1435–1441.

Chen, X. (2009) Small RNAs and their roles in plant development. *Annual Review of Cell Developmental Biology*, 25, 21–44.

Clément, M. et al. (2009) Actin-depolymerizing factor 2-mediated actin dynamics are essential for root-knot nematode infection of Arabidopsis. *The Plant Cell*, 21, 2963–2979.

Combier, J.P. et al. (2006) MtHAP2–1 is a key transcriptional regulator of symbiotic nodule development regulated by microRNA169 in *Medicago truncatula*. *Genes and Development*, 20, 3084–3088.

Cooper, W.R. et al. (2005) Effects of jasmonate-induced defenses on root-knot nematode infection of resistant and susceptible tomato cultivars. *Journal of Chemical Ecology*, 31, 1953–1967.

Danchin, E.G.J. et al. (2010) Multiple lateral gene transfers and duplications have promoted plant parasitism ability in nematodes. *Proceedings of the National Academy of Sciences USA*, 107, 17651–17656.

Das, S. et al. (2010) Transcriptional profiling of root-knot nematode induced feeding sites in cowpea (*Vigna unguiculata* L. *Walp.*) using a soybean genome array. *BMC Genomics*, 11, 480.

Davies, E. et al. (2011) Degradation of the plant cell wall by nematodes. In: J. Jones, G. Gheysen, and C. Fenoll (eds) *Genomics and Molecular Genetics of Plant–Nematode Interactions*, Springer, New York, pp. 255–272.

Davis, E.L. et al. (2008) Parasitism proteins in nematode-plant interactions. *Current Opinion in Plant Biology*, 11, 360–366.

de Almeida Engler, J. et al. (1999) Molecular markers and cell cycle inhibitors show the importance of cell cycle progression in nematode-induced galls and syncytia. *The Plant Cell*, 11, 793–808.

de Almeida Engler, J. et al. (2011) Unravelling the plant cell cycle in nematode induced feeding sites. In: J. Jones, G. Gheysen, and C. Fenoll (eds) *Genomics and Molecular Genetics of Plant–Nematode Interactions*, Springer, New York, pp. 349–368.

de Almeida Engler, J. and Favery, B. (2011) The plant cytoskeleton remodelling in nematode induced feeding sites. In: J. Jones, G. Gheysen, and C. Fenoll (eds) *Genomics and Molecular Genetics of Plant–Nematode Interactions*. Springer, New York, pp. 369–393.

de Almeida Engler, J. et al. (2010) Plant actin cytoskeleton remodeling by plant parasitic nematodes. *Plant Signaling and Behavior*, 5, 213–217.

de Almeida Engler, J. et al. (2004) Dynamic cytoskeleton rearrangements in giant cells and syncytia of nematode-infected roots. *The Plant Journal*, 38, 12–26.

De Meutter, J. et al. (2005) Production of auxin and related compounds by the plant parasitic nematodes *Heterodera schachtii* and *Meloidogyne incognita*. *Communications in Agricultural and Applied Biological Sciences*, 70, 51–60.

De Meutter, J. et al. (2003) Identification of cytokinins produced by the plant parasitic nematodes *Heterodera schachtii* and *Meloidogyne incognita*. *Molecular Plant Pathology*, 4, 271–277.

Devos, S. et al. (2006) A hormone and proteome approach to picturing the initial metabolic events during *Plasmodiophora brassicae* infection on Arabidopsis. *Molecular Plant–Microbe Interactions*, 19, 1431–1443.

Devos, S. et al. (2005) Infection of Chinese cabbage by *Plasmodiophora brassicae* leads to a stimulation of plant growth: impacts on cell wall metabolism and hormone balance. *New Phytologist*, 166, 241–250.

Fahlgren, N. et al. (2007) High-throughput sequencing of Arabidopsis microRNAs: evidence for frequent birth and death of MIRNA genes. *PLoS One*, 2, e219.

Favery, B. et al. (2004) Arabidopsis formin AtFH6 is a plasma membrane-associated protein upregulated in giant cells induced by parasitic nematodes. *The Plant Cell*, 16, 2529–2540.

Favery, B. et al. (2002) The endosymbiosis-induced genes *ENOD40* and *CCS52a* are involved in endoparasitic-nematode interactions in *Medicago truncatula*. *Molecular Plant–Microbe Interactions*, 15, 1008–1013.

Fiers, M. et al. (2005) The 14-amino acid CLV3, CLE19, and CLE40 peptides trigger consumption of the root meristem in Arabidopsis through a CLAVATA2-dependent pathway. *The Plant Cell*, 17, 2542–2553.

Fiers, M. et al. (2004) Mis-expression of the CLV3/ESR-like gene CLE19 in Arabidopsis leads to a consumption of root meristem. *Gene*, 327, 37–49.

Fiers, M. et al. (2007) CLE peptide ligands and their roles in establishing meristems. *Current Opinion in Plant Biology*, 10, 39–43.

Fosu-Nyarko, J. et al. (2009) Functional characterization of transcripts expressed in early-stage *Meloidogyne javanica*-induced giant cells isolated by laser microdissection. *Molecular Plant Pathology*, 10, 237–248.

Fudali, S. et al. (2008) Two tomato alpha-expansins show distinct spatial and temporal expression patterns during development of nematode-induced syncytia. *Physiologia Plantarum*, 132, 370–383.

Fujimoto, T. et al. (2011) Expression profile of jasmonic acid-induced genes and the induced resistance against the root-knot nematode (*Meloidogyne incognita*) in tomato plants (*Solanum lycopersicum*) after foliar treatment with methyl jasmonate. *Journal of Plant Physiology*, 168, 1084–1097.

Fuller, V.L. et al. (2007) Differential gene expression in Arabidopsis following infection by plant-parasitic nematodes *Meloidogyne incognita* and *Heterodera schachtii*. *Molecular Plant Pathology*, 8, 595–609.

Gal, T. et al. (2006) Expression of a plant expansin is involved in the establishment of root knot nematode parasitism in tomato. *Planta*, 224, 155–162.

Gheysen, G. and Fenoll, C. (2002) Gene expression in nematode feeding sites. *Annual Review of Phytopathology*, 40, 191–219.

Gheysen, G. and Mitchum, M.G. (2009) Molecular insights in the susceptible plant response to nematode infection. In: R.H. Berg and C.G. Taylor (eds) *Cell Biology of Plant Nematode Parasitism*, Springer-Verlag, Berlin.

Gheysen, G. and Mitchum, M.G. (2011) How nematodes manipulate plant development pathways for infection. *Current Opinion in Plant Biology*, 14, 415–421.

Goellner, M. et al. (2001) Endo-beta-1,4-glucanase expression in compatible plant–nematode interactions. *The Plant Cell*, 13, 2241–2255.

Goverse, A. et al. (2000) Both induction and morphogenesis of cyst nematode feeding cells are mediated by auxin. *Molecular Plant–Microbe Interactions*, 13, 1121–1129.

Grosshans, H. and Filipowicz, W. (2008) Molecular biology: the expanding world of small RNAs. *Nature*, 451, 414–416.

Grundler, F.M.W. et al. (1998) Formation of cell wall openings in root cells of *Arabidopsis thaliana* following infection by the plant-parasitic nematode *Heterodera schachtii*. *European Journal of Plant Pathology*, 104, 545–551.

Grunewald, W. et al. (2009) Manipulation of auxin transport in plant roots during Rhizobium symbiosis and nematode parasitism. *The Plant Cell*, 21, 2553–2562.

Hamamouch, N. et al. (2011) Expression of Arabidopsis pathogenesis-related genes during nematode infection. *Molecular Plant Pathology*, 12, 355–364.

Hammes, U.Z. et al. (2005) Nematode-induced changes of transporter gene expression in Arabidopsis roots. *Molecular Plant–Microbe Interactions*, 18, 1247–1257.

He, X.F. et al. (2008) Characterization of conserved and novel microRNAs and their targets, including a TuMV-induced TIR-NBS-LRR class R gene-derived novel miRNA in Brassica. *FEBS Letters*, 582, 2445–2452.

Hewezi, T. and Baum, T.J. (2010) Sequence divergences between cyst nematode effector protein orthologs may contribute to host specificity. *Plant Signaling and Behavior*, 5, 187–189.

Hewezi, T. and Baum, T.J. (2012) Complex feedback regulations govern the expression of miRNA396 and its GRF target genes. *Plant Signaling & Behavior*, 7, 749–751.

Hewezi, T. et al. (2008a) Arabidopsis small RNAs and their targets during cyst nematode parasitism. *Molecular Plant–Microbe Interactions*, 21, 1622–1634.

Hewezi, T. et al. (2008b) Cellulose binding protein from the parasitic nematode *Heterodera schachtii* interacts with Arabidopsis pectin methylesterase: cooperative cell wall modification during parasitism. *The Plant Cell*, 20, 3080–3093.

Hewezi, T. et al. (2010) Arabidopsis spermidine synthase is targeted by an effector protein of the cyst nematode *Heterodera schachtii*. *Plant Physiology*, 152, 968–984.

Hewezi, T. et al. (2012) The Arabidopsis microRNA396-*GRF1/GRF3* regulatory module acts as a developmental regulator in the reprogramming of root cells during cyst nematode infection. *Plant Physiology*, 159, 321–335.

Hobe, M. et al. (2003) Loss of CLE40, a protein functionally equivalent to the stem cell restricting signal CLV3, enhances root waving in Arabidopsis. *Development Genes and Evolution*, 213, 371–381.

Horiguchi, G. et al. (2005) The transcription factor AtGRF5 and the transcription coactivator AN3 regulate cell proliferation in leaf primordia of *Arabidopsis thaliana*. *The Plant Journal*, 43, 68–78.

Huang, G. et al. (2006b) Engineering broad root-knot resistance in transgenic plants by RNAi silencing of a conserved and essential root-knot nematode parasitism gene. *Proceedings of the National Academy of Sciences*, 103, 14302–14306.

Huang, G. et al. (2005) Two chorismate mutase genes from the root-knot nematode *Meloidogyne incognita*. *Molecular Plant Pathology*, 6, 23–30.

Huang, G. et al. (2006a) A root-knot nematode secretory peptide functions as a ligand for a plant transcription factor. *Molecular Plant–Microbe Interactions*, 19, 463–470.

Huang, X. et al. (2003) Expression of the Arabidopsis MCM gene PROLIFERA during root-knot and cyst nematode infection. *Phytopathology*, 93, 35–41.

Hutangura, P. et al. (1999) Auxin induction is a trigger for root gall formation caused by root-knot nematodes in white clover and is associated with the activation of the flavonoid pathway. *Australian Journal of Plant Physiology*, 26, 221–231.

Ibrahim, H. et al. (2011) Analysis of gene expression in soybean (*Glycine max*) roots in response to the root knot nematode *Meloidogyne incognita* using microarrays and KEGG pathways. *BMC Genomics*, 12, 220.

Inzé, D. and De Veylder, L. (2006) Cell cycle regulation in plant development. *Annual Review of Genetics*, 40, 77–105.

Ithal, N. and Mitchum, M.G. (2011) Laser capture microdissection of nematode feeding cells. *Plant Immunity*, 712, 227–240.

Ithal, N. et al. (2007a) Parallel genome-wide expression profiling of host and pathogen during soybean cyst nematode infection of soybean. *Molecular Plant–Microbe Interactions*, 20, 293–305.

Ithal, N. et al. (2007b) Developmental transcript profiling of cyst nematode feeding cells in soybean roots. *Molecular Plant–Microbe Interactions*, 20, 510–525.

Jammes, F. et al. (2005) Genome-wide expression profiling of the host response to root-knot nematode infection in Arabidopsis. *The Plant Journal*, 44, 447–458.

Jones, M. (1981) Host cell responses to endoparasitic nematode attack: structure and function of giant cells and syncytia. *Annals of Applied Biology*, 97, 353–372.

Jones, M. and Dropkin, V. (1976) Scanning electron microscopy in nematode-induced giant transfer cells. *Cytobios*, 15, 149–161.

Jones, M. and Payne, H. (1978) Early stages of nematode-induced giant-cell formation in roots of *Impatiens balsamica*. *Journal of Nematology*, 10, 70–84.

Jones-Rhoades, M.W. and Bartel, D.P. (2004) Computational identification of plant MicroRNAs and their targets, including a stress-induced miRNA. *Molecular Cell*, 14, 787–799.

Kageyama, K. and Asano, T. (2009) Life cycle of *Plasmodiophora brassicae*. *Journal of Plant Growth Regulation*, 28, 203–211.

Kakkar, R.K. et al. (2000) Polyamines and plant morphogenesis. *Biologia Plantarum*, 43, 1–11.

Kandoth, P.K. et al. (2011) The soybean Rhg1 locus for resistance to the soybean cyst nematode *Heterodera glycines* regulates the expression of a large number of stress- and defense-related genes in degenerating feeding cells. *Plant Physiology*, 155, 1960–1975.

Karczmarek, A. et al. (2008) Expression of two functionally distinct plant endo-beta-1,4-glucanases is essential for the compatible interaction between potato cyst nematode and its hosts. *Molecular Plant–Microbe Interactions*, 21, 791–798.

Karczmarek, A. et al. (2004) Feeding cell development by cyst and root-knot nematodes involves a similar early, local and transient activation of a specific auxin-inducible promoter element. *Molecular Plant Pathology*, 5, 343–346.

Katan, J. (1996) Interactions of roots with soil-borne pathogens. In: Y. Waisel, A. Eshel, and U. Kafkafi (eds) *Plant Roots: The Hidden Half*. Marcel Dekker, New-York.

Katiyar-Agarwal, S. et al. (2007) A novel class of bacteria-induced small RNAs in Arabidopsis. *Genes and Development*, 21, 3123–3134.

Katiyar-Agarwal, S. et al. (2006) A pathogen-inducible endogenous siRNA in plant immunity. *Proceedings of the National Academy of Sciences USA*, 103, 18002–18007.

Kim, J.H. et al. (2003) The AtGRF family of putative transcription factors is involved in leaf and cotyledon growth in Arabidopsis. *The Plant Journal*, 36, 94–104.

Kim, J.H. and Kende, H. (2004) A transcriptional coactivator, AtGIF1, is involved in regulating leaf growth and morphology in Arabidopsis. *Proceedings of the National Academy of Sciences USA*, 101, 13374–13379.

Kim, J.H. and Lee, B.H. (2006) GROWTH-REGULATING FACTOR4 of *Arabidopsis thaliana* is required for development of leaves, cotyledons, and shoot apical meristem. *Journal of Plant Biology*, 49, 463–468.

Kirik, V. et al. (2001) CPR5 is involved in cell proliferation and cell death control and encodes a novel transmembrane protein. *Current Biology*, 11, 1891–1895.

Klink, V. et al. (2009) A gene expression analysis of syncytia laser microdissected from the roots of the *Glycine max* (soybean) genotype PI 548402 (Peking) undergoing a resistant reaction after infection by *Heterodera glycines* (soybean cyst nematode). *Plant Molecular Biology*, 71, 525–567.

Klink, V. et al. (2007a) Laser capture microdissection (LCM) and comparative microarray expression analysis of syncytial cells isolated from incompatible and compatible soybean (*Glycine max*) roots infected by the soybean cyst nematode (*Heterodera glycines*). *Planta*, 226, 1389–1409.

Klink, V. et al. (2007b) A time-course comparative microarray analysis of an incompatible and compatible response by *Glycine max* (soybean) to *Heterodera glycines* (soybean cyst nematode) infection. *Planta*, 226, 1423–1447.

Klink, V.P. et al. (2010) Syncytium gene expression in *Glycine max* [PI 88788] roots undergoing a resistant reaction to the parasitic nematode *Heterodera glycines*. *Plant Physiology and Biochemistry*, 48, 176–193.

Koltai, H. et al. (2001) Overlapping plant signal transduction pathways induced by a parasitic nematode and a rhizobial endosymbiont. *Molecular Plant–Microbe Interactions*, 14, 1168–1177.

Lambert, K.N. et al. (1999) Cloning and characterization of an esophageal-gland-specific chorismate mutase from the phytoparasitic nematode *Meloidogyne javanica*. *Molecular Plant–Microbe Interactions*, 12, 328–336.

Lee, C. et al. (2011) The novel cyst nematode effector protein 19C07 interacts with the Arabidopsis auxin influx transporter LAX3 to control feeding site development. *Plant Physiology*, 155, 866–880.

Lelandais-Briere, C. et al. (2009) Genome-wide *Medicago truncatula* small RNA analysis revealed novel microRNAs and isoforms differentially regulated in roots and nodules. *The Plant Cell*, 21, 2780–2796.

Li, Y. et al. (2010) Identification of microRNAs involved in pathogen-associated molecular pattern-triggered plant innate immunity. *Plant Physiology*, 152, 2222–2231.

Lohar, D.P. et al. (2004) Cytokinins play opposite roles in lateral root formation, and nematode and rhizobial symbioses. *The Plant Journal*, 38, 203–214.

Lu, C. et al. (2005) Elucidation of the small RNA component of the transcriptome. *Science*, 309, 1567–1569.

Lu, S. et al. (2007) MicroRNAs in loblolly pine (*Pinus taeda* L.) and their association with fusiform rust gall development. *The Plant Journal*, 51, 1077–1098.

Lu, S.-W. et al. (2009) Structural and functional diversity of CLAVATA3/ESR (CLE)-Like genes from the potato cyst nematode *Globodera rostochiensis*. *Molecular Plant–Microbe Interactions*, 22, 1128–1142.

Ludwig-Müller, J. (2009) *Plasmodiophora brassicae*, the causal agent of clubroot disease: a review on molecular and biochemical events in pathogenesis. *Journal of Plant Disease and Plant Protection*, 106, 109–127.

Ludwig-Müller, J. et al. (2009) Metabolism and plant hormone action during clubroot disease. *Journal of Plant Growth Regulation*, 28, 229–244.

Ludwig-Müller, J. and Schuller, A. (2008) What can we learn from clubroots: alterations in host roots and hormone homeostasis caused by *Plasmodiophora brassicae*. *European Journal of Plant Pathology*, 121, 291–302.

Mazarei, M. et al. (2003) Expression of an Arabidopsis phosphoglycerate mutase homologue is localized to apical meristems, regulated by hormones, and induced by sedentary plant-parasitic nematodes. *Plant Molecular Biology*, 53, 513–530.

Mitchum, M. et al. (2004) The promoter of the *Arabidopsis thaliana* cel1 endo-1,4-beta glucanase gene is differentially expressed in plant feeding cells induced by root-knot and cyst nematodes. *Molecular Plant Pathology*, 5, 175–181.

Mitchum, M.G. et al. (2008) Diverse and conserved roles of CLE peptides. *Current Opinion in Plant Biology*, 11, 75–81.

Muller, R. et al. (2008) The receptor kinase CORYNE of Arabidopsis transmits the stem cell-limiting signal CLAVATA3 independently of CLAVATA1. *Plant Cell*, 20, 934–946.

Nahar, K. et al. (2011) The jasmonate pathway is a key player in systemically induced defense against root knot nematodes in rice. *Plant Physiology*, 157, 305–316.

Navarro, L. et al. (2006) A plant miRNA contributes to antibacterial resistance by repressing auxin signaling. *Science*, 312, 436–439.

Navarro, L. et al. (2008) Suppression of the microRNA pathway by bacterial effector proteins. *Science*, 321, 964–967.

Niebel, A. et al. (1996) Induction of cdc2a and cyc1At expression in *Arabidopsis thaliana* during early phases of nematode-induced feeding cell formation. *The Plant Journal*, 10, 1037–1043.

Niebel, A. et al. (1993) Induction patterns of an extensin gene in tobacco upon nematode infection. *The Plant Cell*, 5, 1697–1710.

Oelkers, K. et al. (2008) Bioinformatic analysis of the CLE signaling peptide family. *BMC Plant Biology*, 8, 1.

Olmedo-Monfil, V. et al. (2010) Control of female gamete formation by a small RNA pathway in Arabidopsis. *Nature* 464, 628–632.

Opperman, C.H. et al. (2008) Sequence and genetic map of *Meloidogyne hapla*: a compact nematode genome for plant parasitism. *Proceedings of the National Academy of Sciences USA*, 105, 14802–14807.

Pandey, S.P. et al. (2008) Herbivory-induced changes in the small-RNA transcriptome and phytohormone signaling in *Nicotiana attenuata*. *Proceedings of the National Academy of Sciences USA*, 105, 4559–4564.

Patel, N. et al. (2010) A nematode effector protein similar to annexins in host plants. *Journal of Experimental Botany*, 61, 235–248.

Portillo, M. et al. (2009) Isolation of RNA from laser-capture-microdissected giant cells at early differentiation stages suitable for differential transcriptome analysis. *Molecular Plant Pathology*, 10, 523–535.

Priya, D.B. et al. (2011) Transgenic tobacco plants constitutively expressing Arabidopsis NPR1 show enhanced resistance to root-knot nematode, *Meloidogyne incognita*. *BMC Research Notes*, 4, 231.

Puthoff, D.P. et al. (2007) GeneChip profiling of transcriptional responses to soybean cyst nematode, *Heterodera glycines*, colonization of soybean roots. *Journal of Experimental Botany*, 58, 3407–3418.

Puthoff, D.P. et al. (2003) Arabidopsis gene expression changes during cyst nematode parasitism revealed by statistical analyses of microarray expression profiles. *The Plant Journal*, 33, 911–921.

Pysh, L.D. et al. (1999) The GRAS gene family in Arabidopsis: sequence characterization and basic expression analysis of the SCARECROW-LIKE genes. *The Plant Journal*, 18, 111–119.

Qin, L. et al. (2004) Plant degradation: a nematode expansin acting on plants. *Nature*, 427, 30.

Ramsay, K. et al. (2006) Laser capture microdissection: a novel approach to microanalysis of plant-microbe interactions. *Molecular Plant Pathology*, 7, 429–435.

Ramsay, K. et al. (2004) Using laser capture microdissection to study gene expression in early stages of giant cells induced by root-knot nematodes. *Molecular Plant Pathology*, 5, 587–592.

Replogle, A. et al. (2011) Nematode CLE signaling in Arabidopsis requires CLAVATA2 and CORYNE. *The Plant Journal*, 65, 430–440.

Richardson, L. and Price, N. (1984) Observations on the biology of *Meloidogyne incognita* and the Diageotropica tomato mutant. *Revue de Nématologie*, 7, 97–99.

Ron, M. et al. (2010) Proper regulation of a sperm-specific cis-nat-siRNA is essential for double fertilization in Arabidopsis. *Genes and Development*, 24, 1010–1021.

Sarkar, A.K. et al. (2007) Conserved factors regulate signalling in *Arabidopsis thaliana* shoot and root stem cell organizers. *Nature*, 446, 811–814.

Schaff, J.E. et al. (2007) Comprehensive transcriptome profiling in tomato reveals a role for glycosyltransferase in Mi-mediated nematode resistance. *Plant Physiology*, 144, 1079–1092.

Schwach, F. et al. (2009) Deciphering the diversity of small RNAs in plants: the long and short of it. *Briefings in Functional Genomics and Proteomics*, 8, 472–481.

Siemens, J. et al. (2006) Transcriptome analysis of Arabidopsis clubroots indicate a key role for cytokinins in disease development. *Molecular Plant–Microbe Interactions*, 19, 480–494.

Sobczak, M. et al. (2011) Cell wall modifications induced by nematodes. In: J. Jones, G. Gheysen, and C. Fenoll (eds) *Genomics and Molecular Genetics of Plant–Nematode Interactions*. Springer, New York, pp. 395–422.

Sobczak, M. and Golinowski, W. (2011) Cyst nematodes and syncytia. In: J.D. Jones, G. Gheysen and C. Fenoll (eds) *Genomics and Molecular Genetics of Plant–Nematode Interactions*, Springer, New York, pp. 61–82.

Strabala, T.J. et al. (2006) Gain-of-function phenotypes of many CLAVATA3/ESR genes, including four new family members, correlate with tandem variations in the conserved CLAVATA3/ESR domain. *Plant Physiology*, 140, 1331–1344.

Sunkar, R. et al. (2006) Posttranscriptional induction of two Cu/Zn superoxide dismutase genes in Arabidopsis is mediated by downregulation of miR398 and important for oxidative stress tolerance. *The Plant Cell*, 18, 2051–2065.

Swarup, K. et al. (2008) The auxin influx carrier LAX3 promotes lateral root emergence. *Nature Cell Biology*, 10, 946–954.

Szakasits, D. et al. (2009) The transcriptome of syncytia induced by the cyst nematode *Heterodera schachtii* in Arabidopsis roots. *The Plant Journal*, 57, 771–784.

Trudgill, D.L. and Blok, V.C. (2001) Apomictic, polyphagous root-knot nematodes: exceptionally successful and damaging biotrophic root pathogens. *Annual Review of Phytopathology*, 39, 53–77.

Tucker, M.L. et al. (2007) Gene expression profiles for cell wall-modifying proteins associated with soybean cyst nematode infection, petiole abscission, root tips, flowers, apical buds, and leaves. *Journal of Experimental Botany*, 58, 3395–3406.

Tucker, M.L. et al. (2011) Gene expression profiling and shared promoter motif for cell wall-modifying proteins expressed in soybean cyst nematode-infected roots. *Plant Physiology*, 156, 319–329.

Van de Cappelle, E. et al. (2008) AtCDKA;1 silencing in *Arabidopsis thaliana* reduces reproduction of sedentary plant-parasitic nematodes. *Plant Biotechnology Journal*, 6, 749–757.

Van der Eycken, W. et al. (1996) A molecular study of root-knot nematode-induced feeding sites. *The Plant Journal*, 9, 45–54.

Vanholme, B. et al. (2009) Structural and functional investigation of a secreted chorismate mutase from the plant-parasitic nematode *Heterodera schachtii* in the context of related enzymes from diverse origins. *Molecular Plant Pathology*, 10, 189–200.

Vaucheret, H. (2006) Post-transcriptional small RNA pathways in plants: mechanisms and regulations. *Genes and Development*, 20, 759–771.

Vercauteren, I. et al. (2002) An *Arabidopsis thaliana* pectin acetylesterase gene is upregulated in nematode feeding sites induced by root-knot and cyst nematodes. *Molecular Plant–Microbe Interactions*, 15, 404–407.

Waetzig, G.H. et al. (1999) Localization of hydrogen peroxide during the defence response of *Arabidopsis thaliana* against the plant-parasitic nematode *Heterodera glycines*. *Nematology*, 1, 681–686.

Wang, J. et al. (2010) Dual roles for the variable domain in protein trafficking and host-specific recognition of *Heterodera glycines* CLE effector proteins. *New Phytologist*, 187, 1003–1017.

Wang, J. et al. (2011) Identification of potential host plant mimics of CLAVATA3/ESR (CLE)-like peptides from the plant-parasitic nematode *Heterodera schachtii*. *Molecular Plant Pathology*, 12, 177–186.

Wang, X. et al. (2001) Signal peptide-selection of cDNA cloned directly from the esophageal gland cells of the soybean cyst nematode *Heterodera glycines*. *Molecular Plant–Microbe Interactions*, 14, 536–544.

Wang, X. et al. (2005) A parasitism gene from a plant-parasitic nematode with function similar to CLAVATA3/ESR (CLE) of *Arabidopsis thaliana*. *Molecular Plant Pathology*, 6, 187–191.

Wang, X. et al. (2007) The tobacco Cel7 gene promoter is auxin-responsive and locally induced in nematode feeding sites of heterologous plants. *Molecular Plant Pathology*, 8, 423–436.

Wang, Z. et al. (2003) Differential display analysis of gene expression in the cytoplasm of giant cells induced in tomato roots by *Meloidogyne javanica*. *Molecular Plant Pathology*, 4, 361–371.

Wasson, A.P. et al. (2009) Differing requirements for flavonoids during the formation of lateral roots, nodules and root knot nematode galls in *Medicago truncatula*. *New Phytologist*, 183, 167–179.

Whitford, R. et al. (2008) Plant CLE peptides from two distinct functional classes synergistically induce division of vascular cells. *Proceedings of the National Academy of Sciences USA*, 105, 18625–18630.

Wieczorek, K. et al. (2006) Expansins are involved in the formation of nematode-induced syncytia in roots of *Arabidopsis thaliana*. *The Plant Journal*, 48, 98–112.

Wieczorek, K. and Grundler, F.M.W. (2006) Expanding nematode-induced syncytia: the role of expansins. *Plant Signaling and Behavior*, 1, 223–224.

Wieczorek, K. et al. (2008) Arabidopsis endo-1,4-beta-glucanases are involved in the formation of root syncytia induced by *Heterodera schachtii*. *The Plant Journal*, 53, 336–351.

Wildermuth, M.C. (2010) Modulation of host nuclear ploidy: a common plant biotroph mechanism. *Current Opinion in Plant Biology*, 13, 449–458.

Wubben, M. J., 2nd et al. (2001) Susceptibility to the sugar beet cyst nematode is modulated by ethylene signal transduction in *Arabidopsis thaliana*. *Molecular Plant–Microbe Interactions*, 14, 1206–1212.

Xie, Q. et al. (2000) Arabidopsis NAC1 transduces auxin signal downstream of TIR1 to promote lateral root development. *Genes and Development*, 14, 3024–3036.

11 Genomics of the Root–Actinorhizal Symbiosis

Valérie Hocher, Nicole Alloisio, Laurent Laplaze, Didier Bogusz, and Philippe Normand

Introduction

With the exception of water, nitrogen availability constitutes the main limiting factor of plant growth. The amounts of available nitrogen for plants in soils are for most regions of the world relatively low. The atmosphere, being constituted of 78% of di-nitrogen, potentially represents an infinite source of nitrogen, but the only existing organisms able to exploit this resource are diazotrophic bacteria, which possess the enzymatic complex nitrogenase. Therefore, some plants have evolved the capability to associate with some nitrogen-fixing bacteria to benefit from this source of nitrogen. The most complex of such associations lead to the formation of new root organs called nodules where the bacteria are hosted in specialized plant cells, fed photosynthates, and fix nitrogen. Two root nodule symbioses are known between nitrogen fixing soil bacteria and higher plants: legumes (as well as the nonlegume *Parasponia*) associated with rhizobia and plants collectively designated actinorhizal that interact with *Frankia*. Inside root nodules, the microsymbionts provide the plant with fixed nitrogen and, in exchange, the bacteria are supplied with carbon by the host plant. Legume and actinorhizal nodules differ in their ontogeny and structure (Franche et al. 1998; Wall 2000). However, phylogenetic studies based on *rbcL* gene sequence analysis have shown that all plants able to enter a root nodule symbiosis belong to the same clade, thus suggesting that they share a predisposition for symbiosis (Soltis et al. 1995; Doyle 1998; Kirstner and Parniske 2002). The molecular bases of this predisposition are so far unknown. Although the symbiosis between *rhizobia* and legumes involves more than 1700 plant species of the *Fabaceae* (legumes) family, few model legume species have been the subject of extensive and detailed studies that have resulted in the discovery of the effectors and transducing signal pathways involved in plant–host recognition (Cohn et al. 1998). Conversely, the symbiotic association between *Frankia* and actinorhizal plants is still poorly understood at the molecular level (Franche and Bogusz 2011; Pawlowski et al. 2011). In contrast to Legumes, model plants have not been identified in actinorhizal plants. Majorities are trees or shrubs with long generation times, thus no genetic systems for the study of these symbioses are available. However, molecular approaches have progressed during last decades mainly for the temperate species, *Alnus glutinosa* and *Datisca glomerata*, and the tropical trees of the *Casuarinaceae* family (Obertello et al. 2003; Laplaze et al. 2008). For the latter, transgenic plants have been obtained (Franche et al. 1998) and recent work show the successful application of a RNA silencing system (RNAi) using hairy root (Gherbi et al. 2008a; Markmann et al. 2008). Thus, actinorhizal symbioses emerged recently as simple and original systems of root nodule

Root Genomics and Soil Interactions, First Edition. Edited by Martin Crespi.

initiation and they offer the opportunity to explore developmental strategies significantly different from legumes to form nitrogen-fixing nodules. One of the key goals for understanding actinorhizal root nodule symbiosis is the identification of the network of plant and bacterial determinants whose gene products interact to initiate root nodule formation and to support the nitrogen fixation process. This is made possible by the recent development of genomic technologies. Model legumes have recently been at the forefront of genomics research into root–microbe interactions and genomic programs developed on *Lotus japonicus* and *Medicago truncatula* greatly contributed to the amount of resources actually available on plant nitrogen-fixing symbiosis (Cook 1999; Udvardi et al. 2005; Mathesius and Van Noorden 2011). For actinorhizal species, genomics is now on the way with the recent sequencing of an EST database and the design of DNA chips for two species *A. glutinosa* and *Casuarina glauca*.

After a brief review of the symbiotic partners and a short description of the morphological and cytological symbiotic events leading to the development of actinorhizal nodules, we review the current state of knowledge on the genomics of the actinorhizal symbiotic interaction.

Actinorhizal Symbiosis

Actinorhizal Species

Legumes, *Parasponia,* and actinorhizal plants are flowering plants encompassing families all found within the Eurosid I clade of the eudicots (Soltis et al. 1995). While legumes are within the order of Fabales and are represented by a single family, the Fabaceae, actinorhizal plants are more taxonomically diverse. They are simply defined as a group by their ability to be nodulated by *Frankia* bacteria. They represent about 200 species encompassing 25 genera in 8 different angiosperm families, in 3 different orders (Huss-Danell 1997): the *Betulaceae*, *Casuarinaceae,* and *Myricaceae* of the order Fagales; the *Rosaceae*, *Rhamnaceae,* and *Elaeagnaceae* of the order Rosales; and the *Coriariaceae* and *Datiscaceae* of the order Cucurbitales (Gualtieri and Bisseling 2000; Vessey et al. 2004). In some families, all members are nodulated (*Coriariaceae*, *Elaeagnaceae*, *Datiscaceae,* and *Casuarinaceae*), whereas in others, only a portion of the genera is nodulated (*Betulaceae*, *Myricaceae*, *Rhamnaceae,* and the *Rosaceae*). In at least one case (*Dryas*), nodulation apparently does not extend to all members of a single genus (Kohls et al. 1994). These observations have led to the conclusion that while the predisposition, or *potential*, to form the nitrogen-fixing symbiosis may have evolved only once, the realization of that potential has occurred and/or been lost multiple times (Swensen 1996; Benson and Clawson 2000). Actinorhizal plants are perennial dicotyledonous angiosperms, and are, with the exception of the genera *Dastica*, woody trees or shrubs. Examples of well-known genera include *Alnus* (alder), *Elaeagnus* (autumn olive), *Hippophae* (sea buckthorn), and *Casuarina* (she oak). These plants are widely distributed, found on all continents except Antarctica and are a diverse group of mostly woody dicots. Most members are found in temperate zones, with only a few members being found in tropical environments and in arctic environments (Baker and Schwintzer 1990; Sprent and Parsons 2000). Most actinorhizal plants are capable of high rates of nitrogen fixation comparable to those found in legumes (Torrey 1976). In Egypt, a nitrogen-fixing potential of 288 kg N ha^{-1} has been reported for *Casuarina* (Diem and Dommergues 1990). Ecologically, actinorhizal plants are usually pioneers on nitrogen-poor soils and are frequently found in relatively harsh sites, including glacial till, new volcanic soil, sand dunes, clear cuts, and desert and chaparral (Schwencke and Caru 2001). As a consequence, these plants play an essential role in land stabilization and soil reclamation (Sprent and Parsons 2000). In addition, some actinorhizal species can grow well under a range of environmental stresses, such as high salinity, extreme pH, and the presence of heavy metal (Dawson 1990). Actinorhizal plants have been used in erosion control, soil reclamation, agroforestry, and dune stabilization, as well as in fuel production. For instance *Casuarinaceae* are utilized in stabilizing desert and coastal dunes (i.e., in shelter belts) and in the reclamation of salt-affected soil (Diem and Dommergues 1990), as well as in revegetation of limestone mine spoils in Kenya (Sprent and Parson 2000) and *Alnus* has been used for revegetation of tar sands heaps in the Alberta massive oil fields and hydroelectric dam dykes in the giant Northern Quebec projects (Roy et al. 2007). Thus, actinorhizal plants play extremely important roles in the N cycle of forests and in the revegetation of various landscapes. The properties of actinorhizal plants as pioneer species are linked mainly with the ability of their root to develop original organs, the nodules, containing the symbiotic N-fixing bacteria, in response to low nitrogen conditions.

The Actinomycete Frankia

The microsymbiont *Frankia* is a filamentous, branching, gram-positive actinobacterium and is characterized by a slow growth rate and high G+C DNA content (Simonet et al. 1990; Benson and Silvester 1993). The first successful isolation of *Frankia* in culture was reported only in 1978 after a century of unsuccessful attempts (Callaham et al. 1978), and even now, several actinorhizal plant genera have not yielded a *Frankia* that can fulfill Koch's postulates. In pure culture, *Frankia* presents three major structures: vegetative hyphae (multiplication form), vesicles that are the site of nitrogen fixation under aerobic conditions, and complex sporangia called multilocular that constitute a dissemination and resistance form.

No *Frankia* strain specific to a single host plant species has been described to date (Pawlowski and Sprent 2008). Nevertheless, host specificity is present at different levels and a broad correspondence can be defined between the phylogenies of *Frankia* strains and actinorhizal plants. The phylogeny of the genus *Frankia* has been deduced by comparative sequence analysis of the 16S rRNA gene, the genes for nitrogen fixation (*nif* genes), and of other genes (Benson and Clawson. 2000). All analyses agree that the genus comprises three major groups or clusters (referred to here as groups 1, 2 and 3), each having different and sometimes overlapping plant specificity, physiological properties, and symbiotic interactions. Members of each cluster show distinct host ranges (Normand et al. 1996). Strains belonging to cluster I (="*Alnus*" cluster) were isolated from plants belonging to the Fagale clade and show the higher level of specificity, as they are only able to interact with plants belonging to this clade. Strains belonging to cluster III (="*Elaeagnus*" cluster) have a wider host range and can interact with plants belonging to five families within two distant plant clades, the Rosales and the Fagales. The third group of *Frankia* (= cluster II or "uncultured") has not yet been isolated in pure culture, but cross-inoculation experiments performed with crushed nodules also suggest a broad host range for members of this cluster that nodulate plants belonging to four families within the Rosales and Cucurbitales clades.

Classification of *Frankia* strains into species is still incomplete with only one species described fully, *Frankia alni* (Fernandez et al. 1989), the other remaining as simply genomospecies. Besides, the unisolated strains remain poorly known with a few descriptive works such as that of Becking (1970), who described species on very scant data. The large rise of molecular or genomic approaches may dramatically change this aspect soon, in particular for several *Casuarina*- (Rouvier et al. 1996; Simonet et al. 1999) and *Alnus*-infective strains. *F. alni* and most other *Alnus*-infective strains belong to cluster 1 as defined on the basis of 16S analysis, the same cluster to which *Casuarina*-infective strains belong (Normand et al. 1996). Despite this close genetic proximity with 1% distance in the 16S gene, the two plants constitute two mutually exclusive infectivity groups, besides the *Datisca–Coriaria–Rosaceae* and the *Rhamnaceae–Elaeagnaceae* groups (Normand and Fernandez 2009).

Development of Actinorhizal Nodules

The morphological steps in the development of actinorhizal nodules have been described in detail (Wall 2000; Laplaze et al. 2008).

Infection Process and Nodule Development

Depending on the host plant, two modes of infection of actinorhizal plants by *Frankia* have been described: intercellular root invasion and intracellular root hair infection (Berry and Sunnel 1990; Duhoux et al. 1996; Wall and Berry 2008).

Intracellular infection via root hairs (e.g., of *Casuarina, Alnus, Myrica*) starts with root hair curling induced by an unknown *Frankia* signal. After invagination of growing filaments of *Frankia* into the curled root hairs, infection proceeds intracellularly in the root cortex. *Frankia* hyphae become encapsulated by a cell wall deposit that is believed to consist of xylans, cellulose, and pectins of host origin (Lalonde and Knowles 1975; Berg 1990). At the same time, limited cell divisions occur in the cortex near the invading root hair, leading to the formation of a small external protuberance called the prenodule (Berry and Sunnel 1990). Infection threads consisting of lines of encapsulated *Frankia* hyphae progress intracellularly toward this mitotically active zone and finally invade most cells of the prenodule (Berg 1999). *Frankia* filaments inside the infected plant cells are always surrounded by plant plasma membrane. As the prenodule develops, cell divisions are induced in the pericycle opposite a protoxylem pole and give rise to the nodule primordium. While cortical cell divisions lead to the formation of a nodule primordium in legumes, actinorhizal prenodules do not evolve in nodules. The function of the *C. glauca* prenodule is not yet fully understood but a study of the expression of symbiosis-related genes (*cg12*, *cghb*) coupled to cellular modification (cell wall lignification) indicated that the prenodule displays the same characteristics as the nodule and can thus be considered as a very simple symbiotic organ (Laplaze et al. 2000). The prenodule could thus be a parallel symbiotic organ on its own or the remaining form of a common nodule ancestor for legumes and actinorhizal plants (Gualtieri and Bisseling 2000; Laplaze et al. 2000).

Concerning the intercellular root invasion pathway (e.g., *Discaria, Ceanothus, Elaeagnus, Hippophae*), *Frankia* hyphae penetrate between two adjacent rhizoderm cells and progress apoplastically through cortical cells within an electron-dense matrix secreted into the intercellular spaces (Miller and Baker 1985; Racette and Torrey 1989; Liu and Berry 1991; Valverde and Wall 1999). Unlike the intracellular mode of infection, no prenodule is formed in the root cortex. Once the nodule primordium has developed from the pericycle, intracellular penetration by *Frankia* and the formation of infection threads is initiated acropetally in developing cortical cells of the nodule lobe primordium, following a pattern similar to that described in actinorhizal species invaded through root hairs.

Further development of the primordium gives rise to an indeterminate actinorhizal nodule lobe. New lobes arise continuously to form a coralloid nodule. In each lobe, there is a central vascular bundle surrounded by an endoderm, an expanded cortex, and a periderm. *Frankia* is restricted to the cortical cells. Some species like *Casuarina* or *Myrica* develop a so-called nodule root at the apex of each lobe (Duhoux et al. 1996). This nodule root lacks root hairs, has a reduced root cap and displays negative geotropism. It might be involved in the diffusion of gases, especially oxygen, in and out of the nodule lobe (Silvester et al. 1990).

Nodule Functioning

Four zones have been morphologically defined in studies of nodules originating from intracellular infection (Angulo Carmona 1974; Duhoux et al. 1996) or intercellular root invasion (Valverde and Wall 1999) and gene expression (Ribeiro et al. 1995). (i) The *apical meristem* is free of *Frankia*; (ii) adjacent to the meristem is an *infection zone* where some of the young cortical cells resulting from the meristem activity are infected by *Frankia*. The bacterium starts to proliferate but remains encapsulated in a plant-derived matrix, and the plant cells enlarge; (iii) the subsequent *fixation zone* contains both infected and uninfected cortical cells. Infected cells are hypertrophied and are filled with *Frankia* filaments that differentiate vesicles where nitrogen fixation takes place. The appearance and shape of these vesicles are controlled by the plant. In some species like *Casuarina*, infected cells have a lignified cell wall and there is no vesicle differentiation. Uninfected cells are smaller and in some species contain amyloplasts and phenolic compounds, and might be involved in nitrogen and carbon metabolism. Finally a basal *senescence zone* (iv) is observed in old nodules; plant cells and bacteria degenerate and nitrogen fixation is switched off. More recently, a second level of compartmentation was described in *C. glauca* nodules based on the accumulation of flavans, which occurs in uninfected cells in the endodermis and in the cortex. These cells form layers that delimit *Frankia*-infected compartments in the nodule lobe and may play a role in restricting bacterial infection to certain zones of the nodule (Laplaze et al. 1999).

Molecular Events Associated with Nodulation in Actinorhizal Species

During differentiation of the symbiotic actinorhizal root nodule, a set of genes—called actinorhizal nodulin genes—is activated in the developing nodules (Mullin and Dobritsa 1996, Pawlowski et al. 1997). Similar to legumes, two major types of actinorhizal nodulin genes have been defined by their pattern of expression and function. Early nodulin genes are expressed before the beginning of nitrogen fixation; they are thought to be involved in plant infection or in nodule organogenesis, whereas late nodulin genes comprise sequences involved in different metabolic activities necessary for the functioning of the nodule (Obertello et al. 2003). Many nodule-specific or nodule-enhanced expressed genes have been detected in several actinorhizal plants including *Alnus*, *Datisca*, *Elaeagnus*, and *Casuarina* (Pawlowski and Bisseling 1996; Wall 2000; Laplaze et al. 2008; Hocher et al. 2006). *cg12* is one of the earliest induced nodulin genes isolated from *C. glauca* (Laplaze et al. 2000); it is homologous to *ag12* previously described for *A. glutinosa* (Ribeiro et al. 1995). *cg12/ag12* has a strong homology to the subtilisin-like protease families of several plants. On the basis of the pattern of expression, it has been suggested that it may play a role in the processing of extracellular or cell wall proteins during early stages of *Frankia* infection (Svistoonoff et al. 2003, 2004). Other actinorhizal genes such as *dg93* have been characterized in *D. glomerata* (Okubara et al. 2000) or *ag164* and *agNt84* in *A. glutinosa* (Pawlowski et al. 1997). These last two genes could be involved in transport/storage of metals during the infection process. Plant hemoglobin is the most studied late nodulin and was found in legumes and actinorhizal nodules (Fleming et al. 1987). The *hb* symbiotic genes are induced in young infected cells prior to the detection of *Frankia nifH* mRNA, suggesting that Hb contributes to reducing O_2 tension before *nif* gene expression (Gherbi et al. 1997). Other genes related to N and C metabolisms were also characterized (Guan et al. 1996; Ribeiro et al. 1996; van Ghelue et al. 1996; Kim et al. 1999; Jeong et al. 2004).

The Actinorhizal Nodule, a Modified Lateral Root

Because they originate from cell divisions in the pericycle in front of xylem poles, similar to lateral roots, and because of their root-like structure, actinorhizal nodule lobes have been considered as modified lateral roots (Pawlowski and Bisseling 1996; Péret et al. 2009). The formation of an apical nodule root at the apex of some actinorhizal nodule lobes has somehow reinforced this idea. This nodule root devoid of *Frankia* infection is found at the apex of each nodule lobe in some species like *Casuarina* or *Myrica*. It is a modified root lacking a root cap and root hairs that shows negative geotropism (i.e., growing toward the surface; Péret et al. 2009). The presence of a developed aerenchyma suggests that nodule roots facilitate gas exchanges (Silvester et al. 1990). Nothing is known about the mechanisms leading to this switch in the behavior of the nodule lobe meristem, leading in turn to nodule root formation. Detailed study in *A. glutinosa* showed that actinorhizal nodule formation had no impact on the number or distribution of lateral roots (Angulo Carmona 1974). Actinorhizal nodules are therefore additional lateral organs and do not form from preexisting lateral root primordia (Angulo Carmona 1974). Moreover, there is no increase in the production of lateral roots as a result of *Frankia* infection. This suggests that the formation of these two types of organs is regulated independently. Moreover, in some species such as *Comptonia*, cortical cells are involved in the formation of the nodule lobe primordium while lateral root initiation is restricted to the pericycle (Callaham and Torrey 1977). Consequently, even if actinorhizal nodules have root-like features, it is not clear to what extent nodules and lateral roots share common developmental pathways. Unfortunately, few studies have addressed this important issue.

Some studies have relied on the use of heterogeneous genes as markers. For instance, the *Arabidopsis thaliana cdc2a* encodes a cyclin-dependent kinase regulating the G1 to S and G2 to M cell-cycle transitions (Ferreira et al. 1991; Hemerly et al. 1992). It was introduced in *C. glauca* to study changes in cell cycle induced by *Frankia* infection. This work showed that *Frankia* can quickly induce Pro_{cdc2a}*:GUS* expression in pericycle cells in front of xylem poles (Sy et al. 2007). Therefore, during early stages of actinorhizal symbioses, cells competent for lateral root formation reacquire competence for cell division in response to symbiotic signals from *Frankia* (Sy et al. 2007). The *Nicotiana tabacum* gene *HRGPnt3* encodes a plant cell wall protein expressed at early stages of lateral root development (Keller and Lamb 1989) and is therefore a good marker for lateral root initiation. When *HRGPnt3* promoter fused to the *β-glucuronidase* reporter gene was introduced into transgenic *Allocasuarina verticillata* plants, it was found that *HRGPnt3* was not expressed during lateral root or nodule development in *Allocasuarina verticillata,* suggesting that tissue-specific expression of this marker is not conserved in a heterologous environment (Laplaze, unpublished results).

Auxin is the key phytohormone controlling lateral root initiation, development, and emergence (Péret et al. 2010). Despite its central role during lateral root development, little is known about the role of auxin during actinorhizal nodule development. Unfortunately, molecular markers of auxin response such as soybean *GH3* gene or the synthetic DR5 promoter (Ulmasov et al. 1995) are not functional in transgenic plants of the *Casuarinaceae* family (Perrine-Walker et al. 2010). Nevertheless, auxin accumulates in actinorhizal nodules (Wheeler 1979; Perrine-Walker et al. 2010) and drugs interfering with auxin transport and lateral root development such as 2,3,5-triiodobenzoic acid or naphtoxyacetic acid (NOA) perturb actinorhizal nodule development (Duhoux et al. 1996;

Péret et al. 2007). In *C. glauca*, two genes, *CgAUX1* and *CgLAX3*, encoding auxin influx carriers were identified (Péret et al. 2007). These genes are homologous to *Arabidopsis* AUX1 and LAX3 that are involved in lateral root initiation and emergence, respectively (Péret et al. 2007; Péret et al. 2010). Interestingly, *CgAUX1* is expressed in lateral root primordia in *C. glauca* but not nodule lobe primordia (Péret et al. 2007). This demonstrates differences in the development of these two organs. More recently, we identified *CgIAA7* encoding a transcriptional regulator of auxin response in *C. glauca*. This gene is expressed both in lateral root and nodule lobe primordia (Laplaze et al., in preparation).

Besides, *Alnus*-infective *Frankia* strains have been shown to synthesize auxins in pure culture. Indole-acetic acid is an auxin widely produced by bacteria, the demonstration it was produced by ArI3 (Berry et al. 1989) thus came as not surprising. More unusual was the demonstration that another auxin was synthesized by *F. alni*, phenyl-acetic acid (Hammad et al. 2003). This compound has been shown to be present as part of the vesicle-specific lipid, phenylacetate-bacterio-hopane-tetrol that has been shown to provide protection against oxygen diffusion that would otherwise inactivate nitrogenase (Berry et al. 1993). When applied onto *A. glutinosa* roots at a concentration of 10^{-4} and 10^{-5} M, this compound induced the emergence of numerous stunted lateral roots that resemble nodules.

Taken together these data suggest that part of the developmental program controlling lateral root development was recycled during evolution for actinorhizal nodule development. Recent transcriptomics studies in actinorhizal symbioses show that about 10% of genes are significantly regulated in nodules compared to nitrogen-starved roots (Hocher et al. 2011a). This indicates that a large part of the genes expressed in nodules are also expressed in roots and reinforce the idea that actinorhizal nodules are root-like structures. However, these studies do not have the resolution to compare directly lateral root and actinorhizal nodule development programs. Clearly, further work is needed to understand how actinorhizal nodules develop and to what extent this developmental program is derived from lateral root development. However, the development of microarrays and the advent of new sequencing technologies together with lateral root induction protocols (Himanen et al. 2004; Lucas et al. 2008) pave the way to future comparison of global changes in gene expression in actinorhizal plants during nodule and lateral root development.

Genomic Resources for Studying Actinorhizal Symbiosis

Frankia Genomic Resources

Frankia cannot so far be transformed despite several attempts at developing plasmid transformation, protoplasts regeneration, or electroporation (Cournoyer and Normand 1992), thus genetics cannot be used to identify symbiotic determinants, which was one reason why sequencing of the genomes was undertaken. The genomic resources available are expanding significantly since the 2007 publication of three *Frankia* genome sequences belonging to different host-compatibility groups, *F. alni* that infects *Alnus*, *Frankia* CcI3 that infects *Casuarina*, and *Frankia* EAN1pec that infects *Elaeagnus* (Normand et al. 2007a). A JGI program was later accepted to sequence the genome of a *Frankia* sp. that is an obligate endosymbiont of *Datisca glomerata* and the sequence is now available (NC_015656). There are other genomes underway with EuI1c (NC_014666) and EUN1f (NZ_ADGX00000000) already available, both infective on *Elaeagnus* and several others not yet completed. These genomes cover a large part of the existing diversity of *Frankia*, with three of the four clades recognized (Normand et al. 1996) included and only the #4 grouping comprising

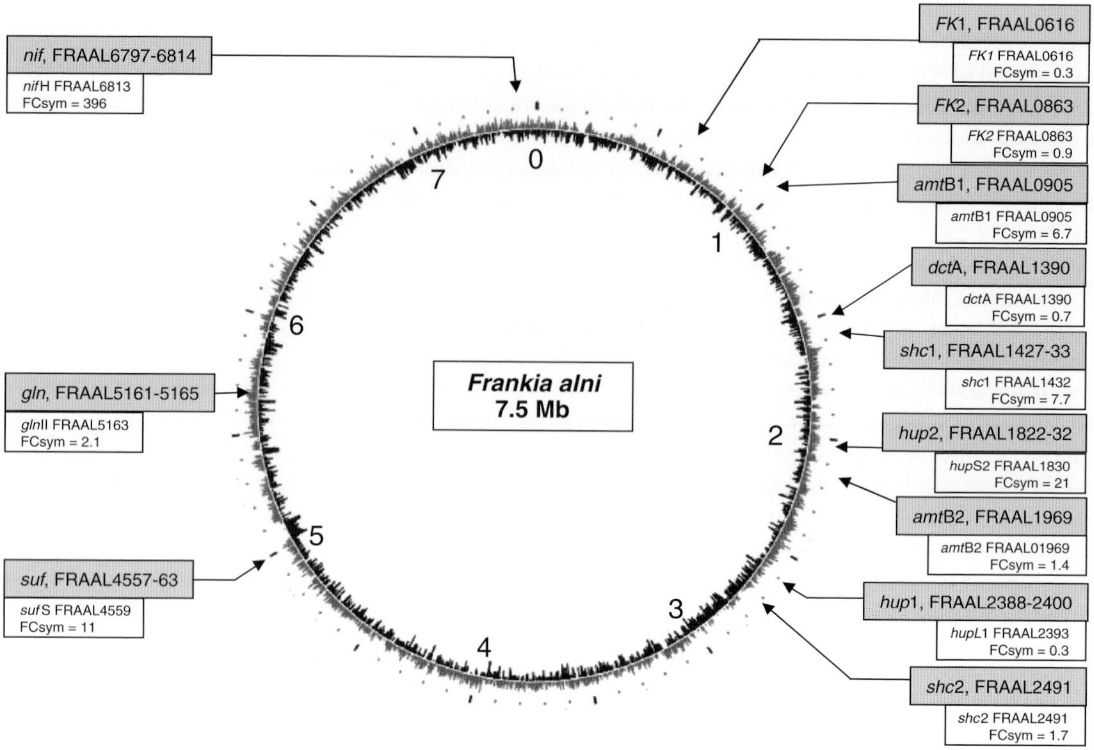

Figure 11.1 Map of the *Frankia alni* genome; G + C% deviations are indicated as well as the most important genes and gene clusters whose functions in symbiosis are known or suggested (indicated by a gray box), derived from the Mage site (Vallenet et al. 2006). Under each gene is indicated the fold change (FC) in nodules relative to growth in a defined medium (BAP+). In the case of clusters, the FC indicated is the most important one. From the top, upregulated clusters are the *amt*B1*glnB1* cluster encoding an ammonium transporter, the *shc*1 cluster that plays a role in synthesis of the hopanoid lipids, the *hup*2 cluster that plays a role in recycling of hydrogen generated by nitrogenase, the *suf* cluster that plays a role in synthesis of iron-sulfur clusters, the *gln* cluster involved in assimilation of ammonium, and the *nif* cluster that codes for nitrogenase. In contrast to their paralogs, the *amt*B2 gene, the *hup*1 cluster, and the *shc*2 gene do not display significant upregulation. The *FK1* and *FK2* genes encoding lectins that play a role in attachment to *Alnus* roots (Pujic et al. submitted) are not upregulated in nodules. Surprisingly, the *dct*A gene encoding C4-dicarboxylate transporter was also not differentially expressed. The most similar homologs to *Rhizobium* spp. *nodB*, *nodC*, and *nodD* genes are also indicated by asterisks and white boxes.

noninfective or nonefficient strains left out. *Frankia* genome analysis revealed the absence of canonical nod genes. Only a few, low similarity *nodB* and *nodC* homologs were detected. Moreover, genes known to be involved in symbiosis such as *nif* (nitrogenase), *shc* (squalene hopene cyclase), *hup* (hydrogenase uptake), and *suf* (sulfur-iron cofactor synthesis) are scattered over the genomes away from the distant *nod* homologs (Figure 11.1). This lack of a symbiotic island is in sharp contrast to the situation of rhizobia where *nif* genes are clustered with *nod* and ancillary genes, as a likely sequel of recent lateral transfers (Amadou et al. 2008). A consequence of this absence of symbiotic island is that the proximity of genes to the *nif* genes cannot be construed as an indication that they are involved in symbiosis. There are also several lectin genes that have been shown to increase the number of nodules formed with limited numbers of microbial cells, but these lectin genes appear to be specific of *Alnus*-infective strains (Pujic et al. submitted).

The genome sequencing of *F. alni* strain ACN14a allows the investigation of symbiosis by large-scale tools such as transcriptomics. Expression profiling of 6607 predicted genes were investigated using microarrays (Alloisio et al. 2010). Transcript levels of these genes were monitored under nitrogen-replete free-living condition and in *A. glutinosa* young nodules (harvested 3–4 weeks after inoculation) to identify genes differentially expressed in symbiosis. The data obtained confirmed that *F. alni* do not contain a symbiotic island consisting of numerous nodule-induced genes as it was found in some rhizobia such as *Bradyrhizobium japonicum* (Pessi et al. 2007) and *Mesorhizobium loti* (Uchiumi et al. 2004). Among the 53 genes displaying a fold-change (FC) value > 10 in symbiosis vs. free-living condition, the most highly induced genes were the *nif* genes, especially the structural *nif*HDK genes encoding nitrogenase (FC = 396; Figure 11.1). Several other genes known to be related to nitrogen fixation were also found in the top-regulated group: genes encoding uptake hydrogenase 2 (*hup2*) that converts hydrogen produced by nitrogenase into protons and electrons, *suf* genes involved in iron-sulfur cluster formation, which is strongly required for key nitrogen fixation-related enzymes such as nitrogenase and uptake hydrogenase 2, and genes belonging to the squalene-hopene cyclase 1 (*shc*1) cluster containing the major genes, leading to hopanoid biosynthesis. Besides these results pointing out that *F. alni* is a nitrogen-fixing machine in symbiosis, data related to nitrogen assimilation revealed the inhibition of *gln*II activation (FC = 2.1) in contrast with the known strong induction upon nitrogen-fixing free-living conditions (FC = 15). The *amt*B1 gene encoding ammonium transport protein was also significantly up-regulated. Thus, it can be suggested that, in nodules, bacterial ammonium assimilation is limited in order to increase ammonium diffusion from the vesicles to host-infected cells. Surprisingly, the *dct*A gene whose product could be involved in the uptake of the host photosynthetic compounds (C4-dicarboxylates) was not differentially expressed. Numerous induced genes encoding proteins that may play a role during symbiosis in transcriptional regulation, signaling processes, protein drug export, protein secretion, lipopolysaccharide, and peptidoglycan biosynthesis were also identified.

All these genomic data will help to understand the bacterial physiological changes occurring in nodules. Following this first global gene expression pattern of *F. alni* in nodules, an investigation of the early stages of infection, which are of particular interest in the absence of known signal molecules initiating symbiosis, will be performed using large-scale cDNA sequencing. Recent review by Benson et al. 2011 addresses the current status of molecular biology and genomics.

Plant Genomic Resources

Between 2006 and 2011, funding from Genopole (Montpellier, France) and Genoscope (Evry, France) allowed the creation of the first actinorhizal EST database for two actinorhizal species *A. glutinosa* and *C. glauca* (Hocher et al. 2006, 2011a). Starting from cDNA libraries obtained from noninoculated roots (controls), inoculated roots (2, 4, and 7 days postinoculation (dpi)), nodules (3 weeks postinoculation), and mycorrhizae (only for *C. glauca*), two sets of high-quality ESTs were sequenced. For both actinorhizal species, the ESTs database contains a total of about 35,000 valid ESTs. Each set of ESTs sequence data was processed and annotated using a multimodule custom pipeline as previously described in Hocher et al. (2006) and further clustered using TGICL (Pertea et al. 2003); http://compbio.dfci.harvard.edu/tgi/software/) to generate a set of nonredundant gene sequences. The clustering resulted in about 15,000 unigenes for *A. glutinosa* and for *C. glauca*. All resulting data (sequences, clustering results, and BLAST results) were automatically integrated into a relational database, searchable *via* a local Web browser-based interface.

To analyze transcriptional changes occurring during symbiosis in *A. glutinosa* and *C. glauca*, a 15K Agilent custom oligonucleotide chip was designed for each species by Imaxio (Clermont Ferrand, France; http://www.imaxio.com/index.php). Two biological conditions were compared for both plants: noninoculated roots and 3-week-old nodules and resulted in the identification of sets of differentially expressed genes for *C. glauca* and *A. glutinosa* (Hocher et al. 2011a).

All these data are currently accessible through public database (Hocher et al. 2011a) and directly upon request from V. Hocher and N. Alloisio. In order to handle all actinorhizal plants and *Frankia* genomics data available, a WEB platform will soon be generated with the aim to integrate and distribute knowledge related to actinorhizal symbioses worldwide

What Did We Learn from Actinorhizal Genomics?

A Similar Genetic Program Is Recruited During Symbiosis of Actinorhizal Species

The global analysis of our data revealed that 11–14% of the 15,000 unigenes were shown to be regulated or specifically induced during nodule development for *A. glutinosa* and *C. glauca*, respectively. A majority of them are linked to transport, metabolism, protein synthesis machinery, cell wall, defense, and response to stress, indicating that nodulation is associated with a high metabolic activity. This situation was found to be very close to that described in different transcriptomic studies performed in model legumes (Asamizu et al. 2000; Journet et al. 2002; Colebatch et al. 2004; El Yahyaoui et al. 2004; Maunoury et al. 2010). Among the regulated genes, all the different nodulins already described in *A. glutinosa* and/or *C. glauca* (Goetting-Minesky and Mullin 1994; Ribeiro et al. 1995, 1996; Guan et al. 1996, 1997; van Ghelue et al. 1996; Pawlowski et al. 1997, 2003; Svistoonoff et al. 2003; Jeong et al. 2004) were identified in both species with a similar regulation pattern (Hocher et al. 2011a).

The results of this global analysis represent an important step toward the understanding of actinorhizal symbiosis. First, for the first time, a similar global behavior is found during nodulation of two actinorhizal species, and this behavior is found very similar to what is described for legumes, thus suggesting similar genetic programs for root nodule symbioses. Second, the regulated genes represent a reservoir of putative new actinorhizal nodulins, and functional studies are now needed for their further characterization.

Flavonoid: An Initial Signal in Actinorhizal Species?

As in the nitrogen-fixing symbiosis between legumes and rhizobia, the actinorhizal symbiotic interaction starts by the exchange of symbiotic signals in the soil between the plant and the bacteria. In legumes, flavonoids are involved as plant chemotactic signals to rhizobia and as activators of *nod* gene expression, which leads to the biosynthesis of lipochitooligosaccharides (Nod Factors), the bacterial symbiotic signals (Oldroyd and Downie 2008). However, very little is known about the plant signaling pathways leading to actinorhizal symbioses.

If several studies have reported the implication of flavonoids during actinorhizal plants–*Frankia* interaction, direct evidence of flavonoids as early signals between plant and *Frankia* is still lacking (Perrine-Walker et al. 2011). The recent development of genomics in *C. glauca* brings new insights into a putative role of flavonoids during actinorhizal symbiosis. *C. glauca* transcriptome analysis allowed to identify eight genes implicated in flavonoids biosynthetic pathway (Hocher et al. 2006;

Auguy et al. 2011), and a kinetic study of their expression during a nodulation time-course revealed a higher accumulation of isoflavone reductase transcripts as soon as 12 h after inoculation of the plant by the bacteria. These data suggest that flavonoids could act as plant signals in actinorhizal symbioses and might determine the symbiotic specificity in actinorhizal symbioses (Hocher et al. 2011b). In order to test this hypothesis, we are currently using a RNAi strategy (Gherbi et al. 2008a) to generate plants impaired in flavonoid biosynthesis through downregulation of a gene encoding chalcone synthase, the first enzyme in this biosynthetic pathway, and look at the effects on nodulation.

On the bacterial side, *Frankia* Nod factor equivalents have not been characterized yet. This was confirmed by the recent sequencing of three *Frankia* genomes that revealed the lack of the canonical Nod factor genes (Normand et al. 2007a, 2007b). A partial purification of a root hair deforming factor from the supernatant of *F. alni* cultures was achieved using root hair deformation assay on *Alnus*. This *Frankia* root hair deforming factor was found to be of different nature than Nod factors (Ceremonie et al. 1999). Several approaches are presently underway to characterize further the root hair deforming factor (Perrine-Walker et al. 2011).

Signal Transduction

In legumes, unraveling the signal transduction pathways required for successful symbiosis has been accelerated in recent years through the identification and study of key genes involved in symbiosis. Recent work using model species as *Medicago truncatula* and *Lotus japonicus* elucidated the role of many genes essential for the different nodulation steps (Oldroyd et al. 2009). A very important discovery was that a genetic overlap exists between legumes root nodule symbiosis and the more ancient arbuscular mycorrhizal symbiosis referred to as the "SYM" pathway (Capoen et al. 2009; Markmann and Parniske 2009).

Aside from the recent demonstration of *SymRK* as a linchpin in *C. glauca* and *Datisca glomerata* (Gherbi et al. 2008b; Markmann et al. 2008), very little is known about symbiotic determinants of actinorhizal plants. The genomic program developed on *A. glutinosa* and *C. glauca* allowed to lift the veil on the possibly involved gene candidates (Hocher et al. 2011a). Most legume symbiosis genes have orthologs in the two actinorhizal plants, which covered all members of the symbiosis pathway from the signal perception to the nodulation process *via* the bacterial and fungal common part. Six of them (*Lys6, CCaMK, Hap2–1, IIMG, Cyp2,* and *Cyp4*) were identified in both species. More interestingly, expression analysis of these symbiosis orthologs in *A. glutinosa* and *C. glauca* revealed that transcript accumulation in nodules versus uninfected roots was not only comparable between the two actinorhizal species but also similar to those found in legumes (Hocher et al. 2011a). Taken all together, these results support for the first time a role for these orthologs in the actinorhizal-*Frankia* symbiosis, and second, they confirm a conserved signaling pathway for endosymbiosis. The fact that a series of well-characterized symbiotic genes in legumes exhibit similar expression patterns in actinorhizal plants further support the existence of a common "SYM" pathway for endosymbioses and, for the first time, points to the possibility of a similar "NOD" pathway between root nodule symbioses (Figure 11.2). The overlapping of legume and actinorhizal root nodule symbioses reinforces the hypothesis of a common genetic ancestor of the nodulating clade with a genetic predisposition for nodulation (Soltis et al. 1995).

To confirm these hypotheses, functional studies are now developed on several candidates like *CCaMK* and *NIN* using RNA interference. Moreover, the identification of common genes implicated

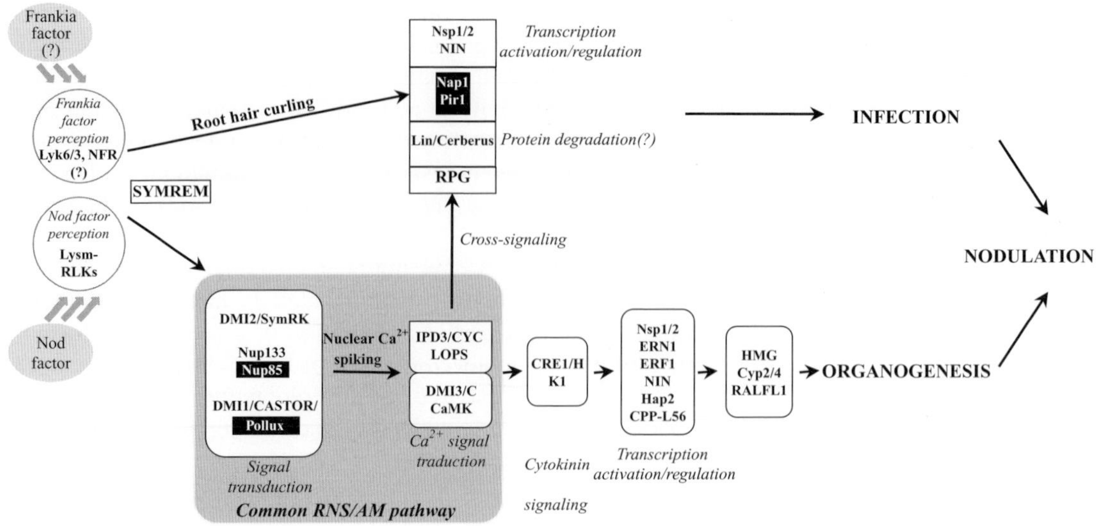

Figure 11.2 Proposed model for a common symbiotic signaling pathway between actinorhizal plants and legumes. The common genes identified in *A. glutinosa* and/or *C. glauca* were assigned to infection and organogenesis pathways. Black boxes indicate legume genes not yet identified in *A. glutinosa* or *C. glauca*. The common pathway between RNS and AM is indicated by gray box. (From Hocher et al. 2011a).

in signaling during mycorrhizal symbiosis is on the way using the *C. glauca* chip and hybridizations with cDNAs from mycorrhized roots.

Conclusion and Future Directions

In conclusion, having genomic resources for actinorhizal plant species and *Frankia* is a very important step that should allow us to progress in identification of key symbiosis determinants.

On the bacterial side, the priority is now to search for symbiotic signal molecules and the isolation of the corresponding genes. Characterization of *Alnus* root hair deforming factor(s) is in progress (van Ghelue 1997; Ceremonie et al. 1999). Whatever the signal molecules isolated are, their implication in symbiosis must be demonstrated. The development of a bioassay using symbiotic genes as markers is also underway (S. Svistoonoff, personal communication).

On the plant side, comparison of the genetic programs that control actinorhizal and rhizobium-legume symbioses must continue (Hocher et al. 2011b). This should lead to the identification of a "core" symbiotic program essential for root nodulation.

The basic knowledge obtained on legume–rhizobium and actinorhizal symbioses may help define strategies to expand the capacity of nodulation to nonnodulating crop plants.

Acknowledgments

We gratefully acknowledge support from IRD, CNRS (Project EC2CO), Genoscope, Genopole of Montpellier, and Agence Nationale de la Recherche (Projects NewNod AN-06-BLAN-0095 and SESAM 2010 BLAN 1708 01).

References

Alloisio, N. et al. (2010) The *Frankia alni* symbiotic transcriptome. *Molecular Plant–Microbe Interaction*, 23, 593–607.

Amadou, C. et al. (2008) Genome sequence of the beta-rhizobium *Cupriavidus taiwanensis* and comparative genomics of rhizobia. *Genome Research*, 18, 1472–1483.

Angulo Carmona, A.F. (1974) La formation des nodules fixateurs d'azote chez *Alnus glutinosa* (L.). *Acta Botanica Neerlandica*, 23, 257–303.

Asamizu, E. et al. (2000) Generation of 7137 non-redundant expressed sequence tags from a legume, *Lotus japonicus*. *DNA Research*, 7, 127–130.

Auguy, F. et al. (2011) Isoflavonoids pathway activation in actinorhizal symbioses. *Functional Plant Biology Special Issue: Research Front on Actinorhizal Plants*, 38, 690–696.

Baker, D.D. and Schwintzer, C.R. (1990) Introduction. In: C.R. Schwintzer and J.D. Tjepkema (eds) *The Biology of Frankia and Actinorhizal Plants*. Academic Press, New York, pp. 157–176.

Becking, J.H. (1970) *Frankiaceae* fam. nov. (Actinomycetales) with one new combination and six new species of the genus *Frankia* Brunchorst 1886. *International Journal of Systematic and Evolutionary Microbiology*, 20, 201–220.

Benson, D. et al. (2011) The biology of *Frankia* sp. strains in the post-genome era. *Molecular Plant–Microbe Interaction*, DOI: 10.1094/MPMI-06–11-0150, in press.

Benson, D.R. and Clawson, M.L. (2000) Evolution of the actinorhizal plant symbioses. In: E.W. Triplett (ed) *Prokaryotic Nitrogen Fixation: A Model System for Analysis of Biological Process*. Horizon Scientific Press, Wymondham, UK, pp. 207–224.

Benson, D.R. and Silvester, W.B. (1993) Biology of *Frankia* strains, actinomycete symbionts of actinorhizal plants. *Microbiological Reviews*, 57, 293–319.

Berg, R.H. (1990) Cellulose and xylans in the interface capsule in symbiotic cells of aetinorhizae. *Protoplasma*, 159, 35–43.

Berg, R.H. (1999) Cytoplasmic bridge formation in the nodule apex of actinorhizal root nodules. *Canadian Journal of Botany*, 77, 1327–1333.

Berry, A.M. et al. (1989) Identification of indole compounds secreted by Frankia HFPArI3 in defined culture medium. *Plant and Soil*, 118, 205–209.

Berry, A.L. and Sunnel, L.A. (1990) The infection process and nodule development. In: C.R. Schwintzer and J.D. Tjepkema (eds) *The Biology of Frankia and Actinorhizal Plants*. Academic Press, New York, pp. 61–81.

Berry, A.M. et al. (1993). Hopanoid lipids compose the Frankia vesicle envelope, presumptive barrier of oxygen diffusion to nitrogenase. *Proceedings of the National Academy of Sciences of the USA*, 90, 6091–6094.

Callaham, D., and Torrey, J.G. (1977) Prenodule formation and primary nodule development in roots of Comptonia (Myricaceae). *Canadian Journal of Botany*, 55, 2306–2318.

Callaham, D. et al. (1978) Isolation and cultivation in vitro of the actinomycete causing root nodulation in Comptonia. *Science*, 199, 899–902.

Capoen, W. et al. (2009) Calcium spiking patterns and the role of the calcium/calmodulin-dependent kinase CCaMK in lateral root base nodulation of *Sesbania rostrata*. *The Plant Cell*, 21, 1526–1540.

Ceremonie, H. et al. (1999) Structural and functional comparison of *Frankia* root hair deforming factor and rhizobia Nod factor. *Canadian Journal of Botany*, 77, 1293–1301.

Cohn, J. et al. (1998) Legume nodule organogenesis. *Trends in Plant Science*, 3, 105–110.

Colebatch, G. et al. (2004) Global changes in transcription orchestrate metabolic differentiation during symbiotic nitrogen fixation in *Lotus japonicus*. *The Plant Journal*, 39, 487–512.

Cook, D. (1999) *Medicago truncatula*—a model in the making! *Current Opinion in Plant Biolology*, 2, 301–304.

Cournoyer, B. and Normand, P. (1992) Electropermeabilization of Frankia intact cells to plasmid DNA. *Acta Oecologica*, 13, 369–378.

Dawson, J.O. (1990) Interactions among actinorhizal and associated species. In: C.R. Schwintzer and J.D. Tjepkema (eds) *The Biology of Frankia and Actinorhizal Plants*. Academic Press, New York, pp. 299–316.

Diem, H.G. and Dommergues, Y.D. (1990) Current and potential uses and management of Casuarinaceae in the tropics and subtropics. In: C.R. Schwintzer and J.D. Tjepkema (eds) *The Biology of Frankia and Actinorhizal Plants*. Academic Press, New York, pp. 317–342.

Doyle, J.J. (1998) Phylogenetic perspectives on nodulation: evolving views of plants and symbiotic bacteria. *Trends in Plant Science*, 3, 473–478.

Duhoux, E. et al. (1996) Le nodule actinorhizien. *Acta Botanica Gallica*, 143, 593–608.

El Yahyaoui, F. et al. (2004) Expression profiling in *Medicago truncatula* identifies more than 750 genes differentially expressed during nodulation, including many potential regulators of the symbiotic program. *Plant Physiology*, 136, 3159–3176

Fernandez, M.P. et al. (1989) Deoxyribonucleic acid relatedness among members of the genus *Frankia*. *International Journal of Systematic Bacteriology*, 39, 424–429.

Ferreira, R.C.G. et al. (1991). The Arabidopsis functional homolog of the p34cdc2 protein kinase. *The Plant Cell*, 3, 531–540.

Fleming, A.I. et al. (1987). The purification, characterization and ligand-binding kinetics of hemoglobins from root nodules of the non-leguminous *Casuarina* glauca-Frankia symbiosis. *Biochemistry Biophysic Acta*, 911, 209–220.

Franche, C. and Bogusz, D. (2011) Signalling and communication in the actinorhizal symbiosis. In: S. Perroto and F. Baluska (eds) *Signaling and Communication in Plant Symbiosis*. Spinger-Verlag, Berlin Heidelberg. Doi 10. 1007/978/978–3-642–20966-6–4, in press.

Franche, C. et al. (1998) Actinorhizal symbioses: recent advances in plant molecular and genetic transformation studies. *Critical Review in Plant Science*, 17, 1–28.

Gherbi, H. et al. (1997) Cloning of a full-length symbiotic hemoglobin cDNA and *in situ* localization of the corresponding mRNA in *Casuarina glauca* root nodule. *Physiologia Plantarum*, 99, 608–616.

Gherbi, H. et al. (2008b) SymRK defines a common genetic basis for plant root endosymbioses with arbuscular mycorrhiza fungi, rhizobia, and Frankia bacteria. *Proceedings of the National Academy of Sciences of the USA*, 105, 4928–4932.

Gherbi, H. et al. (2008a) Post-transcriptional gene silencing in the root system of the actinorhizal tree *Allocasuarina verticillata*. *Molecular Plant–Microbe Interaction*, 21, 518–524.

Goetting-Minesky, M.P. and Mullin, B.C. (1994) Differential gene expression in an actinorhizal symbiosis: evidence for a nodule-specific cysteine proteinase. *Proceedings of the National Academy of Sciences of the USA*, 91, 9891–9895.

Gualtieri, G. and Bisseling, T. (2000) The evolution of nodulation. *Plant Molecular Biology*, 42, 181–194.

Guan, C. et al. (1997) *ag13* is expressed in *Alnus glutinosa* nodules in infected cells during endosymbiont degradation and in the nodule pericycle. *Physiologia Plantarum*, 99, 601–607.

Guan, C. et al. (1996) Nitrogen metabolism in actinorhizal nodules of *Alnus glutinosa*: expression of glutamine synthetase and acetylornithine transaminase. *Plant Molecular Biology*, 32, 1177–1184.

Hammad, Y. et al. (2003) A possible role for phenyl acetic acid (PAA) on *Alnus glutinosa* nodulation by Frankia. *Plant and Soil*, 254, 193–206.

Hemerly, A. et al. (1992). Genes regulating the plant cell cycle: isolation of a mitotic-like cyclin from *Arabidopsis thaliana*. *Proceedings of the National Academy of Science of the USA*, 89, 3295–3299.

Himanen, K. et al. (2004) Transcript profiling of early lateral root initiation. *Proceedings of the National Academy of Science of the USA*, 101, 5146–5151.

Hocher, V. et al. (2011a) Transcriptomics of actinorhizal symbioses reveals homologs of the whole common symbiotic signaling cascade. *Plant Physiology*, 156, 700–711.

Hocher, V. et al. (2011b) Early signaling in actinorhizal symbioses. *Plant Signaling and Behavior*, 6, 1377–1379.

Hocher, V. et al. (2006) Expressed sequence tag analysis in *Casuarina glauca* actinorhizal nodule and root. *New Phytologist*, 169, 681–688.

Huss-Danell, K. (1997) Actinorhizal symbioses and their N_2 fixation. *New Phytologist*, 136, 375–405.

Jeong, J. et al. (2004) A nodule-specific dicarboxylate transporter from alder is a member of the peptide transporter family. *Plant Physiology*, 134, 969–978.

Journet, E.P. et al. (2002) Exploring root symbiotic programs in the model legume *Medicago truncatula* using EST analysis. *Nucleic Acids Research*, 30, 5579–5592.

Keller, B. and Lamb, C.J. (1989). Specific expression of a novel cell hydroxyproline-rich glycoprotein gene in lateral root initiation. *Genes Development*, 3, 1639–1646.

Kim, H.B. et al. (1999) Isolation and characterization of a cDNA clone encoding asparagine synthetase from root nodules of *Elaeagnus umbellata*. *Plant Science*, 149, 85–94.

Kistner, C. and Parniske, M. (2002) Evolution of signal transduction in intracellular symbiosis. *Trends in Plant Science*, 7, 511–518.

Kohls, S.J. et al. 1994. Nodulation patterns of actinorhizal plants in the family Rosaceae. *Plant and Soil*, 162, 229–239.

Lalonde, M. and Knowles, R. (1975) Ultrastructure, composition and biogenesis of the encapsulation material surrounding the endophyte of *Alnus crispa* var. *mollis* root nodules. *Canadian Journal of Botany*, 5, 1951–1971.

Laplaze, L. et al. (2008) Molecular Biology of Actinorhizal Symbioses. In: K. Pawlowski and W.E. Newton (eds) *Nitrogen-Fixing Actinorhizal Symbioses*, Vol. 6. Springer Verlag, Dordrecht, The Netherlands, pp. 235–253.

Laplaze, L. et al. (2000) *Casuarina glauca* prenodule cells display the same differentiation as the corresponding nodule cells. *Molecular Plant–Microbe Interaction*, 13, 107–112.

Laplaze, L. et al. (1999) Flavan-containing cells delimit *Frankia*-infected compartments in *Casuarina glauca* nodules. *Plant Physiology*, 121, 113–122.

Liu, Q. and Berry, A.M. (1991) The infection process and nodule initiation in the *Frankia-Ceanothus* root nodule symbiosis: a structural and histochemical study. *Protoplasma*, 163, 82–92.

Lucas, M. et al. (2008) An auxin transport-based model of root branching in *Arabidopsis thaliana*. *PLoS One*, 3, e3673.

Markmann, K. and Parniske, M. (2009) Evolution of root endosymbiosis with bacteria: how novel are nodules? *Trends in Plant Science*, 14, 77–86.

Markmann, K. et al. (2008) Functional adaptation of a plant receptor-kinase paved the way for the evolution of intracellular root symbioses with bacteria. *PLoS Biology*, 6, e68.

Mathesius, and Van Noorden, (2011) Genomics of root-microbe interactions. In: A. Costa de Oliveira and R.K. Varshney (eds) *Root Genomics*. Springer-Verlag, Berlin Heidelberg, pp. 73–97.

Maunoury, N. et al. (2010) Differentiation of symbiotic cells and endosymbionts in *Medicago truncatula* nodulation are coupled to two transcriptome-switches. *PLoS One*, 5, e9519.

Miller, I.M. and Baker, D.D. (1985) The initiation, development and structure of root nodules in *Elaeagnus angustifolia* L. (*Elaeagnaceae*). *Protoplasma*, 128, 107–119.

Mullin, B.C. and Dobritsa, S.V. (1996) Molecular analysis of actinorhizal symbiotic systems: progress to date. *Plant and Soil*, 186, 9–20.

Normand, P. and Fernandez, M. (2009) Evolution and diversity of Frankia. In: K. Pawlowski (ed.) *Prokaryotic Symbionts in Plants, Microbiology Monographs*, Vol. 8. (A. Steinbüchel (Series ed.)). Springer Verlag, Berlin Heidelberg, pp. 103–125.

Normand, P. et al. (2007a) Genome characteristics of facultatively symbiotic *Frankia* sp. strains reflect host range and host plant biogeography. *Genome Research*, 17, 7–15.

Normand, P. et al. (1996) Molecular phylogeny of the genus Frankia and related genera and emendation of the family *Frankiaceae*. *International Journal of Systematic Bacteriology*, 46, 1–9.

Normand, P. et al. (2007b) Exploring the genomes of *Frankia* sp. *Physiologia Plantarum*, 13, 331–343.

Obertello, M. et al. (2003) Actinorhizal nitrogen fixing nodules: infection process, molecular biology and genomics. *African Journal of Biotechnology*, 2, 528–538.

Okubara, P.A. et al. (2000) *Dg93*, a nodule-abundant mRNA of Datisca glomerata with homology to a soybean early nodulin gene. *Plant Physiology*, 122, 1073–1079.

Oldroyd, G.E. and Downie, J.A. (2008) Coordinating nodule morphogenesis with rhizobial infection in legumes. *Annual Review of Plant Biology*, 59, 519–546.

Oldroyd, G.E. et al. (2009) Reprogramming plant cells for endosymbiosis. *Science*, 324, 753–754.

Pawlowski, K., and Sprent, J. (2008) Comparison between Actinorhizal and legume symbiosis. In: K. Pawlowski and W.E. Newton (eds) *Nitrogen-Fixing Actinorhizal Symbioses*, Vol. 6. Springer Verlag, Dordrecht, The Netherlands, pp. 261–281.

Pawlowski, K., and Bisseling, T. (1996) Rhizobial and actinorhizal symbioses: what are the shared features? *The Plant Cell*, 8, 1899–1913.

Pawlowski, K. et al. (2011) Progress on research on actinorhizal plants. *Functional Plant Biology, Special Issue: Research Front on Actinorhizal Plants*, 38, 633–638.

Pawlowski, K. et al. (2003) Distinct patterns of symbiosis-related gene expression in actinorhizal nodules from different plant families. *Molecular Plant–Microbe Interaction*, 16, 796–807.

Pawlowski, K. et al. (1997) A nodule-specific gene family from *Alnus glutinosa* encodes glycine- and histidine-rich proteins expressed in the early stages of actinorhizal nodule development. *Molecular Plant–Microbe Interaction*, 10, 656–664.

Péret, B. et al. (2007) Auxin influx activity is associated with *Frankia* infection during actinorhizal nodule formation in *Casuarina glauca*. *Plant Physiology*, 144, 1852–1862.

Péret, B. et al. (2010) When plants socialize: symbioses and root development. In: T. Beeckman (ed) *Root Development. Annual Plant Reviews*, Vol. 37. Wiley-Blackwell, New Jersey, pp. 209–238.

Péret, B. et al. (2009) Arabidopsis lateral root development: an emerging story. *Trends in Plant Science*, 14, 399–408.

Perrine-Walker, F. et al. (2010) Auxin carriers localization drives auxin accumulation in plant cells infected by *Frankia* in *Casuarina glauca* actinorhizal nodules. *Plant Physiology*, 154, 1372–1380.

Perrine-Walker, F. et al. (2011) Symbiotic signalling in actinorhizal symbioses. *Current Protein and Peptide Science*, 12, 156–164.

Pertea, G. et al. (2003) TIGR Gene Indices clustering tools (TGICL): a software system for fast clustering of large EST datasets. *Bioinformatics/Computer Applications in The Biosciences—Bioinformatics*, 19, 651–652.

Pessi, G. et al. (2007) Genome-wide transcript analysis of Bradyrhizobium japonicum bacteroids in soybean root nodules. *Molecular Plant–Microbe Interaction*, 20:1353–1363.

Racette, S. and Torrey, J.G. (1989) Root nodule initiation in *Gymnostoma (Casuarinaceae)* and *Sheperdia* (*Elaeagnaceae*) induced by *Frankia* strain HFPGpI1. *Canadian Journal of Botany*, 67, 2873–2879.

Ribeiro, A. et al. (1995) A nodule-specific gene encoding a subtilisin-like protease is expressed in early stages of actinorhizal nodule development. *The Plant Cell*, 7, 785–794.

Ribeiro, A. et al. (1996) Identification of *agthi1*, whose product is involved in biosynthesis of the thiamine precursor thiazole, in actinorhizal nodules of *Alnus glutinosa*. *The Plant Journal*, 10, 361–368.

Rouvier, C. et al. (1996) Genetic diversity among *Frankia* strains nodulating members of the family *Casuarinaceae* in Australia revealed by PCR and restriction fragment length polymorphism analysis with crushed root nodules. *Applied and Environmental Microbiology*, 62, 979–985.

Roy, S. et al. (2007) Review: combining alders, frankiae, and mycorrhizae for soil remediation and revegetation. *Canadian Journal of Botany*, 85, 237–251.

Schwencke, J. and Caru, M. (2001) Advances in actinorhizal symbiosis: host plant–*Frankia* interactions, biology and applications in arid land reclamation. A review. *Arid Land Research Management*, 15, 285–327.

Silvester, W.B. et al. (1990) Oxygen regulation and hemoglobin. In: C.R. Schwintzer and J.D. Tjepkema (eds) *The Biology of Frankia and Actinorhizal Plants*. Academic Press, New York, pp. 157–176.

Simonet, P. et al. (1999) Co-evolution between Frankia populations and host plants in the family Casuarinaceae and consequent patterns of global dispersal. *Environnemental Microbiology*, 1, 525–533.

Simonet, P. et al. (1990) The genetics of the Frankia–actinorhizal symbiosis. In: P.M. Gresshoff (ed.) *Molecular Biology of Symbiotic Nitrogen Fixation*. CRC Press, Bocaraton, pp. 77–109.

Soltis, D.E. et al. (1995) Chloroplast gene sequence data suggest a single origin of the predisposition for symbiotic nitrogen fixation in angiosperms. *Proceedings of the National Academy of Sciences of the USA*, 92, 2647–2651.

Sprent, J.I. and Parsons, R. (2000) Nitrogen fixation by legume and non-legume trees. *Field Crops*, 65, 183–196.

Svistoonoff, S. et al. (2004) Infection-related activation of the *cg12* promoter is conserved between actinorhizal and legume-rhizobia root nodule symbioses. *Plant Physiology*, 136, 3191–3197.

Svistoonoff, S. et al. (2003) *cg12* expression is specifically linked to infection of root hairs and cortical cells during *Casuarina glauca* and *Allocasuarina verticillata* actinorhizal nodule development. *Molecular Plant–Microbe Interaction*, 16, 600–607.

Swensen, S.M. (1996) The evolution of actinorhizal symbioses: evidence for multiple origins of the symbiotic association. *American Journal of Botany*, 83, 1503–1512.

Sy, M.O. et al. (2007) The cell-cycle promoter cdc2aAt from *Arabidopsis thaliana* is induced in the lateral roots of the actinorhizal tree *Allocasuarina verticillata* during the early stages of the symbiotic interaction with *Frankia*. *Physiologia Plantarum*, 130, 409–417.

Torrey, J.G. (1976) Initiation and development of root nodules of Casuarina (*Casuarinaceae*). *American Journal of Botany*, 63, 335–345.

Uchiumi, T. et al. (2004) Expression islands clustered on the symbiosis island of the *Mesorhizobium loti* genome. *Journal of Bacteriology*, 186, 2439–2448.

Udvardi, M.K. et al. (2005) *Lotus japonicus*: legume research in the fast lane. *Trends in Plant Science*, 10, 222–228.

Ulmasov, T. et al. (1995) The soybean *GH2/4* gene that encodes a glutathione *S*-transferase has a promoter that is activated by a wide range of chemical agents. *Plant Physiology*, 108, 919–927.

Vallenet, D. et al. (2006) MaGe: a microbial genome annotation system supported by synteny results. *Nucleic Acids Research*, 34, 53–65.

Valverde, C. and Wall, L.G. (1999) Time course of nodule development in the *Discaria trinervis* (Rhamnaceae)—*Frankia* symbiosis. *New Phytologist*, 141, 345–354.

van Ghelue, M. et al. (1997) Early interaction between *Alnus glutinosa* and *Frankia* strain ArI3. Production and specificity of root hair deformation factor(s). *Physiologia Plantarum*, 99, 579–587.

van Ghelue, M. et al. (1996) Sucrose synthase and enolase expression in actinorhizal nodules of *Alnus glutinosa*: comparison with legume nodules. *Molecular and General Genetics*, 250, 437–446.

Vessey, J.K. et al. (2004) Root-based N_2-fixing symbioses: legumes, actinorhizal plants, *Parasponia* sp. and cycads. *Plant and Soil*, 266, 205–230.

Wall, L.G. (2000) The actinorhizal symbiosis. *Journal of Plant Growth Regulation*, 19, 167–182.

Wall, L.G. and Berry, A.L. (2008) Early interactions, infection and nodulation. In: K. Pawlowski and W.E. Newton (eds) *Nitrogen-Fixing Actinorhizal Symbioses*, Vol. 6, Springer Verlag, Dordrecht, The Netherlands, pp. 147–166.

Wheeler, C.T. et al. (1979) Hormones in plants bearing actinomycete nodules. *Botanical Gazette*, 140, 52–57.

12 Plant Growth Promoting Rhizobacteria and Root Architecture

Thais L.G. Carvalho, Paulo C.G. Ferreira, and Adriana S. Hemerly

Introduction

Many soil bacteria interact with plants, are able to multiply, and colonize root niches at all stages of plant development even in the presence of a competitive microflora. In some of these interactions, bacteria are said to have a neutral effect, not affecting plant growth. On the other hand, other soil bacteria have the potential to affect plant growth and crop yield significantly, exerting positive or negative effects.

Negative regulators of plant growth are known as deleterious bacteria. They inhibit growth through production of harmful metabolites—such as cyanide (Alström and Burns 1989; Kremer and Souissi 2001) and phytotoxins (Schippers 1988; Alström and Burns 1989; Bolton et al. 1989) or production of high concentrations of phytohormones (Arshad and Frankenberger 1998; Persello-Cartieaux et al. 2003). Moreover, they can also increase the negative effects of pathogens (Suslow and Schroth 1982; Preston 2004) or attenuate the beneficial effects of mycorrhiza (Schippers et al. 1987; Nehl et al. 1997).

Unlike deleterious bacteria, beneficial bacteria can induce plant growth by increasing the availability of nutrients, by production of phytohormones or by reducing the negative effects of pathogens (Bloemberg and Lugtenberg 2001; Vessey 2003). These organisms are called plant growth-promoting rhizobacteria (PGPR) (Dobbelaere and Okon 2007). The PGPRs are commonly known as biofertilizers, when they promote plant growth specifically by increasing nutrient availability or improving plant access to nutrients, such as by increasing the surface area of roots (Somers et al. 2004). When bacteria promote growth through the control of harmful organisms, they are commonly referred as biological control agents or biopesticides (Holgin 1998; Somers et al. 2004). Because of their positive effects on plant growth, PGPRs could be used as a tool for maximizing gains in agricultural productivity.

Most of the knowledge about PGPRs comes from identification of beneficial bacteria and their growth promotion effects on plants. Lately, advances are emerging on the molecular mechanisms involved in these interactions and in growth promotion. So far, few studies used genomic approaches to investigate this type of interaction. In this scenario, in this chapter, we will uncover developmental and genetic controls by which PGPRs promote changes in root growth and architecture. Understanding how genetic programs regulate plant physiological and biochemical changes in response to PGPRs will provide an integrated view of how PGPRs can improve this association, opening new perspectives to enhance agriculture.

Root Genomics and Soil Interactions, First Edition. Edited by Martin Crespi.

Different Root Niches for PGPR Colonization

The beneficial effects of PGPR on plant growth depend on an intimate association between the bacteria and the host plant. However, the degree of relationship can vary depending on where and how PGPR colonizes the host plant, and it can be categorized into two levels of complexity: rhizospheric and endophytic.

Rhizospheric Niches

The rhizosphere is the compartment of the soil around the roots that is under plant influence (Kennedy 2005). Some scientists adopt a more expanded definition for rhizospheric PGPRs, including bacteria from the rhizoplane, which is the surface of roots and soil particles strongly adhered to it (Antoun and Prévost 2005; Figure 12.1).

Root growth causes several changes in physical and chemical properties of the soil, which can affect the ability of PGPRs to colonize the rhizosphere. It includes changes in water potential, pH and salinity, partial pressure of O_2, and mineral and organic composition due to plant exudation (Hasegawa et al. 2005). While most of the soil is poor in nutrients, the rhizosphere represents a region of high nutritional composition (de Weger et al. 1995; Jones et al. 2003; Somers et al. 2004).

The main component of root secretion is the mucilage, which contains hydrated polysaccharides, organic acids, vitamins and amino acids, therefore an excellent substrate for microbial growth. Mucilage adheres to water, helping to form a highly hydrated environment for roots and rhizospheric microorganisms. All this rhizodeposition promotes microbial abundance and activity in the

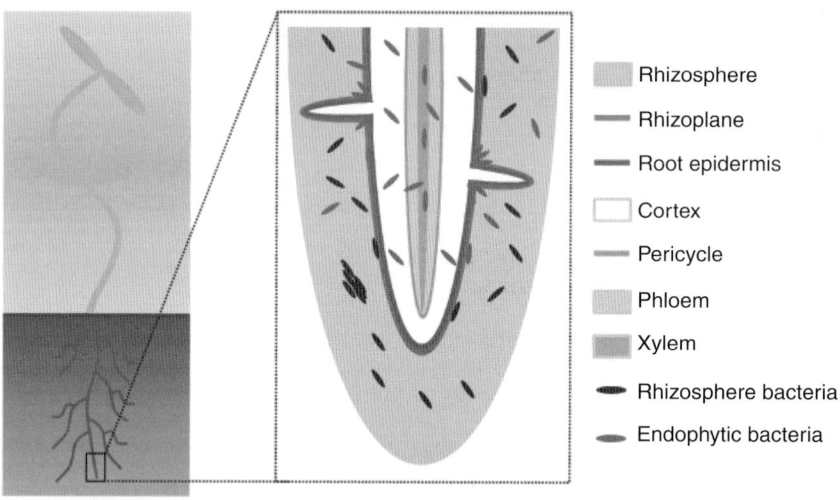

Figure 12.1 Root niches for PGPR colonization. Rhizospheric PGPRs (garnet cells) colonize rhizosphere soil area and roots surface (rhizoplane), but they cannot invade internal plant tissues. Endophytic PGPRs (blue cells) colonize any region within the epidermis of the plant root, and they can reside in apoplastic intercellular spaces and xylem vessel apoplast. In general, the endophytes invade the internal plant tissues through sites of injury in the epidermis, root tips, and root cracks formed at the sites of lateral root emergence. Some endophytic PGPRs can spread through distinct plant organs (stem, leaves, seeds, fruits). (For a color version of this figure, see the color plate section.)

rhizosphere, which can be characterized as the most active microbial habitat of the soil (Burdman et al. 2000; Lucy et al. 2004; Smalla et al. 2006; Dobbelaere and Okon 2007; Hartmann et al. 2008).

Many interactions between plant, soil, and microorganisms occur in the rhizosphere (Brencic and Winans 2005). Several biochemical interactions and molecular signaling trades between plants and soil microorganisms have been described (Pinton et al. 2001). These interactions can significantly influence plant growth and yield.

Endophytic Niches

Endophytes are bacteria that are present within plants. In 1926, endophytic growth was recognized as a particular stage in the life of bacteria, described as an advanced stage of infection and as having a close relationship with mutualistic symbiosis (Perotti 1926). Since then, endophytes have been defined as microorganisms that could be isolated from surface-sterilized plant organs (Henning and Villforth 1940). In agronomy, this concept has been further broadened to encompass all bacteria that can be isolated from surface-sterilized plant tissues and do not visibly harm host plants (Hallmann et al. 1997). In accordance with their life strategies, bacterial endophytes can be classified as "obligate" or "facultative" (Hardoim et al. 2008). Obligate endophytes are strictly dependent on the host plant for their growth and survival. Transmission to other plants occurs vertically or via vectors. Facultative endophytes have a stage in their life cycle in which they subsist outside host plants.

Endophytic sites include any region within the plant, although the vascular system is sometimes considered separately (Figure 12.1). Depending on the host plant and the bacteria strain, PGPRs may be found in other plant tissues than roots, such as seeds, stems, leaves, or fruits (Lugtenberg and Kamilova 2009). Bacteria enter plants through sites of injury in the epidermis, root tips, and lateral roots (Garbeva et al. 2001; Hallman 2001; Carvalho et al. 2011; Figure 12.1). In endophytic relationships, PGPR resides within apoplastic spaces inside the host plant. There is evidence of endophytes living in the apoplastic intercellular space within parenchyma tissue (Dong et al. 1994; Dong et al. 1997) and/or the xylem vessel apoplast (Fuentes-Ramirez et al. 1999; James et al. 2001). Microscopic analysis revealed that PGPRs do not live within healthy host cells. The few reports that suggest an intracellular localization are controversial and could characterize a saprophytic colonization (Bellone et al. 1997; Cocking et al. 2006). The microorganisms that reach the intercellular regions must compete with the plant defense system, which is activated when bacteria enters the plant. Remarkably, bacteria colonize tissues of most organs of the infected plant, without causing visible disease symptoms (Hardoim et al. 2008; Carvalho et al. 2011). In the endophytic places, bacteria can access nutrients and water more easily than on the plant surface and are less affected by environmental fluctuations (Vessey 2003). Different from the symbiotic association between rhizobium and legumes, new anatomical structures have not been observed in endophytic PGPR associations (Vessey 2003).

PGPR Recognition by Plants

PGPR colonize plants, living on nutrients obtained from plant cells. The highest accomplishment for plant-colonizing bacteria lies in fitting in the host environment and being able to tolerate or avoid plant defense responses. Success for plants lies in being able to distinguish beneficial bacteria from pathogenic ones and in using induced defense responses to protect against dangerous pathogens at minimum cost.

Plants evolved the capacity to recognize and respond to a wide range of generic microbial molecules known as pathogens-associated molecular patterns (PAMP) (Gómez-Gómez and Boller 2002). Defense responses are elicited by these general elicitors in plants. PAMP-elicited defense responses may contribute to restriction of endophytic growth and to systemic-induced resistance elicited by PGPRs. Two of the most widely studied PAMPs produced by PGPRs are flagellins, subunits of the polar flagella produced by motile PGPRs, and lipopolysaccharides (LPSs), constituents of bacterial envelope (Figure 12.2).

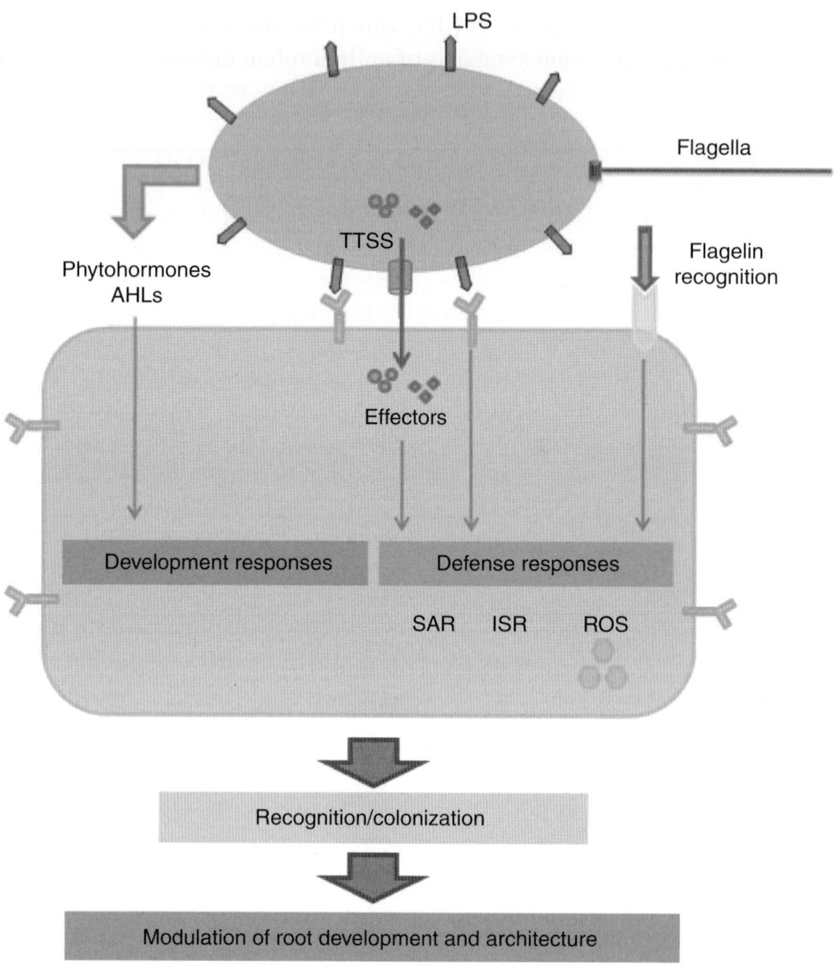

Figure 12.2 Overview of mechanisms involved in PGPRs recognition by plant cells. The figure illustrates a range of factors described to be produced by PGPRs, which can be recognized by plant cells, triggering specific responses. Lipopolysaccharides (LPS), constituents of bacterial envelope, are recognized by plant hosts. Flagellins, subunits of the polar flagella, are produced by motile PGPRs and are recognized by FLS2, a membrane-associated kinase with an extracellular LRR domain. PGPRs type III protein secretion system (TTSS) is used to deliver effector proteins directly into cytoplasm of host cells. The recognition process and the complex signal transduction pathways lead to activation of defense responses such as ISR, SAR, and ROS production. Host plant colonization depends on plant–bacteria recognition. This colonization together with phytohormones production by PGPR will result in root growth promotion and modulation of root architecture. (For a color version of this figure, see the color plate section.)

Flagellin recognition in plants is mediated by FLS2, a membrane-associated kinase with an extracellular leucine-rich repeat (LRR) domain. FLS2 is a member of the Toll family of receptor kinases, which has been linked to pathogen recognition in plants (Gómez-Gómez and Boller 2002). Purified flagellin elicits an oxidative burst, callose deposition, and synthesis of antimicrobial proteins in plant cells (Felix et al. 1999; Gómez-Gómez et al. 1999). Flagellin recognition by plants is host and strain specific. One unresolved question with regard to flagellin recognition is whether flagella are expressed at all stages of plant colonization. Flagella are important for initial colonization of roots but not for endophytic multiplication (Haefele and Lindow 1987). Regulation of flagella expression could be an additional mechanism used to avoid plant recognition.

A second commonly recognized factor is LPS, which has been mostly studied in the context of plant pathogens. LPS can induce plant synthesis of antimicrobial factors and suppress programmed cell death associated with the hypersensitive response (HR), an effect referred to as "localized induced resistance" or "localized induced response" (LIR; Dow et al. 2000; Newman et al. 2002). Induction of LIR by PGPR's LPSs may enhance local defense responses to plant pathogens, but perhaps the most important role of LPS in PGPR–plant interactions may be in priming systemic expression of plant defense responses (Dow et al. 2000).

Another system that can play an important role in modulation of host defense responses by PGPRs is the type III protein secretion system (TTSS) (Figure 12.2). This system is used by bacteria to deliver effector proteins into cytoplasm of host cells (Buttner and Bonas 2002; Greenberg and Vinatzer 2003). TTSS effectors are highly diverse, but their collective function appears to be to render the host more susceptible to colonization and to promote bacterial multiplication in host tissue (Gabriel 1999; Abramocitch et al. 2003). Regulatory, structural, and effector genes of PGPRs TTSS are closely related to those of pathogenic bacteria (Preston et al. 2001; Wolfgang et al. 2003). Therefore, it seems likely that PGPRs TTSSs promote colonization of susceptible hosts in much the same way as in plant pathogens. Modulation of host responses or host-specific recognition of effectors secreted by PGPRs could have a significant impact on induction of local and systemic defense mechanisms and on the ability of PGPRs to live endophytically in plant tissues. It is possible that the use of TTSS effectors imposes host-specificity on plant–PGPR interactions, as it has been observed for *Rhizobium* (Preston et al. 2004). Current evidence clearly suggests that plant cells can and do receive TTSS-secreted effectors from a wide range of plant-colonizing bacteria, including PGPRs, but further analyses are needed to address the role of TTSSs in the ecology of plant-colonizing bacteria (Preston et al. 2004).

Many recent studies of plant-PGPR interactions have focused on the ability of PGPR to induce systemic defense responses such as induced systemic resistance (ISR) or systemic acquired resistance (SAR) in host plants (Figure 12.2). Induced resistance is defined as active resistance dependent on physical or chemical barriers of the host plant, activated by biotic or abiotic agents (Press et al. 1997). Induced resistance arising from plant–PGPR interactions may be linked to ISR, SAR, or both, in combination with other effects of PGPRs on plants and plant-associated microbes (Preston et al. 2004). There is substantial communication between the two pathways, and they can act synergistically or antagonistically (Conrath et al. 2002; Kunkel and Brooks 2002).

An important stress signal used by plants is the generation of reactive oxygen species (ROS) (Figure 12.2). ROS affect plant cells in two main ways: as a cause of stress through oxidative damage to plant molecules and as signaling intermediates (Finkel 2003; Foreman et al. 2003). ROS may also act directly as antimicrobial factors when they are produced during plant defense responses; and active ROS detoxification by microbes might alter plant stress responses. The role of plant-derived ROS in plant–microbe interactions has been studied for many years. However, it has only recently become clear that bacteria-derived ROS may also affect plant–PGPR interactions. De Meyer et al.

(1999) suggested that ISR by plant–bacteria interactions could be due to SA biosynthesis. However, Audenaert et al. (2002) showed that the generation of ROS makes an equally important contribution toward the induced resistance observed in these interactions.

One recent and still controversial question concerning plant–bacteria interactions is the role and impact of acylated homoserine lactones (AHLs). Many PGPRs use AHLs to monitor the external environment and the proximity of other bacteria (Bodman et al. 2003; Loh et al. 2002). Mathesius et al. (2003) used proteomics to show that the model legume *Medicago truncatula* responds to AHLs produced by its symbiotic bacteria *Sinorhizobium meliloti*. However, it is not a consensus whether AHLs constitute another class of molecule involved in modulating local and systemic plant responses to bacteria. Some authors have proposed the role for these AHLs in root growth promotion by PGPRs (Figure 12.2).

Modulation of Root Growth and Architecture by PGPRs

It has been demonstrated that plant growth is strongly regulated by various environmental conditions, including association with microorganisms (Lucy et al. 2004). Analysis of the effects on plant growth caused by inoculation with PGPRs demonstrated that bacteria association has a major role both in promotion of root development and in modulating root architecture (Vessey 2003; Lucy et al. 2004; Table 12.1) . The increase in root surface mediated by PGPRs improves nutrient absorption by the plant (Kapulnik et al. 1985; Lifshitz et al. 1987), then raising nutrient levels, and consequently enhance plant growth as a whole.

Increase in the total root system, quantified as root dry weight, is the most common reported plant response mediated by inoculation with different PGPR formulations in various plant species (Table 12.1). Comprehensive field trial experiments were performed in Florida to evaluate the effects of different formulations of PGPRs on several tomato and pepper growth parameters (Kokalis-Burelle et al. 2002). PGPR-treated tomato and pepper transplants showed significant increases in root weight, which correlates with the increases in yield observed in these experiments. A study conducted with sugar beet analyzed 2 years of field trials of inoculation experiments with several strains of *Bacillus, Burkholderia, and Pseudomonas* (Cakmakci et al. 2001). The results indicated that PGPRs significantly affected yield and quality parameters, increasing root weight and sugar yield.

Changes in root architecture mediated by inoculation with PGPRs have also been analyzed (Table 12.1). Some studies demonstrated that PGPR association can increase primary root length (Vessey 2003; Lucy et al. 2004; López-Bucio et al. 2007). Treatment of clipped soybean roots with *Azospirillum brasilense* Sp7 caused a 63% increase in root dry weight, more than a 6-fold increase in specific root length (root length per unit root dry weight) and more than a 10-fold increase in total root length (Molla et al. 2001). Inoculation of canola seedlings with *Pseudomonas putida* GR12–2 resulted in roots significantly longer than those of control plants (Lifshitz et al. 1987). Cucumber plants also showed increased root length when colonized by *Pythium aphanidermatum* (Uthede et al. 1999).

The major impact of PGPR on root morphogenesis is the overproduction of lateral roots and root hairs, which are commonly associated with the enhanced performance of plants in response to challenging environment (Persello-Cartieaux et al. 2003; López-Bucio et al. 2007). Kapulnik et al. (1985) demonstrated that inoculation with *Azospirillum* increased root elongation and total root surface by promotion of both lateral root and root hair formation. They also observed that higher numbers of associated bacteria result in negative effects on root growth, suggesting the existence of

Table 12.1 Examples of effects of PGPR inoculation on root development of several plant species.

Bacteria	Result of Bacteria Inoculation	Plant	References
Azospirillum amazonense	Increase in root dry weight	Sugarcane	Oliveira et al. 2002
Azospirillum brasilense	Increase in lateral root number Increase in root fresh weights Increase in lateral root number, root length, and root dry weight	Sorghum, Bean, Wheat, Maize, Chickpeas, Faba beans	Sarig et al. 1992; Vedder-Weiss et al. 1999; Dobbelare et al. 2001; Hamaoui et al. 2001
Azospirillum lipoferum	Increase in primary root length and root fresh weight Larger root systems	Maize	Jacoud et al. 1998; Dobbelaere et al. 2001
Azospirillum chroococcum	Increase in the extension of the growing root	Barley	Harper et al. 1979
Azotobacter sp.	Increase in root length	Rice	Alam et al. 2001
Bacillus sp.	Increase in root length	Rice	Alam et al. 2001
Bacillus polymyxa	Increase in root yield	Sugar beet	Cakmakci et al. 2001
Bacillus amyloliquefaciens	Increase in weight of roots	Tomato, pepper	Kokalis-Burelle et al. 2002
Bacillus pumilus	Increase in weight of roots	Tomato, pepper	Kokalis-Burelle et al. 2002
Bacillus subtilis	Increase in weight of roots Increase in root length	Tomato, pepper, cucumber	Kokalis-Burelle et al. 2002 Uthede et al. 1999
Bacillus cereus	Increase in weight of roots Increase in root length	Tomato, pepper, cucumber	Kokalis-Burelle et al. 2002 Uthede et al. 1999
Burkholderia sp.	Increase in root yield Increase in root dry weight	Sugar beet, sugarcane	Cakmakci et al. 2001 Oliveira et al. 2002
Burkholderia vietnamiensis	Increase in root weight	Rice	Tran Van et al. 2000
Enterobacter sp.	Increase in root length	Rice	Alam et al. 2001
Enterobacter aerogenes	Increase in root length	Cucumber	Uthede et al. 1999
Enterobacter agglomerans	Increase in root length	Cucumber	Uthede et al. 1999
Gluconacetobacter diazotrophicus	Increase in root dry weight	Sugarcane	Oliveira et al. 2002
Herbaspirillum rubrisubalbicans	Increase in root dry weight	Sugarcane	Oliveira et al. 2002
Herbaspirillum seropedicae	Increase in root dry weight	Sugarcane	Oliveira et al. 2002
Pseudomonas sp.	Increase in root yield Increase in root dry weight Increase in root weight	Sugar beet, canola, lettuce, cucumber, tomato	Cakmakci et al. 2001 Bertrand et al. 2001 Van Peer and Schippers 1998
Pseudomonas cepacia	Increase in root weight	Winter wheat	De Freitas and Germida 1991
Pseudomonas putida	Increase in root weight Increase in root length Root elongation Root elongation	Winter wheat, cucumber, canola, lettuce, tomato, barley, wheat	De Freitas and Germida 1991 Uthede et al. 1999 Lifshitz et al. 1987 Hall et al. 1996
Phyllobacterium sp.	Increase in root dry weight	Canola	Bertrand et al. 2001
Variovorax sp.	Increase in root dry weight	Canola	Bertrand et al. 2001
Xanthobacter sp.	Increase in root length	Rice	Alam et al. 2001

a critical extent of colonization by *Azospirillum* to promote root growth. Inoculation of *Arabidopsis thaliana* with *Bacillus megaterium* showed diverse root-architecture alterations (López-Bucio et al. 2007). These PGPRs caused inhibition of primary root growth followed by an increase in lateral root number, lateral root growth, and root hair length. Fallik and Okon (1994) reported that inoculation of maize with *A. brasilense* resulted in proliferation of root hairs, which could have dramatic effects on increasing root surface area.

Altogether, these observations demonstrate that PGPR can significantly regulate plant performance in agricultural and marginal soils by positively affecting growth, particularly root development and architecture. The alterations on root architecture lead to an increase in total absorptive root surface area, improving nutrient and water uptake, which may have positive effects on crop yield.

Mechanisms of Plant Growth Promotion by PGPRs

PGPRs can promote plant growth by direct and indirect mechanisms (Glick et al. 1998; Glick et al. 1999; Persello-Cartieaux et al. 2003; Mantelin and Touraine 2004; Spaepen et al. 2009). They indirectly benefit plant development by suppressing deleterious microorganisms that inhibit plant growth. Some of the indirect mechanisms are competition for nutrients and colonization sites with plant harmful microorganisms, regulation of pathogens colonization by induction of ISR, and production of antibiotic compounds (Lugtenberg and Kamilova 2009).

There are several ways bacteria can directly facilitate development of their host plants. Direct promotion usually is a result of the provision of a compound that is synthesized by the bacterium, which facilitates the uptake of nutrients by the plant. Production of phytohormones and improvement in plant nutrition are the two prominent direct mechanisms (Brown 1974; Davision 1988; Kloepper et al. 1989; Lambert and Joos 1989; Patten and Glick 1996; Glick et al. 1999).

Root morphogenesis and growth are under the control of both environmental stimuli and endogenous developmental programs. In both cases, plant hormones coordinate adaptive changes in cell division and differentiation that lead to changes in root development and architecture (López-Bucio et al. 2003, 2005). Phytohormones and nutrients are some of the compounds provided to the plant by PGPRs, and both are known regulators of root morphogenesis. They are considered the most important candidates to mediate PGPRs effect on root growth, and they will be discussed in more detail.

Plant Growth Promotion through Hormone Production

The production of substances that promote plant growth by bacteria has been reported for many bacterial species, and the hypothesis that it contributes to the plant growth-promoting effects has been launched about 50 years ago (Barea et al. 1976). Since then, several studies provided evidences that different strains of PGPR promote plant growth due to the production of different phytohormones (Vessey 2003; Beattie et al. 2006; Spaepen et al. 2008).

Auxins

About 80% of bacteria from rhizosphere are able ro produce indole acetic acid (IAA), indicating a possible role in the interaction with the plant (Patten and Glick 1996; Khalid et al. 2003). Auxins are an important class of hormones controlling many aspects of root development and architecture, such as primary root growth, lateral root formation, and root hair development (Benkóva and Hejátko 2009; Fukaki and Tasaka 2009).

The contribution of this phytohormone in promoting plant growth has been demonstrated for *A. brasilense*, *Aeromonas veronii*, *Agrobacterium* spp., *Alcaligenes piechaudii*, *Bradyrhizobium* spp. *Comamonas acidovorans*, *Enterobacter* spp., and *Rhizobium leguminosarum* (Vessey 2003), among others, and occurs through the increase in root growth and through proliferation and elongation of root hairs (Vessey 2003).

There are different IAA biosynthesis pathways in bacteria, which are very similar to those found in plants (Costacurta and Vanderleyden 1995). Most of the IAA biosynthetic routes start from the main precursor tryptophan (trp). The routes are named according to the intermediate molecules produced. It is important to note that not all routes are characterized to the same extent and that there may be multiple routes in the same organism.

The importance of auxin production in PGPR ability to promote plant growth has been demonstrated through inoculation studies with bacteria mutants (Barbieri et al. 1986; Barbieri and Galli 1993). So far, all mutants in specific genes of IAA biosynthetic pathways still show a residual production of IAA, although the levels can reach 90–99% of reduction. All attempts to isolate a null mutant failed, indicating redundancy of IAA biosynthetic pathways in bacteria (Barbieri et al. 1986; Prinsen et al. 1993; Carreno-Lopez et al. 2000; Spaepen et al. 2007). A direct link between the production of IAA by *Azospirillum* and changes in the morphology of wheat roots was demonstrated by Dobbelaere et al. (1999). These results were corroborated by a recent study of Spaepen et al. (2008), in which inoculation with the wild strain *A. brasilense* shortened the primary root and increased formation of root hairs while the mutant strain in IAA biosynthesis no longer induces these effects.

Cytokinin

In bacteria, cytokinin synthesis was first described in pathogens as an important virulence factor when produced in high levels (Costacurta and Vanderleyden 1995; Barash and Manulis-Sasson 2007). Several reports showed that PGPRs also have the ability to produce cytokinins (Barea et al. 1976; Frankenberger and Arshad 1995; De Salamone et al. 2001). The mechanisms of action of cytokinins in plant–PGPR interactions are poorly understood, mainly due to the lack of mutants to allow quantification of the contribution of bacterial cytokinins to plant growth promotion.

Cytokinin is essential for cell division despite that it has negative effects on root development (Benkóva and Hejátko 2009; Fukaki and Tasaka 2009). Cytokinin negatively regulates root growth by reducing the relative elongation rate and by blocking the increase of meristem size through inhibition of root meristem cell division (Beemster and Baskin 2000). Cytokinin also inhibits auxin transport, which perturbs the establishment of an auxin gradient during LR initiation, thereby inhibiting the asymmetric cell division for LR initiation (Benkóva and Hejátko 2009). Many bacteria produce auxins and cytokinins, therefore the effect of these PGPRs on plant growth will depend on the balance between these two phytohormones (Spaepen et al. 2009).

Gibberellins

Gibberellins (GAs) perform an important role in mediating the effects of environmental stimuli on plant development. A positive role of GAs in root growth and branching was demonstrated in studies with plant mutants. GA-deficient pea mutant lines have fewer lateral roots (Ferguson et al. 2005) and *Arabidopsis* GA-deficient mutants have reduced primary root growth (Fu and Harberd 2003). These results indicate that GAs are positive regulators of root growth and branching.

Some PGPRs, as *Bacillus pumilus*, *Bacillus licheniformis*, and some strains of *Azospirillum* spp and *Pseudomonas* spp, synthesize gibberellins in culture medium (Gaudin et al. 1994; Gutierrez-Manero et al. 2001). In addition to auxin, *Gluconacetobacter diazotrophicus* produces GA1 and GA3 (Sevilla et al. 2001). It has been speculated that the induction of sugarcane growth by inoculation with a *G. diazotrophicus* mutant strain unable to fix nitrogen (*nif-*) could be due to release of those phytohormones (Sevilla et al. 2001). Nevertheless, there are still insufficient data to understand the role of these compounds in the promotion of plant growth mediated by PGPRs and information about GA phytohormone production by other bacteria are still very infrequent (De Salamone et al. 2001).

Ethylene

Among several effects on plant growth and development, the phytohormone ethylene can inhibit root growth and lateral root development (Benkóva and Hejátko 2009; Fukaki and Tasaka 2009). It has been proposed that some PGPRs might reduce ethylene levels in plants by cleaving the ethylene precursor aminocyclopropane-1-carboxylate (ACC) by the action of the bacteria enzyme ACC deaminase (Glick et al. 1998; Grichko and Glick 2001; Penrose et al. 2001). This enzyme competes with ACC oxidase and catalyzes the cleavage of ACC to α-ketobutyrate and ammonia, which is then used as a nitrogen source by the bacteria (Glick et al. 1998). In a study by Duan et al. (2006), 27 of 233 new isolates of *Rhizobium* spp from various locations in Saskatchewan, Canada, showed ACC deaminase activity. In another study, ACC deaminase activity was found in several bacterial isolates, including *Azospirillum* spp., *Rhizobium* spp., *Agrobacterium* spp., *Achromobacter* sp., *Burkholderia* spp., *Ralstonia* sp., *Pseudomonas* spp., and *Enterobacter* spp. (Blaha et al. 2006). In addition, 62 of 88 strains of *Pseudomonas* isolated worldwide contained ACC deaminase (Wang et al. 2001). This downregulation of ethylene levels by PGPRs, in combination with the action of auxins, can cause substantial positive effects on root growth and development.

On the other hand, gas chromatography revealed high levels of ethylene in tissues of some plants colonized by PGPRs, which may be related to the production of ethylene by inoculated bacteria (Weingart and Volksch 1997; Weingart et al. 2001). This ethylene production might be important for activation of various defense responses in plants and it is a necessary component to induce systemic resistance to diseases caused by bacteria (van Loon et al. 2006). Also, the positive effect of ethylene on root hair formation is well described (Libault et al. 2010), which is a growth promotion result of PGPR colonization. This dual function of PGPRs on ethylene production makes it difficult to define the exact role of ethylene produced by different species of PGPRs in promoting root development (van Loon et al. 2006).

ABA

The production of abscisic acid (ABA) by some PGPRs in culture medium has been demonstrated for *A. brasilense, Helianthus annuus, B. megaterium, Pseudomonas vulgaris*, and some strains of *Bradyrhizobium japonicum*; (Karadeniz et al. 2006; Boiero et al. 2007; Forchetti et al. 2007; Cohen and Bandurski 2008).

Since ABA is a positive regulator of primary root growth and inhibits lateral root formation (Benkóva and Hejátko 2009, Fukaki and Tasaka 2009), it could take part in the regulatory network controlling root growth promotion by PGPRs. In addition, it has already been proposed that ABA-producing bacteria can increase plant growth by interfering with plant cytokinin concentration, since ABA inhibits the synthesis of this phytohormone (Miernyk 1979).

In addition to the evidences of phytohormone production by several PGPRs in culture media, important information about these substances production is coming from genomic analyses. Genomic approaches identified several genes involved in the biosynthetic pathways of these compounds for various PGPRs, such as *Azospirillum* spp., *B. amyloliquefaciens, B. japonicum, Burkholderia spp., Enterobacter cloaceae, G. diazotrophicus, and Herbaspirillum spp.* (Steenhoudt and Vanderleyden 2000; Pedrosa 2005; Idris et al. 2007; Ryu and Patten 2008; Bertalan et al. 2009; Morrone et al. 2009).

Phytohormone regulatory effects on root development are summarized in Figure 12.3. Auxin is a positive regulator of root development in general, inducing primary root growth, lateral root

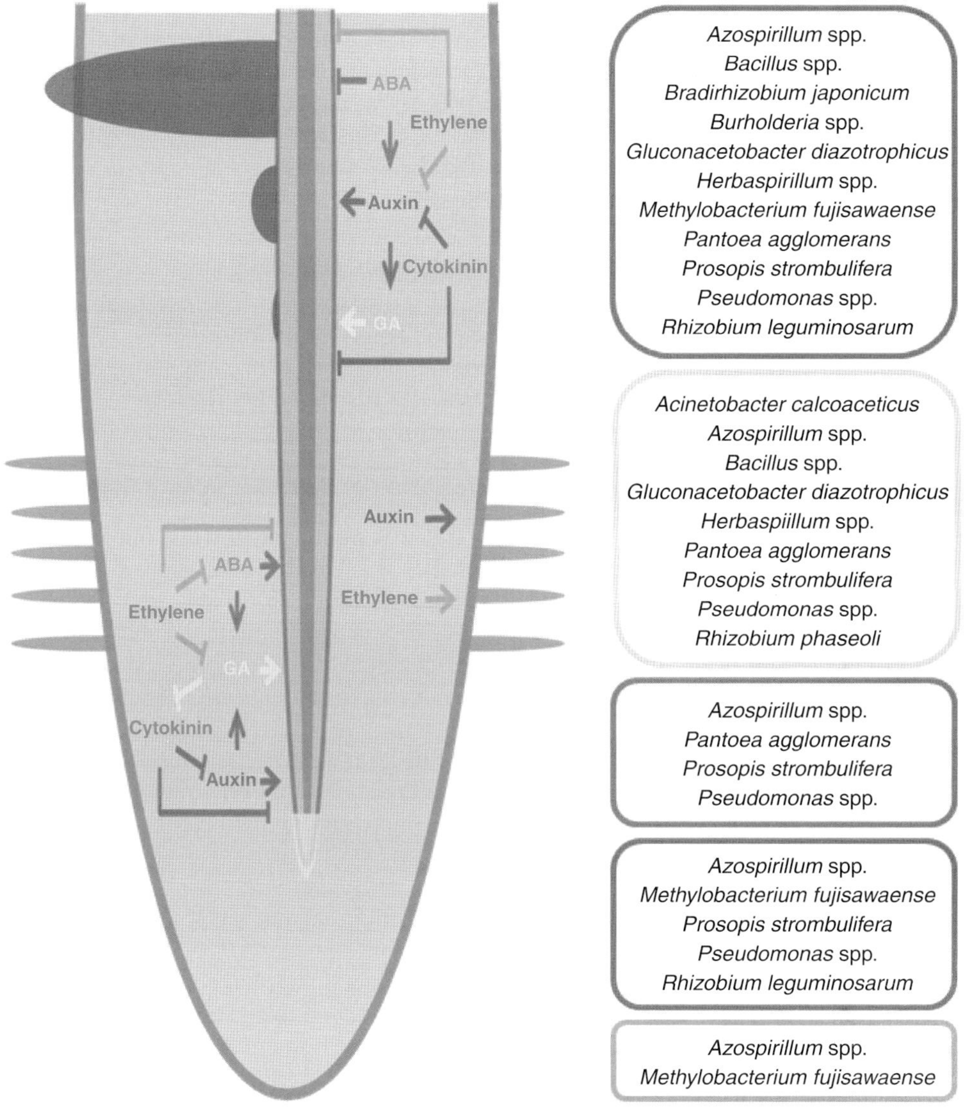

Figure 12.3 Proposed model for different mechanisms regulating root architecture by PGPRs. PGPR inoculations regulate different aspects of root architecture, such as primary root growth, lateral root formation, and root hair development. These processes are controlled by intricate regulatory networks signaled by phytohormones. Several PGPRs produce plant growth regulators such as auxin (red), ethylene (green), gibberellins (yellow), cytokinin (purple), and abscisic acid (blue). Each balloon contains the PGPRs already described as phytohormone producers, and the colors of the balloons represent the different phytohormones. (For a color version of this figure, see the color plate section.)

formation, and root hair development. GAs also appears as a positive regulator of root growth. ABA, on the other hand, shows a positive effect on primary root growth but a negative role in lateral root formation. Cytokinin is a negative regulator of root growth. Ethylene inhibits primary root growth and lateral root formation, and is a positive regulator of root hairs. Ethylene levels may be reduced by PGPR ACC deaminase activity.

Phytohormone control of root growth and architecture is a complex process, governed by integrated regulatory networks. There is an intricate cross talk among all different hormone response pathways, and the final root architecture might be regulated by the balance of several hormone levels in specific tissues and root developmental processes. In addition, different hormones levels are provided during specific host-bacteria associations. In this context, it is remarkable to note that root growth is a major outcome from most plant–PGPR associations.

Plant Growth Promotion through Increase in Nutrients Availability

In addition to plant growth promotion through the production of phytohormones, some PGPRs can improve plant nutrition by providing specific nutrients for plants, especially nitrogen, phosphorus, and iron (Glick et al. 1999; Podile and Kishore 2006). As discussed below, some of these nutrients are known signals regulating root growth and architecture.

Iron

Iron (Fe) is an essential micronutrient for plants and it is abundant in soils, but is found mostly in the form of ferric hydroxide, unavailable to plants. Bacteria have developed a strategy for a more efficient uptake of Fe through the production and secretion of organic iron chelators, called siderophores (Buyer et al. 1993). These compounds act outside of the cell membrane, capturing molecules of Fe^{+3} in solution. The siderophore–Fe^{+3} complexes are absorbed, thus making the Fe absorbed available for microbial or plant metabolism (Neilandjs and Leong 1986; Raaijmakers et al. 1995).

The role of siderophores in increasing the plant Fe nutrition was well established through analysis of production of these molecules in mutants (Kloepper et al. 1991). PGPRs mutant strains, deficient in the production of siderophores, lost the ability to promote growth, while complementation of the production of these molecules restored its activity. Treatment of maize seeds with strains producing siderophores of *Pseudomonas* spp. GRP3A, PRS9, and *Pseudomonas chlororaphis* promoted an increase in germination, length of roots and aerial part, and in the dry weight of seedlings (Sharma and Johri 2003).

Low Fe availability induces morphological changes in root epidermal cells, increasing its differentiation into trichoblasts (López-Bucio et al. 2003). It seems that increase in root growth promoted by PGPRs is not a direct effect of Fe availability on root development and might be caused by the enhancement of plant nutrition as a whole.

Phosphorus

Phosphorus (P) is the second mineral macronutrient that limits plant growth, after nitrogen. However, even in P-rich soils, the majority occurs in an insoluble form, unavailable to plants (Stevenson and Cole 1999). In addition, three quarters of phosphate fertilizers applied to the soil reprecipitate into insoluble forms (Goldstein 1986), requiring large doses of phosphate fertilizers to increase crop productivity.

PGPRs can stimulate metabolic processes that are effective in the solubilization and mineralization of P from poorly available forms of inorganic and organic phosphorus. Solubilization of inorganic phosphates may occur through the release of organic acids that acidify the soil and release monobasic ($H_2PO_4^-$) and dibasic (HPO_4^{-2}) soluble ions (Kucey et al. 1989). The generation of these ions increases the available form of phosphorus for plants (Gyaneshwar et al. 2002).

In addition to the inorganic phosphorus present in soil, its organic form can be mineralized by PGPRs, making it available to plants as soluble phosphate. Mineralization occurs through the acid and alkaline phosphatase of PGPRs, involving dephosphorylation through hydrolysis of phosphodiester and phospho-anhydrous bonds (Rodriguez and Fraga 1999; Gyaneshwar et al. 2002). Phytate (myo-inositol hexaquifosfato) represents about 20–50% of soil organic phosphorus, and phytase (myo-inositol hexafosfato fosfohidrolase) is an enzyme belonging to the group of acid phosphatase that hydrolyzes phytate to myo-inositol and orthophosphoric acid (Vohra and Satyanarayana 2003). This enzyme have been purified and characterized in some strains of PGPRs (Greiner et al. 1997; Kerovuo et al. 1998).

Soil availability of P is a known signal-regulating root growth and architecture. Low P availability increases root hair numbers and length and induces lateral root formation (López-Bucio et al. 2003). On the other hand, increase in P supply promotes primary root elongation (López-Bucio et al. 2003). Hence, this increase in P availability in the soil by PGPRs may represent an important mechanism for root growth modulation by these bacteria. However, P released by these microorganisms, in general, is not sufficient to promote a substantial plant growth. Undoubtedly, the knowledge of those mechanisms, and the overcoming of a series of obstacles encountered in field conditions, might improve the use of PGPRs in crop inoculation, notably to increase the absorption of solubilized phosphates (Rodriguez and Fraga 1999).

Nitrogen

Nitrogen (N) is an essential nutrient to plants, which can be easily lost in the soil by leaching or by bacterial denitrification (Vance 2001). An important feature of some PGPRs is the ability to supply N to plants through biological nitrogen fixation (BNF). This is one of the most important biological processes required for plant growth and is being performed only for prokaryotic microorganisms, called diazotrophic bacteria, which are able to reduce atmospheric N into a form available to plants.

It is important to understand and quantify the positive effects of PGPRs on plant growth mediated by the provision of nitrogen fixed by the bacteria to the host plant. There is compelling evidence that plants are able to obtain large and significant contributions of N from plant-associated BNF. Large contributions to the plant N supply from N_2-fixing PGPR have been reported in sugarcane (Urquiaga et al. 1992; Mirza et al. 2001) and in mangrove (Bashan and Houlguin 1998). ^{15}N isotope dilution and ^{15}N natural abundance studies have shown significant contributions of N_2 fixation to some Brazilian sugarcane varieties, indicating that BNF can supply up to 70% of the plant N requirements and provide over 150 Kg N fixed hay^{-1} per year (Urquiaga et al. 1992; Boddey et al. 1995). Furthermore, gene reporter fusions demonstrated that the Nif genes of *Azospirillum* and *Azoarcus rhizobacteria* strains are induced when associated with the roots of wheat and rice, respectively (Vande-Broek et al. 1993; Egener et al. 1998, 1999). This "biofertilization" hypothesis is attractive, since N availability is the main yield-limiting factor in many agricultural situations and the leaching of N fertilizers into groundwater causes environmental problems.

N signaling has a direct role on regulation of root growth and architecture (López-Bucio et al. 2003). Increasing N availability reduces primary root elongation, and lateral root density remains constant across a range of N concentrations. However, in low nitrate soils, patches of high N have a localized stimulatory effect on lateral root development in many plant species (Zhang and Forde 1998; López-Bucio et al. 2003). Carbon : nitrogen ratios also control root system architecture and a low sucrose : nitrate ratio induces lateral root formation, likely through increase in nitrate availability (Nibau et al. 2008). Taken all together, it is possible to propose that BNF is an important mechanism of nutrient availability regulation by some PGPRs and this N supply to the plant might be one mechanism involved in modulating root growth and architecture.

Plant Genetic Programs Controlling Modulation of Root Growth and Architecture by PGPRs

Although physiological mechanisms controlling modulation of root growth and architecture by PGPRs are being unraveled, genomic studies on the identification of the plant genetic programs involved are still incipient.

Cartieaux et al. (2003) monitored gene expression of approximately 14,300 genes of the *A. thaliana* genome by microarray analysis to provide insights into physiological and biochemical changes mediated by *Pseudomonas thivervalensis*. Approximately 50% of reduction in primary root length and enhanced lateral root number was observed in plants inoculated with *P. thivervalensis*, accompanied by an increase in hypocotyl length (Persello-Cartieaux et al. 2001). Expression analysis revealed that only nine transcripts were reduced in inoculated roots, while in shoots, the level of 63 transcripts was modified in response to inoculation. It includes two unknown proteins: a homolog of the tomato *DEFECTIVE CHLOROPLASTS AND LEAVES* gene that is required for development of protoplastids and the *low temperature and salt-responsive protein*, LTI6B, a gene associated with stress responses. Although genetic and biochemical experiments demonstrated a key role of bacterially derived auxin in root modifications (Persello-Cartieaux et al. 2001), none of the auxin-signaling pathway genes was differentially expressed in inoculated roots. This data was unexpected because of the strong effects on roots of *P. thivervalensis* colonized plants. Possibly, activation of root branching is restricted to few cells of the pericycle, and such localized signal may be too diluted in whole root samples to be detected by microarray analysis.

Verhagen et al. (2004) used microarray analysis to determine the transcript profile of *A. thaliana* roots in response to inoculation with *P. fluorescens* WCS417r. *Arabidopsis* inoculated plants were larger than control, as observed for other plant species (van Peer and Schippers 1989). Ninety-seven genes showed consistent changes in expression in roots of inoculated plants. Among them, 18% were annotated as "unclassified proteins," 18% were predicted to be involved in cell rescue and defense, 14% in metabolism, 15% in regulating gene transcription, and 7% in cellular communication and signal transduction. Within this set of differentially expressed genes, members of ethylene response pathway, such as ethylene response factor (ERF) 10, ethylene responsive element binding factors (EREBP) 1, and EREBP2, were downregulated during colonization. Another noteworthy point is the upregulation of a transcription factor of lateral organ boundaries (LOB) domain family, which has members with important role in root development (Inukai et al. 2005; Okushima et al. 2007) and also linked to auxin signaling, such as targets of AUXIN RESPONSIVE FACTORs (Okushima et al. 2007).

Another study used microarray to analyze *A. thaliana* gene expression during endophytic colonization by a different *P. fluorescens* strain, the FPT9601-T5 (Wang et al. 2005). Analyses of total plant fresh weight demonstrated that root colonization by this PGPR promoted plant growth 3 weeks after inoculation. The majority of the *P. fluorescens* regulated genes belonged to the functional categories "metabolism," "transcription," "cellular communication/signal transduction mechanism," "cell rescue, defense, and virulence," "development" (including systemic and biogenesis of cellular components), and "unclassified proteins" (including classification not yet clear-cut). Some genes involved in plant growth promotion were upregulated, such as auxin-induced genes, NAM-like genes, and those involved in C-compound and carbohydrate metabolism, some of which are shown to be regulated by auxin in other plant species (Catala et al. 1997; Taguchi et al. 2001). The two largest functional categories (except for unclassified protein) of the downregulated genes were metabolism and transcription. The downregulated genes that belong to the transcription category comprise two major transcription factor families, ERF (ethylene-response factor)/AP2 and Myb, which are known to regulate plant development (Martin and Paz-Ares 1997).

Advances on the understanding of sugarcane genes regulated during association with endophytic diazotrophic bacteria have been achieved mainly by studies of gene expression profile (Carvalho et al. 2011). These PGPRs can promote sugarcane root development and increase plant biomass and productivity, possibly by BNF and other mechanisms such as plant hormone production and plant nutrition optimization (Fuentes-Ramirez et al. 1993; Sevilla et al. 2001; Oliveira et al. 2002; Oliveira et al. 2005; Baldani and Baldani 2005). By analyzing a EST database of sugarcane plantlets inoculated with *G. diazotrophicus* strain PAL5 or *H. rubrisubalbicans* strain HCC103, candidate genes exclusively and preferentially expressed in PGPR-inoculated sugarcane plants were identified (Nogueira et al. 2001; Vargas et al. 2003; Carvalho et al. 2011). Exclusive or preferential expression of nitrate transporter, asparagine synthetase, peptide transporter, and other carbon metabolism genes was observed, suggesting that sugarcane nitrogen assimilation apparatus can play a role in the incorporation of the nitrogen compound released by the diazotrophic bacteria during association (Nogueira et al. 2001). Differential expression of glutamine synthetase genes was observed in sugarcane genotypes with contrasting BNF efficiencies, which may explain, at least in part, the capacity of high BNF genotypes to grow in low nitrogen addition (Nogueira et al. 2005). Also, members of phytohormone biosynthesis, transport, and response pathways, including auxin, gibberellin, and ethylene, are preferentially or exclusively expressed in inoculated plants (Nogueira et al. 2001; Vargas et al. 2003). Gene expression analysis by qRT-PCR revealed that ethylene receptor SCER1 and the transcription factor SCERF2 are induced in sugarcane inoculated plants, while SCERF1 is repressed. These genes showed an opposite expression profile in response to pathogens, suggesting that specific components of ethylene signaling pathway may identify a beneficial endophytic association from pathogenic ones, modulating plant response to diazotrophic endophytes (Cavalcante et al. 2007). In addition, several genes involved in plant growth and development that are regulated by hormones were identified as preferentially expressed in sugarcane plants inoculated with the endophytic diazotrophs, such as specific genes that control cell division and expansion (Nogueira et al. 2001). It indicates that some basic mechanisms controlling plant growth and development might participate in this association.

Taking all together, genomic approaches used to study plant gene expression during PGPR colonization revealed the regulation of several candidate genes that could be involved in modulating root architecture. Root growth results basically from cell divisions and elongation, and the observed induction of genes involved in these processes by PGPRs may represent an important mechanism regulating root growth promotion by PGPRs. Upregulation of transcription factors, such as members of NAM and LOB families, seems to be another response to PGPR colonization involved in root architecture modulation. Remarkably, members of ethylene and auxin signaling pathways, two of the major regulators of root development, responded to PGPRs inoculation in the different systems investigated. The upregulation of auxin response genes corroborates that this phytohormone might mediate root growth induced by PGPR colonization. Gene expression profiles also showed that some ethylene pathway members are downregulated, which may be important for root growth promotion, as this phytohormone negatively regulates root growth and branching. Moreover, it cannot be ruled out that ethylene signaling regulation may be an important mechanism regulating bacteria colonization and defense resistance against PGPRs.

Conclusions

PGPRs inoculation can be considered an important approach for increasing crop yield in agriculture. Plant growth promotion, particularly of roots, represents a central feature to enhance plant performance in agricultural and marginal soils. Beneficial effects of many PGPR strains can be observed at low bacterial numbers, in successful experiments conducted in laboratory culture, greenhouse, or in the field. The lack of consistency in PGPR inoculation results in the field remains a major challenge for the use of these bacteria as biofertilizers. Advances in the comprehension of the molecular basis of these interactions may help in the establishment of successful plant–PGRPs associations.

Environmental conditions influence PGPR association efficiency; hence, further studies may be necessary to define what bacterial traits are useful and necessary for different environmental conditions and plants, in order to select or construct optimal bacterial strains. Also, it would be very useful to have a better understanding of how different bacterial strains work together as the synergistic promotion of plant growth may further enhance crop yield under field conditions.

The understanding of molecular mechanisms involved in root growth and architecture regulation by PGPRs is just starting. Transcriptome profile analysis identified differentially expressed plant genes during PGPR colonization. Regulatory networks involved in various plant processes such as plant–microorganism recognition, defense, plant hormone signaling, plant growth, and nitrogen metabolism were reported to be responsive to PGPR bacteria colonization. Comprehension of how various plant regulatory mechanisms are coordinated and connected to promote root growth and branching is still an important challenge. Nevertheless, the knowledge that is being accumulated in different systems may guide studies to improve plant association with PGPRs and, possibly, enhance its impact in crop yields and agriculture.

Acknowledgments

The authors thank INCT (Instituto Nacional de Ciência de Tecnologia) in Biological Nitrogen Fixation, FAPERJ (Fundação de Amparo à Pesquisa do Estado do Rio de Janeiro), CNPq (Conselho Nacional de Desenvolvimento Científico e Tecnológico), and CAPES (Coordenação de Aperfeiçoamento de Pessoal de Nível Superior) for the financial support.

References

Abramovitch et al. (2003) Pseudomonas type III effector AvrPtoB induces plant disease susceptibility by inhibition of host cell death. *EMBO Journal*, 22, 60–69.

Alam, M.S. et al. (2001) Grain yield and related physiological characteristics of rice plants (*Oryza sativa* L.) inoculated with free-living rhizobacteria. *Plant Production Science*, 4, 125–130.

Alström, S. and Burns, R.G. (1989) Cyanide production by rhizobacteria as a possible mechanism of plant growth inhibition. *Biology and Fertility of Soils*, 7, 232–238.

Antoun, H. and Prévost, D. (2005) Ecology of plant growth promoting rhizobacteria. In: Z.A. Siddiqui (ed.) *PGPR: Biocontrol and Biofertilization*, First edition. Springer, Dordrecht, The Netherlands, pp. 1–38.

Arshad, M. and Frankenberger, W.T. (1998) Plant growth-regulating substances in the rhizosphere: microbial production and functions. *Advances in Agronomy*, 62, 45–151.

Audenaert, K. et al. (2002) Induction of systemic resistance to Botrytis cinerea in tomato by *Pseudomonas aeruginosa* 7NSK2: role of salicylic acid, pyochelin and pyocyanin. *Molecular Plant Microbe Interaction*, 15, 1147–1156.

Baldani, J.I. and Baldani, V.L.D. (2005) History on the biological nitrogen fixation research in graminaceous plants: special emphasis on the Brazilian experience. *Anais da Academia Brasileira de Ciências*, 77, 549–579.

Barash, I. and Manulis-Sasson, S. (2007) Virulence mechanism and host specificity of gall-forming *Pantoea agglomerans*. *Trends in Microbiology*, 15, 538–545.

Barbieri, P. and Galli, E. (1993) Effect on wheat root development of inoculation with an *Azospirillum brasiliense* mutant altered indole-3-acetic acid production. *Research in Microbiology*, 144, 69–75.

Barbieri, P. et al. (1986) Wheat inoculation with *Azospirillum brasiliense* Sp6 and sme mutants altered in nitrogen fixation and indole-3-acetic acid production. *FEMS Microbiology Letters*, 36, 87–90.

Barea, J.M. et al. (1976) Production of plant growth-regulators by rhizosphere phosphate-solubilizing bacteria. *Journal of Applied Bacteriology*, 40, 129–134.

Bashan, Y. and Houlguin, G. (1998) Proposal for the division of plant growth-promoting rhizobacteria into two classifications: biocontrol-PGPB (plant growth-promoting bacteria) and PGPB. *Soil Biology and Biochemistry*, 30, 1225–1228.

Beattie, G.A., (2006). Plant associated bacteria: survey, molecular phylogeny, genomics and recent advances. In: S.S. Gnanamanickam (ed.) *Plant-Associated Bacteria*. Springer, The Netherlands, pp. 1–56.

Beemster, G.T.S. and Baskin, T.I. (2000) STUNTED PLANT 1 mediates effects of cytokinin, but not of auxin, on cell division and expansion in the root of Arabidopsis. *Plant Phisyology*, 124, 1718–1727.

Bellone, C.H. et al. (1997) Cell colonization and infection thread formation in sugar cane roots by *Acetobacter diazotrophicus*. *Soil Biology and Biochemistry*, 29, 965–967.

Benková, E. and Hejátko, J. (2009) Hormone interactions at the root apical meristem. *Plant Molecular Biology*, 69, 383–396.

Bertalan, M. et al. (2009) Complete genome sequence of the sugarcane nitrogen-fixing endophyte *Gluconacetobacter diazotrophicus* Pal5. *BMC Genomics*, 10, 450–467.

Bertrand, H. et al. (2001) Isolation and identification of the most efficient plant growth-promoting bacteria associated with canola (*Brassica napus*). *Biology and Fertility of Soils*, 33, 152–156.

Blaha, D. et al. (2006) Phylogeny of the 1-aminocyclopropane-1-carboxylic acid deaminase-encoding gene acdS in phytobeneficial and pathogenic Protobacteria and relation with strain biogeography. *FEMS Microbial Ecology*, 56, 455–470.

Bloemberg, G.V. and Lugtenberg, B.J.J. (2001) Molecular basis of plant growth promotion and biocontrol by rhizobacteria. *Current Opinion in Plant Biology*, 4, 343–350.

Boddey, R.M. et al. (1995) Biological nitrogen-fixation associated with sugarcane and rice—contributions and prospects for improvement. *Plant and Soil*, 174, 195–209.

Bodman, S.B. et al. (2003) Quorum sensing in plant-pathogenic bacteria. *Annual Review of Phytopathology*, 41, 455–482.

Boiero, L. et al. (2007) Phytohormone production by strains of *Bradyrhizobium japonicum* and possible physiological and technological implications. *Applied Microbiology and Biotechnology*, 74, 874–880.

Bolton, H. et al. (1989) Characterization of a toxin produced by a rhizobacterial *Pseudomonas* sp. that inhibits wheat growth. *Plant and Soil*, 114, 279–287.

Brencic, A. and Winans, S.C. (2005) Detection of and response to signlas involved in host–microbe interactions by plant-associated bacteria. *Microbial Molecular Biology Reviews*, 69, 155–194.

Brown, M.E. (1974) Seed and root bacterizations. *Annual Reviews in Phytopathology*, 12, 181–197.

Burdman, S. et al. (2000) Recent advances in the use of plant growth promoting rhizobacteria (PGPR) in agriculture. In: N.S. Subba and Y.R. Dommergues (eds) *Microbial Interactions in Agriculture and Forestry*, First edition, Vol. 2. Science Publishers, Enfield, NH, pp. 229–250.

Buttner, D. and Bonas, U. (2002) Getting across—bacterial type III effector proteins on their way to the plant cell. *EMBO Journal*, 21, 5313–5322.

Buyer, J.S. et al. (1993) A method for detection of pseudobactin, the siderophore produced by a plant-growth-promoting *Pseudomonas* strain, in the barley rhizosphere. *Applied and Enviromental Microbiology*, 59, 677–681.

Cakmakci, R. et al. (2001). Effect of N_2-fixing bacterial inoculations on yield of sugar beet and barley. *Journal of Plant Nutrition and Soil. Science*, 164, 527–531.

Carreño-Lopez, R. et al. (2000) Physiological evidence for different regulated tryptophan-dependent pathways for índole-3-acetic acid synthesis in *Azospirillum brasilense*. *Molecular and General Genetics*, 264, 521–530.

Cartieaux, F. et al. (2003) Transcriptome analysis of *Arabidopsis* colonized by a plant-growth promoting rhizobacterium reveals a general effect on disease resistance. *Plant Journal*, 36, 177–188.

Carvalho, T.L.G. et al. (2011) Sugarcane genetic controls involved in the association with beneficial endophytic nitrogen fixing bacteria. *Tropical Plant Biology*, 4, 31–41.

Catala, C. et al. (1997) Auxin regulation and spatial localization of an endo-1,4-*D*-glucanase and a xyloglucan endotransglycosylase in expanding tomato hypocotyls. *Plant Journal*, 12, 417–426

Cavalcante, J.J. et al. (2007) Members of ethylene signaling pathway are regulated in sugarcane during the association with nitrogen-fixing endophytic bacteria. *Journal of Experimental Botany*, 58, 673–686.

Cocking, E.C. et al. (2006) Intracellular colonization of roots of *Arabidopsis* and crop plants by *Gluconacetobacter diazotrophicus*. *Vitro Cell Development Biology Plan*, 42, 74–82.

Cohen, J.D. and Bandurski, R.S. (1982 ou 2008) Chemistry and physiology of the bound auxins. *Annual Review of Plant Physiology*, 33, 403–430.

Conrath, U., Pieterse, C.M.J., and Mauch-Mani, B. (2002) Priming in plant–pathogen interactions. *Molecular Plant Microbe Interaction*, 7, 1360–1385.

Costacurta, A. and Vandeleyden, J. (1995) Synthesis of phytohormones by plant-associated bacteria. *Critical Reviews in Microbiology*, 21, 1–18.

Davison, J. (1988) Plant beneficial bacteria. *Biotechnology*, 6, 282–286.

De Freitas, J.R. and Germida, J.J. (1991) *Pseudomonas cepacia* and *Pseudomonas putida* as winter wheat inoculants for biocontrol of *Rhizoctonia solani*. *Canadian Journal Microbiology*, 37, 780–784.

De Meyer et al. (1999) Nanogram amounts of salicylic acid produced by the rhizobacterium *Pseudomonas aeruginosa* 7NSK2 activate the systemic acquired resistance pathway in bean. *Molecular Plant Microbe Interaction* 12, 450–458.

De Salamone, I.E.G. et al. (2001) Cytokinin production by plant growth promoting rhizobacteria and selected mutants. *Canadian Journal of Microbiology*, 47, 404–411.

De Weger, L.A. et al. (1995) Colonization of the rhizosphere of crop plants by plant-beneficial pseudomonads. *FEMS Microbiology Ecology*, 17, 221–227.

Dobbelaere, S. et al. (1999) Phytostimulatory effect of *Azospirillum brasilense* wild type and mutant strains altered in IAA production on wheat. *Plant and Soil*, 212, 155–164.

Dobbelaere, S. et al. (2001) Responses of agronomically important crops to inoculation with *Azospirillum*. *Australian Journal of Plant Physiology*, 28, 871–879.

Dobbelaere, S. and Okon, Y. (2007) The plant growth-promoting effect and plant responses. In: C. Elmerich and W.E. Newton (eds) *Associative and Endophytic Nitrogen-Fixing Bacteria an Cyanobacterial Associations*, First edition. Springer Verlag, New York, pp. 145–170.

Dong, Z. et al. (1994) A nitrogen fixing endophyte of sugarcane stems (a new role for the apoplast). *Plant Physiology*, 105, 1139–1147.

Dong, Z. et al. (1997) Does *Acetobacter diazotrophicus* live and move in the xylem of sugarcane stems? Anatomical and physiological data. *Annals of Botany*, 80, 147–158.

Dow, M. et al. (2000) The induction and modulation of plant defense responses by bacterial lipopolysaccharides. *Annual Reviews in Phytopathology*, 38, 241–261.

Duan, J. et al. (2006) *1-Aminocyclopropane-1-carboxylate (ACC) deaminase genes in Rizhobia: isolation, characterization and regulation*. Proceedings in the 7th International PGPR Workshop, Amsterdan.

Egener, T. et al. (1999) Endophytic expression of nif genes of Azoarcus sp. strain BH72 in rice roots. *Molecular Plant–Microbe Interaction*, 12, 813–819.

Egener, T. et al. (1998) Use of green fuorescent protein to detect expression of nif genes of *Azoarcus* sp. BH72, a grass-associated diazotroph, on rice roots. *Molecular Plant–Microbe Interaction*, 11, 71–75.

Fallik, E. and Okon, Y. (1996). Inoculants of *Azospirillum brasilense:* biomass production, survival and growth promotion of Setaria italica and Zea mays. *Soil Biology and Biochemistry*, 28, 123–126.

Felix, G. et al. (1999) Plants recognize bacteria through the most conserved domain of flagellin. *Plant Journal*, 18, 265–276.

Ferguson, B.J. et al. (2005) Nodulation phenotypes of gibberellin and brassinosteroid mutants of pea. *Plant Physiology*, 138, 2396–2405.

Finkel, T. (2003) Oxidant signals and oxidative stress. *Current Opinion in Cell Biology*, 15, 247–254.

Foreman, J. et al. (2003) Reactive oxygen species produced by NADPH oxidase regulate plant cell growth. *Nature*, 422, 442–446.

Forchetti, G. et al. (2007) Endophytic bacteria in sunflower (*Helianthus annuus* L.): isolation, characterization, and production of jasmonates and abscisic acid in culture medium. *Applied Microbiology and Biotechnology*, 76, 1145–1152.

Frankenberger, W.T.J. and Arshad, M. (eds) (1995) *Phytohormones in Soils: Microbial Production and Function*, Marcel Dekker, New York.

Fu, X. and Harberd, N.P. (2003) Auxin promotes *Arabidopsis* root growth by modulating gibberellin response. *Nature*, 421, 740–743.

Fuentes-Ramirez, L.E. et al. (1993) *Acetobacter diazotrophicus*, an indolacetic acid-producing bacterium isolated from sugarcane cultivars in Mexico. *Plant and Soil*, 154, 145–150.

Fuentes-Ramirez, L.E. et al. (1999) Colonization of sugarcane by *Gluconacetobacter diazotrophicus* is inhibited by high N fertilization. *FEMS Microbiology Ecology*, 29, 117–128.

Fukaki,H. and Tasaka,M. (2009) Hormone interactions during lateral root formation. *Plant Molecular Biology*, 69, 437–449.

Gabriel, D. W. (1999) Why do pathogens carry avirulence genes? *Physiological and Molecular Plant Pathology*, 55, 205–214.

Garbeva, P. et al. (2001) Analysis of endophytic bacterial communities of potato by plating and denaturing gradient gel electrophoresis (DGGE) of 16s rDNA based PCR fragments. *Microbial Ecology*, 41, 369–383.

Gaudin, V. et al. (1994) Bacterial genes modifying hormonal balance in plant. *Plant Physiology an Biology*, 32, 11–29.

Gómez-Gómez, L. and Boller, T. (2002) Flagellin perception: a paradigm for innate immunity. *Trends in Plant Science*, 7, 251–256.

Gómez-Gómez, L. et al. (1999) A single locus determines sensitivity to bacterial flagellin in *Arabidopsis thaliana*. *Plant Journal*, 18, 277–284.

Gutierrez-Manero, F.J. et al. (2001) The plant growth-promoting rhizobacteria *Bacillus pumilus* and *Bacillus licheniformis* produce high amounts of physiologically active gibberellins. *Physiologia Plantarum*, 111, 206–211.

Glick, B.R. et al. (eds) (1999) *Biochemical and Genetic Mechanism Used by Plant Growth Promoting Bacteria*. Imperial College Press, London.

Glick, B.R. et al. (1998) A model for the lowering of plant ethylene concentrations by plant growth-promoting bacteria. *Journal od Theoretical Biology*, 190, 63–68.

Goldstein, A.H. (1986) Bacterial mineral phosphate solubilization: historical perspectives and future prospects. *American Journal of Alternative Agriculture*, 1, 57–65.

Greenberg, J.T. and Vinatzer, B. A. (2003) Identifying type III effectors of plant pathogens and analyzing their interaction with plant cells. *Current Opinion in Microbiology*, 6, 20–28.

Greiner, R. et al. (1997) Purification and characterization of a phytase from *Klebsiella terrigena*. *Archieves of Biochemistry and Biophysics*, 341, 201–206.

Grichko, V.P. and Glick, B.R. (2001) Amelioration of flooding stress by ACC deaminase-contaning plant growth-promoting bacteria. *Plant Physiology and Biochemistry*, 39, 11–17.

Gyaneshwar, P. et al. (2002) Role of soil microorganisms in improving P nutrition of plants. *Plant and Soil*, 245, 83–93.

Haefele, D.M. and Lindow, S. E. (1987) Flagellar motility confers epiphytic fitness advantages to *Pseudomonas syringae*. *Applied Environmental Microbioogy*, 53, 2528–2533.

Hallmann, J. et al. (1997) Bacterial endophytes in agricultural crops. *Canadian Journal of Microbiology*, 43, 895–914.

Hallmann, J. (2001) Plant interactions with endophytic bacteria. In: M.J. Jeger and N.J. Spence (eds), *Biotic Interactions in Plant-Pathogen Associations*, pp. 87–119. CAB International, USA.

Hamaoui, B. et al. (2001) Effects of inoculation with *Azospirillum brasilense* on chickpeas (*Cicer arietinum*) and faba beans (*Vicia faba*) under different growth conditions. *Agronomie*, 21, 553–560.

Hardoim, P.R. et al. (2008) Properties of bacterial endophytes and their proposed role in plant growth. *Trends in Microbiology*, 16, 463–471.

Harper, S.H.T. and Lynch, J.M., (1979) Effects of *Azotobacter chroococcum* on barley seed germination and seedling development. *Journal of Genetic Microbiology*, 112, 45–51.

Hartmann, A. et al. (2008) Lorenz Hiltner, a pioneer in rhizosphere microbial ecology and soil bacteriology research. *Plant and Soil*, 312, 7–14.

Henning, K. and Villforth, F. (1940) Experimentelle untersuchungen zur frage der bacteriesymbiose in höheren pflanzen und ihre beeinflussung durch "Leitemente." *Biochem Zeitschr*, 305, 299–309.

Idris, E.E. et al. (2007) Tryptophan-dependent production of indole-3-acetic acid (IAA) affects level of plant growth promotion by *Bacillus amyloliquefaciens* FZB42. *Molecular Plant–Microbe Interaction*, 20, 619–626.

Inuaki, Y. et al. (2005) Crown rootless 1, which is essential for crown root formation in rice, is a target of an auxin response factor in auxin signaling. *The Plant Cell*, 17, 1387–1396.

Jacoud, C. et al. (1998) Development of a strain-specific probe to follow inoculated *Azospirillum lipoferum* CRT1 maize root development by inoculation. *FEMS Microbial Ecology*, 27, 43–51.

James, E.K. et al. (2001) Further observations on the interaction between sugar cane and *Gluconacetobacter diazotrophicus* under laboratory and greenhouse conditions. *Journal of Experimental Botany*, 52, 747–760.

Jones, D.L. et al. (2003) Associative nitrogen fixation and root exudation: what is theoretically possible in the rhizosphere? *Symbiosis*, 35, 19–38.

Kapulnik, Y. et al. (1985) Changes in root morphology of wheat caused by *Azospirillum* inoculation. *Canadian Journal of Microbiology*, 31, 881–887.

Karadeniz, A. et al. (2006) Auxin, gibberellin, cytokinin and abscisic acid production in some bacteria. *World Journal of Microbiology and Biotechnology*, 22, 1061–1064.

Kennedy, A.C. (2005) Rhizosphere. In: D.M. Sylvia, J.J. Fuhrmann, P.G. Hartel, and D.A. Zuberer (eds) *Principles and Applications of Soil Microbiology*, Second edition. Pearson, Prentice Hall, New Jersey, pp. 242–262.

Kerovuo, J. et al. (1998) Isolation, characterization, molecular gene cloning, and sequencing of a novel phytase from *Bacilus subtilis*. *Applied and Environmental Microbiology*, 64, 2079–2085.

Khalid, A. et al. (2003) Growth and yield responses of wheat to inoculation with auxin producing PGPR. *Pakistan Journal of Botany*, 35, 483–498.

Kloepper, J.W. et al. (1991) Plant growth promotion mediated by bacterial rhizosphere colonizers. In: D.L. Keister and P.B. Cregan (eds) *The Rhizosphere and Plant Growth*, First Edition, pp. 315–326. Kluwer Academic Publisher, Netherlands.

Kloepper, J.W. et al. (1989) Free-living bacterial inocula for enhancing crop productivity. *Trends in Biotechnology*, 7, 39–43.

Kokalis-Burelle, N. et al. (2002) Field evaluation of plant growth-promoting rhizobacteria amended transplant mixes and soil solarization for tomato and pepper production in Florida. *Plant and Soil*, 238, 257–266.

Kremer, R.J. and Souissi, T. (2001) Cyanide production by rhizobacteria and potential for suppression of weed seedling growth. *Current Microbiology*, 43, 182–186.

Kucey, R.M. et al. (1989) Microbially mediated increases in plant-available phosphorus. *Advances in Agronomy*, 42, 199–227.

Kunkel, B.N. and Brooks, D.M. (2002) Cross talk between signaling pathways in pathogen defense. *Current Opinion in Plant Biology*, 5, 325–331.

Lambert, B. and Joos, H. (1989) Fundamental aspects of rhizobacterial plant growth promotion research. *Trends in Biotechnology*, 7, 219–219.

Libault, M. et al. (2010) Complete transcriptome of the soybean root hair cell, a single-cell model, and its alteration in response to *Bradyrhizobium japonicum* infection. *Plant Physiology*, 152, 541–552.

Lifshitz, R. et al. (1987) Growth promotion of canola (rapeseed) seedlings by a strain of *Psedomonas putida* under gnotobiotic conditions. *Canadian Journal of Microbiology*, 33, 390–395.

Loh, J. et al. (2002) Quorum sensing in plant-associated bacteria. *Current Opinion in Plant Biology*, 5, 1–6.

López-Bucio, J. et al. (2007) *Bacillus megaterium* rhizobacteria promote growth and alter root root-system architecture through na auxin- and ethylene-independent signaling mechanism in *Arabidopsis thaliana*. *Molecular Plant–Microbe Interaction*, 20, 207–217.

López-Bucio, J. et al. (2005) An auxin transport independent pathway is involved in phosphate stress-induced root architectural alterations in *Arabidopsis*. Identification of BIG as a mediator of auxin in pericycle cell activation. *Plant Physiology*, 137, 681–691.

López-Bucio, J. et al. (2003) The role of nutrient availability in regulating root architecture. *Current Opinion in Plant Biology*, 6, 280–287.

Lucy, M. et al. (2004) Applications of free living plant growth-promoting rhizobacteria. *Antonie Leeuwenhoek*, 86, 1–25.

Lugtenberg, B. and Kamilova, F. (2009) Plant growth promoting rhizobacteria. *Annual Review of Microbiology*, 63, 541–556.

Mantelin, S. and Touraine, B. (2004) Plant growth-promoting bacteria and nitrate availability: impact of development and nitrate uptake. *Journal of experimental Botany*, 55, 27–34.

Martin, C. and Paz-Ares, J. (1997) MYB transcription factors in plants. *Trends in Genetics*, 13, 67–73.

Mathesius, U. et al. (2003) From the cover: extensive and specific responses of a eukaryote to bacterial quorum-sensing signals. *Proceedings of the National Academy of Sciences USA*, 100, 1444–1449.

Miernyk, J.A. (1979) Abscisic acid inhibition of kinetin nucleotide formation in germinating lettuce seeds. *Physiologia Plantarum*, 45, 63–66.

Mirza, M.S. et al. (2001) Isolation, partial characterization, and the effect of plant growth-promoting bacteria (PGPB) on micro-propagated sugarcane in vitro. *Plant and Soil*, 237, 47–54.

Molla, A.H. et al. (2001) Potential for enhancement of root growth and nodulation of soybean co-inoculated with *Azospirillum* and *Bradyrhizobium* in laboratory systems. *Soil Biology and Biochemistry*, 33, 457–463.

Morrone, D. et al. (2009) Gibberellin biosynthesis in bacteria: separate ent-copalyl diphosphate and ent-kaurene synthases in *Bradyrhizobium japonicum*. *FEBS Letters*, 583, 475–480.

Nehl, D.B. et al. (1997) Deleterious rhizosphere bacteria: an integrating perspective. *Applied Soil Ecology*, 5, 1–20.

Neilandjs, B. and Leongs, A. (1986) Siderophores in relation to plant growth and disease. *Annual Review of Plant Physiology*, 31, 187–208.

Newman, M.A. et al. (2002) Prior exposure to lipopolysaccharide potentiates expression of defenses in response to bacteria. *Plant Journal*, 29, 487–495.

Nibau, C. et al. (2008) Branching out in new directions: the control of root architecture by lateral root formation. *New Phytologist*, 179, 595–614.

Nogueira, E.M. et al. (2005) Characterization of gluthamine synthetase genes in sugarcane genotypes with different rates of biological nitrogen fixation. *Plant Science*, 169, 819–832.

Nogueira, E.M., Vinagre, F. and Masuda, H.P. (2001) Expression of sugarcane genes induced by inoculation with *Gluconacetobacter diazotrophicus* and *Herbaspirillum rubrisubalbicans*. *Genetic and Molecular Biology*, 24, 199–206.

Oliveira, A.L.M. et al. (2002) The effect of inoculating endophytic N_2-fixing bacteria on micropropagated sugarcane plants. *Plant and Soil*, 242, 205–215.

Oliveira, A.L.M. et al. (2003) Response of micropropagated sugarcane varieties to inoculation with endophytic diazotrophic bacteria. *Brazilian Journal of Microbiology*, 34, 59–61.

Okushima, Y. et al. (2007) ARF7 and ARF19 regulate lateral root formation via direct activation of *LBD/ASL* genes in *Arabidopsis*. *The Plant Cell*, 19, 118–130.

Patten, C.L. and Glick, B.R. (1996) Bacterial biosynthesis of indole-3-acetic acid. *Canadian Journal of Microbiology*, 42, 207–220.

Pedrosa, F. (2005) A report on the genome of *Herbaspirillum seropedicae* strain Z78. *Current Plant Science and Biotechnology Agriculture*, 41, 111–114.

Penrose, D.M. et al. (2001) Determination of 1-aminocycopropane-1-carboxylic acid (ACC) to acess the effects of ACC deaminase-contaning bacteria on roots of canola seedlings. *Canadian Journal of Microbiology*, 47, 77–80.

Perotti, R. (1926) On the limits of biological inquiry on soil science. *Proceeding of International Society of Soil Science*, 2, 146–161.

Persello-Cartieaux, F. et al. (2001) Utilization of mutants to analyze the interaction between *Arabidopsis thaliana* and its naturally root-associated *Pseudomonas*. *Planta*, 212,190–198.

Persello-Cartieaux, F. et al. (2003) Tales from the underground: molecular plant–rhizobia interactions. *Plant, Cell and Environment*, 26, 189–199.

Pinton, R. et al. (2001) (eds) *The Rhizosphere. Biochemistry and Organic Substances at the Soil–Plant Interface*. Marcel Dekker, New York.

Podile, A.R. and Kishore, G.K. (2006) Plant growth-promoting rhizobacteria. In: S.S. Gnanamanickam (ed) *Plan-Associated Bacteria*, First edition. Springer, Dordrecht, The Netherlands, pp. 195–230.

Press, C.M. et al. (1997) Salicylic acid produced by *Serratia marcescens* is not the primary determinant of induced resistance in cucumber or tobacco. *Molecular Plant Microbe Interaction*, 10, 761–768.

Preston, G.M. (2004) Plant perceptions of plant growth-promoting *Pseudomonas*. *Philosophical Transactions of the Royal Society Lond B Biological Sciences*, 359, 907–918.

Preston, G.M. (2004) Plant Perceptions of plant growth-promoting *Pseudomonas*. Philosophical Transactions of the Royal Society London, Series B. *Biological Sciences* 359, 907-918.

Preston, G.M. et al. (2001) Type III secretion in plant growth-promoting *Pseudomonas fluorescens* SBW25. *Molecular Microbiology*, 41, 999–1014.

Prinsen, E. et al. (1993) *Azospirillum brasilense* índole-3-acetic acid biosynthesis: evidence for a non-tryptophan dependent pathway. *Molecular Plant–Microbe Interactions*, 6, 609–615.

Raaijmakers, J.M. et al. (1995) Dose response relationships in biological control of *Fusarium* wilt of radish by *Pseudomonas* spp. *Phytopathology*, 85, 1075–1081.

Rodriguez, H. and Fraga, R. (1999) Phosphate solubilizing bactéria and their role in plant growth promotion. *Biotechnology Advances*, 17, 319–339.

Ryu, R.J. and Patten, C.L. (2008) Aromatic amino acid-dependent expression of indole-3-pyruvate decarboxylase is regulated by TyrR in *Enterobacter cloacae* UW5. *Journal of Bacteriology*, 190, 7200–7208.

Sarig, S. et al. (1992) Effect of *Azospirillum brasilense* inoculation on growth dynamics and hydraulic conductivity of *Sorghum bicolor* roots. *Journal of Plant Nutrition*, 15, 805–819.

Schippers, B. (1988) Biological control of pathogens with rhizobacteria. *Royal Society of London Philosophical Transactions, London*, 318, 283–293.

Schippers, B. et al. (1987) Interactions of deleterious and beneficial rhizosphere microorganisms and the effect of cropping practices. *Annual Review of Phytopathology*, 25, 339–358.

Sevilla, M. et al. (2001) Comparison of benefit to sugarcane plant growth and $^{15}N_2$ incorporation following inoculation of sterile plants with *Acetobacter diazotrophicus* wild-type and *nif* – mutant strains. *Molecular Plant–Microbe Interactions*, 14, 358–366.

Sharma, A. and Johri, B.N. (2003) Growth promoting influence of siderophore-producing *Pseudomonas* strains GRP3 and PRS9 in maize (*Zea mays* L.) under iron limiting conditions. *Microbiological Research*, 158, 243–248.

Smalla, K. et al. (2006) The rhizosphere: soil compartment influenced by root. *FEMS Microbial Ecology*, 56, 165–172.

Somers, E. et al. (2004) Rhizosphere bacterial signaling: a love parade beneath our feet. *Critical Reviews in Microbiology*, 30, 205–240.

Spaepen,S. et al. (2009) Plant growth-promoting actions of rhizobacteria. *Advances in Botanical Research*, 51, 283–320.

Spaepen, S. et al. (2008) Effects of *Azospirillum brasilense* indole-3-acetic acid production on inoculated wheat plants. *Plant and Soil*, 312, 15–23.

Spaepen,S. et al. (2007) Indole-3-acetic acid in microbial and microorganism-plant signaling. *FEMS Microbiology Reviews*, 31, 425–448.

Steenhoudt, O. and Vanderleyden, J. (2000) *Azospirillum*, a free-living nitrogen-fixing bacterium closely associated with grasses: genetic, biochemical and ecological aspects. *FEMS Microbiology Reviews*, 24, 487–506.

Stevenson, F.J. and Cole, M.A. (1999) *Cycles of soils: carbon, nitrogen, phosphorus, sulfur, micronutrients*, Second edition. Wiley, New York.

Suslow, T.V. and Schroth, M.N. (1982) Role of deleterious rhizobacteria as minor pathogens in reducing crop growth. *Phytopathology*, 72, 111–115.

Taguchi, G. et al. (2001) Molecular cloning and heterologous expression of novel glucosyltransferases from tobacco cultured cells that have broad substrate specificity and are induced by salicylic acid and auxin. *European Journal of Biochemistry*, 268, 4086–4094.

Tran Van, V. et al. (2000) Repeated beneficial effects of rice inoculation with a strain with a strain of *Burkholeria vietnamiensis* on early and late yield components in low fertility sulphate acid soils of Vietnam. *Plant and Soil* 281, 273–284.

Urquiaga, S. et al. (1992) Contribution of nitrogen fixation to sugarcane: nitrogen-15 and nitrogen balance estimates. *Soil Science Society of American Journal*, 56, 105–114.

UthedeR.S. et al. (1999) Rhizobacterial growth and yield promotion of cucumber plants inoculated with *Pythium aphanidermatum*. *Canadian Journal of Plant Pathology*, 21, 265–271.

van Loon,LC. et al. (2006) Significance of inducible defense-related proteins in infected plants. *Annual Reviews in Phytopathology*, 44, 135–162.

Van Peer, R. and Schippers, B. (1988) Plant growth responses to bacterization with selected *Pseudomonas* spp. strains and rhizosphere microbial development in hydroponic cultures. *Canadian Journal of Microbiology*, 35, 456–463.

Vance, C.P. (2001) Symbiotic nitrogen fixation and phosphorus acquisition: lant nutrition in a world of declining renewable resources. *Plant Physiology*, 127, 390–397.

Vande-Broek, A. et al. (1993) Spatial-temporal colonization patterns of *Azospirillum brasilense* on the wheat root surface and expression of the bacterial *nifH* gene during association. *Molecular Plant–Microbe Interaction*, 6, 592–600.

Vargas, C. et al. (2003) Signaling pathways mediating the association between sugarcane and endophytic diazotrophic bacteria: a genomic approach. *Symbiosis*, 35, 159–180.

Vedder-Weiss, D. et al. (1999) Root growth, respiration and beta-glucosidase activity in maize (*Zea mays)* and common bean (*Phaseolus vulgaris)* inoculated with *Azospirillum brasilense*. *Symbiosis*, 26, 363–377.

Verhagen, B.W.M. et al. (2004) The transcriptome of rhizobacteria-induced systemic resistance in *Arabidopsis*. *Molecular Plant–Microbe Interaction*, 17, 895–908.

Vessey, J.K. (2003) Plant growth promoting rhizobacteria as biofertilizers. *Plant and Soil*, 255, 571–586.

Vohra, A. and Satyanarayana, T. (2003) Phytases: microbial sources, production, purification, and potential biotechnological applications. *Critical Reviews in Biotechnology*, 23, 29–60.

Wang,C. et al. (2001) Cosmopolitan distribution of *phlD*-containing dicotyledonous crop-associated biocontrol pseudomonads of worldwide origin. *FEMS Microbiology Ecology*, 37, 105–116.

Wang, Y.Q. et al. (2005) Microarray analysis of the gene expression profile induced by the endophytic plant growth-promoting rhizobacteria, *Pseudomonas fluorescens* FPT9601-T5 in Arabidopsis. *Molecular Plant–Microbe Interaction*, 18, 385–396.

Weingart, H. and Völksch, B. (1997) Ethylene production by *Pseudomonas syringae* pathovars *in vitro* and *in planta*. *Applied Environmental Microbiology*, 63, 156–161.

Weingart, H. et al. (2001) The role of ethylene production in virulence of *Pseudomonas syringae* pvs. *glycinea* and *phaseolicola*. *Phytopathology*, 91, 511–518.

Wolfgang, M.C. et al. (2003) Coordinate regulation of bacterial virulence genes by a novel adenylate cyclase-dependent signaling pathway. *Developmental Cell*, 4, 253–263.

Zhang, H. and Forde, B.G. (1998) An *Arabidopsis* MADS box gene that controls nutrient-induced changes in root architecture. *Science*, 279, 407–409.

13 Translational Root Genomics for Crop Improvement

Reyazul Rouf Mir, Mahendar Thudi, Siva K. Chamarthi,
L. Krishnamurthy, Pooran M. Gaur, and Rajeev K. Varshney

Introduction

Breeding of crop plants in the 21st century needs consciousness and awareness of climate change. For instance, both biotic and abiotic stresses need attention of breeders in order to breed crops showing resistance/tolerance to these stresses in a changing climate scenario. In developing countries, drought is the major abiotic stress and is already limiting crop productivity in several species. This tendency is going to worsen the situation in the years to come (Varshney et al. 2011a). Enhancing crop productivity in resource-poor dry land conditions is a formidable challenge. Conserving resources through management practices and engineering plants for superior extraction of these resources coupled with an increased efficiency of resource utilization deserve emphasis. Though resource conservation through management practices are equally important, development of superior resource use efficiency as a seed-based technology always has greater acceptance and adaptability.

Roots, which are underground hidden parts of plants, are considered as first plant organs to be exposed, as well as to respond in stresses. However, among several parts of the plant, roots have received little attention of researchers so far despite their major role in plant–soil interactions (Sheshshayee et al. 2011). Thus, they seem to hold the key for the next plant breeding revolution, leading to improved crop productivity in environmentally challenged situations. Recent advancements in plant genomics are certainly of help in crop improvement efforts (Varshney et al. 2005) and genomics-assisted breeding applications such as marker-assisted selection (MAS) have been already used in developing superior cultivars in several crop species (Varshney et al. 2006, 2010; Gupta et al. 2010). The field of root genomics is an exciting and promising field of research and some technical advancements in plant-omics are believed to generate some useful data for pursuing translational research. Whole genome approaches such as microarrays or next-generation sequencing (NGS)-based transcript profiling or functional genomics approaches may provide some useful information about the type and nature of structural and functional genes involved in various aspects of root growth, development, and water and nutrient uptake. In addition, on the basis of whole genome sequence available for several crop species (for references, see Jackson et al. 2011; Morrell et al. 2012) or gene information available in model plant species, use of comparative genomics or bioinformatics approaches may provide interesting candidate genes. Once the details of these genes are known, some other approaches like overexpression, TILLLING (Targeting-Induced Local Lesions IN Genomes; Varshney et al. 2011b), qRT-PCR, gene-knockout mutant analysis may be used to further validate their utility in root trait improvement.

Root Genomics and Soil Interactions, First Edition. Edited by Martin Crespi.

In this article, we highlight the importance of root genomics research in the context of drought stress and discuss recent developments in genetic and molecular dissection of root traits. A special emphasis has been given on translational root genomics for developing superior varieties through molecular breeding for root traits.

Root Research for Crop Improvement

Studying roots in higher plants assumes importance because they provide firm anchorage of the plant in their soil substratum and helps in absorption and effective supply of water and nutrients to the shoot. These are the most important roles of the root system (Varshney et al. 2011b). In addition, roots assume importance for plants by producing a number of growth hormones, including cytokinins and ABA. From an ecological point of view, roots play some role in weathering of rocks, thus leading to the formation of soil. Sometimes, tiny roots spread out by growing horizontally and form a thick mat-like network; they may prevent soil erosion and through their symbiotic association, roots with rhizobia, mycorrhiza, and other organisms can fix atmospheric nitrogen or enrich mineral soil content. However, keeping in view the current scenario of importance of drought in view of changing climate, the water mining capacity of roots from deeper soil profiles is considered as one of the important adaptive strategies evolved by plants to survive water-scarce conditions. With an objective of paying more attention on root research at genetic, physiological, and molecular level, a *Plant Root Genomics Consortium* was developed by University of Missouri in collaboration of other partners from USA (http://rootgenomics.rnet.missouri.edu/prgc/index.html). The primary goal of this consortium is to develop understanding of molecular mechanisms used by plant roots to get water and nutrition from soil and to find out the possible role of roots in adaptation to drought conditions and further transfer this knowledge for crop improvement through various breeding and biotechnological approaches. Discussions are currently underway to extend this consortium into an International Root Genomics Consortium.

The role of roots is very significant to enhance drought tolerance. In simpler terms, drought tolerance is the ability of a plant to avoid or tolerate stress at organism level (Levitt 1972; Blum 2005). Drought is often regarded as a major threat to ecosystems, as water stress limits crop yield more than all other biotic and abiotic factors combined (Lambers et al. 2008). In these situations, the ability of the plants to explore water resource by reaching roots deep in the soil and extracting water has great relevance in maintaining water relation as well as carbon assimilation. Therefore, genotypes with long deep-roots have been found more tolerant to drought (Li et al. 2005a; Kashiwagi et al. 2005, 2006; Reynolds and Tuberosa 2008). In addition, some other related traits have been suggested to help plants to perform better in water-limited conditions. For instance, low leaf conductance under nonlimited water conditions during the vegetative stage, higher fraction of transpirable soil water (FTSW) thresholds that reduce transpiration, thus avoiding rapid soil water depletion, and low leaf expansion rate when soil moisture is still nonlimiting for plant growth have been suggested in chickpea (Zaman-Allah et al. 2011).

Genetic Dissection of Root Traits

Quantitative Trait Locus (QTL) Discovery for Root Traits

In simple terms, a QTL is a segment of DNA that affects a quantitative trait or the region within genome that contains genes associated with a particular quantitative trait. The plant root system is considered highly dynamic and responds to changes in environmental parameters, including stresses such as drought, nutrient deficiencies, water logging, and salinity. However, despite their essential role in plant growth, adaptation, and mineral nutrient acquisition, the root system has remained unexplored, owing to the difficulty in screening techniques until recently. Once the considerable understanding of root growth and development, both at the whole plant level and at the molecular level, is achieved, the next step is to devise strategies for identification of important genes/QTLs associated with various root traits followed by their validation and subsequently introgression into crops through molecular breeding approaches (Varshney et al. 2011b). In terms of genetic control, root traits are believed to be complex controlled by a number of genes/QTLs. Therefore, understanding the genetic control of root development and functions of component root traits is considered inherently important for breeding improved cultivars for root traits that are well adapted to variable climates. For QTL mapping studies, generating large-scale and precise phenotyping data on mapping populations is very critical. In recent years, significant progress has been made toward improving phenotyping capabilities (Manschadi et al. 2008; Gregory et al. 2009; Hund et al 2009; Nagel et al. 2009; Yazdanbakhsh and Fisahn 2009; Chen et al. 2011) as well as marker genotyping capacities (see Mir et al., in press). As a result, marker-trait association (MTA) studies are gaining importance to gain new insights into the genetic control of root system architecture (see de Dorlodot et al. 2007; Courtois et al. 2009; Hochholdinger and Tuberosa 2009; Coudert et al. 2010; Kell 2011; Gowda et al. 2011; Tuberosa et al. 2011).

In general, there are two important approaches for the study of MTA: (i) QTL interval mapping commonly called linkage mapping/linkage analysis-based QTL mapping and (ii) linkage disequilibrium-based association mapping (Figure 13.1). Both approaches have their own advantages and disadvantages and have been already used for discovering QTLs/genes for a variety of traits in all important crop plants (see Myles et al. 2009; Rafalski 2010). The various steps involved in both approaches have been discussed elsewhere (see Varshney et al. 2011b; Chamarthi et al. 2011; Figure 13.1). By using one of these two approaches, QTLs for root traits have been identified in about 15 plant species (Kalliokoski et al. 2008). However, rice, maize, and wheat among cereal crops and soybean, common bean and chickpea among legumes have dominated root trait QTL studies. Some of these important QTL studies have been summarized in Table 13.1.

Rice is probably the crop species in which a maximum number of QTL studies for root traits have been targeted (see Courtois et al. 2009; Priya et al. 2009). More than 900 QTLs related to root traits such as maximum length, number, thickness, volume, and their effects on yield under varying moisture regimes have been studied in rice (Kamoshita et al. 2002a; MacMillan et al. 2006; Steele et al. 2006, 2007; Yue et al. 2006). Since precise phenotyping is considered crucial in the study of MTA for root traits, a range of phenotyping platforms and methods including green house (Komoshita et al. 2002b), hydroponic cultures (Obara et al. 2010), basket method (Uga et al. 2011), poly vinyl chloride (PVC) cylinders (Qu et al. 2008), and field conditions (Ikeda et al. 2007; Yue et al. 2008) have been used. In terms of genetic localization, QTLs for maximum root length have been identified almost on all 12 chromosomes (Hemamalini et al. 2000; Zhang et al. 2001; Kamoshita et al. 2002a, 2002b; Courtois et al. 2003; Horii et al. 2006; MacMillan et al. 2006; Yue et al. 2006). A major QTL for root length (*qRL6.1*) based on phenotyping of seedlings grown under hydroponic conditions has been identified on chromosome 6 (Obara et al. 2010) and delimited to a 337 kb region of the Nipponbare genome. Another major QTL controlling RDR (for ratio of deep rooting), which explains 66.6% total phenotypic variation (PV) was detected on chromosome 9 by using RIL populations derived from a cross IR64 $\times$ Kinandang Patong (Uga et al. 2011).

In the case of maize, QTLs for various root traits have been identified based on phenotyping using a range of soil moisture regimes (Tuberosa et al. 2002a, 2003; Landi et al. 2007). QTLs of root pulling resistance at flowering time were identified based on phenotyping in field conditions (Giuliani et al.

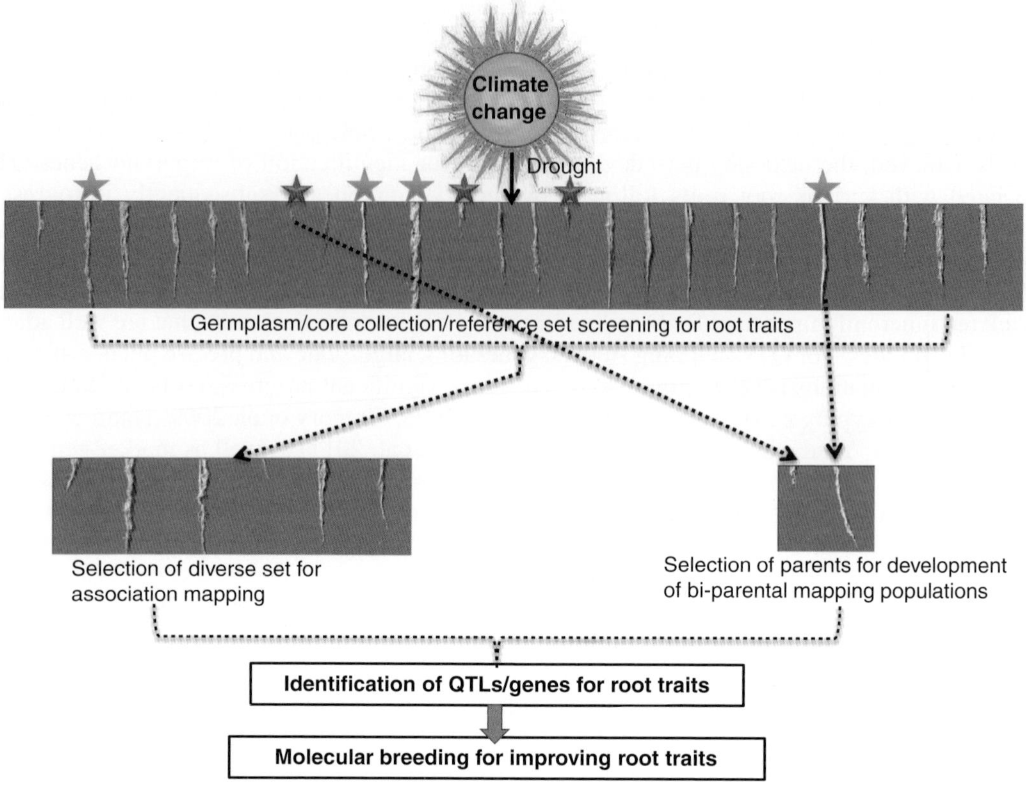

Figure 13.1 A schematic figure showing discovery and translation of root trait QTLs for crop improvement. The figure shows screening of germplasm under natural conditions for root traits. While a diverse set of germplasm showing a variation of root traits is selected and used for association mapping, contrasting lines comprising of higher root, and lower root biomass are selected for developing biparental mapping populations and linkage mapping-based QTL analysis. Once major QTLs are identified, molecular breeding for these QTLs can lead to development of superior lines for improved root traits. Red stars indicated lines with short roots (mostly drought susceptible), while green lines indicated lines with long roots (mostly drought tolerant). (For a color version of this figure, see the color plate section.)

2000). Recently, Ruta et al. (2010) identified 13 QTLs for six seedling traits: elongation rates of axile roots, the rate constant of lateral root elongation (kLat), the final respective lengths LAx and LLat, and the ratios kLat/ERAx and LLat/LAx. In this study, QTLs for the elongation rates of axile roots responded more clearly to water stress compared to root length. In addition, many QTLs for root traits and their responses to drought and phosphorus deficiency have been reported in maize (Lebreton et al. 1995; Guingo et al. 1998; Landi et al. 2002; Tuberosa et al. 2002b; Hund et al. 2004; Mano et al. 2005; Zhu et al. 2005a, 2005b, 2006). However, QTL analysis for root traits in response to low N stress is not well studied.

In the case of wheat, Ma et al. (2005) found a QTL for root-growth rate under Al treatment. QTLs of root traits (primary/lateral root length and number, root dry matter) under control conditions and during nitrogen deficiency were also identified (Laperche et al. 2006). Several studies identified QTLs for early root growth in wheat (An et al. 2006; Laperche et al. 2006; Li et al. 2007; Sanguineti et al. 2007; Sharma et al. 2011). In a recent study, a total of 15 QTLs, including 6 additive and 9

Table 13.1 Root trait QTLs identified in some important cereals and legume crop plants.

Crop	Mapping Population	Traits	Environment	QTLs	Phenotypic Variation Explained (PVE; %)	Reference
Rice	IR64 × INRC10192 (RIL)	Shoot length, maximum root length, shoot dry weight, root dry weight, and root to shoot dry weight ratio	PEG-induced stress conditions	19	7.9–29.8	Srividhya et al. 2010
	IR64 × Kinandang Patong (RIL)	Ratio of deep rooting	Basket method	1	66.6	Uga et al. 2011
	Koshihikari × Kasalath (CSSL)	Root length	Hydroponic culture	1	21.1	Obara et al. 2010
	IR64 × Kinandang Patong (RIL)	Stele transverse area, total area, and number of late metaxylem vessels	Upland field; well watered	10	18.7–23.9	Uga et al. 2008
	IRAT109 × Yuefu (RIL)	Basal root thickness, root number, maximum root length, root fresh weight, root dry weight, and root volume	PVC pipes at seedling, tillering, heading, grain filling, and mature stages	84 additive and 86 epistatic	19 and 10	Qu et al. 2008
	Otomemochi × Yumenohatamochi	Number of new roots, maximum new root length, root, and shoot ratio	Low land field, well-watered and drought stress	20	7.0–31.2	Ikeda et al. 2007
	Akihikari × IRAT109 (BIL)		Solution culture	5	8–24	Horii et al. 2006
	Bala × Azucena (RIL)	Maximum root length, maximum root thickness, root mass below 50 cm, total plant dry mass, % root mass, and shoot length	Soil box method	28	5.4–13.5	MacMillan et al. 2006
	IRAT 109 × Yuefu (DH)	Basal root thickness, total root number, maximum root length, root fresh weight, root dry weight, ratio of root fresh weight to shoot fresh weight, ratio of root dry weight to shoot dry weight	PVC cylinders, low land field, well watered, and drought	51	1.1–25.6	Li et al. 2005a
	IR1552 × Azucena (RIL)	Seminal root length, adventitious root number, lateral root length, and lateral root number	Pot culture, well watered, and drought stress	23	11.4–20.0	Zheng et al. 2003
	IAC165 × Co39 (RIL)	Root length, root thickness, and root dry weight	PVC cylinders, aerobic	29	6.3–24.4	Courtois et al. 2003
	IR64 × Azucena (DH)		PVC cylinders, well watered, and drought stress	3	12.9–30.7	Venuprasad et al. 2002
	CT9993-5-10-1-M × IR62266-42-6-2	Root thickness	Green house	8	6.8–51.8	Kamoshita et al. 2002a

(Continued)

Table 13.1 (*Continued*)

Crop	Mapping Population	Traits	Environment	QTLs	Phenotypic Variation Explained (PVE; %)	Reference
	CT9993 × IR62266 (DH)	–	Wax-petrolatum layer system and field	36	8.0–37.0	Zhang et al. 2001
	ZYQ8 × JX17	Maximum root length, dry root weight, root/shoot ratio of dry weight	Solution culture	5	10.4–19.9	Xu et al. 2001
	IR64 × Azucena (DH)	PVC cylinders	PVC cylinders; drought stress	29	11.9–26.7	Hemamalini et al. 2000
Maize	Lo964 × Lo1016 (RIL)	Elongation rates of axile roots, rate constant of lateral root elongation	Roots were grown in pouches under well-watered conditions or at low water potential induced by the osmolyte (PEG 8000)	13	2.3–21.9	Ruta et al. 2010
	Z3 × 87-1	Number of axial roots, average axial root length, length of lateral root, root length	Hydroponics	17	11–43.7	Liu et al. 2008
	B73 × *Zea luxurians* (F_2)	Root angle	Greenhouse	4	0.123–0.396	Omori and Mano 2007
	B73 × Mo17 (RIL)	Fractal dimension (FD)1, FD2, FD change over time	–	47	24.6–46.8	Bohn et al. 2006
	B73 and Mo17 (RIL)	Seminal root length and seminal root number		5	11–25.4	Zhu et al. 2006
	B73 and Mo17 (RIL)	Root hair length, RHL under high fertility, root hair length under low phosphorus	Cigar roll culture system	5	9.6–31.9	Zhu et al. 2005a
	B73 × Mo17 (RIL)	Lateral root length, lateral root number under high fertility, lateral root number under low fertility	Cigar roll culture system	8	3.5–18.9	Zhu et al. 2005b
	B64 × Na4 (F_2)	Adventurous root formation	Flooding condition		–	Mano et al. 2005
	Lo964 × Lo1016 ($F_{2:4}$)	Root weight, root length, diameter of primary and seminal roots		40	–	Hund et al. 2004
	Lo964 × Lo1016 (F_3 families)	Root length, primary root diameter, primary root weight, weight of adventitious seminal roots	Hydroponics	37	6.8–36.2	Tuberosa et al. 2002a
	Lo964 × Lo1016 (F_3 families)	Vertical root pulling resistance	Hydroponics	19	–	Landi et al. 2002

Wheat	Xiaoyan 54 × Jing 411 (RIL)	Maximum root length, primary root length, lateral root length, total root length, and root tip, and number of seminal roots	Hydroponics	22	4.2–68	Ren et al. 2011
	Pavon 76 × Pavon 1RS.1BL	Root length and root weight	Green house, natural photoperiod	15	–	Sharma et al. 2011
	Huapei 3 × Yumai 57, (immortalized F_2)	Root total length, root surface area, root average diameter, root volume, root tips, maximum root length, shoot dry weight, root dry weight, and ratio of root dry weight to shoot dry weight per plant	Hydroponics	19	0.01–11.91	Li et al. 2010
	Creso × Pedroso (RIL)	Length, ProjArea, SurfArea, average diameter, LenPerVol, volume	Hydroponics	–	60	Petrarulo et al. 2009
		Root length, root number, and root dry matter		30	10	Li et al. 2007
	Hanxuan 10 × Lumai 14	Shoot dry weight (SDW), root dry weight (RDW), tiller number (TN), and NUP (total plant N uptake) per plant	Hydroponics, field	17, 17	4.3–9.6, 6–21.9	An et al. 2006
Soybean	Essex × Forrest	Basal root thickness, maximum root length, lateral root number, root dry weight, root fresh weight, shoot fresh weight, shoot dry weight (SDW), RDW/SDW, and RFW/SFW	–	19	12–34	Williams et al. 2012
	PI416937 × Benning	Fibrous roots	Field	5	51	Abdel-Haleem et al. 2011
	Hartwig × Flyer (RIL)	Basal root thickness, maximum root length, lateral root number, root dry weight, root fresh weight, shoot fresh weight, shoot dry weight (SDW), RDW/SDW, and RFW/SFW	Greenhouse	43	–	Ouertani et al. 2010
	BD 2 × BX10 (RIL)	Total root length, total root surface area, root width, ratio of root/shoot, root dry weight, root P content, plant P-content, shoot P content, plant dry weight, shoot dry weight, seed number per plant, seed weight per plant	Field	31	10.9–18.7	Liang et al. 2010
	Essex × Forrest	Basal root thickness, lateral root number, maximum root length, root fresh weight, root dry weight, shoot fresh weight, shoot dry weight	Green house and field	89	–	Washington et al. 2009
	Kefeng 1 × Nannong 1138-2	Dry root weight, total root length and root volume	Green house	13	3.6–24.7	Gai et al. 2007

(Continued)

Table 13.1 (*Continued*)

Crop	Mapping Population	Traits	Environment	QTLs	Phenotypic Variation Explained Explained (PVE; %)	Reference
Common bean	DOR364 × G19833 and G2333 × G198339 (RILs)	Number of basal root whorls	Growth pouch	2	25–58	Miguel et al. 2009
	G19833 × AND696	Root growth, seed yield, seed P content, and P use efficiency	Field	29	18–33	Cichy et al. 2009b
	DOR 364 × G19833	Root length, specific root length, P-accumulation, tap root length, tap root diameter	Field and green house	26	0.08–0.51	Beebe et al. 2006
	G2333 × G19839	Number of adventitious, adventitious dry weight, adventitious length, adventitious specific root length, shoot dry weight.	Green house and field	19	19–61 field condition and 18–39 green house	Ochoa et al. 2006
	DOR364 × G19833 (RIL)	Total root hair length for basal roots, hair length for tap roots	Solution culture	19	0.33–0.5	Yan et al. 2004
	DOR364 × G19833 (RIL)	Basal root angle, shallow basal root length, phosphorous root acquisition	Growth pouch and field	16	9.3–20.3	Liao et al. 2004
Chickpea	ICC 4958 × ICC 1882	Root length, root volume, root length density, root surface area, root dry weight, shoot length, shoot dry weight	Field	52	35	Varshney et al. (unpublished)
	ICC 283 × ICC 8261	Root length, root volume, root length density, root surface area, root dry weight, shoot length, shoot dry weight	Field	45	35	Varshney et al. (unpublished)

epistatic QTLs have been detected for different traits of root length and root weight on 1RS segment using a high-resolution chromosome arm-specific mapping population (Sharma et al. 2011). In the case of sorghum, nodal root angle (qRA) that influences vertical and horizontal root distribution in the soil profile is considered one of the most important traits that may provide new opportunities for improving drought adaptation mechanisms. Mace et al. (2012) have reported four QTLs for qRA in addition to three QTLs for root dry weight: two for shoot dry weight and three for plant leaf area. Further, qRA QTL explained 58.2% of the phenotypic variance was also validated across a range of diverse inbred lines.

Among legumes, QTLs for root fresh weight, maximum root length, basal root thickness, lateral root number, root dry weight, total root surface area, root width, ratio of root/shoot, root dry weight, root P content, plant P content, and shoot P content have been identified in soybean under field conditions (Liang et al. 2010; Abdel-Haleem et al. 2011) as well as under greenhouse conditions (Gai et al. 2007; Washington et al. 2009). On the other hand, a large number of QTLs for P efficiency have been identified in soybean (Li et al. 2005b; Zhang et al. 2009). In the case of common bean, a large number of QTLs for root morphology and physiology as related to P nutrition have been reported (Beebe et al. 2006; Li et al. 2007; Chen et al. 2009; Cichy et al. 2009a; Li et al. 2009). More interestingly, QTLs for basal root growth and development were shown tightly linked with the QTLs for P uptake efficiency under field conditions (Beebe et al. 2006). In the case of chickpea, based on phenotyping of root traits under PVC cylinder and SSR genotyping on two mapping populations (ICC 4958 × ICC 1882 and ICC 283 × ICC 8261), a genomic region, harboring QTLs for various drought tolerance related root traits, has been identified on linkage group 4, which explained more than 35% phenotypic variation (unpublished data).

Meta-QTL Analysis for Root Traits

As different QTL studies employ different mapping populations like recombinant inbred lines (RILs), doubled haploids (DHs), backcross introgression lines, and near isogenic lines (NILs) that segregate for different root traits, it is difficult to exploit the varietal difference for improved root traits by MAS and for identification of concerned alleles. In this context, meta-analysis of QTLs is considered one of the best approaches that is conducted when hundreds of QTLs based on large number of studies involving a number of mapping populations are available in the published literature. This approach is a useful reductionist approach to bring down the number of genuine and real QTLs for their efficient use in MAS. Meta-analysis of QTLs provides narrow confidence intervals for meta-QTLs (MQTLs), permitting easier positional candidate gene identification. Meta-analysis of QTLs is usually applied to multiple mapping populations, but can be applied to one also (Khowaja et al. 2009).

For comparing root traits across experiments and developmental stages, the International Society of Root Research (ISRR, www.rootresearch.org) proposed a general nomenclature for roots (Zobel and Weisel 2010) to compare root types in different crop species. Recently, a database "QlicRice" was developed by Smita et al. (2011), which serves as a web interface and platform for data mining of abiotic stress responsive QTLs and their corresponding sequenced gene loci, which help researchers to retrieve association among different agronomic traits. Meta-analysis of drought-related QTLs in the Bala × Azucena mapping population of rice from 13 experiments and 25 independent screens resulted in three meta-QTLs within a space of 35 cM on chromosome 9 (Khowaja et al. 2009). Swamy et al. (2011) identified that the MQTL for grain yield under drought coincided with at least one of the MQTL identified for root and leaf morphology traits under drought in earlier reports.

A rational database "Rootbrowse" has also been constructed using QTLs, markers, and genome sequence information (Priya et al. 2009).

In the case of maize, MQTL analysis based on 15 independent QTL studies of 9 mapping populations resulted in the identification of seven MQTL for root traits that are colocalized with grain yield and other drought responsive traits in the field. Furthermore, on the basis of MQTL analysis, Landi et al. (2010) inferred that one single locus, root-yield-1.06, has a major constitutive effect on root traits, plant vigor and productivity across water regimes, genetic backgrounds, and inbreeding levels.

Molecular Dissection of Root Traits

Candidate Genes Based on Genome Sequence Information

As evident from Table 13.1, the root trait QTLs have been studied extensively in rice than any other crop. It is also known that the rice root system is complex, consisting of seminal, adventitious, and lateral roots. The availability of the rice genome sequence has opened some new avenues to look at the large number of QTLs already identified to control root development and to identify candidate genes in those QTL regions. In this context, a majority of root QTLs identified in rice were integrated computationally into the *japonica* genome and it was observed that a maximum number of QTLs for root development is present on chromosome 1 and fewer QTLs on some other chromosomes (Ulaganathan et al. 2007; http://www.kulab.org/research1.htm). As expected, several QTLs reported in different studies utilizing different mapping populations have been found to possess QTLs in the same genomic regions, thus showing that many QTL regions are common across different mapping populations. Some of these genomic regions possessing several overlapping QTLs could also be resolved into multiple QTLs. More interestingly, the putative candidate genes governing root development in these QTLs regions were extracted and their putative functions were computationally analyzed. This led to identification of small number of genes (200–300 genes) in some of the QTL regions, while a large number of genes (600–800 genes) were present in larger QTLs/genomic regions. In summary, computational analysis of the genes revealed and confirmed the polygenic and complex nature of root development in rice.

These genes identified in genomic regions for root development mostly belongs to categories of (a) transcription factors, (b) auxin metabolism and transport genes, (c) auxin-responsive genes, (d) auxin-related proteasome pathway genes, (e) environmental sensors, and (f) biotic and abiotic stress tolerance genes (http://www.kulab.org/research1.htm). However, it will be essential to validate these genes using functional genomics approaches like quantitative RT-PCR or reverse genetic approaches like knockout mutant analysis and TILLING. Nevertheless, availability of the genome sequence in several crop species like maize (Schnable et al. 2009), soybean (Schmutz et al. 2010), sorghum (Paterson et al. 2009), pigeonpea (Varshney et al. 2012), and others offer the possibility to identify the candidate genes underlying the QTL regions identified in these crop species.

QTL Cloning for Root Traits

Most of the QTL mapping studies involves crossing of varieties with contrasting root characteristics, development of mapping populations, and identification of number of QTLs on different chromosomes. Since most of these experiments involve the use of different sets of markers and localization of QTLs on long genomic regions, thus making it difficult to exploit the varietal difference for improved drought tolerance by MAS and for identification of the concerned alleles. Therefore, one would like to move close to the target QTL or even use perfect functional markers for the introgression of respective QTL/gene for a root trait through molecular breeding. Therefore, efforts may be made to clone all the major and important QTL for a root trait. The positional cloning of a major QTL requires (i) large mapping population (>2000 plants) obtained after crossing two NILs for the target QTL, (ii) identifying the genomic region covering the QTL region, and (iii) validating of effect of candidate gene(s) on phenotype (see Salvi and Tuberosa 2005; Tuberosa and Salvi 2006).

Several reports are available on mapping of QTLs for different traits (see Salvi and Tuberosa 2007); however, only few reports are available where efforts have been made to clone the root trait QTLs. The major obstacle, in this context, is precise phenotyping for root traits in the thousands of plants. Nevertheless, some efforts have been made to clone the QTLs for root traits. For instance in maize two major QTLs on chromosome bins 1.06 and 2.04 (*root-ABA1*) affecting root architecture and a number of agronomic traits, including grain yield, have been targeted (Tuberosa and Salvi 2007). Similarly, efforts are underway for fine mapping/cloning of "*root-ABA1*" QTL that is responsible for root architecture and leaf ABA concentration in pearl millet (Kholova et al. 2010a, 2010b). In future, newer genomics approaches like association mapping (Rafalski 2010) and next generation sequencing (Varshney et al. 2009) are expected to facilitate cloning of QTLs for root traits.

Molecular Breeding for Root Traits

MTA studies identify important genomic regions, explaining a significant proportion of PV for root traits. Once major QTLs are identified, molecular markers associated with these QTLs need to be validated on a range of germplasm to select the candidate QTLs as well as the elite cultivars for introgression of the QTL using marker-assisted backcross (MABC) approach. This approach has been successfully used in several crops for improving different traits (Gupta et al. 2010; Varshney et al. 2010; Chamarthi et al. 2011; Kulwal et al. 2011). However, the power and efficiency of this approach has been limited, especially for root traits. For instance, Steele et al. (2006, 2007) demonstrated the power of MABC by releasing the first novel upland rice variety Birsa Vikas Dhan 111 (PY 84) through molecular breeding in the Jharkhand state of India. The variety was bred through MABC involving introgression of three genomic regions carrying root growth QTL on chromosomes 2 (root length), 9 (root thickness), and 11 (root penetration) from the donor Philippines variety Azucena into the recurrent parent Kalinga III. The target QTLs were first identified by Adam Price (now at Aberdeen University, UK) and Birgitte Courtois (CIRAD, France/IRRI, Philippines; http://www.cazs.bangor.ac.uk/ccstudio/WhatsNew/cazsWhatsNew2.php?ID=14; Cairns et al. 2009; Courtois et al. 2009; Norton and Price 2009). The variety developed in a collaborative partnership between Centre for Arid Zone Studies (CAZS) Natural Resources, Bangor University, UK; Gramin Vikas Trust, Ranchi, Jharkhand, India; and Birsa Agricultural University, Ranchi, Jharkhand, India, showed improved root growth, thus leading to improved performance under drought conditions. Similarly, Shen et al. (2001) also showed the effectiveness of QTL-based transfer of root traits, by involving the transfer of Azucena allele at four QTL alleles for deeper roots (on chromosomes 1, 2, 7, and 9) from selected DH lines into IR64. Thus, they demonstrated that NILs with an improved root system permit testing the importance of root depth for water-limited environments. In the case of chickpea also, a hotspot harboring several QTLs for drought tolerance-related traits (with >30% PV) is being introgressed into a leading chickpea variety JG11 through MABC approach (unpublished results). Using both foreground and background selection with SSR/AFLP markers, BC_3F_4 progenies have been derived and are being evaluated for their agronomic performance under rainfed and irrigated conditions in India, Kenya, Ethiopia, and Tanzania. These are some of the nice examples of translational genomic research for crop improvement.

Summary and Outlook

Roots are considered primary targets in view of current scenario of importance of drought in changing climate. Therefore, it is important to understand genetic and molecular mechanisms involved in conferring drought tolerance through root traits. While a linkage mapping-based approach has been used extensively to identify QTLs for a range of root traits in some crop species, association mapping has also started to be used in some cases for identification of QTLs. Some efforts have also been made toward cloning of QTLs for roots traits in species like maize and pearl millet. In terms of translation, a few reports have become available on molecular breeding through MABC and superior varieties have been developed. MABC approach for translating root genomics is possible only in the case where QTLs contribute to higher PV. In the cases where root traits are controlled by many and small-effect QTLs, the MARS approach that involves intermating selected individuals in each selection cycle (Eathington et al. 2007; Ribaut and Ragot 2007; Bernardo 2008) seems to be a better approach. Genome-wide selection or genomic selection approach is the other approach that involves selection of genotypes based on genomic estimated breeding values, estimated using genome-wide marker profile data for making the crosses (Bernardo and Yu 2007). Availability of genome sequences, NGS technologies, high-throughput genotyping, as well as phenotyping facilities are expected to accelerate root genomics research, especially for translation to crop improvement.

Acknowledgments

Authors are thankful to CGIAR Generation Challenge Programme (GCP) for providing Theme-Leader Discretionary grant and Tropical Legume I (TLI) project and Department of Biotechnology (DBT) of Government of India for Centre of Excellence (CoE) grant that has made it possible to write this article.

References

Abdel-Haleem, H. et al. (2011) Identification of QTLs for increased fibrous roots in soybean. *Theoretical and Applied Genetics*, 122, 935–946.

An, D.G. et al. (2006) Mapping QTLs for nitrogen uptake in relation to the early growth of wheat (*Triticum aestivum* L.). *Plant Soil* 284, 73–84.

Beebe, S.E. et al. (2006) Quantitative trait loci for root architecture traits correlated with phosphorus acquisition in common bean. *Crop Science*, 46, 413–423.

Bernardo, R. (2008) Molecular markers and selection for complex traits in plants: learning from the last 20 years. *Crop Science*, 48, 1649–1664.

Bernardo, R. and Yu, J. (2007) Prospects for genome wide selection for quantitative traits in maize. *Crop Science*, 47, 1082–1090.

Blum, A. (2005) Drought resistance, water-use efficiency and yield potential—are they compatible, dissonant or mutually exclusive? *Australian Journal of Agricultural Research* 56, 1159–1168.

Bohn, M. et al. (2006) Genetic evaluation of root complexity in maize. *Acta Agronomica Hungarica,* 54, 291–303.

Cairns, J. et al. (2009) Mapping quantitative trait loci associated with root growth in upland rice (*Oryza sativa* L.) exposed to soil water-deficit in fields with contrasting soil properties. *Field Crops Research,* 114, 108–118.

Chamarthi, S.K. et al. (2011) Trait mapping and molecular breeding. In: A. Pratap and J. Kumar (eds) *Biology and Breeding of Food Legumes*. CABI International, Oxford shire, pp. 296–313.

Chen, J.Y. et al. (2009) Identification of QTLs for phosphorus utilization efficiency in maize (*Zea mays* L.) across P levels. *Euphytica*, 167, 245–252.

Chen, Y.L. et al. (2011) Development of a novel semi-hydroponic phenotyping systemfor studying root architecture. *Functional Plant Biology*, 38, 355–363.

Cichy, K.A. et al. (2009a) QTL analysis of root architecture traits and low phosphorus tolerance in an Andean bean population. *Crop Science*, 49, 59–68.

Cichy et al. (2009b). QTL analysis of seed iron, zinc, and phosphorus levels in an Andean bean population. *Crop Science*, 49, 1742–1750.

Coudert, Y. et al. (2010) Genetic control of root development in rice, the model cereal. *Trends in Plant Science*, 15, 219–226.

Courtois, B. et al. (2003) Locating QTLs controlling constitutive root traits in the rice population IAC165 × Co39. *Euphytica*, 134, 335–345.

Courtois, B. et al. (2009) Rice root genetic architecture: meta-analysis from a drought QTL database. *Rice*, 2, 115–128.

de Dorlodot, S. et al. (2007) Root system architecture: opportunities and constraints for genetic improvement of crops. *Trends in Plant Science*, 12, 474–481.

Eathington, S.R. et al. (2007) Molecular markers in a commercial breeding program. *Crop Science*, 47, S154–S163.

Gai, J. et al. (2007) Identification, inheritance and QTL mapping of root traits related to tolerance to rhizospheric stresses in soybean (*G. max* (L.) *Merr.*). *Frontiers of Agriculture in China*, 1, 119–128.

Giuliani, M.M. et al. (2000) Comparative QTL analysis in maize for vertical root pulling resistance in the field and root traits in hydroponics. *International Conference on Plant & Animal Genome VIII*, January 9–12, San Diego, California, p. 114.

Gowda, V.R.P. et al. (2011) Root biology and genetic improvement for drought avoidance in rice. *Field Crop Research* 122, 1–13.

Gregory, P.J. et al. (2009) Root phenomics of crops: opportunities and challenges. *Functional Plant Biology*, 36, 922–929.

Guingo, E. et al. (1998) Genetic analysis of root traits in maize. *Agronomie*, 18, 225–235.

Gupta, P.K. et al. (2010) Marker assisted selection as a component of conventional plant breeding. *Plant Breeding Reviews*, 33, 145–217.

Hemamalini, G.S. et al. (2000) Molecular marker assisted tagging of morphological and physiological traits under two contrasting moisture regimes at peak vegetative stage in rice (*Oryza sativa* L.). *Euphytica*, 112, 69–76.

Hochholdinger, F. and Tuberosa, R. (2009) Genetic and genomic dissection of maize root development and architecture. *Current Opinion in Plant Biology*, 12, 172–177.

Horii, H. et al. (2006) Quantitative trait loci for adventitious and lateral roots in rice. *Plant Breeding*, 125, 198–200.

Hund, A. et al. (2009) Growth of axile and lateral roots of maize: I. Development of a phenotyping platform. *Plant Soil*, 325, 335–349.

Hund, A. et al. (2004) QTL controlling root and shoot traits of maize seedlings under cold stress. *Theoretical and Applied Genetics*, 109, 618–629.

Ikeda, H. et al. (2007) Genetic analysis of rooting ability of transplanted rice (*Oryza sativa* L.) under different water conditions. *Journal of Experimental Botany*, 58, 309–318.

Jackson, S.A. et al. (2011) Sequencing crop genomes: approaches and applications. *The New Phytologist*, 191, 915–925.

Kalliokoski, T. et al. (2008) Coarse root architecture of three boreal tree species growing in mixed stands. *Silva Fenn*, 42, 189–210.

Kamoshita, A. et al. (2002a) Mapping QTL for root morphology of a rice population adapted to rainfed lowland conditions. *Theoretical and Applied Genetics*, 104, 880–893.

Kamoshita, A. et al. (2002b) Effects of phenotyping environment on identification of quantitative trait loci for rice root morphology under anaerobic conditions. *Crop Science*, 42, 255–265.

Kashiwagi, J. et al. (2005) Genetic variability of drought-avoidance root traits in the mini-core germplasm collection of chickpea (*Cicer arietinum* L.). *Euphytica*, 146, 213–222.

Kashiwagi, J. et al. (2006) Variability of root length density and its contributions to seed yield in chickpea (*Cicer arietinum* L) under terminal drought stress. *Field Crop Research*, 95, 171–181.

Kell, D.B. (2011) Breeding crop plants with deep roots: their role in sustainable carbon, nutrient and water sequestration. *Annals of Botany*, 108, 407–418.

Kholová, J. et al. (2010b) Constitutive water conserving mechanisms are correlated with the terminal drought tolerance of pearl millet [*Pennisetum glaucum* (L.) R. Br.]. *Journal of Experimental Botany*, 61, 369–377.

Kholová, J. et al. (2010a) Terminal drought-tolerant pearl millet [*Pennisetum glaucum* (L.) R. Br.] have high leaf ABA and limit transpiration at high vapour pressure deficit. *Journal of Experimental Botany*, 61, 1431–1440.

Khowaja, F.S. et al. (2009) Improved resolution in the position of drought-related QTLs in a single mapping population of rice by meta-analysis. *BMC Genomics*, 10, 276.

Kulwal, P.L. et al. (2011) *Genomics interventions in crop breeding for sustainable agriculture. In: Encyclopedia of Sustainability Science and Technology*. Springer, New York, DOI 10.1007/978-1-4419-0851-3.

Lambers, H. et al. (2008) *Plant Physiological Ecology*, Second edition. Springer, New York.

Landi, P. et al. (2007) Root-ABA1 QTL affects root lodging, grain yield, and other agronomic traits in maize grown under well-watered and water-stressed conditions. *Journal of Experimental Botany*, 58, 319–326.

Landi, P. et al. (2002) Detection of QTLs for vertical root pulling resistance in maize and overlap with QTLs for root traits in hydroponics and for grain yield under different water regimes. *Maydica*, 47, 233–243.

Landi, P. et al. (2010) Characterization of root-yield-1.06, a major constitutive QTL for root and agronomic traits in maize across water regimes. *Journal of Experimental Botany*, 61, 3553–3562.

Laperche, A. et al. (2006) A simplified conceptual model of carbon/nitrogen functioning for QTL analysis of winter wheat adaptation to nitrogen deficiency. *Theoretical and Applied Genetics,* 113, 1131–1146.

Lebreton, C. et al. (1995) Identification of QTL for drought responses in maize and their use in testing causal relationships between traits. *Journal of Experimental Botany*, 46, 853–865.

Levitt, J. (1972) *Responses of Plants to Environmental Stresses*, Academic Press, New York.

Li, Y.D. et al. (2005b) QTL mapping of phosphorus deficiency tolerance in soybean (*Glycine max* L. Merr.). *Euphytica*, 142, 137–142.

Li, Z. et al. (2005a) QTL mapping of root traits in a doubled haploid population from a cross between upland and lowland japonica rice in three environments. *Theoretical and Applied Genetics*, 110, 1244–1252.

Li, J.Z. et al. (2009). Root and shoot traits responses to phosphorus deficiency and QTL analysis at seedling stage using introgression lines of rice. *Journal of Genetics and Genomics*, 36, 173–183.

Li, Z.K. et al. (2010) Analysis of QTLs for root traits at seedling stage using an immortalized F_2 population of wheat. *Acta Agronomica Sinica* , 36, 442–448.

Li, Z.X. et al. (2007) Molecular mapping of QTLs for root response to phosphorus deficiency at seedling stage in wheat (*Triticum aestivum* L.). *Proceedings of the National Academy of Sciences of the United States of America*, 17, 1177–1184.

Liang, Q. et al. (2010) QTL analysis of root traits as related to phosphorus efficiency in soybean. *Annals of Botany*, 106, 223–234.

Liao, H. et al. (2004) Genetic mapping of basal root gravitropism and phosphorus acquisition efficiency in common bean. *Functional Plant Biology*, 31, 959–970.

Liu, J. et al. (2008) Mapping QTLs for root traits under different nitrate levels at the seedling stage in maize (*Zea mays* L.). *Plant Soil*, 305, 253–265.

Ma, H.X. et al. (2005) Molecular mapping of a quantitative trait locus for aluminium tolerance in wheat cultivar Atlas 66. *Theoretical and Applied Genetics*, 112, 51–57.

Mace, E.S. et al. (2012) QTL for nodal root angle in sorghum (*Sorghum bicolor* L. Moench) co-locate with QTL for traits associated with drought adaptation. *Theoretical and Applied Genetics,* 12, 97–109.

MacMillan, K. et al. (2006) Assessing the importance of genotype × environment interaction for root traits in rice using a mapping population. I: a soil-filled box screen. *Theoretical and Applied Genetics,* 113, 977–986.

Mano, Y. et al. (2005) QTL mapping of adventitious formation under flooding conditions in tropical maize (*Zea mays* L.) seedlings. *Breeding Science,* 55, 343–347.

Manschadi, A.M. et al. (2008) Genotypic variation in seedling root architectural traits and implications for drought adaptation in wheat (*Triticum aestivum* L.). *Plant Soil,* 303, 115–129.

Miguel, M.A. et al. (2009) QTL mapping of basal root whorl number in common bean (*Phaseolus vulgaris* L). ASA-CSSA-SSAA International Annual Meeting, November 1–5, 2009, Pittsburg, PA. http://a-c-s.confex.com/crops/2009am/webprogram/Paper55496.html.

Mir, R.R. et al. (in press) Evolving molecular marker technologies for plant genetics research and breeding applications. In: T. Lübberstedt and R.K. Varshney (eds) *Diagnostics in Plant Breeding*. Springer, The Netherlands.

Morrell, P.L. et al. (2012) Crop genomics: advances and applications. *Nature Reviews Genetics,* 13, 85–96.

Myles, S. et al. (2009) Association mapping: critical considerations shift from genotyping to experimental design. *Plant Cell*, 21, 2194–2204.

Nagel, K.A. et al. (2009) Temperature responses of roots: impact on growth, root system architecture and implications for phenotyping. *Functional Plant Biology,* 36, 947–959.

Norton, G.J. and Price, A.H. (2009) Mapping of quantitative trait loci for seminal root morphology and gravitropic response in rice. *Euphytica*, 166, 229–237.

Obara, M. et al. (2010) Fine-mapping of qRL6.1, a major QTL for root length of rice seedlings grown under a wide range of NH_4^+ concentrations in hydroponic conditions. *Theoretical and Applied Genetics,* 121, 535–547.

Ochoa, I.E. et al. (2006) QTL analysis of adventitious root formation in common bean under contrasting phosphorus availability. *Crop Science*, 46, 1609–1621.

Omori, F. and Mano, Y. (2007) QTL mapping of root angle in F_2 populations from maize "B73" × teosinte "*Zea luxurians*." *Plant Root*, 1, 57–65.

Ouertani, K. et al. (2010) Quantitative trait loci (QTL) that underlie root and shoot traits in the soybean [*Glycine max* (L.) *Merr.*] "Hartwig" by "Flyer" recombinant inbred line (RIL) population. *Internation Conference on Plant & Animal Genomes XVIII*, January 9–13, 2010, Town and Country Convention Center, San Diego, California. http://www.intl-pag.org/18/abstracts/P05f_PAGXVIII_395.html.

Paterson, A.H. et al. (2009) The *Sorghum bicolor* genome and the diversification of grasses. *Nature*, 457, 551–556.

Petrarulo, M. et al. (2009) Mapping QTLS for root morphological traits in durum wheat. *International Symposium on Root Research and Applications*, 2–4 September, Boku, Vienna, Austria. http://asrr.boku.ac.at/fileadmin/files/RRcd/session07/poster/195.pdf

Priya, P.S. et al. (2009) Genome-wide analyses of rice root development QTLs and development of an online resource, Root browse. *Bioinformation*, 3, 279–281.

Qu, Y. et al. (2008) Mapping QTLs of root morphological traits at different growth stages in rice. *Genetica*, 133, 187–200.

Rafalski, A. (2010) Association genetics in crop improvement. *Current Opinion in Plant Biology*, 13, 174–180.

Ren, Y. et al. (2011) Major quantitative trait loci for seminal root morphology of wheat seedlings. *Molecular Breeding*, 30, 139–148.

Reynolds, M. and Tuberosa, R. (2008) Translational research impacting on crop productivity in drought prone environments. *Current Opinion in Plant Biology*, 11, 171–179.

Ribaut, J.M. and Ragot, M. (2007) Marker-assisted selection to improve drought adaptation in maize: the backcross approach, perspectives, limitations, and alternatives. *Journal of Experimental Botany*, 58, 351–360.

Ruta, N. et al. (2010) QTLs for the elongation of axile and lateral roots of maize in response to low water potential. *Theoretical and Applied Genetics*, 120, 621–631.

Salvi, S. and Tuberosa, R. (2005) To clone or not to clone plant QTLs: present and future challenges. *Trends in Plant Science*, 10, 297–304.

Salvi, S. and Tuberosa, R. (2007) Cloning QTLs in plants. In: R.K. Varshney and R. Tuberosa (eds) *Genomics-Assisted Crop Improvement*, Vol. I. Springer, New York, pp. 207–226.

Sanguineti, M.C. et al. (2007) Genetic dissection of seminal root architecture in elite durum wheat germplasm. *Annals of Applied Biology*, 151, 291–305.

Schmutz, J. et al. (2010) Genome sequence of the palaeopolyploid soybean. *Nature*, 463, 178–183.

Schnable, P.S. et al. (2009) The B73 maize genome: complexity, diversity, and dynamics. *Science*, 326, 1112–1115.

Sharma, S. et al. (2011) Dissection of QTL effects for root traits using a chromosome arm-specific mapping population in bread wheat. *Theoretical and Applied Genetics*, 122, 759–769.

Shen, L. et al. (2001) Evaluation of near-isogenic lines of rice introgressed with QTLs for root depth through marker-aided selection. *Theoretical and Applied Genetics*, 103, 427–437.

Sheshshayee, M.S. et al. (2011) Phenotyping for root traits and their improvement through biotechnological approaches to sustaining crop productivity. In: A.C. de Oliveira and R.K. Varshney (eds) *Root Genomics*. Springer-Verlag, Berlin Heidelberg, pp. 205–232.

Smita, S. et al. (2011) QlicRice: a web interface for abiotic stress responsive QTL and loci interaction channels in rice. *Database*, Doi: 10.1093/database/bar037.

Srividya, A. et al. (2010). Identification and mapping of landrace derived QTL associated with yield and its components in rice under different nitrogen levels and environments. *International Journal of Plant Breeding and Genetics*, 4, 210–227.

Steele, K.A. et al. (2006) Marker-assisted selection to introgress rice QTLs controlling root traits into an Indian upland rice variety. *Theoretical and Applied Genetics*, 112, 208–221.

Steele, K.A. et al. (2007) Field evaluation of upland rice lines selected for QTLs controlling root traits. *Field Crop Research*, 101, 180–186.

Swamy, M.B.P. et al. (2011) Meta-analysis of grain yield QTL identified during agricultural drought in grasses showed consensus. *BMC Genomics*, 12, 319.

Tuberosa, R. and Salvi, S. (2006) Genomics approaches to improve drought tolerance in crops. *Trends in Plant Science*, 11, 405–412.

Tuberosa, R. and Salvi, S. (2007) From QTLs to genes controlling root traits in maize. In: J.H.J. Spiertz, P.C. Struik, and H.H. van Laar (eds) *Scale and Complexity in Plant Systems Research: Gene Plant Crop Relations*. Springer, Dordrecht, The Netherlands, pp. 15–24.

Tuberosa, R. et al. (2002a) Identification of QTLs for root characteristics in maize grown in hydroponics and analysis of their overlap with QTLs for grain yield in the field at two water regimes. *Plant Molecular Biology*, 48, 697–712.

Tuberosa, R. et al. (2003) Searching for quantitative trait loci controlling root traits in maize: a critical appraisal. *Plant Soil*, 255, 35–54.

Tuberosa, R. et al. (2002b) Mapping QTLs regulating morpho-physiological traits and yield: case studies, short comings and perspectives in drought-stressed maize. *Annals of Botany*, 89, 941–963.

Tuberosa, R. et al. (2011) Genomics of root architecture and functions in maize. In: A. Costa de Oliveira and R.K. Varshney (eds) *Root Genomics*. Springer, Berlin Heidelberg, pp. 179–204.

Uga, Y. et al. (2008) QTL underlying natural variation in stele and xylem structures of rice root. *Breeding Science*, 58, 7–14.

Uga, Y. et al. (2011) *Dro*1, a major QTL involved in deep rooting of rice under upland field conditions. *Journal of Experimental Botany*, 62, 2485–2494.

Ulaganathan, K. et al. (2007) Physical integration of root QTLs with the *japonica* genome and prediction of probable candidate genes for drought tolerance in rice. *Molecular Plant Breeding*, 5, 213–214.

Varshney, R.K. et al. (2005) Genomics-assisted breeding for crop improvement. *Trends in Plant Science*, 10, 621–630.

Varshney, R.K. et al. (2006) Advances in cereal genomics and applications in crop breeding. *Trends in Biotechnology*, 24, 490–499.

Varshney, R.K. et al. (2011a) Agricultural biotechnology for crop improvement in a variable climate: hope or hype? *Trends in Plant Science*, 16, 363–371.

Varshney, R.K. et al. (2011b) Genomics and physiological approaches for root trait breeding to improve drought tolerance in chickpea (*Cicer arietinum* L.). In: A.D. de Oliveira and R.K. Varshney (eds) *Root Genomics*. Springer, Berlin Heidelberg, pp. 233–250.

Varshney, R.K. et al. (2010) Legume genomics and breeding. *Plant Breeding Reviews*, 33, 257–304.

Varshney, R.K. et al. (2009) Next-generation sequencing technologies and their implications for crop genetics and breeding. *Trends in Biotechnology*, 27, 522–530.

Varshney, R.K. et al. (2012) Draft genome sequence of pigeonpea (*Cajanus cajan*), an orphan legume crop of resource-poor farmers. *Nature Biotechnology*, 30, 83–89.

Venuprasad, R. et al. (2002) Tagging quantitative trait loci associated with grain yield and root morphological traits in rice (*Oryza sativa* L.) under contrasting moisture regimes. *Euphytica*, 128, 293–300.

Washington, E. et al. (2009) A genetic analysis of root traits in soybean plants grown in the greenhouse and in the field. *International Conference on Plant & Animal Genomes XVII*, January 10–14, 2009, Town & Country Convention Center, San Diego, California. http://www.intl-pag.org/17/abstracts/P05f_PAGXVII_374.html.

Williams, B. et al. (2012) Genetic analysis of root and shoot traits in the "Essex" by "Forrest" recombinant inbred line (RIL) population of soybean [*Glycine max* (L.) *Merr*.]. *Journal of Plant Genome Sciences*, 1, 1–9.

Xu, J.C. et al. (2001) QTL mapping of the root traits in rice seedling. *Acta Genet Sin*, 28, 433–438.

Yan, X. et al. (2004) QTL mapping of root hair and acid exudation traits and their relationship to phosphorus uptake in common bean. *Plant Soil*, 265, 17–29.

Yazdanbakhsh, N. and Fisahn, J. (2009) High throughput phenotyping of root growth dynamics, lateral root formation, root architecture and root hair development enabled by PlaRoM. *Functional Plant Biology*, 36, 938–946.

Yue, B. et al. (2008) Identification of quantitative trait loci for four morphologic traits under water stress in rice (*Oryza sativa* L.). *Journal of Genetics and Genomics* 35, 569–575.

Yue, B. et al. (2006) Genetic basis of drought resistance at reproductive stage in rice: separation of drought tolerance from drought avoidance. *Genetics*, 172, 1213–1228.

Zaman-Allah, M. et al. (2011) A conservative pattern of water use, rather than deep or profuse rooting, is critical for the terminal drought tolerance of chickpea. *Journal of Experimental Botany*, 62, 4239–4252.

Zhang, D. et al. (2009) Detection of quantitative trait loci for phosphorus deficiency tolerance at soybean seedling stage. *Euphytica*, 167, 313–322.

Zhang, J. et al. (2001) Locating genomic regions associated with components of drought resistance in rice: comparative mapping within and across species. *Theoretical and Applied Genetics*, 103, 19–29.

Zheng, B.S. et al. (2003) Mapping QTLs and candidate genes for rice root traits under different water-supply conditions and comparative analysis across three populations. *Theoretical and Applied Genetics*, 107, 1505–1515.

Zhu, J.M. et al. (2005a) Mapping of QTL controlling root hair length in maize (*Zea mays* L.) under phosphorus deficiency. *Plant Soil*, 270, 299–310.

Zhu, J.M. et al. (2005b) Mapping of QTLs for lateral root branching and length in maize (*Zea mays* L.) under differential phosphorus supply. *Theoretical and Applied Genetics*, 111, 688–695.

Zhu, J.M. et al. (2006) Detection of quantitative trait loci for seminal root traits in maize (*Zea mays* L.) seedlings grown under differential phosphorus levels. *Theoretical and Applied Genetics*, 113, 1–10.

Zobel, R.W. and Weisel, Y. (2010) A plant root system architectural taxonomy: a framework for root nomenclature. *Plant Biosystems*, 144, 507–512.

Index

Root Genomics and Soil Interactions, First Edition. Edited by Martin Crespi.